Texts and Monographs in Physics

Springer
Berlin
Heidelberg
New York
Barcelona
Budapest
Hong Kong
London
Milan
Paris
Santa Clara
Singapore
Tokyo

Texts and Monographs in Physics

Series Editors: R. Balian W. Beiglböck H. Grosse E. H. Lieb N. Reshetikhin H. Spohn W. Thirring

From Microphysics to Macrophysics I + II Methods and Applications of Statistical Physics By R. Balian

Variational Methods in Mathematical Physics A Unified Approach By P. Blanchard and E. Brüning

Quantum Mechanics: Foundations and Applications 3rd enlarged edition By A. Böhm

The Early Universe Facts and Fiction 3rd corrected and enlarged edition By G. Börner

Operator Algebras and Quantum Statistical Mechanics I + II 2nd edition By O. Bratteli and D. W. Robinson

Geometry of the Standard Model of Elementary Particles By A. Derdzinski

Scattering Theory of Classical and Quantum *N*-Particle Systems By J. Dereziński and C. Gérard

Random Walks, Critical Phenomena, and Triviality in Quantum Field Theory By R. Fernández, J. Fröhlich and A. D. Sokal

Quantum Relativity A Synthesis of the Ideas of Einstein and Heisenberg By D. R. Finkelstein

Quantum Mechanics I + II By A. Galindo and P. Pascual

The Elements of Mechanics By G. Gallavotti

Local Quantum Physics Fields, Particles, Algebras 2nd revised and enlarged edition By R. Haag

Supersymmetric Methods in Quantum and Statistical Physics By G. Junker

Elementary Particle Physics Concepts and Phenomena By O. Nachtmann

Inverse Schrödinger Scattering in Three Dimensions By R. G. Newton

Scattering Theory of Waves and Particles 2nd edition By R. G. Newton

Quantum Entropy and Its Use By M. Ohya and D. Petz

Generalized Coherent States and Their Applications By A. Perelomov

Essential Relativity Special, General, and Cosmological Revised 2nd edition By W. Rindler

Path Integral Approach to Quantum Physics An Introduction 2nd printing By G. Roepstorff

Advanced Quantum Theory and Its Applications Through Feynman Diagrams 2nd edition By M. D. Scadron

Finite Quantum Electrodynamics The Causal Approach 2nd edition By G. Scharf

From Electrostatics to Optics A Concise Electrodynamics Course By G. Scharf

Large Scale Dynamics of Interacting Particles By H. Spohn

General Relativity and Relativistic Astrophysics By N. Straumann

The Mechanics and Thermodynamics of Continuous Media By M. Šilhavý

The Dirac Equation By B. Thaller

The Theory of Quark and Gluon Interactions 2nd completely revised and enlarged edition By F. J. Ynduráin

Relativistic Quantum Mechanics and Introduction to Field Theory By F. J. Ynduráin

Jan Dereziński Christian Gérard

Scattering Theory of Classical and Quantum *N*-Particle Systems

With 15 Figures

Springer

Jan Dereziński

Department of Mathematical Methods
in Physics, Warsaw University
Hoża 74
PL-00-682 Warsaw, Poland

Christian Gérard

Centre de Mathématiques
Ecole Polytechnique
F-91128 Palaiseau Cedex, France

Editors

Library of Congress Cataloging-in-Publication Data

Dereziński, Jan, 1957–
Scattering Theory of Classical and Quantum N-Particle Systems / J. Dereziński, C. Gérard.
p. cm. – (Texts and monographs in physics, ISSN 0172-5998) Includes bibliographical references.
ISBN 3-540-62066-4 (alk. paper)
1. Scattering (Physics) 2. Scattering (Mathematics) 3. Quantum theory. 4. Mathematical physics – Asymptotic theory. I. Gérard, Christian, 1960– . II. Title. III. Series.
QC20.7.S3D47 1997 539.7'58'0151–dc21 96-46757 CIP

ISSN 0172-5998
ISBN 3-540-62066-4 Springer-Verlag Berlin Heidelberg New York

Typesetting: Camera-ready copy from the authors
Cover design: *design & production* GmbH, Heidelberg
SPIN: 10481305 55/3144-5 4 3 2 1 0 - Printed on acid-free paper

We dedicate this book to our beloved
Ida, Line, Marion, Michał and Pierre

Acknowledgments

One of us (Jan Dereziński) would like to express his gratitude for the hospitality and the financial support of the following institutions: Centre de Mathématiques of Ecole Polytechnique, Schrödinger Institute, Aarhus University, IHES, Institute for Mathematics and its Applications at the University of Minnesota, Mittag Leffler Institute, Université Paris Sud, Caltech, Université Paris VII. He also acknowledges grants from Komitet Badań Naukowych.

Both of us profited from discussions with our colleagues working on scattering theory. We use this opportunity to express our gratitude to all of them, in particular to V. Enss, R. Froese, V. Georgescu, G. M. Graf, G. Hagedorn, I. Herbst, H. Isozaki, A. Jensen, S. Nakamura, E. Mourre, P. Perry, I. Sigal, E. Skibsted, A. Soffer, X. P. Wang, D. Yafaev, K. Yajima and L. Zieliński.

Contents

0. Introduction

A system of N non-relativistic classical particles interacting with pair potentials is described by a Hamiltonian of the form

$$H(x_1,\ldots,x_N,\xi_1,\ldots,\xi_N) := \sum_{j=1}^{N}\frac{1}{2m_j}\xi_j^2 + \sum_{1\le i<j\le N} V_{ij}(x_i - x_j). \tag{0.0.1}$$

This Hamiltonian generates a flow $\phi(t)$ on the phase space $\mathbb{R}^{3N}\times\mathbb{R}^{3N}$.

An analogous system of N quantum particles is described by a Hamiltonian of the form

$$H := -\sum_{j=1}^{N}\frac{1}{2m_j}\Delta_j + \sum_{1\le i<j\le N} V_{ij}(x_i - x_j). \tag{0.0.2}$$

It generates the unitary evolution e^{-itH} on the Hilbert space $L^2(\mathbb{R}^{3N})$. The aim of this book is to describe the asymptotic behavior of the dynamics $\phi(t)$ and e^{-itH} for $t\to\pm\infty$. (The cases $t\to-\infty$ and $t\to+\infty$ are completely analogous; for notational convenience we will restrict ourselves to the case $t\to+\infty$).

Roughly speaking, we will show that the evolution for large time is simpler. It is close to the free evolution in certain variables; in other variables it resembles a bounded motion. This observation, which was made by physicists a long time before the first rigorous results about this subject, is the basis for the interpretation of many physical experiments. In fact, in many experiments the particles are initially far apart, then they scatter from one another and at the end they move away. Experimental physicists measure the probability of various outcomes of such experiments – the so-called scattering cross-sections. These scattering cross-sections can be computed from the so-called scattering operator. Note, however, that in our monograph we will not study the scattering operator itself – which relates the asymptotics $t\to-\infty$ with the asymptotics $t\to+\infty$. We will concentrate on the questions related to the limits $t\to+\infty$ (which means the study of wave transformations/operators and asymptotic observables).

Usually scattering theory is considered in the context of quantum mechanics. As a matter of fact, the results for quantum scattering theory are usually far more satisfactory than their classical counterparts. The theorem about the asymptotic completeness of long-range N-body scattering gives a very satisfactory understanding of the problem in the quantum case. Such an understanding, at least for more than 2 particles, is lacking in the classical case. Nevertheless, we think that it is instructive to study classical and quantum scattering in a parallel way. Moreover, in order to thoroughly understand long-range quantum scattering one needs first to study its classical analog. One can argue that scattering is a semi-classical phenomenon and a lot of scattering properties of classical and quantum systems are analogous. We try to stress this aspect in our presentation.

As we indicated earlier, the main results of scattering theory have a fundamental importance for understanding some problems of non-relativistic physics. The Hamiltonians (0.0.1) and (0.0.2) are very often used to describe physical systems. Therefore, the main results of our monograph are usually very close to the real physical world. There is also, however, another aspect of scattering theory that we would like to stress. It is indeed a very elegant, natural and deep mathematical theory. Its main results should be appealing to any mathematician, even if he or she is not interested in its physical content. Their proofs are usually very elegant and intuitive, at the same time they are tricky and technical.

We will always suppose that the 2-body potentials $V_{ij}(x)$ decay at infinity. Roughly speaking, the general form of the assumptions that we will impose will be the following:

$$|\partial_x^\alpha V_{ij}(x)| \leq C_\alpha (1+|x|)^{-\mu-|\alpha|}, \quad |\alpha| \leq n. \tag{0.0.3}$$

In the quantum case, we can also accommodate local singularities of the potentials. We will not do so in the classical case. If $\mu > 1$ and $n = 0$ in (0.0.3), then this is the so-called *short-range case.* If $\mu > 0$ and $n = 2$ in (0.0.3), then this is the *long-range case.* The dimension of the one-particle configuration space will always be arbitrary.

The short-range case is much simpler to study. For short-range potentials, it is possible to compare the full dynamics with a dynamics that is, at least partially, free. Unfortunately, the physically most important class of potentials are Coulomb potentials, which are long-range.

Our monograph would be much simpler, shorter and less interesting if we restricted ourselves to short-range potentials. In fact, long-range scattering is the central subject of our monograph. We study it under very general conditions, which are motivated by the mathematical structure of the problem. If we were interested just in the physically relevant case, that is in Coulomb potentials, we would not make some of the constructions that we need to do to handle the general case. Nevertheless, we believe that exploring the problem in its natural mathematical generality gives a lot of insight that will be useful for future applications. In order to construct modified wave operators in the long-range case with $\mu > 0$, we are forced to study in detail the corresponding classical system and to understand the relationship between the classical and quantum dynamics quite deeply. For Coulomb potentials, or more generally, for potentials satisfying $\mu > 1/2$ and $n = 1$ in (0.0.3), it would be sufficient to use the so-called Dollard modifiers, which is a cheap way of avoiding some of the work that we do.

If in Hamiltonians (0.0.1) and (0.0.2) we have just one particle, then there is no interaction whatsoever and the dynamics is free. If we have 2 particles, then after separating the center-of-mass motion we obtain reduced Hamiltonians

$$H(x,\xi)=\frac{1}{2}\xi^2+V(x) \tag{0.0.4}$$

in the classical case and

$$H=-\frac{1}{2}\Delta+V(x) \tag{0.0.5}$$

in the quantum case, where the potential $V(x)$ decays in all directions. This is the most often studied and the best understood case of scattering theory. The first four chapters are devoted to the 2-body case.

The technical difficulties of proving the asymptotic completeness of 2-body systems can be divided into two phases. First one needs to show that scattering trajectories (in the classical case) and scattering states (in the quantum case) move away from the origin as C_0t with $C_0>0$. Then one can introduce wave transformations (in the classical case) and wave operators (in the quantum case). In the first phase, one does not see the difference between the short-range and the long-range case. Only in the second phase, the difference becomes important. In fact, if we already accomplished the first phase, the existence and completeness of wave transformations/operators in the short-range case is very easy, whereas in the long-range case, at least for a very slow decay, there is still a lot of work to do. A pedagogical device that helps to describe the second phase in a clean way is studying Hamiltonians of the form

$$H(x,\xi)=\frac{1}{2}\xi^2+V(t,x) \tag{0.0.6}$$

in the classical case and

$$H=-\frac{1}{2}\Delta+V(t,x) \tag{0.0.7}$$

in the quantum case, where instead of the spatial decay (0.0.3) we have the time decay, typically,

$$|\partial_x^\alpha V(t,x)|\le C_\alpha(1+t)^{-\mu-|\alpha|},\quad |\alpha|\le n. \tag{0.0.8}$$

We call the assumption (0.0.8) $\mu>1$, $n=0$ the *fast-decaying case* – it is the analog of the short-range case for time-independent potentials. The assumption (0.0.8) with $\mu>0$, $n=2$ is called the *slow-decaying case* – it is the analog of the long-range case. We devote Chaps. 1 and 3 to the study of scattering for time-decaying potentials.

The asymptotic completeness of 2-body systems can be shown in many ways, and the literature on the subject is very rich. We do not intend to review all the techniques that were used in this context (see e.g. [RS, vol III and IV], [Hö2, vol II and IV] and [Yaf4]).

In the case of systems of $N\ge 3$ and especially $N\ge 4$ particles, scattering theory becomes much more difficult than in the 2-body case. The literature on the subject is much more limited.

The case of 3-body systems is intermediate between the 2-body case and the general N-body case. The asymptotic completeness of the 3-body systems, both in the short- and long-range case with $\mu > \sqrt{3}-1$, was first proven by Enss [E5, E6]. Note, however, that the method of Enss does not seem to generalize to the case of general N-body systems.

To our understanding, there are essentially two approaches to proving asymptotic completeness for N-body systems. The first approach uses the so-called local decay estimate

$$\int_0^\infty \|\langle x\rangle^{-\frac{1}{2}-\epsilon}\chi(H)\mathrm{e}^{-itH}\phi\|^2 \mathrm{d}t < \infty, \qquad (0.0.9)$$

where $\epsilon > 0$ and χ is a cutoff supported away of bound states and thresholds of H. The estimate (0.0.9) is usually proven by considering the boundary values of the resolvent. Its proof is based on the ideas of Mourre [Mo1, Mo2, PSS]. The proof of asymptotic completeness based on the first approach uses then time-independent observables, for instance homogeneous functions of degree zero on the configuration space. Error terms that arise due to the commuting of observables are usually $O(\langle x\rangle^{-2})$ and can be handled by the local decay estimate (0.0.9). This approach was used in the first proof of the asymptotic completeness of N-body short-range systems of Sigal and Soffer [SS1] and in the proof of Yafaev [Yaf5].

The second approach is fully time-dependent. One does not need to know (0.0.9) nor to study the resolvent. The proof uses time-dependent observables, for instance of the form $J(\frac{x}{t})$, where J is a compactly supported function. The error terms are typically of the order $O(t^{-2})$ and can be estimated in norm. This approach was used in the work of Graf [Gr], which contained the first reasonably simple proof of the asymptotic completeness of N-body short-range systems and in [De8] where the first proof of the asymptotic completeness of long-range N-body systems with $\mu > \sqrt{3}-1$ was given.

Throughout our monograph (also when considering 2-body systems), we will stick to the second approach. We believe that this is the most natural and the simplest method of handling the problem of asymptotic completeness. In particular, we do not need to study the resolvent. Note, however, that the first approach, involving time-independent observables and a study of the resolvent, has its advantages. It yields results about the properties of eigenfunction expansions and scattering matrices that are of a considerable interest and that seem to be inaccessible to a purely time-dependent approach.

We made an effort to make our monograph self-contained and accessible to a reader with a modest mathematical background. The mathematical tools that we use are quite limited, they include some elements of the real analysis, basic properties of operators in Hilbert spaces and the simplest classes of pseudo-differential and Fourier integral operators. We included a number of appendices that provide the reader with an introduction to the mathematical tools that we use.

1. Classical Time-Decaying Forces

1.0 Introduction

The motion of a non-relativistic free particle in Euclidean space is described by the Hamiltonian

$$H_0(x,\xi) = \frac{1}{2m}\xi^2.$$

A free particle moves along straight lines with a constant velocity.

To describe the motion of a particle in an external potential, one uses Hamiltonians of the form

$$H(x,\xi) = \frac{1}{2m}\xi^2 + V(x), \tag{1.0.1}$$

which yield the equations of motion

$$m\partial_t^2 x(t) = -\nabla_x V(x(t)). \tag{1.0.2}$$

If the potential decays as $x \to \infty$, it is natural to expect that trajectories that escape to infinity sufficiently fast should resemble free trajectories. Comparison of the dynamics generated by the "full Hamiltonian" $H(x,\xi)$ and the "free Hamiltonian" $H_0(x,\xi)$ is the main subject of classical scattering theory, which we would like to present in the first two chapters of this book.

A system of two particles interacting through a pair potential is described by a Hamiltonian of the form

$$H(x_1,\xi_1,x_2,\xi_2) = \frac{1}{2m_1}\xi_1^2 + \frac{1}{2m_2}\xi_2^2 + V(x_1 - x_2). \tag{1.0.3}$$

In such systems, it is convenient to separate the motion of the center of mass. To this end, we change the variables by setting $y_{\mathrm{cm}} := (m_1x_1+m_2x_2)/(m_1+m_2)$ and $y := x_1 - x_2$. We denote the corresponding momenta by η_{cm} and η. We also introduce the total mass $m_{\mathrm{cm}} := m_1 + m_2$ and the reduced mass $m := 1/\left(m_1^{-1} + m_2^{-1}\right)$. Then (1.0.3) becomes

$$H(y_{\mathrm{cm}},\eta_{\mathrm{cm}},y,\eta) = \frac{1}{2m_{\mathrm{cm}}}\eta_{\mathrm{cm}}^2 + \frac{1}{2m}\eta^2 + V(y). \tag{1.0.4}$$

Thus the motion of the system separates into two independent parts: the motion of the center of mass, which is free, and the relative motion of the pair of particles, which is described by a Hamiltonian of the form (1.0.1). Therefore it is common to call (1.0.1) a *two-body Hamiltonian* (although the name *one-body Hamiltonian* is also justified and used).

Our exposition of classical scattering is divided into two chapters. Only in the second chapter we treat Hamiltonians of the form (1.0.1). The first chapter develops scattering theory for somewhat different systems.

In the first chapter we study the motion of a particle subject to a time-dependent force $F(t,x)$. The equation of motion is the well known Newton's equation

$$\partial_t^2 x(t) = F(t, x(t)). \tag{1.0.5}$$

The force $F(t,x)$ and its derivatives with respect to x are assumed to decay in time, but no assumptions on the decay in space are imposed.

Usually, in this chapter, we will not assume that the force is conservative. If the force is conservative, then there exists a potential $V(t,x)$ such that

$$F(t,x) = -\nabla_x V(t,x) \tag{1.0.6}$$

and the dynamics is generated by the time-dependent Hamiltonian

$$H(t,x,\xi) = \frac{1}{2}\xi^2 + V(t,x). \tag{1.0.7}$$

This assumption, although customary in the literature, has nothing to do with scattering theory, therefore in this chapter we consider the more general equations (1.0.5) without assuming (1.0.6).

Dynamics with time-decaying forces are probably not so physically motivated as those generated by (1.0.2). Nevertheless, from the mathematical point of view, systems with the dynamics of the form (1.0.5) constitute a very natural class to study scattering theory. Most of the results about scattering for systems that belong to this class can be formulated and developed in a particularly clean and simple way. One can remark in parentheses that this class is invariant with respect to translations and Galilean transformations (changes of coordinates from one reference system to another one that moves with a constant velocity). Scattering theory for time-independent systems presented in Chap. 2 is somewhat more complicated (in particular, it is not invariant with respect to Galilean transformations).

In practice, if one is studying a time-independent system, it is very convenient to reduce the problem to a time-dependent one. This is why we are considering time-decaying forces in Chap. 1 and the time-independent ones in Chap. 2. We also try to concentrate most of our discussions of various fine points of two-body scattering theory in the first chapter.

In order to give the reader a rough idea what kind of assumptions can appear in a study of scattering theory for time-decaying forces, let us consider the following condition:

$$\partial_x^\alpha F(t,x) \in O(t^{-1-\mu-|\alpha|}), \text{ for } |\alpha| = 0, 1. \tag{1.0.8}$$

If in the above condition we assume that $\mu > 1$, then we say that our system is fast-decaying. If $1 \geq \mu > 0$ then the system is said to be slow-decaying. (Note that outside of the introduction we will use better, more general assumptions on the forces than (1.0.8)). Let us briefly describe the content of Chap. 1.

In Sects. 1.1 and 1.2 we introduce the basic notation, a large part of which will be used throughout some of the next chapters.

In Sect. 1.3 we introduce the *asymptotic momentum* $\xi^+(y,\eta)$, a basic asymptotic quantity that exists both in the slow- and fast-decaying case. For every initial conditions (y,η), it is defined as the limit of the momentum $\xi(t,y,\eta)$. It exists under the condition

$$F(t,x) \in O(t^{-1-\mu}) \quad \text{for some} \quad \mu > 0. \tag{1.0.9}$$

Section 1.4 is devoted to an exposition of scattering theory for fast-decaying forces. If $\mu > 1$ in (1.0.9) (which is essentially the fast-decaying condition) then we can introduce the asymptotic position $x_{\mathrm{fd}}^+(y,\eta)$ defined as

$$\lim_{t\to\infty} \left(x(t,y,\eta) - t\xi^+(y,\eta) \right).$$

If $\phi(s,t)$ denotes the evolution of our system and $\phi_0(t)$ denotes the free flow, then

$$\lim_{t\to\infty} \phi_0(-t)\phi(t,0)(y,\eta) = (x_{\mathrm{fd}}^+(y,\eta), \xi^+(y,\eta)). \tag{1.0.10}$$

The mapping

$$X \times X' \ni (y,\eta) \mapsto (x_{\mathrm{fd}}^+(y,\eta), \xi^+(y,\eta)) \in X \times X'$$

allows us to label in a natural way all the trajectories of a given system with trajectories of the free system. This fundamental property goes traditionally under the name of the *asymptotic completeness* of the wave transformations.

If we assume, in addition, that

$$\nabla_x F(t,x) \in O(t^{-2-\mu}) \quad \text{for some} \quad \mu > 0, \tag{1.0.11}$$

then

$$\lim_{t\to\infty} \phi(0,t)\phi_0(t) \tag{1.0.12}$$

exists. We call (1.0.12) the *wave transformations* and denote it by $\mathcal{F}_{\mathrm{fd}}^+$. Clearly, (1.0.10) is its inverse. Moreover, if $(y,\eta) = \mathcal{F}_{\mathrm{fd}}^+(x,\xi)$, one has

$$\lim_{t\to\infty} (\phi(t,0)(y,\eta) - \phi_0(t)(x,\xi)) = 0. \tag{1.0.13}$$

Note that (1.0.13) does not follow directly from (1.0.12). The property (1.0.13) means that, for any trajectory of the free system, there exists a trajectory of the perturbed system that is asymptotic to it. This property is traditionally known as the *existence* of the wave transformations.

Scattering in the slow-decaying case is much more complicated than scattering for fast-decaying forces. We devote to it the rest of this chapter.

In Sect. 1.5 we begin with a study of solutions $\tilde{y}(s, t_1, t_2, x, \xi)$ of the equation of motion with the following boundary conditions: at the initial time t_1 we fix the position x and at the final time t_2 we fix the momentum ξ. It turns out that if $\mu > 0$ in (1.0.8), $T \leq t_1 \leq t_2$ and T is large enough, then we can solve uniquely this problem. (The final time t_2 can be put even to ∞).

The above result is used to prove that the distance of any two trajectories with the same asymptotic momentum converges to a limit as $t \to \infty$. Therefore, the union of all trajectories with a given asymptotic momentum is naturally isomorphic to the configuration space in the sense of *affine spaces*. It is a somewhat weaker statement than in the fast-decaying case, where one had a natural isomorphism in the sense of *vector spaces*.

In general, in the slow-decaying case wave transformations $\mathcal{F}_{\text{fd}}^{+}$ are not well defined. Fortunately, there exists a substitute. To define this substitute, we fix T big enough and, for every momentum ξ, we denote by $s \mapsto Y(s, \xi)$ the trajectory that has the asymptotic momentum ξ and that starts at the time T at the origin. Now we can give our first definition of the modified asymptotic position $x_{\text{sd}}^{+}(y, \eta)$, given by the limit

$$\lim_{t \to \infty} \left(x(t, y, \eta) - Y(t, \xi^{+}(y, \eta)) \right),$$

and also our first definition of the modified wave transformation $\mathcal{F}_{\text{sd}}^{+}$, defined as the inverse of the map

$$(y, \eta) \mapsto (x_{\text{sd}}^{+}(y, \eta), \xi^{+}(y, \eta)).$$

A second definition of these concepts is given in Sect. 1.6. It is based on a comparison of the flow generated by the equations (1.0.5) with the so-called *modified free evolution* $\phi_{\text{sd}}(t)$. The modified free flow has the form

$$\phi_{\text{sd}}(t)(x, \xi) = (x + \tilde{Y}(t, \xi), \xi),$$

where the function $\tilde{Y}(t, \xi)$, called a *modifier*, is defined to be the position of the trajectory that at time T is at the origin and at time t has the momentum ξ. It resembles the free flow, but it also has some large-time corrections to the free motion reflecting the influence of slow-decaying forces.

The second definition of the modified wave transformation is the following:

$$\mathcal{F}_{\text{sd}}^{+} := \lim_{t \to \infty} \phi(0, t) \phi_{\text{sd}}(t).$$

The second definition of the asymptotic position is

$$x_{\text{sd}}^{+}(y, \eta) := \lim_{t \to \infty} \left(x(t, y, \eta) - \tilde{Y}(t, \xi(t, y, \eta)) \right).$$

The definitions of the asymptotic position and the modified wave transformation from Sects. 1.5 and 1.6 are equivalent.

In most of Chap. 1 we do not assume that the force is the gradient of a potential and therefore we do not use the Hamiltonian formalism. It is mainly in Sect. 1.8 where we add the assumption that the force is conservative and therefore the flow is symplectic. In this section we construct solutions of the Hamilton-Jacobi and eikonal equations. They will be important in the chapters on quantum scattering theory. Solutions of the Hamilton-Jacobi equation will be also used in Chap. 2 devoted to time-independent Hamiltonian systems.

From the constructions of Sect. 1.8 it easily follows that if the force is conservative, then the modified wave transformation $\mathcal{F}_{\mathrm{sd}}^{+}$ is symplectic. In fact, let $S(t,\xi)$ be the solution of the Hamilton-Jacobi equation

$$\partial_t S(t,\xi) = \frac{1}{2}\xi^2 + V(t, \nabla_\xi S(t,\xi))$$

with the initial condition $S(T,\xi) = 0$. Then it is easy to see that $\tilde{Y}(t,\xi) = \nabla_\xi S(t,\xi)$. Therefore $S(t,\xi) + \langle x, \xi\rangle$ is the generating function of the transformation $\phi_{\mathrm{sd}}(t)$. This clearly implies that $\phi_{\mathrm{sd}}(t)$ is a canonical transformation.

In Sect. 1.8 we also construct certain solutions of the eikonal equation

$$-\partial_s \Phi^+(s,x,\xi) = \frac{1}{2}(\nabla_x \Phi^+(s,x,\xi))^2 + V(s,x).$$

These solutions turn out to be generating functions of the time translated wave transformations $\phi(s,0)\circ\mathcal{F}_{\mathrm{fd}}^{+}$ and $\phi(s,0)\circ\mathcal{F}_{\mathrm{sd}}^{+}$.

In the slow-decaying case, before introducing the asymptotic position and the modified wave transformation one has to make certain arbitrary choices. One can say that one has to fix the gauge, which is best described by a choice of the family of reference trajectories $Y(t,\xi)$. Of course, the choice that we describe in Sects. 1.5 and 1.6 is quite arbitrary, there is nothing special about the time $t_1 = T$ and the position $x = 0$ that were fixed to define $Y(t,\xi)$ and $\tilde{Y}(t,\xi)$. It is natural to ask what happens if we change this "gauge". We discuss these things in Sect. 1.9.

In Sect. 1.10 we study the smoothness of basic quantities defined in the preceding section. Roughly speaking, we use the following condition:

$$\partial_x^\alpha F(t,x) \in O(t^{-1-|\alpha|-\mu}), \quad \text{for all } \alpha \in \mathbb{N}^n. \tag{1.0.14}$$

where $\mu > 0$ is a smooth version of the slow-decaying condition, and $\mu > 1$ is a smooth version of the fast-decaying condition. Results of Sect.1.10 will be used in our study of quantum scattering.

In Sect. 1.11 we give conditions under which one can define a relative wave transformation for a pair of time-decaying forces.

In Sect. 1.12 we discuss various possible modified free dynamics, such as the so-called Dollard dynamics, which works in the case $\mu > 1/2$, and its improved versions due to Buslaev and Matveev. These dynamics have the advantage that they are easier to calculate than the dynamics $\phi_{\mathrm{sd}}(t)$ that we introduced in Sect. 1.5. On the other hand, with the exception of the Dollard dynamics, they require much more smoothness on the potentials.

In the literature, classical scattering theory was seldom treated as the end in itself. The main papers investigating explicitly this subject by methods similar to the ones used here were [Sim1, Pro] in the short-range case and [He] in the long-range case. Another approach to this subject, more inspired by classical statistical mechanics, was used earlier in [Co2, Hu2, Hu3].

Most efforts were devoted to quantum scattering theory. Nevertheless, many results about classical scattering were used as intermediate steps in quantum scattering theory (especially in the long-range case). Many ideas were first discovered in the setting of quantum scattering theory, and only afterwards their classical counterparts were formulated – which from the logical point of view is rather unnatural. So the list of contributions to classical scattering theory contains many papers on quantum scattering.

Most of Sect. 1.3 is based on [Sim1].

In the long-range case, the paper that is usually quoted as the first is that of Dollard [Do1]. This paper dealt with the Coulomb quantum case. Its ideas were further developed in [BuMa, AlKa, Hö1]. All of them were devoted primarily to the quantum case. The first (and almost the only) paper that dealt specifically with classical long-range scattering was due to I. Herbst [He]. In fact, most of the material of Sects. 1.5 and 1.6 can be found in [He]. In particular, I. Herbst noticed that there are two ways of defining a modified wave operator: starting from a classification of trajectories as in Sect. 1.5 or using a modified free flow as in Sect. 1.6. Note, however, that our presentation of these results uses weaker assumptions on the forces than I. Herbst's.

The boundary value problem considered at the beginning of Sect. 1.5 in Theorem 1.5.1 was considered by many authors that studied long-range scattering, e.g. in [He, Hö1, KiYa1]. These authors first solved the Cauchy problem and only then applied the inverse function theorem to obtain solutions of the boundary value problem. The idea to use an integral equation to solve this boundary problem in one step was used in [De6].

The arguments used to show the existence of the modified asymptotic position contained in the the proof of Lemma 1.9.7 are a classical adaptation of an argument that I. Sigal used in the quantum case in [Sig2]. The use of solutions of the Hamilton-Jacobi equations in long-range scattering theory is probably due to L. Hörmander. Before, people used functions that solved the Hamilton-Jacobi equation only approximately, such as in [BuMa] (see Sect. 1.5). The use of solutions of the eikonal equation in scattering theory (see Propositions 1.8.3) seems to have first appeared in [IK1] in the quantum context. Many estimates on the derivatives of classical trajectories and other objects constructed in classical scattering can be found in various papers on quantum long-range scattering, such as [Ki5, KiYa1, Hö1, IK1].

Let us remark that, strictly speaking, many of the above-mentioned papers treated time-independent Hamiltonians. Nevertheless, most of their methods apply to the case of time-decaying non-conservative forces as well.

1.1 Basic Notation

X will denote a finite dimensional Euclidean vector space, which plays the role of the configuration space. Its elements will be denoted by x, y, sometimes z. The Euclidean norm of x will be denoted by $|x|$.

X' will stand for the space dual to X – the momentum space. Its elements will be denoted by ξ and η. (X and X' are naturally isomorphic, but still it will be usually useful to make a distinction between them). $X \times X'$ will play the role of the phase space of our problem. If $x \in X$ and $\xi \in X'$, then $\langle x, \xi \rangle$ denotes the duality bracket of x and ξ.

Time will be denoted by t, s and sometimes u. We will usually restrict ourselves to positive times.

$\mathbb{N}$ denotes the set of natural numbers. We assume that $0 \in \mathbb{N}$. Multi-indices, that is elements of $\mathbb{N}^n$ ($n = \dim X$), will be denoted by α, β, etc. $|\alpha|$ will denote the length of the multi-index α. As a rule, derivatives will be understood in the distributional sense.

$\nabla_x^k f(x)$ will denote the kth differential of f at the point x (which is a k linear functional on X). $\partial_x^\alpha f(x)$ will denote the αth partial derivative of f at x.

If Θ is an open subset of $\mathbb{R}^n$, then $C^k(\Theta, \mathbb{R}^m)$ will denote the space of k times continuously differentiable functions from Θ to $\mathbb{R}^m$.

Let us note that, for any function $\Theta \ni x \mapsto f(x)$, we have the following identity:

$$\|\nabla f\|_{L^\infty(\Theta)} = \sup_{x,y \in \Theta,\ x \neq y} \frac{|f(x) - f(y)|}{|x - y|}. \tag{1.1.1}$$

Functions for which (1.1.1) is finite are called uniformly Lipschitz. By writing $f \in C^{0,1}(\Theta, \mathbb{R}^m)$, we will mean that the function f is locally Lipschitz, that is, for every compact $\Theta_1 \subset \Theta$,

$$\inf_{x,y \in \Theta_1,\ x \neq y} \frac{|f(x) - f(y)|}{|x - y|} < \infty.$$

We will write $f \in C^{k,1}(\Theta, \mathbb{R}^m)$ if and only if $\nabla_x^{k-1} f \in C^{0,1}(\Theta, \mathbb{R}^m)$. By writing

$$|\partial_x^\alpha f(x)| \leq C, \quad |\alpha| \leq k,$$

we will mean that $f \in C^{k-1,1}$ and $\|\nabla^k f\|_\infty \leq C$.

If Y is a locally compact space and E a Banach space, $C_\infty(Y, E)$ stands for the space of functions $f \in C(Y, E)$ that satisfy $\lim_{y \to \infty} f(y) = 0$. The space $C_\infty(Y, \mathbb{C})$ will be simply denoted by $C_\infty(Y)$.

We will use the symbols $\langle t \rangle := \sqrt{1 + t^2}$, $\langle x \rangle := \sqrt{1 + x^2}$, etc.

The open ball with center at x_0 and radius $r > 0$ is denoted by

$$B(x_0, r) := \{x \mid |x - x_0| < r\}.$$

If $\Theta \subset \mathbb{R}$, then $\mathbb{1}_\Theta(s)$ will denote the characteristic function of Θ.

C will be the generic name for constants that appear in various estimates. Often they will be different, even though denoted with the same letter. We will usually omit the expressions "for all $\alpha \in \mathbb{N}^n$" and "there exists a $C \in \mathbb{R}$ such that"; the reader will easily figure out where they should be added to make a rigorous mathematical statement.

By writing

$$f(t) \in o(t^k),$$

we will mean that

$$\lim_{t \to \infty} \frac{f(t)}{t^k} = 0.$$

Sometimes we will need to describe estimates that, apart from $t \in \mathbb{R}$, involve some auxiliary variables, say $x \in X$. Let $g(t,x)$ be a positive function and $\Theta \subset \mathbb{R} \times X$. Then by writing "uniformly for $(t,x) \in \Theta$ we have

$$f(t,x) \in o(t^k) g(t,x)\text{"} \tag{1.1.2}$$

we will mean the following: if we set

$$f(t) := \sup_{\{x \mid (t,x) \in \Theta\}} \frac{|f(t,x)|}{g(t,x)},$$

then

$$f(t) \in o(t^k).$$

Likewise, by writing "uniformly for $(t,x) \in \Theta$ we have

$$f(t,x) \in g(t,x) L^1(\mathrm{d}t)\text{"} \tag{1.1.3}$$

we will mean that

$$f(t) \in L^1(\mathrm{d}t).$$

It is convenient to introduce certain families of Banach spaces. These Banach spaces will be used as technical tools in various proofs of this and the next chapter.

For $m \geq 0$, $T > 0$, we put

$$Z_T^m := \left\{ z \in C^0([T, \infty[, X) \mid \sup_{s \in [T,\infty[} \frac{|z(s)|}{|s-T|^m} < \infty \right\} \tag{1.1.4}$$

with the norm

$$\|z\|_{Z_T^m} := \sup_{s \in [T,\infty[} \frac{|z(s)|}{|s-T|^m}.$$

We will also need to consider the closed subspace $Z_{T,\infty}^m$ of Z_T^m defined by

$$Z_{T,\infty}^m := \left\{ z \in Z_T^m \mid \lim_{s \to \infty} \frac{|z(s)|}{|s-T|^m} = 0 \right\}. \tag{1.1.5}$$

1.2 Newton's Equation

Let us now introduce basic definitions related to Newton's equations. We will assume that

$$\mathbb{R}^+ \times X \ni (t,x) \mapsto F(t,x) \in X'$$

is a measurable function that, for all $T_1 \leq T_2 < \infty$, satisfies the following conditions:

$$\begin{aligned} &\textstyle\int_{T_1}^{T_2} |F(t,0)| \mathrm{d}t < \infty, \\ &\textstyle\int_{T_1}^{T_2} \|\nabla_x F(t,\cdot)\|_\infty \mathrm{d}t < \infty. \end{aligned} \tag{1.2.1}$$

Throughout this chapter we consider the equations of motion in the presence of a time-dependent force $F(t,x)$

$$\begin{cases} \partial_t x(t) = \xi(t), \\ \partial_t \xi(t) = F(t, x(t)). \end{cases} \tag{1.2.2}$$

It follows from Proposition A.2.4 that the condition (1.2.1) guarantees the existence and the uniqueness of the solution of (1.2.2) for all positive times. These solutions will be called *phase space trajectories.*

We will denote by $[0,\infty[\ni t \mapsto (x(t,s,y,\eta), \xi(t,s,y,\eta))$ the solution of the equations (1.2.2) with initial conditions

$$\begin{cases} x(s,s,y,\eta) = y, \\ \xi(s,s,y,\eta) = \eta. \end{cases}$$

We will also write $(x(t,y,\eta), \xi(t,y,\eta))$ for $(x(t,0,y,\eta), \xi(t,0,y,\eta))$. Often, we will drop y, η and write simply $(x(t), \xi(t))$.

We denote by $\phi(t,s)$ the evolution generated by the equations (1.2.2), that is,

$$\phi(t,s)(y,\eta) = (x(t,s,y,\eta), \eta(t,s,y,\eta)).$$

$\phi_0(t)$ will stand for the flow generated by the Newton equation with a force equal to 0. The flow $\phi_0(t)$ is simply given by

$$\phi_0(t)(x,\xi) \;=\; (x + t\xi, \xi).$$

Clearly, instead of equations (1.2.2), one can study the Newton equation

$$\partial_t^2 x(t) = F(t, x(t)). \tag{1.2.3}$$

Solutions of (1.2.3) will be called *configuration space trajectories.* We will use interchangeably both phase space and configuration space trajectories (which contain the same amount of information) and we will usually call them just *trajectories*, which should not lead to ambiguities.

Sometimes we will assume that the force $F(t,x)$ is conservative. This means that $F(t,x) = -\nabla_x V(t,x)$ for some time-dependent potential $V(t,x)$. In the conservative case, the equations (1.2.2) are the Hamilton equations for the Hamiltonian

$$H(t,x,\xi) = \frac{1}{2}\xi^2 + V(t,x).$$

1.3 Asymptotic Momentum

We will start our exposition of scattering theory for (1.2.2) with a construction of the so-called *asymptotic momentum.*

Theorem 1.3.1

Assume that

$$\int_0^\infty \|F(t,\cdot)\|_\infty \mathrm{d}t < \infty. \tag{1.3.1}$$

Then for every (y,η)*, there exists the uniform limit*

$$\lim_{t\to\infty} \xi(t,y,\eta) =: \xi^+(y,\eta). \tag{1.3.2}$$

Moreover, the function $\xi^+(\cdot,\cdot)$ *has the following properties:*
(i)

$$\lim_{t\to\infty} \frac{x(t,y,\eta)}{t} = \xi^+(y,\eta);$$

(ii) the function

$$X \times X' \ni (y,\eta) \mapsto \xi^+(y,\eta) \in X'$$

is continuous;
(iii)

$$|\xi^+(y,\eta) - \eta| \le C.$$

Proof. Clearly, one has

$$\xi(t) - \eta = \int_0^t F(u,x(u))\mathrm{d}u,$$

from which we obtain that $\xi(t)$ has the Cauchy property near ∞. This shows the existence of the limit (1.3.2). $\xi^+(\cdot,\cdot)$ is a uniform limit of continuous functions, hence it is continuous.

To prove *(i)*, we write

$$\frac{\mathrm{d}}{\mathrm{d}t}(x(t) - t\xi^+) = \xi(t) - \xi^+ = -\int_t^\infty F(u,x(u))\mathrm{d}u.$$

Hence,

$$x(t) - t\xi^+ = y - \int_0^t uF(u,x(u))\mathrm{d}u - t\int_t^\infty F(u,x(u))\mathrm{d}u. \tag{1.3.3}$$

Thus,

$$\frac{x(t)}{t} - \xi^+ = \frac{y}{t} - \int_0^t \frac{u}{t}F(u,x(u))\mathrm{d}u - \int_t^\infty F(u,x(u))\mathrm{d}u. \tag{1.3.4}$$

By the Lebesgue dominated convergence theorem, (1.3.4) converges to zero. □

The limit (1.3.2) will be called the *asymptotic momentum* along the trajectory $(x(t), \xi(t))$.

Under the assumptions of Theorem 1.3.1, we can extend the definition of $\xi(t,s,y,\eta)$ to $t = \infty$ by setting

$$\xi(\infty,s,y,\eta) := \lim_{t\to\infty} \xi(t,s,y,\eta).$$

The following proposition will be useful in Chaps. 2 and 5:

Proposition 1.3.2
Assume (1.3.1). Then we have, uniformly for $0 \le s \le t < \infty$,

$$x(t,s,y,\eta) - y - (t-s)\eta \in |t-s|o(s^0),$$
$$\xi(t,s,y,\eta) - \eta \in o(s^0).$$

Proof. We use

$$x(t,s,y,\eta) - y - (t-s)\eta) = \textstyle\int_s^t (t-u)F(u,x(u))\mathrm{d}u,$$
$$\xi(t,s,y,\eta) - \eta = \textstyle\int_s^t F(u,x(u))\mathrm{d}u.$$

□

1.4 Fast-Decaying Case

The asymptotic momentum can be defined under very broad conditions on the forces, which include both the fast- and slow-decaying case. Next we would like to impose a more restrictive condition on the decay of the force. We are going to study scattering theory in the fast-decaying case. Roughly speaking this means that the force $F(t,\cdot)$ decays in t like $\langle t\rangle^{-1-\mu}$ for some $\mu > 1$.

Theorem 1.3.1 gives a partial classification of all trajectories. Namely, they are classified by the limit (1.3.2). One would also like to classify all trajectories corresponding to a given asymptotic momentum. In the fast-decaying case, it is possible to classify them in a natural way with elements of the configuration space X. This is described in the following theorem.

Theorem 1.4.1
Assume that

$$\int_0^\infty \langle t\rangle \|F(t,\cdot)\|_\infty \mathrm{d}t < \infty. \tag{1.4.1}$$

Then for any (y,η)*, there exists a uniform limit*

$$\lim_{t\to+\infty}(x(t,y,\eta) - t\xi^+(y,\eta)) =: x^+_{\mathrm{fd}}(y,\eta). \tag{1.4.2}$$

Moreover, the following is true:
(i) the function

$$X \times X' \ni (y,\eta) \mapsto x^+_{\mathrm{fd}}(y,\eta) \in X$$

is continuous;
(ii)

$$|x^+_{\mathrm{fd}}(y,\eta) - y| \leq C;$$

(iii)

$$\lim_{t\to\infty}(x(t,y,\eta) - t\xi(t,y,\eta)) = x^+_{\mathrm{fd}}(y,\eta).$$

Proof. By Lebesgue's dominated convergence theorem, the limit of (1.3.3) exists when $t \to \infty$ and is equal to

$$\lim_{t\to\infty}(x(t) - t\xi^+) = y - \int_0^\infty uF(u,x(u))\mathrm{d}u.$$

$x^+_{\mathrm{fd}}(\cdot,\cdot)$ is continuous as the uniform limit of continuous functions. □

The function $x^+_{\mathrm{fd}}(\cdot,\cdot)$ will be called the *asymptotic position*. The subscript fd stands for "fast-decaying".

Our next theorem shows that in the fast-decaying case, under an additional condition on the force, the mapping from the initial conditions to the scattering data is bijective. We first introduce some notation. For $t \in [0,\infty[$ and $(x,\xi) \in X \times X'$, we denote by $[0,t] \ni s \mapsto (y_{\mathrm{fd}}(s,t,x,\xi), \eta_{\mathrm{fd}}(s,t,x,\xi))$ the solution of

$$\begin{cases} \partial_s y_{\mathrm{fd}}(s,t,x,\xi) = \eta_{\mathrm{fd}}(s,t,x,\xi), \\ \partial_s \eta_{\mathrm{fd}}(s,t,x,\xi) = F(s, y_{\mathrm{fd}}(s,t,x,\xi)), \\ y_{\mathrm{fd}}(t,t,x,\xi) = x + t\xi, \qquad \eta_{\mathrm{fd}}(t,t,x,\xi) = \xi. \end{cases} \tag{1.4.3}$$

We will use the following convention to extend these trajectories for large time: for $s \geq t$, instead of using the flow of Newton's equations, we put

$$(y_{\mathrm{fd}}(s,t,x,\xi), \eta_{\mathrm{fd}}(s,t,x,\xi)) := (x + s\xi, \xi). \tag{1.4.4}$$

Theorem 1.4.2
Suppose that (1.4.1) is true and

$$\int_0^\infty \langle t\rangle \|\nabla_x F(t,\cdot)\|_\infty \mathrm{d}t < \infty. \tag{1.4.5}$$

Then the trajectory $(y_{\rm fd}(s,t,x,\xi), \eta_{\rm fd}(s,t,x,\xi))$ *converges as* $t \to \infty$, *uniformly in* $\mathbb{R}_s \times X \times X'$, *to a trajectory* $(y_{\rm fd}(s,\infty,x,\xi), \eta_{\rm fd}(s,\infty,x,\xi))$, *which satisfies*

$$\begin{cases} \lim\limits_{s\to\infty} (y_{\rm fd}(s,\infty,x,\xi) - x - s\xi) = 0, \\ \lim\limits_{s\to\infty} (\eta_{\rm fd}(s,\infty,x,\xi) - \xi) = 0. \end{cases} \tag{1.4.6}$$

Moreover, the following statements are true:
(i) the trajectory $(y_{\rm fd}(s,\infty,x,\xi), \eta_{\rm fd}(s,\infty,x,\xi))$ *is the only one that satisfies (1.4.6);*
(ii) the mapping

$$\begin{aligned} &[0,\infty] \times X \times X' \ni (t,x,\xi) \\ &\mapsto (y_{\rm fd}(s,t,x,\xi) - x - s\xi,\ \eta_{\rm fd}(s,t,x,\xi) - \xi) \in C_\infty(\mathbb{R}_s, X \times X') \end{aligned}$$

is continuous.

Proof. For simplicity of notation, we will usually suppress the parameters (t,x,ξ) in $y_{\rm fd}(s,t,x,\xi)$. Clearly, $y_{\rm fd}(s)$ satisfies the following integral equation:

$$y_{\rm fd}(s) = x + s\xi + \int_s^t (u-s)F(u, y_{\rm fd}(u))\mathrm{d}u. \tag{1.4.7}$$

We will also set

$$z(s) := y_{\rm fd}(s) - x - s\xi.$$

For any $(t,x,\xi) \in [0,\infty] \times X \times X'$, we define $\mathcal{P}$ as a mapping on $Z^0_{T,\infty}$ by the formula

$$\mathcal{P}(z)(s) := \int_s^t (u-s)F(u, z(u) + x + u\xi)\mathrm{d}u. \tag{1.4.8}$$

Now we can rewrite (1.4.7) as

$$z = \mathcal{P}(z). \tag{1.4.9}$$

We will show that, for large enough T, there exists a unique solution of the equation (1.4.9) in $Z^0_{T,\infty}$.

Note that $\mathcal{P}$ is well defined as a map of $Z^0_{T,\infty}$ into itself. Moreover, for $z \in Z^0_{T,\infty}$, $\mathcal{P}(z)$ is continuous with respect to (t,x,ξ) on $[0,\infty] \times X \times X'$. Using (1.4.5), we obtain that $\mathcal{P}$ is a contraction on $Z^0_{T,\infty}$ uniformly with respect to $(t,x,\xi) \in [0,\infty] \times X \times X'$, provided we take T large enough. In fact, one has

$$\|\mathcal{P}(z_1) - \mathcal{P}(z_2)\|_{Z^0_T} \le \int_T^\infty \langle u\rangle \|\nabla_y F(u,\cdot)\|_\infty \|z_1 - z_2\|_{Z^0_T} \mathrm{d}u.$$

By the fixed point theorem (see Appendix A.2), there exists a unique solution in $Z^0_{T,\infty}$ of the equation (1.4.9), which depends continuously on (t, x, ξ). This gives a solution $y_{\rm fd}(s, \infty, x, \xi)$ defined for $s \geq T$. We extend it for $s \in [0, T]$, using the existence and uniqueness for the flow defined by the equations (1.2.2). □

Note the identities

$$\phi(s,t)\phi_0(t)(x,\xi) = (y_{\rm fd}(s,t,x,\xi),\ \eta_{\rm fd}(s,t,x,\xi)),$$

$$\phi_0^{-1}(t)\phi(t,s)(y,\eta) = (x(t,s,y,\eta) - t\xi(t,s,y,\eta),\ \xi(t,s,y,\eta)).$$

One can interpret the Theorems 1.4.1 and 1.4.2 in a more conventional way by introducing the *wave transformations.*

Theorem 1.4.3
(i) Assume (1.4.1). Then there exists the limit

$$\lim_{t\to\infty} \phi_0^{-1}(t)\phi(t,0) \tag{1.4.10}$$

uniformly on $X \times X'$. The limit is a continuous map from $X \times X'$ into itself.
(ii) Assume, in addition, (1.4.5). Then there exists the limit

$$\lim_{t\to\infty} \phi(0,t)\phi_0(t), \tag{1.4.11}$$

uniformly on $X \times X'$. The mapping

$$\mathcal{F}^+_{\rm fd} : X \times X' \to X \times X'$$

defined by the limit (1.4.11) is called the wave transformations. *It is continuous and bijective. Moreover, (1.4.10) is equal to $(\mathcal{F}^+_{\rm fd})^{-1}$.*
(iii) If $(y,\eta) = \mathcal{F}^+_{\rm fd}(x,\xi)$, then one has

$$\lim_{t\to\infty} (\phi(t,0)(y,\eta) - \phi_0(t)(x,\xi)) = 0. \tag{1.4.12}$$

Note the following identity:

$$(\mathcal{F}^+_{\rm fd})^{-1}(y,\eta) = (x^+_{\rm fd}(y,\eta), \xi^+(y,\eta)).$$

Remark. Let us note that the statement *(iii)* of the above theorem does not follow from the statements *(i)* and *(ii)* (or the other way around). In fact, the relation described in *(iii)* can be used to give another definition of the wave transformation $\mathcal{F}^+_{\rm fd}$ (beside the one given by the limit (1.4.11)). Fortunately, under the assumptions of Theorem 1.4.3 both definitions are equivalent.

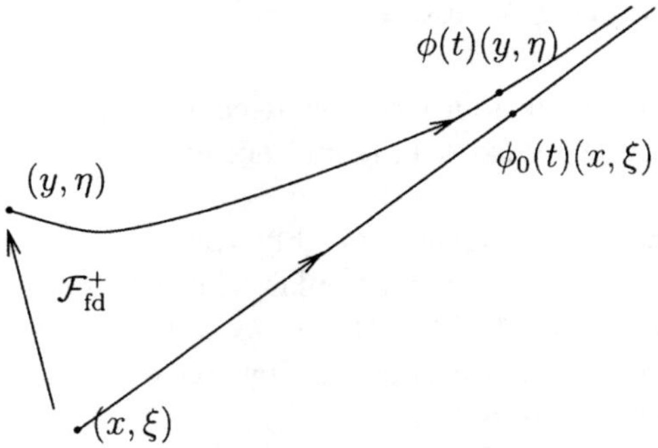

Fig. 1.1. Fast-decaying wave transformation

Remark. In the definition stated after (1.2.2), we use letters (y, η) to denote the initial conditions and (x, ξ) to denote trajectories. In the definition before Theorem 1.4.2, it is the other way around: (y, η) denote trajectories and initial conditions are expressed in terms of (x, ξ). At a first sight this may seem inconsistent, in reality this is quite a natural convention: in both cases (y, η) are the position and the momentum at the time s and the letters (x, ξ) are used to denote the data related to the time $t \to \infty$. We will try to conform to these conventions throughout this and the next chapter, but it will not always be easy to be consistent. In particular, in Sect. 1.5, where we study the boundary value problem related to the slow-decaying case, we will not stick to this convention.

The problem (1.4.6) can be viewed as a kind of a Cauchy problem with the initial conditions set at infinity. In the usual Cauchy problem, where we put the initial conditions at a given point, a typical assumption used to guarantee the uniqueness of the solution is the Lipschitz condition. In the case of (1.4.6), the assumption (1.4.5) on the gradient of the force is an analog of the Lipschitz condition. Below we will give an example of the non-uniqueness for the problem (1.4.6) in the absence of condition (1.4.5). This example (or, more precisely, its time-independent version) is due to Simon [Sim1].

Example 1.4.4
Take a cutoff function $\chi \in C_0^\infty(\mathbb{R})$ such that $\chi(s) = 1$ for $|s| \leq 2$ and $\chi(s) = 0$ for $|s| \geq 3$. Consider the potential

$$V(t,x) = -\frac{\rho^2}{2}|x|^{(2\rho+2)/\rho}\chi(xt^\rho),$$

for $x \in \mathbb{R}$ and $\rho > 0$. It is easy to check that the force $-\nabla_x V(t,x)$ decays like $t^{-2-\rho}$, but the gradient of the force $-\nabla_x^2 V(t,x)$ decays only like t^{-2}. The following two solutions are asymptotic to the same free solution when $t \to \infty$:

$$x_1(t) := 0, \quad x_2(t) := t^{-\rho}.$$

1.5 Slow-Decaying Case I

In this section we begin our study of slow-decaying forces. Our aim is to classify all the trajectories in this case, taking into account their asymptotic behavior as $t \to \infty$.

The slow-decaying case means roughly that the force $F(t,x)$ decays like $\langle t\rangle^{-1-\mu}$ for some $0 < \mu \leq 1$. In particular, the field felt by a particle moving freely in the Coulomb potential has the decay of this type with $\mu = 1$.

Most of the time, when studying the slow-decaying case, we will assume the following conditions on the forces:

$$\int_0^\infty \langle t\rangle^{|\alpha|} \|\partial_x^\alpha F(t,\cdot)\|_\infty \mathrm{d}t < \infty, \quad |\alpha| = 0, 1. \tag{1.5.1}$$

Scattering theory in the slow-decaying case is more difficult than in the fast-decaying case. In general, the usual wave transformation does not exist and the problem (1.4.3) that was used to define the trajectories $(y_{\mathrm{fd}}(s), \eta_{\mathrm{fd}}(s))$ is of no use. Instead, it is more natural to consider a mixed problem, where the boundary conditions are the initial position and the final momentum. Hence we will start with a rather detailed study of this problem, which is the subject of the next theorem.

Theorem 1.5.1
Assume (1.5.1). Then there exists T such that if $T \leq t_1 < t_2 \leq \infty$ and $(x,\xi) \in X \times X'$, there exists a unique trajectory

$$[t_1, t_2] \ni s \mapsto (\tilde{y}(s, t_1, t_2, x, \xi), \tilde{\eta}(s, t_1, t_2, x, \xi))$$

satisfying

$$\begin{cases} \partial_s \tilde{y}(s, t_1, t_2, x, \xi) = \tilde{\eta}(s, t_1, t_2, x, \xi), \\ \partial_s \tilde{\eta}(s, t_1, t_2, x, \xi) = F(s, \tilde{y}(s, t_1, t_2, x, \xi)), \\ \tilde{y}(t_1, t_1, t_2, x, \xi) = x, \qquad \tilde{\eta}(t_2, t_1, t_2, x, \xi) = \xi. \end{cases} \tag{1.5.2}$$

($\tilde{\eta}(\infty, t_1, \infty, x, \xi) = \xi$ means of course $\lim_{s\to\infty} \tilde{\eta}(s, t_1, \infty, x, \xi) = \xi$). Moreover, the solution satisfies the following estimates, uniformly for $T \leq t_1 \leq s \leq t_2 \leq \infty$ and $(x,\xi) \in X \times X'$:

$$\partial_\xi^\beta (\tilde{y}(s, t_1, t_2, x, \xi) - x - (s - t_1)\xi) \in o(s^0)|s - t_1|, \quad |\beta| \leq 1, \tag{1.5.3}$$

$$\partial_x^\alpha (\tilde{y}(s, t_1, t_2, x, \xi) - x - (s - t_1)\xi) \in o(t_1^0), \quad |\alpha| = 1, \tag{1.5.4}$$

$$\partial_x^\alpha \partial_\xi^\beta (\tilde{\eta}(s, t_1, t_2, x, \xi) - \xi) \in o(s^{-|\alpha|}), \quad |\alpha| + |\beta| \leq 1, \tag{1.5.5}$$

$$\partial_x^\alpha (\tilde{\eta}(s, t_1, t_2, x, \xi) - \xi) \in L^1(\mathrm{d}s), \quad |\alpha| = 1, \tag{1.5.6}$$

$$\tilde{y}(s, t_1, t_2, x, \xi) - \tilde{y}(s, t_1, \infty, x, \xi) \in o(t_2^0)|s - t_1|, \tag{1.5.7}$$

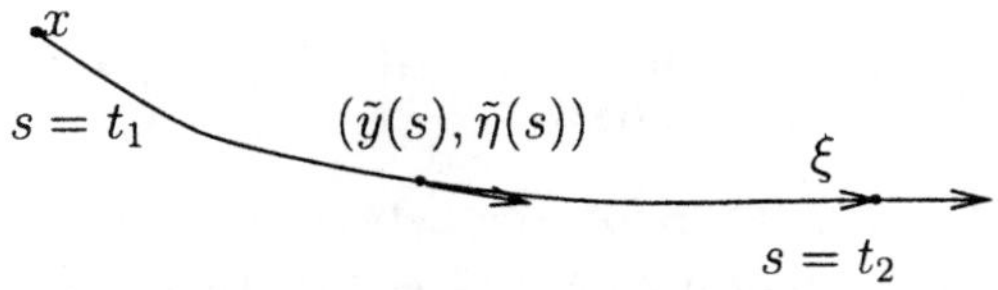

Fig. 1.2. Boundary value problem for slow-decaying scattering

$$\tilde{\eta}(s, t_1, t_2, x, \xi) - \tilde{\eta}(s, t_1, \infty, x, \xi) \in o(t_2^0). \tag{1.5.8}$$

Proof. We will suppress parameters t_1, t_2, x and ξ when possible, to simplify the notation. An easy computation shows that $\tilde{y}(s)$ has to satisfy the following integral equation:

$$\begin{aligned}\tilde{y}(s) \ &= x + (s - t_1)\xi - \textstyle\int_{t_1}^{s} (u - t_1) F(u, \tilde{y}(u))\mathrm{d}u \\ &-(s - t_1) \textstyle\int_{s}^{t_2} F(u, \tilde{y}(u))\mathrm{d}u.\end{aligned} \tag{1.5.9}$$

We will set

$$\tilde{z}(s) := \tilde{y}(s) - x - (s - t_1)\xi$$

and introduce the following function:

$$\zeta_{t_1,s}(u) := \begin{cases} 0, & u \le t_1, \\ u - t_1, & t_1 \le u \le s, \\ s - t_1, & s \le u. \end{cases}$$

We extend $\tilde{z}(s)$ by setting $\tilde{z}(s) := \tilde{z}(t_2)$ for $t_2 \le s \le \infty$. We define

$$\mathcal{P}(\tilde{z})(s) := - \textstyle\int_{t_1}^{t_2} \zeta_{t_1,s}(u) F(u, x + (u - t_1)\xi + \tilde{z}(u))\mathrm{d}u. \tag{1.5.10}$$

where the map $\mathcal{P}$ depends on the parameters t_1, t_2, x, ξ. Now we can write the equation (1.5.9) in the form

$$\tilde{z} = \mathcal{P}(\tilde{z}). \tag{1.5.11}$$

As in the proof of Theorem 1.4.2, we will apply the fixed point theorem to solve (1.5.11), but now we choose for our Banach space the space $Z_{t_1,\infty}^1$. We have

$$|s - t_1|^{-1} |\mathcal{P}(\tilde{z})(s)| \le \int_{t_1}^{\infty} |s - t_1|^{-1} \zeta_{t_1,s}(u) \|F(u, \cdot)\|_\infty \mathrm{d}u. \tag{1.5.12}$$

Note that

$$0 \le |s - t_1|^{-1} \zeta_{t_1,s}(u) \le 1,$$

for any u, we have the pointwise convergence

$$\lim_{s\to\infty}\left(\sup_{\{t_1 \mid T\le t_1\le s\}} |s-t_1|^{-1}\zeta_{t_1,s}(u)\right)=0,$$

and $\|F(u,\cdot)\|_\infty \in L^1(\mathbb{R}^+)$. Therefore, by Lebesgue's dominated convergence theorem, (1.5.12) is $o(s^0)$, uniformly for $T\le t_1\le s$. Hence, $\mathcal{P}$ is bounded on $Z^1_{t_1,\infty}$ with norm $o(t_1^0)$.

Similarly we estimate:

$$\begin{aligned}&|s-t_1|^{-1}|\nabla_{\tilde z}\mathcal{P}(\tilde z)v(s)|\\ &\le \textstyle\int_{t_1}^\infty |s-t_1|^{-1}\zeta_{t_1,s}(u)|u-t_1|\|\nabla_{\tilde y}F(u,\cdot)\|_\infty\|v\|_{Z^1_{t_1,\infty}}\mathrm{d}u.\end{aligned}\tag{1.5.13}$$

Therefore,

$$\|\nabla_{\tilde z}\mathcal{P}(\tilde z)\|_{B(Z^1_{t_1,\infty})}\le\int_{t_1}^\infty |u-t_1|\|\nabla_{\tilde y}F(u,\cdot)\|_\infty \mathrm{d}u\in o(t_1^0).$$

We fix T such that

$$\int_T^\infty |u-T|\|\nabla_{\tilde y}F(u,\cdot)\|_\infty\mathrm{d}u<1.$$

Clearly, if $T\le t_1$ then the map $\mathcal{P}$ is a contraction on $Z^1_{t_1,\infty}$. So, by the fixed point theorem, there exists a unique solution of (1.5.2).

Let us now prove that $\tilde z(s)$ satisfies the estimates (1.5.3) – (1.5.6).

The fact that $|\tilde z(s)|\in o(s^0)|s-t_1|$ is immediate by the properties of the range of $\mathcal{P}$ that we described above.

Let us now prove (1.5.3) with $|\beta|=1$. We use the identity

$$(1-\nabla_{\tilde z}\mathcal{P}(\tilde z))\nabla_\xi\tilde z=\nabla_\xi\mathcal{P}(\tilde z).\tag{1.5.14}$$

Let us first check that $\nabla_\xi\mathcal{P}(\tilde z)$ belongs to the Banach space $Z^1_{t_1,\infty}$. Indeed, we see that

$$|s-t_1|^{-1}|\nabla_\xi\mathcal{P}(\tilde z)(s)|\le\int_{t_1}^\infty |s-t_1|^{-1}\zeta_{t_1,s}(u)|u-t_1|\|\nabla_{\tilde y}F(u,\cdot)\|_\infty\mathrm{d}u.$$

By (1.5.1) and Lebesgue's dominated convergence theorem, this quantity goes to 0 when t_1 and $s-t_1$ go to ∞, which proves that $\nabla_\xi\mathcal{P}(\tilde z)\in Z^1_{t_1,\infty}$ and $\|\nabla_\xi\mathcal{P}(\tilde z)\|_{Z^1_{t_1}}\in o(t_1^0)$. We already know that $\nabla_{\tilde z}\mathcal{P}(\tilde z)$ is contracting on $Z^1_{t_1,\infty}$, for $t_1\ge T$. Hence, using (1.5.14), we get that $\nabla_\xi\tilde z\in Z^1_{t_1,\infty}$ and $\|\nabla_\xi\tilde z\|_{Z^1_{t_1}}\in o(t_1^0)$, which implies (1.5.3) with $|\beta|=1$.

Let us prove (1.5.4). First note that $\nabla_x\mathcal{P}(\tilde z)$ belongs to $Z^0_{t_1}$. Indeed,

$$|\nabla_x\mathcal{P}(\tilde z)(s)|\le\int_{t_1}^\infty\zeta_{t_1,s}(u)\|\nabla_{\tilde y}F(u,\cdot)\|_\infty\mathrm{d}u\in o(t_1^0).\tag{1.5.15}$$

Moreover,

$$\|\nabla_{\tilde z}\mathcal{P}(\tilde z)\|_{B(Z^0_{t_0})}\le\int_{t_1}^\infty\zeta_{t_1,s}(u)\|\nabla_{\tilde y}F(u,\cdot)\|_\infty\mathrm{d}u\in o(t_1^0).$$

Thus, for $t_1 \geq T$, the map $\mathcal{P}$ is a contraction on $Z^0_{t_1}$, and using

$$(1 - \nabla_{\tilde{z}}\mathcal{P}(\tilde{z}))\nabla_x \tilde{z} = \nabla_x \mathcal{P}(\tilde{z}), \tag{1.5.16}$$

we obtain (1.5.4).

The fact that $\eta(s) - \xi = \dot{\tilde{z}}(s) \in o(s^0)$ follows from the formula

$$\dot{\tilde{z}}(s) = -\int_s^{t_2} F(u, \tilde{y}(u))\mathrm{d}u. \tag{1.5.17}$$

To prove (1.5.5) with $|\beta| = 1$, we differentiate (1.5.17) with respect to ξ and obtain

$$\begin{aligned} |\nabla_\xi \dot{\tilde{z}}(s)| &\leq C \textstyle\int_s^\infty \|\nabla_{\tilde{y}} F(u, \cdot)\|_\infty |\nabla_\xi \tilde{y}(u)| \mathrm{d}u \\ &\leq C_1 \textstyle\int_s^\infty \|\nabla_{\tilde{y}} F(u, \cdot)\| |u - t_1| \mathrm{d}u \in o(s^0). \end{aligned}$$

Next, we differentiate the identity (1.5.17) with respect to x. We get

$$\begin{aligned} |\nabla_x \dot{\tilde{z}}(s)| &\leq C \textstyle\int_s^\infty \|\nabla_{\tilde{y}} F(u, \cdot)\|_\infty |\nabla_x \tilde{y}(u)| \mathrm{d}u \\ &\leq C_1 s^{-1} \textstyle\int_s^\infty u \|\nabla_x F(u, \cdot)\|_\infty \mathrm{d}u \in o(s^{-1}). \end{aligned} \tag{1.5.18}$$

This shows (1.5.5) with $|\alpha| = 1$.

Using

$$\begin{aligned} \textstyle\int_T^\infty |\nabla_x \dot{\tilde{z}}(s)| \mathrm{d}s &\leq C \textstyle\int_T^\infty \mathrm{d}s \int_s^\infty \|\nabla_{\tilde{y}} F(u, \cdot)\|_\infty \mathrm{d}u \\ &= C \textstyle\int_T^\infty (s - T) \|\nabla_{\tilde{y}} F(s, \cdot)\|_\infty \mathrm{d}s < \infty, \end{aligned}$$

we obtain (1.5.6). (1.5.7) follows from the continuity of $\mathcal{P}$ on $Z^1_{T,\infty}$ with respect to t_2. (1.5.8) follows from (1.5.17) and (1.5.7). □

Let us remark that the above theorem immediately implies that ξ^+ maps X' onto X'.

If in Theorem 1.5.1 we fix t_1 and put $t_2 = \infty$, we get a complete classification of trajectories using the initial (at time t_1) position x and the asymptotic momentum ξ. This classification is not satisfactory from the point of view of scattering theory, because x is not an asymptotic quantity and because it depends on the choice of t_1.

In the fast-decaying case, all the trajectories with a given asymptotic momentum were classified with the function $x^+_{\mathrm{fd}}(\cdot, \cdot)$, which was a natural asymptotic quantity. Therefore, they were labeled with elements of the configuration space X. In the slow-decaying case, they can be classified in a natural way by elements of the affine space X. Unlike in the fast-decaying case, in general we cannot replace "affine" by "vector". This classification is the subject of the next theorem, which was first proven by I. Herbst [He] under more restrictive conditions on the forces.

Theorem 1.5.2
Assume (1.5.1).
(i) Let $(y_1(s), \eta_1(s))$ *and* $(y_2(s), \eta_2(s))$ *be two trajectories such that*

$$\lim_{s\to\infty} \eta_1(s) = \lim_{s\to\infty} \eta_2(s).$$

Then there exists

$$\lim_{s\to\infty} (y_1(s) - y_2(s)).$$

Moreover,

$$\eta_1(s) - \eta_2(s) \in o(s^{-1}). \qquad (1.5.19)$$

(ii) Let $(y_1(s), \eta_1(s))$ *be a trajectory and* $x \in X$*. Then there exists a unique trajectory* $(y_2(s), \eta_2(s))$ *such that*

$$\begin{cases} \lim_{s\to\infty} \eta_1(s) = \lim_{s\to\infty} \eta_2(s), \\ \lim_{s\to\infty} (y_1(s) - y_2(s)) = x. \end{cases}$$

Proof. Let us first show *(i)*. We choose T large enough. Let $x_i := y_i(T)$ and $\xi := \lim_{s\to\infty} \eta_i(s)$, $i = 1, 2$. Using the notation introduced in Theorem 1.5.1, we note that $\eta_i(s) = \tilde{\eta}(s, T, \infty, x_i, \xi)$. Therefore,

$$\eta_1(s) - \eta_2(s) = \int_0^1 (x_1 - x_2)\nabla_x \tilde{\eta}(s, T, \infty, \tau x_1 + (1-\tau)x_2, \xi)\mathrm{d}\tau.$$

Using estimates (1.5.6), (1.5.5) of Theorem 1.5.1 we get that $\eta_1(s) - \eta_2(s) \in L^1(\mathrm{d}s) \cap o(s^{-1})$, which proves *(i)*.

Let us now prove *(ii)*. We will use the following integral equation:

$$y_2(s) - y_1(s) = x - \int_s^\infty (u - s)(F(u, y_2(u)) - F(u, y_1(u)))\mathrm{d}u. \qquad (1.5.20)$$

We set

$$z(s) := y_2(s) - y_1(s) - x$$

and

$$\mathcal{P}(z)(s) := -\int_s^\infty (u - s)(F(u, y_1(u) + x + z(u)) - F(u, y_1(u)))\mathrm{d}u.$$

We see that z has to satisfy

$$z = \mathcal{P}(z). \qquad (1.5.21)$$

By the now standard argument, we see that $\mathcal{P}$ is a contraction on the Banach space $Z^0_{T,\infty}$, for T large enough. Therefore, by the fixed point theorem, the equation (1.5.21) possesses a unique solution, which ends the proof of the theorem. □

Theorem 1.5.2 means that the space of trajectories with a given asymptotic momentum has the structure of an affine space. To turn it into a vector space, we have to fix an "origin" in this space. Unlike in the fast-decaying case, where we could use the free trajectory $W_1(t,\xi) := t\xi$ to fix the origin in a natural way, in the slow-decaying case, we have to make an arbitrary choice. Namely, for each $\xi \in X'$, we need to choose a certain trajectory $Y(t,\xi)$ with the asymptotic momentum ξ.

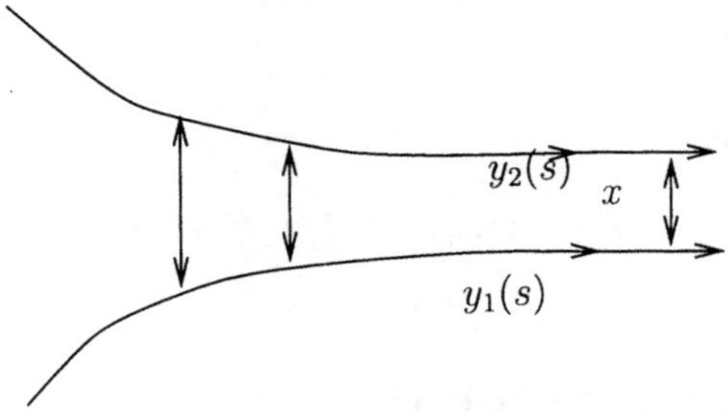

Fig. 1.3. Affine space structure

One way to do this is to fix T large enough such that there exist the trajectories

$$Y(t,\xi) := \tilde{y}(t,T,\infty,0,\xi), \quad E(t,\xi) := \tilde{\eta}(t,T,\infty,0,\xi)$$

constructed in Theorem 1.5.1. (Let us stress that the choice of $t_1 = T$, and $x = 0$ is completely arbitrary).

From Theorem 1.5.2 and its proof we see that the following theorem is true:

Theorem 1.5.3
Assume the hypotheses of Theorem 1.5.1. Then for any (y,η), there exists

$$\lim_{t\to\infty}(x(t,y,\eta) - Y(t,\xi^+(y,\eta))) =: x^+_{\mathrm{sd}}(y,\eta).$$

Moreover, the following statements are true:
(i) the function

$$X \times X' \ni (y,\eta) \mapsto x^+_{\mathrm{sd}}(y,\eta) \in X$$

is continuous;
(ii)

$$|x^+_{\mathrm{sd}}(y,\eta)| \le C|x(T,y,\eta)|;$$

(iii)

$$x(t,y,\eta) - Y(t,\xi^+(y,\eta)) - x^+_{\mathrm{sd}}(y,\eta) \in o(t^0)|x(T,y,\eta)|;$$

(iv) the map

$$X \times X' \ni (y,\eta) \mapsto (x^+_{\mathrm{sd}}(y,\eta), \xi^+(y,\eta)) \in X \times X' \tag{1.5.22}$$

is bijective.

We will call $x_{\rm sd}^+(\cdot,\cdot)$ the *modified asymptotic position.*

Theorem 1.5.3 leads to our first definition of the modified wave transformation.

Definition 1.5.4
The inverse of the map (1.5.22) will be called the modified wave transformations *and will be denoted* $\mathcal{F}_{\rm sd}^+$.

Clearly, we have

$$(\mathcal{F}_{\rm sd}^+)^{-1}(y,\eta) = (x_{\rm sd}^+(y,\eta), \xi^+(y,\eta)).$$

1.6 Slow-Decaying Case II

So far, scattering theory for slow-decaying forces seems to have little resemblance to the fast-decaying case. Now we would like to present a construction of modified wave transformations that is more parallel to the one we gave in the fast-decaying case.

Let T be the number fixed before Theorem 1.5.3. Denote, for $t \geq T$,

$$\tilde{Y}(t,\xi) := \tilde{y}(t,T,t,0,\xi). \tag{1.6.1}$$

The function $\tilde{Y}(t,\xi)$ is called a *modifier.* Let us note its properties:

$$\begin{aligned} &\tilde{Y}(T,\xi) = 0, \\ &\partial_\xi^\beta(\tilde{Y}(t,\xi) - t\xi) \in o(t), \quad |\beta| = 0,1, \\ &Y(t,\xi) = \tilde{Y}(t, E(t,\xi)), \\ &x(t,s,\tilde{Y}(s,\eta),\eta) = \tilde{Y}(t,\xi(t,s,\tilde{Y}(s,\eta),\eta)). \end{aligned} \tag{1.6.2}$$

If we differentiate the last identity with respect to t and plug in $s = t$ then we obtain the following equation

$$\partial_t\tilde{Y}(t,\xi) = \xi - \nabla_\xi\tilde{Y}(t,\xi)F(t,\tilde{Y}(t,\xi)). \tag{1.6.3}$$

(1.6.3) is closely related to the Hamilton-Jacobi equation, which we will consider in the conservative case. The following theorem gives an alternative way to construct the function $x_{\rm sd}^+(\cdot,\cdot)$ using $\tilde{Y}(t,\xi)$ instead of $Y(t,\xi)$.

Theorem 1.6.1
Assume that the hypotheses of Theorem 1.5.1 hold. Then

$$x(t,y,\eta) - \tilde{Y}(t,\xi(t,y,\eta)) - x_{\rm sd}^+(y,\eta) \in o(t^0)|x(T,y,\eta)|.$$

Proof. Set

$$\xi^+ := \xi^+(y,\eta).$$

Note that

$$\xi(t) = \tilde{\eta}(t,T,\infty,x(T),\xi^+).$$

The difference of $x(t,y,\eta) - Y(t,\xi^+(y,\eta))$ and $x(t,y,\eta) - \tilde{Y}(t,\xi(t,y,\eta))$ equals

$$\begin{aligned}
&\tilde{Y}(t,\xi(t)) - \tilde{Y}(t,E(t,\xi^+)) \\
&= \tilde{y}(t,T,t,0,\tilde{\eta}(t,T,\infty,x(T),\xi^+)) - \tilde{y}(t,T,t,0,\tilde{\eta}(t,T,\infty,0,\xi^+)) \\
&= \textstyle\int_0^1 \nabla_\xi \tilde{y}(t,T,t,0,\tau\tilde{\eta}(t,T,\infty,x(T),\xi^+) + (1-\tau)\tilde{\eta}(t,T,\infty,0,\xi^+))\mathrm{d}\tau \\
&\times \textstyle\int_0^1 \nabla_x \tilde{\eta}(t,T,\infty,\sigma x(T),\xi^+)x(T)\mathrm{d}\sigma \\
&\in O(t)o(t^{-1})|x(T)| = o(t^0)|x(T)|.
\end{aligned}$$

(At the end we used estimates (1.5.3) and (1.5.5)). □

The identity

$$x^+_{\mathrm{sd}}(y,\eta) = \lim_{t\to\infty}(x(t,y,\eta) - \tilde{Y}(t,\xi(t,y,\eta))) \tag{1.6.4}$$

obtained in Theorem 1.6.1 can be viewed as the second definition of $x^+_{\mathrm{sd}}(y,\eta)$.

For $t \in [T,\infty[$, we denote by $[0,t] \ni s \mapsto (y_{\mathrm{sd}}(s,t,x,\xi), \eta_{\mathrm{sd}}(s,t,x,\xi))$ the unique trajectory such that

$$\begin{cases} y_{\mathrm{sd}}(t,t,x,\xi) = x + \tilde{Y}(t,\xi), \\ \eta_{\mathrm{sd}}(t,t,x,\xi) = \xi. \end{cases} \tag{1.6.5}$$

Our next result is an analog of Theorem 1.4.2.

Theorem 1.6.2
Assume that the hypotheses of Theorem 1.5.1 hold.

Then the trajectory

$$(y_{\mathrm{sd}}(s,t,x,\xi), \eta_{\mathrm{sd}}(s,t,x,\xi))$$

converges as $t \to \infty$ *to a trajectory* $(y_{\mathrm{sd}}(s,\infty,x,\xi), \eta_{\mathrm{sd}}(s,\infty,x,\xi))$, *which satisfies*

$$\begin{cases} \lim\limits_{s\to\infty}(y_{\mathrm{sd}}(s,\infty,x,\xi) - x - Y(s,\xi)) = 0, \\ \lim\limits_{s\to\infty}(\eta_{\mathrm{sd}}(s,\infty,x,\xi) - \xi) = 0. \end{cases} \tag{1.6.6}$$

Moreover, the following statements are true:
(i) the trajectory $(y_{\mathrm{sd}}(s,\infty,x,\xi), \eta_{\mathrm{sd}}(s,\infty,x,\xi))$ *is the only one that satisfies (1.6.6);*
(ii) the mapping

$$[0,\infty]\times X\times X'\ni(t,x,\xi)$$
$$\mapsto(y_{\rm sd}(s,t,x,\xi)-x-y_{\rm sd}(s,t,0,\xi),\ \eta_{\rm sd}(s,t,x,\xi)-\xi)\in C^0_\infty(\mathbb{R}_s,X\times X')$$

is continuous;
(iii) we have, uniformly for $0\le s\le t$, $T\le t$, $(x,\xi)\in X\times X'$,

$$y_{\rm sd}(s,t,x,\xi)-y_{\rm sd}(s,\infty,x,\xi)\in o(t^0)(\langle x\rangle+\langle s\rangle), \tag{1.6.7}$$

$$\eta_{\rm sd}(s,t,x,\xi)-\eta_{\rm sd}(s,\infty,x,\xi)\in o(t^0)(\langle s\rangle^{-1}\langle x\rangle+1). \tag{1.6.8}$$

Proof. First note that we already can solve (1.6.5) for $x=0$. Namely, by (1.6.1)

$$(y_{\rm sd}(s,t,0,\xi),\eta_{\rm sd}(s,t,0,\xi))=(\tilde y(s,T,t,0,\xi),\tilde\eta(s,T,t,0,\xi)). \tag{1.6.9}$$

By (1.5.7) and (1.5.8), there exists a limit of (1.6.9) as $t\to\infty$ that satisfies

$$\tilde y(s,T,t,0,\xi)-\tilde y(s,T,\infty,0,\xi)\in o(t^0)\langle s\rangle, \tag{1.6.10}$$

$$\tilde\eta(s,T,t,0,\xi)-\tilde\eta(s,T,\infty,0,\xi)\in o(t^0). \tag{1.6.11}$$

Next consider the case of a general x. To simplify the notation, we will write $(y_{\rm sd}(s,x),\eta_{\rm sd}(s,x))$ instead of $(y_{\rm sd}(s,t,x,\xi),\eta_{\rm sd}(s,t,x,\xi))$.

Now $y_{\rm sd}(s,x)$ satisfies the following integral equation:

$$\begin{aligned}&y_{\rm sd}(s,x)\\&=x+\tilde Y(t,\xi)+\xi(s-t)-\textstyle\int_s^t(u-s)F(u,y_{\rm sd}(u,x))du.\end{aligned} \tag{1.6.12}$$

We set

$$z(s):=y_{\rm sd}(s,x)-y_{\rm sd}(s,0)-x. \tag{1.6.13}$$

If we subtract (1.6.12) with an arbitrary x from (1.6.12) with $x=0$, we obtain that $z(s)$ satisfies the integral equation

$$z=\mathcal{P}(z), \tag{1.6.14}$$

where

$$\mathcal{P}(z)(s):=\int_s^t(u-s)(F(u,y_{\rm sd}(u,0)+x+z(u))-F(u,y_{\rm sd}(u,0)))du.$$

Using hypothesis (1.5.1) and the now standard argument, we see that $\mathcal{P}$ is a contraction of $Z^0_{T,\infty}$ for T large enough. Moreover, using Theorem 1.5.1 and Lebesgue's dominated convergence theorem, we see that $\mathcal{P}$ depends continuously on the parameters $(t,x,\xi)\in[0,\infty]\times X\times X'$. Therefore, there exists a unique solution $z\in Z^0_{T,\infty}$ of the fixed point equations, which depends continuously on $(t,x,\xi)\in[0,\infty]\times X\times X'$. This proves that $y_{\rm sd}(s,\infty,x,\xi)$ exists, satisfies (1.6.6), and also the statements *(i)* and *(ii)* are true.

Next we remark that the inequality

$$|z(s)| - \int_s^t u\|\nabla_y F(u,\cdot)\|_\infty |z(u)| \mathrm{d}u \le \int_s^t u\|\nabla_y F(u,\cdot)\|_\infty |x| \mathrm{d}u$$

and the Gronwall inequality imply

$$|z(s)| \le C|x|. \tag{1.6.15}$$

This estimate will be useful in the proof of *(iii)*, which we are going to give now.

To keep track the dependence on the parameters, it will be convenient to denote (1.6.13) by $z(s,t,x,\xi)$ and the mapping $\mathcal{P}$ by $\mathcal{P}_{t,x,\xi}$. Note that, for any $z \in Z^0_{T,\infty}$, we have uniformly in t, x, ξ the following estimate:

$$\|\mathcal{P}_{t,x,\xi}(z) - \mathcal{P}_{\infty,x,\xi}(z)\|_{Z^0_{T,\infty}} \in o(t^0)(\langle x\rangle + |z|).$$

If we take into account (1.6.15), then we see that

$$\|\mathcal{P}_{t,x,\xi}(z(\cdot,\infty,x,\xi)) - \mathcal{P}_{\infty,x,\xi}(z(\cdot,\infty,x,\xi))\|_{Z^0_{T,\infty}} \in o(t^0)\langle x\rangle.$$

Therefore, from the proof of Proposition A.2.2 we obtain

$$\|z(\cdot,t,x,\xi) - z(\cdot,\infty,x,\xi)\|_{Z^0_{T,\infty}} \in o(t^0)\langle x\rangle.$$

This together with (1.6.10) ends the proof of (1.6.7)

From (1.6.7) and from

$$\dot{z}(s,t,x,\xi) = -\int_s^t (F(u, y_{\mathrm{sd}}(u,t,x,\xi)) - F(u, y_{\mathrm{sd}}(u,t,0,\xi)))\mathrm{d}u$$

we obtain easily

$$\dot{z}(s,t,x,\xi) - \dot{z}(s,\infty,x,\xi) \in o(t^0)(\langle s\rangle^{-1}\langle x\rangle + 1).$$

This, together with (1.6.11) yields (1.6.8). □

As in the section on the fast-decaying case, we will now interpret Theorems 1.6.1 and 1.6.2 using wave transformations. We start by defining an appropriate *modified free flow.*

Definition 1.6.3
We define the modified free flow $\phi_{\mathrm{sd}}(t)$ *by*

$$X \times X' \ni (x,\xi) \mapsto \phi_{\mathrm{sd}}(t)(x,\xi) := (x + \tilde{Y}(t,\xi), \xi) \in X \times X'.$$

Note that, in general, the above introduced modified free flow is defined only for $t \ge T$.

Note the following identities:

$$\phi(s,t)\phi_{\mathrm{sd}}(t)(x,\xi) = (y_{\mathrm{sd}}(s,t,x,\xi), \eta_{\mathrm{sd}}(s,t,x,\xi)),$$

$$\phi_{\mathrm{sd}}^{-1}(t)\phi(t,s)(y,\eta) = (x(t,s,y,\eta) - \tilde{Y}(t,\xi(t,s,y,\eta), \ \xi(t,s,y,\eta)).$$

We have the following corollary of Theorems 1.6.1 and 1.6.2.

Theorem 1.6.4
Assume that the hypotheses of Theorem 1.5.1 hold. Then the limits

$$\lim_{t\to\infty} \phi(0,t)\phi_{\mathrm{sd}}(t), \tag{1.6.16}$$

$$\lim_{t\to\infty} \phi_{\mathrm{sd}}^{-1}(t)\phi(t,0) \tag{1.6.17}$$

exist and are equal to $\mathcal{F}_{\mathrm{sd}}^{+}$ *and* $(\mathcal{F}_{\mathrm{sd}}^{+})^{-1}$ *respectively. The convergence of the momentum component is uniform and the convergence of the position component is of the type* $o(t^0)|x(T,y,\eta)|$ *in (1.6.17). The convergence of (1.6.16) is of the type* $o(t^0)\langle x\rangle$.

We have the following identity:

$$(\mathcal{F}_{\mathrm{sd}}^{+})^{-1}(y,\eta) = (x_{\mathrm{sd}}^{+}(y,\eta), \xi^{+}(y,\eta)).$$

Remark. Let us mention a difference between the slow-decaying and the fast-decaying case, first pointed out by Herbst [He]. Namely, if $(y,\eta) = \mathcal{F}_{\mathrm{sd}}^{+}(x,\xi)$, then in general it is *false* that

$$\lim_{t\to\infty} (\phi(t,0)(y,\eta) - \phi_{\mathrm{sd}}(t)(x,\xi)) = 0. \tag{1.6.18}$$

The correct analog of (1.4.12) from the fast-decaying case is (1.6.6) or, in other words,

$$\lim_{t\to\infty} (\phi(t,0)(y,\eta) - (x + Y(t,\xi),\xi)) = 0. \tag{1.6.19}$$

Of course, the momentum component of (1.6.18) converges to zero. But, in general, the position component is divergent. Below we will give an example to illustrate this statement. But first let us give a heuristic argument why the existence of (1.6.18) should not be expected.

Let us subtract the x-component of (1.6.19) from the x-component of (1.6.18). We obtain

$$\tilde{Y}(t,\xi) - Y(t,\xi) = \tilde{y}(t,T,t,0,\xi) - \tilde{y}(t,T,t,0,\tilde{\eta}(t,T,\infty,0,\xi)). \tag{1.6.20}$$

We know by Theorem 1.5.1 that $\tilde{\eta}(t,T,\infty,0,\xi) - \xi \in o(t^0)$ and $\nabla_\xi \tilde{y}(t,T,t,0,\xi) \in O(t)$. This yields a bound $o(t)$ on (1.6.20) and not $o(t^0)$.

Example 1.6.5 Let us give an example illustrating the above remark. We will consider the flow generated by the 1−dimensional force

$$F(t,x) = t^{-1-\mu}.$$

It is easy to solve exactly the equations of motion for this problem. In particular, we have

$$\tilde{y}(s,t_1,t_2,x,\xi)$$
$$= \begin{cases} x+(s-t_1)\xi+\left(\frac{1}{1-\mu}+\frac{1}{\mu}\right)\left(s^{1-\mu}-t_1^{1-\mu}\right)-\frac{1}{\mu}(s-t_1)t_2^{-\mu}, & 0<\mu<1, \\ x+(s-t_1)\xi+\log s-\log t_1-(s-t_1)t_2^{-1}, & \mu=1. \end{cases}$$

We can choose $T=1$, and then we obtain

$$Y(s,\xi)=\begin{cases} (s-1)\xi+\left(\frac{1}{1-\mu}+\frac{1}{\mu}\right)(s^{1-\mu}-1), & 0<\mu<1, \\ (s-1)\xi+\log s, & \mu=1, \end{cases}$$

and

$$\tilde{Y}(s,\xi)=\begin{cases} (s-1)\xi+\left(\frac{1}{1-\mu}+\frac{1}{\mu}\right)(s^{1-\mu}-1)-\frac{1}{\mu}(s-1)s^{-\mu}, & 0<\mu<1, \\ (s-1)\xi+\log s-(s-1)s^{-1}, & \mu=1. \end{cases}$$

Thus, for $0<\mu\leq 1$,

$$\lim_{s\to\infty}(Y(s,\xi)-\tilde{Y}(s,\xi))$$

exists only in the case $\mu=1$, and then it is non-zero.

1.7 Boundary Conditions for Wave Transformations

Let $\mathcal{F}^+$ denote $\mathcal{F}^+_{\mathrm{fd}}$ or $\mathcal{F}^+_{\mathrm{sd}}$. Consider the equation

$$(y,\eta)=\mathcal{F}^+(x,\xi).$$

It turns out that it is often useful to express (η,x) in terms of (y,ξ). Unfortunately, in general, this is possible only if we replace (y,η) with $\phi(s,0)(y,\eta)$ for s big enough. In this section we look more closely at $\phi(s,0)\circ\mathcal{F}^+$ with such boundary condition.

First assume (1.5.1). Let T be given by Theorem 1.5.1. For any $x,\xi\in X\times X'$ and $s\geq T$, we define

$$\zeta^+(s,x,\xi):=\tilde{\eta}(s,s,\infty,x,\xi).$$

Next we assume the fast-decaying condition (1.4.1) and (1.4.5). We set

$$z^+_{\mathrm{fd}}(s,x,\xi):=x^+_{\mathrm{fd}}(\phi(0,s)(x,\zeta^+(s,x,\xi))).$$

We clearly have

$$\phi(s,0)\circ\mathcal{F}^+_{\mathrm{fd}}(z^+_{\mathrm{fd}},\xi)=(x,\zeta^+).$$

where we write $(z^+_{\mathrm{fd}},\zeta^+)$ instead of $(z^+_{\mathrm{fd}}(s,x,\xi),\zeta^+(s,x,\xi))$.

Proposition 1.7.1
The following identities are true:

$$\begin{aligned} z_{\mathrm{fd}}^{+}(s,x,\xi) &= \lim_{t\to\infty}(\tilde{y}(t,s,\infty,x,\xi)-t\xi) \\ &= \lim_{t\to\infty}(\tilde{y}(t,s,t,x,\xi)-t\xi), \end{aligned}$$

$$z_{\mathrm{fd}}^{+}(s,x,\xi)-x+s\xi=\int_s^\infty(\tilde{\eta}(t,s,\infty,x,\xi)-\xi)\mathrm{d}t. \tag{1.7.1}$$

Assume now the slow-decaying assumption (1.5.1). We set

$$z_{\mathrm{sd}}^{+}(s,x,\xi):=x_{\mathrm{sd}}^{+}(\phi(0,s)(x,\zeta^{+}(s,x,\xi))).$$

We clearly have

$$\phi(s,0)\circ\mathcal{F}_{\mathrm{sd}}^{+}(z_{\mathrm{fd}}^{+},\xi)=(x,\zeta^{+}).$$

where we write $(z_{\mathrm{sd}}^{+},\zeta^{+})$ instead of $(z_{\mathrm{sd}}^{+}(s,x,\xi),\zeta^{+}(s,x,\xi))$.

Proposition 1.7.2
The following identities are true:

$$\begin{aligned} z_{\mathrm{sd}}^{+}(s,x,\xi) &= \lim_{t\to\infty}(\tilde{y}(t,s,\infty,x,\xi)-Y(t,\xi)) \\ &= \lim_{t\to\infty}(\tilde{y}(t,s,t,x,\xi)-\tilde{Y}(t,\xi)), \end{aligned}$$

$$z_{\mathrm{sd}}^{+}(s,x,\xi)-x+Y(s,\xi)=\int_s^\infty(\tilde{\eta}(t,s,\infty,x,\xi)-\tilde{\eta}(t,s,\infty,Y(s,\xi),\xi))\mathrm{d}t.$$

Moreover, uniformly for $x,\xi\in X\times X'$ and $T\leq s$, we have the following estimates.

$$\begin{aligned} &z_{\mathrm{sd}}^{+}(s,x,\xi)-x+Y(s,\xi)\in o(s^0)|x-Y(s,\xi)|, \\ &\partial_\xi^\beta(z_{\mathrm{sd}}^{+}(s,x,\xi)-x+Y(s,\xi))\in o(s^0)(|x-Y(s,\xi)|+\langle s\rangle),\quad |\beta|\leq 1, \\ &\partial_x^\alpha(z_{\mathrm{sd}}^{+}(s,x,\xi)-x+Y(s,\xi))\in o(s^0),\quad |\alpha|=1, \\ &\partial_x^\alpha\partial_\xi^\beta(\zeta^{+}(s,x,\xi)-\xi)\in o(s^{-|\alpha|}),\quad |\alpha|+|\beta|\leq 1, \\ &\partial_x^\alpha(\zeta^{+}(s,x,\xi)-\xi)\in L^1(\mathrm{d}s),\quad |\alpha|=1. \end{aligned} \tag{1.7.2}$$

Proof. The first identity follows immediately from the Definition 1.5.4. Let us show the second identity.

Set

$$z_{\mathrm{sd}}(t):=\tilde{y}(t,s,t,x,\xi)-\tilde{Y}(t,\xi),\quad \zeta(t):=\tilde{\eta}(s,s,t,x,\xi).$$

Then

$$\phi_{\mathrm{sd}}^{-1}(t)\phi(t,s)(x,\zeta(t))=(z_{\mathrm{sd}}(t),\xi).$$

Clearly,

$$\lim_{t\to\infty} \zeta_{\rm sd}(t) = \zeta_{\rm sd}^+.$$

Moreover, $\phi_{\rm sd}^{-1}(t)\phi(t,s)$ converges uniformly on compact sets to $(\mathcal{F}_{\rm sd}^+)^{-1}\phi(0,s)$. Hence

$$\lim_{t\to\infty} z_{\rm sd}(t) = z_{\rm sd}^+.$$

The three last estimates of (1.7.2) follow immediately from Theorem 1.5.1. The first two estimates follow from the identity

$$\begin{aligned} &z_{\rm sd}^+(s,x_0,\xi) - x_0 + Y(s,\xi) \\ &= (x_0 - Y(s,\xi)) \textstyle\int_s^\infty \int_0^1 \nabla_x \tilde\eta(t,s,\infty,\tau x_0 + (1-\tau)Y(s,\xi),\xi){\rm d}t{\rm d}\tau. \end{aligned} \tag{1.7.3}$$

□

1.8 Conservative Forces

In addition to the hypotheses (1.5.1), let us assume that the force is conservative, that is, there exist a real potential $V(t,x)$ such that

$$F(t,x) = -\nabla_x V(t,x).$$

Under this assumption, Theorem 1.5.1 can be used to solve the Hamilton-Jacobi equation. We refer to Appendix A.3 for some general facts about the Hamilton-Jacobi equation.

With the notation of Theorem 1.5.1, for $T \le t_1 \le t_2 < \infty$, we put

$$\begin{aligned} S(t_1,t_2,x,\xi) \ &:= \langle \xi, \tilde y(t_2,t_1,t_2,x,\xi)\rangle \\ &\quad - \textstyle\int_{t_1}^{t_2} \left(\frac{1}{2}\tilde\eta^2(s,t_1,t_2,x,\xi) - V(s,\tilde y(s,t_1,t_2,x,\xi))\right){\rm d}s. \end{aligned}$$

By Appendix A.3, we know that $S(t_1,t_2,x,\xi)$ satisfies two Hamilton-Jacobi equations. Let us describe the properties of this function in the following proposition.

Proposition 1.8.1
(i) The function $S(t_1,t_2,x,\xi)$ is the only $C^{1,1}(\mathbb{R}^n_\xi)$ solution of the problem

$$\begin{cases} \partial_{t_2} S(t_1,t_2,x,\xi) = \frac{1}{2}\xi^2 + V(t_2, \nabla_\xi S(t_1,t_2,x,\xi)), \\ S(t,t,x,\xi) = \langle x,\xi\rangle. \end{cases}$$

(ii) The function $S(t_1,t_2,x,\xi)$ is the only $C^{1,1}(\mathbb{R}^n_x)$ solution of the problem

$$\begin{cases} -\partial_{t_1} S(t_1,t_2,x,\xi) = \frac{1}{2}(\nabla_x S(t_1,t_2,x,\xi))^2 + V(t_1,x), \\ S(t,t,x,\xi) = \langle x,\xi\rangle. \end{cases}$$

(iii) $\nabla_x S(t_1,t_2,x,\xi) = \tilde{\eta}(t_1,t_1,t_2,x,\xi)$; $\nabla_\xi S(t_1,t_2,x,\xi) = \tilde{y}(t_2,t_1,t_2,x,\xi)$;
(iv) The following estimates are true, uniformly for $T \le t_1 \le t_2$, $(x,\xi) \in X \times X'$:

$$\partial_\xi^\beta \left(S(t_1,t_2,x,\xi) - \langle x,\xi\rangle - \tfrac{1}{2}(t_2-t_1)\xi^2\right) \in |t_2-t_1| o(t_2^0), \quad |\beta| = 1,2;$$
$$\partial_x^\alpha \partial_\xi^\beta \left(S(t_1,t_2,x,\xi) - \langle x,\xi\rangle - \tfrac{1}{2}(t_2-t_1)\xi^2\right) \in o(t_1^{1-|\alpha|}), \quad |\alpha| \ge 1,\ |\alpha|+|\beta| \le 2;$$
$$\partial_x^\alpha (S(t_1,t_2,x,\xi) - \langle x,\xi\rangle - \tfrac{1}{2}(t_2-t_1)\xi^2) \in L^1(\mathrm{d}t_1), \quad |\alpha| = 2.$$

Proof. *(i) – (iii)* follow immediately from Appendix A.3. The estimates *(iv)* follow immediately from the equations *(iii)* and Theorem 1.5.1. □

The Hamilton-Jacobi equation containing x-derivatives is sometimes called the eikonal equation. Below we construct certain solutions of the eikonal equation. We start with the fast-decaying case.

Proposition 1.8.2
Assume (1.4.1) and (1.4.5) and that the force is conservative. Then the wave transformation $\mathcal{F}_{\mathrm{fd}}^+$ *is symplectic. Moreover, the following limit exists:*

$$\lim_{t\to\infty} \left(S(s,t,x,\xi) - \frac{1}{2}t\xi^2\right) =: \Phi_{\mathrm{fd}}^+(s,x,\xi).$$

It satisfies the eikonal equation

$$-\partial_s \Phi_{\mathrm{fd}}^+(s,x,\xi) = \frac{1}{2}(\nabla_x \Phi_{\mathrm{fd}}^+(s,x,\xi))^2 + V(s,x).$$

It is a generating function of $\phi(s,0) \circ \mathcal{F}_{\mathrm{fd}}^+$, *that is,*

$$\nabla_x \Phi_{\mathrm{fd}}^+(s,x,\xi) = \zeta^+(s,x,\xi), \quad \nabla_\xi \Phi_{\mathrm{fd}}^+(s,x,\xi) = z_{\mathrm{fd}}^+(s,x,\xi).$$

Next let us return to the assumption (1.5.1). Recall that, in order to define the family of trajectories $Y(t,\xi)$, we fixed a time T. With the same T, let us set

$$S(t,\xi) := S(T,t,0,\xi). \tag{1.8.1}$$

The function $[T,\infty[\times X' \ni (t,\xi) \mapsto S(t,\xi)$ is the unique solution of the Hamilton-Jacobi equation with the zero initial condition at time T:

$$\begin{cases} \partial_t S(t,\xi) = \frac{1}{2}\xi^2 + V(t, \nabla_\xi S(t,\xi)), \\ S(T,\xi) = 0. \end{cases}$$

Moreover, by (1.6.1),

$$\nabla_\xi S(t,\xi) = \tilde{Y}(t,\xi),$$

and the modified free flow can be written as

$$\phi_{\rm sd}(t)(x,\xi) = (x + \nabla_\xi S(t,\xi), \xi). \tag{1.8.2}$$

We can also construct a solution of the eikonal equation in the slow-decaying case, as described in the following proposition. Consequently, the following fact is true:

Proposition 1.8.3
Assume (1.5.1) and that the force is conservative. Then the modified wave transformation is symplectic. Moreover, the following limit exists:

$$\lim_{t\to\infty} (S(s,t,x,\xi) - S(t,\xi)) =: \Phi^+_{\rm sd}(s,x,\xi).$$

It satisfies the eikonal equation

$$-\partial_s \Phi^+_{\rm sd}(s,x,\xi) = \frac{1}{2}(\nabla_x \Phi^+_{\rm sd}(s,x,\xi))^2 + V(s,x).$$

It is a generating function of $\phi(s,0) \circ \mathcal{F}^+_{\rm sd}$, that is,

$$\nabla_x \Phi^+_{\rm sd}(s,x,\xi) = \zeta^+(s,x,\xi), \quad \nabla_\xi \Phi^+_{\rm sd}(s,x,\xi) = z^+_{\rm sd}(s,x,\xi).$$

Uniformly for $T \le s,\ (x,\xi) \in X \times X'$, we have

$$\begin{aligned}
&\partial_\xi^\beta(\Phi^+_{\rm sd}(s,x,\xi) - \langle x,\xi\rangle + S(s,\xi)) \in o(s^0)(|x - Y(s,x)| + \langle s\rangle), \qquad 1 \le |\beta| \le 2,\\
&\partial_x^\alpha\partial_\xi^\beta(\Phi^+_{\rm sd}(s,x,\xi) - \langle x,\xi\rangle + S(s,\xi)) \in o(s^{1-|\alpha|}), \qquad |\alpha| \ge 1, \quad |\alpha| + |\beta| \le 2,\\
&\partial_x^\alpha(\Phi^+_{\rm sd}(s,x,\xi) - \langle x,\xi\rangle + S(s,\xi)) \in L^1({\rm d}s), \qquad |\alpha| = 2.
\end{aligned}$$

Proof. (1.8.2) means that the flow $\phi_{\rm sd}(t)$ is symplectic and generated by the function $S(t,\xi) + \langle x,\xi\rangle$. This implies that $\mathcal{F}^+_{\rm sd}$ is symplectic.

Next let us compute:

$$\begin{aligned}
&\lim_{t\to\infty} (S(s,t,x,\xi) - S(t,\xi))\\
&= \lim_{t\to\infty} (S(s,t,x,\xi) - S(s,t,0,\xi)) + \lim_{t\to\infty} (S(s,t,0,\xi) - S(T,t,0,\xi))\\
&= x\textstyle\int_0^1 \tilde\eta(s,s,\infty,\tau x,\xi){\rm d}\tau - \int_T^s(\frac{1}{2}(\tilde\eta^2(u,u,\infty,0,\xi) + V(u,0)){\rm d}u.
\end{aligned}$$

Thus, $\Phi^+_{\rm sd}(s,x,\xi)$ is well defined. □

1.9 Gauge Invariance of Wave Transformations

Throughout this section we assume the conditions of Theorem 1.5.1. Note that, in this section, it will be convenient to denote $\mathcal{F}^+_{\rm sd}$, $x^+_{\rm sd}$, $\Phi^+_{\rm sd}$ by $\mathcal{F}^+$, x^+, Φ^+.

We have seen in the preceding sections that, in slow-decaying scattering theory, one has to make some arbitrary choices. For example, the definition of the reference trajectories $Y(t, \xi)$ was just a convention. In this section we examine how the different objects we introduced are modified when we make a different choice of reference trajectories.

Let us first examine what happens when we change the family of reference trajectories in Theorem 1.5.3. Instead of considering $(Y(t, \xi), E(t, \xi))$, let us fix a rather arbitrary family of trajectories $(Y_1(t, \xi), E_1(t, \xi))$ that satisfy

$$\lim_{t\to\infty} E_1(t, \xi) = \xi.$$

By replacing $Y(t, \xi)$ with $Y_1(t, \xi)$ in Theorem 1.5.3, one can define another asymptotic position x_1^+ and another modified wave transformation $\mathcal{F}_1^+$ by

$$\begin{aligned}&\lim_{t\to\infty}(x(t, y, \eta) - Y_1(t, \xi^+(y, \eta))) =: x_1^+(y, \eta),\\ &(\mathcal{F}_1^+)^{-1}(y, \eta) = (x_1^+(y, \eta), \xi^+(y, \eta)).\end{aligned} \tag{1.9.1}$$

One has

$$(\mathcal{F}^+)^{-1}(y, \eta) - (\mathcal{F}_1^+)^{-1}(y, \eta) = (f^+(\xi^+(y, \eta)), 0), \tag{1.9.2}$$

where f^+ is an arbitrary function. The function $f^+(\cdot)$ can be computed from the formula

$$f^+(\xi) = \lim_{t\to\infty}(Y(t, \xi) - Y_1(t, \xi)). \tag{1.9.3}$$

(The above limit exists in the sense of the uniform convergence on compact sets and f^+ is continuous if $Y_1(\cdot, \xi)$ depends continuously on ξ).

Note that if the force $F(t, x)$ is conservative, then it is natural to require that a modified wave transformation be symplectic. This is not always the case for the above defined $\mathcal{F}_1^+$. Therefore, in the conservative case, one prefers a more restrictive class of definitions of modified wave transformations.

One might ask what are the natural quantities in the slow-decaying theory (since $\mathcal{F}_{\rm sd}^+$ is not naturally defined). It is easier to describe them using inverse modified wave transformations. The basic natural objects are the asymptotic momentum

$$\xi^+(y, \eta)$$

and the derivative of an asymptotic position along level sets of the asymptotic momentum:

$$\nabla x^+(y, \eta)\Big|_{\mathrm{Ker}\nabla\xi^+(y,\eta)}.$$

It is easy to see from (1.9.2) that these quantities do not depend on the choice of a modified wave transformation and are uniquely determined by the system itself.

The derivative of a modified wave transformation with respect to x of order $|\alpha| \geq 1$ can be interpreted in an invariant way as a function on the *affine space* X: namely, if $\mathcal{F}^+$ and $\mathcal{F}_1^+$ are related by (1.9.2) then, for any $|\alpha| \geq 1$,

$$\partial_x^\alpha \mathcal{F}^+(x,\xi) = \partial_x^\alpha \mathcal{F}_1^+(x - f^+(\xi), \xi).$$

Note also that if we are given a wave transformation $\mathcal{F}_1^+$, then we can retrieve the family of trajectories $Y_1(\cdot, \xi)$ by the formula

$$Y_1(t,\xi) = x(t, \mathcal{F}_1^+(0,\xi)).$$

Sections 1.5 and 1.6 present two different, although equivalent definitions of modified wave transformations. The definition from Sect. 1.5 is based on a comparison of the flow with a family of *reference trajectories*. The identity (1.6.16) from Sect. 1.6 can be viewed as an alternative definition of $\mathcal{F}^+$, which is based on a comparison of the flow with a *modified free dynamics*. Let us note that the second definition, especially its quantum-mechanical analog, is probably more common in the literature.

We have seen in Sect. 1.6 that we can associate to the family of reference trajectories $Y(t,\xi)$ a modifier $\tilde{Y}(t,\xi)$, which can be used in an alternative definition of the wave transformation $\mathcal{F}^+$. This modifier satisfies the equation

$$Y(t,\xi) = \tilde{Y}(t, E(t,\xi)).$$

Let us assume that we have fixed a different family of reference trajectories $(Y_1(t,\xi), E_1(t,\xi))$ leading to a different modified wave transformation $\mathcal{F}_1^+$. One might ask if there exists a modifier $\tilde{Y}_1(t,\xi)$ satisfying

$$Y_1(t,\xi) = \tilde{Y}_1(t, E_1(t,\xi)), \tag{1.9.4}$$

such that the modified free flow defined with $\tilde{Y}_1(t,\xi)$ can be used to construct the wave transformation $\mathcal{F}_1^+$ as in Theorem 1.6.4.

In general, the answer is negative. The second definition of $\mathcal{F}_1^+$ is possible only if we assume that the reference trajectories satisfy certain regularity properties.

A natural class of reference trajectories, for which the two approaches are equivalent, is described in the following definition.

Definition 1.9.1
Let $(Y_1(t,\xi), E_1(t,\xi))$ be a family of trajectories satisfying

$$\lim_{t\to\infty} E_1(t,\xi) = \xi. \tag{1.9.5}$$

We say that it is a regular family of reference trajectories *if, uniformly in ξ, we have*

$$\begin{aligned} &\partial_\xi^\beta (Y_1(t,\xi) - t\xi) \in o(t), \quad |\beta| = 1, \\ &\partial_\xi^\beta (E_1(t,\xi) - \xi) \in o(t^0), \quad |\beta| = 1. \end{aligned} \tag{1.9.6}$$

(1.9.6) is equivalent to a much simpler condition.

Lemma 1.9.2
A family of reference trajectories satisfying (1.9.5) is regular if and only if there exists $T_1 > T$ and C such that

$$|\partial_\xi^\beta Y_1(T_1, \xi)| \le C, \quad |\beta| = 1. \tag{1.9.7}$$

Proof. The conditions (1.9.6) follow from the identities

$$Y_1(t,\xi) = \tilde{y}(t, T_1, \infty, Y_1(T_1,\xi), \xi),$$
$$E_1(t,\xi) = \tilde{\eta}(t, T_1, \infty, Y_1(T_1,\xi), \xi),$$

(1.9.7) and Theorem 1.5.1. □

Theorem 1.9.3
Suppose that $(Y_1(t,\xi), E_1(t,\xi))$ is a regular family of trajectories. Then for t large enough, there exists a unique solution $\tilde{Y}_1(t,\xi)$ of the equation (1.9.4). It has the following properties:

$$\partial_t \tilde{Y}_1(t,\xi) = \xi - \nabla_\xi \tilde{Y}_1(t,\xi) F(t, \tilde{Y}_1(t,\xi)), \tag{1.9.8}$$

$$\partial_\xi^\beta (\tilde{Y}_1(t,\xi) - t\xi) \in o(t), \quad |\beta| = 1. \tag{1.9.9}$$

The theorem follows from the following lemma.

Lemma 1.9.4
The map

$$X' \ni \xi \mapsto E_1(t,\xi) \in X' \tag{1.9.10}$$

is invertible for t large enough, and the inverse mapping

$$X' \ni \eta \mapsto \Xi_1^+(t,\eta) \in X'$$

satisfies for large t

$$\partial_\eta^\beta (\Xi_1^+(t,\eta) - \eta) \in o(t^0), \quad |\beta| = 0, 1. \tag{1.9.11}$$

Proof. We know that, for $T > T_1$,

$$|\nabla_\xi (E_1(t,\xi) - \xi)| \le \frac{1}{2}. \tag{1.9.12}$$

This proves the global invertibility of the map (1.9.10) for $t \ge T_1$. The estimates (1.9.11) follow from (1.9.12) and (1.5.5). □

Proof of Theorem 1.9.3. We set

$$\tilde{Y}_1(t,\eta) := Y_1(t, \Xi_1^+(t,\eta)).$$

Following the proof of (1.6.2) we can check the following identity

$$x(t,s,\tilde{Y}_1(s,\xi),\xi) = \tilde{Y}_1(t,\xi(t,s,\tilde{Y}_1(s,\xi),\xi)).$$

Differentiating this identity with respect to t and plugging in $s = t$, we obtain (1.9.8). Using the estimates (1.5.3), (1.5.4) of Theorem 1.5.1, we obtain the estimate (1.9.9). □

Definition 1.9.5
A function $[T_1,\infty[\times X' \ni (t,\xi) \mapsto X$ *is called a* regular modifier *if it satisfies*

$$\begin{aligned} &\partial_t \tilde{Y}_1(t,\xi) = \xi - \nabla_\xi \tilde{Y}_1(t,\xi) F(t,\tilde{Y}_1(t,\xi)), \\ &\partial_\xi^\beta \tilde{Y}_1(t,\xi) \in O(t), \quad |\beta| = 1. \end{aligned} \tag{1.9.13}$$

The modified free dynamics associated with $\tilde{Y}_1$ *is defined as*

$$\phi_1(t)(x,\xi) := (x + \tilde{Y}_1(t,\xi), \xi). \tag{1.9.14}$$

Theorem 1.9.6
Suppose that $\tilde{Y}_1(t,\xi)$ *is a regular modifier. Then there exists a unique regular family of reference trajectories* $(Y_1(t,\xi), E_1(t,\xi))$ *such that*

$$Y_1(t,\xi) = \tilde{Y}_1(t, E_1(t,\xi)), \qquad \lim_{t\to\infty} E_1(t,\xi) = \xi.$$

Moreover, the following limits exist uniformly on compact sets:

$$\begin{aligned} &\lim_{t\to\infty} \phi(0,t)\phi_1(t) = \mathcal{F}_1^+, \\ &\lim_{t\to\infty} \phi_1^{-1}(t)\phi(t,0) = (\mathcal{F}_1^+)^{-1}, \\ &\lim_{t\to\infty} \left(\tilde{Y}_1(t,\xi) - \tilde{Y}(t,\xi)\right) = f^+(\xi), \end{aligned}$$

where $\mathcal{F}_1^+$, f^+ *where defined at the beginning of this section.*

The proof of the above theorem is divided into a series of steps.

Lemma 1.9.7
There exists

$$\tilde{x}_1^+(y,\eta) := \lim_{t\to\infty} (x(t,y,\eta) - \tilde{Y}_1(t,\xi(t,y,\eta))).$$

Consequently, there exists

$$(\tilde{\mathcal{F}}_1^+)^{-1} := \lim_{t\to\infty} \phi_1^{-1}(t)\phi(t,0) = (\tilde{x}_1^+, \xi^+).$$

(We will see later on that $\tilde{x}_1^+ = x_1^+$ and $\tilde{\mathcal{F}}_1^+ = \mathcal{F}_1^+$).

Proof. We have, using (1.9.13),

$$\begin{aligned} &\tfrac{\mathrm{d}}{\mathrm{d}t}(x(t,y,\eta) - \tilde{Y}_1(t,\xi(t,y,\eta))) \\ &= \nabla_\xi \tilde{Y}_1(t,\xi)(F(t,\tilde{Y}_1(t,\xi)) - F(t,x(t,y,\eta))). \end{aligned} \tag{1.9.15}$$

Consequently, if we set

$$k(t) := |x(t,y,\eta) - \tilde{Y}_1(t,\xi(t,y,\eta))|,$$

then $k(t)$ satisfies

$$\left|\frac{\mathrm{d}}{\mathrm{d}t}k(t)\right| \leq f(t)k(t), \quad \text{for some} \quad f \in L^1(\mathrm{d}t).$$

By the Gronwall inequality, $f(t)$ is bounded. Applying again (1.9.15) we see that the limit of $x(t,y,\eta) - \tilde{Y}_1(t,\xi(t,y,\eta))$ exists. □

Lemma 1.9.8
There exists the limit

$$\lim_{t\to\infty}(\tilde{Y}_1(t,\xi(t,y,\eta)) - \tilde{Y}(t,\xi(t,y,\eta))) \tag{1.9.16}$$

For any $\xi \in X'$, the limit (1.9.16) does not depend on (y,η) as long as $\xi = \xi^+(y,\eta)$.

Proof. The existence of the limit (1.9.16) follows immediately from Lemma 1.9.7.

If $\xi^+(y_1,\eta_1) = \xi^+(y_2,\eta_2)$, then by (1.5.19)

$$\xi(t,y_1,\eta_1) - \xi(t,y_2,\eta_2) \in o(t^{-1}).$$

Hence

$$\begin{aligned} &\tilde{Y}_1(t,\xi(t,y_1,\eta_1)) - \tilde{Y}_1(t,\xi(t,y_2,\eta_2)) \in o(t^0), \\ &\tilde{Y}(t,\xi(t,y_1,\eta_1)) - \tilde{Y}(t,\xi(t,y_2,\eta_2)) \in o(t^0). \end{aligned}$$

Therefore, the limit (1.9.16) is the same for (y_1,η_1) and (y_2,η_2). □

Set $\tilde{f}^+(\xi)$ to be the limit (1.9.16). Define

$$\phi_{\tilde{f}^+}(x,\xi) = (x + \tilde{f}^+(\xi), \xi).$$

Lemma 1.9.9
We have

$$\tilde{x}_1^+(y,\eta) = x^+(y,\eta) + \tilde{f}^+(\xi^+(y,\eta))$$

and, consequently,

$$(\tilde{\mathcal{F}}_1^+)^{-1} = \phi_{\tilde{f}^+}^{-1} \circ (\mathcal{F}^+)^{-1}. \tag{1.9.17}$$

Proof. Using (1.6.4), we obtain

$$\begin{aligned} \tilde{x}_1^+(y,\eta) &= \lim_{t\to\infty}(x(t,y,\eta) - \tilde{Y}(t,\xi(t,y,\eta))) \\ &+ \lim_{t\to\infty}(\tilde{Y}(t,\xi(t,y,\eta)) - \tilde{Y}_1(t,\xi(t,y,\eta))) = x^+(y,\eta) - \tilde{f}^+(y,\eta). \end{aligned}$$

□

Lemma 1.9.10
We have

$$\tilde{f}^+(\xi) = \lim_{t\to\infty}(\tilde{Y}_1(t,\xi) - \tilde{Y}(t,\xi)).$$

Proof. We will write $\phi(t)$ instead of $\phi_{\rm sd}(t)$. Note that

$$\phi_1^{-1}(t)\phi(t)(x,\xi) = (x + \tilde{Y}(t,\xi) - \tilde{Y}_1(t,\xi), \xi).$$

Using (1.9.17), we obtain

$$\begin{aligned} \phi_{\tilde{f}^+}^{-1}(x,\xi) = (\tilde{\mathcal{F}}_1^+)^{-1} \circ \mathcal{F}^+(x,\xi) &= \lim_{t\to\infty} \phi_1^{-1}(t) \circ \phi(t)(x,\xi) \\ &= \lim_{t\to\infty}(x + \tilde{Y}(t,\xi) - \tilde{Y}_1(t,\xi), \xi). \end{aligned}$$

□

We define

$$(Y_1(t,\xi), E_1(t,\xi)) := \phi(t,0) \circ \tilde{\mathcal{F}}_1^+(0,\xi). \tag{1.9.18}$$

Lemma 1.9.11
(1.9.18) is the unique trajectory satisfying (1.9.4).

Proof. Recall from the proof of Lemma 1.9.7 that if we set

$$k(t) := |x(t,y,\eta) - \tilde{Y}_1(t,\xi(t,y,\eta))|,$$

then $k(t)$ satisfies

$$\left|\frac{\mathrm{d}}{\mathrm{d}t}k(t)\right| \leq f(t)k(t), \quad f \in L^1(\mathrm{d}t). \tag{1.9.19}$$

If $\tilde{x}_1^+(y,\eta) = 0$, then it means that

$$\lim_{t\to\infty} k(t) = 0.$$

By the Gronwall inequality (Proposition A.1.1 applied backward in time), this implies $k(t) = 0$ for all $t \geq T_1$. This means that $(Y_1(t,\xi), E_1(t,\xi))$ satisfies (1.9.4).

To prove the uniqueness, let $(x(t,y,\eta), \xi(t,y,\eta))$ be a trajectory and assume that, for some $t_0 \geq T_1$, $x_0 = x(t_0, y, \eta)$ and $\xi_0 = \xi(t_0, y\eta)$, we have

$$x_0 = \tilde{Y}_1(t_0, \xi_0).$$

Hence $k(t_0) = 0$. By (1.9.19) and the Gronwall inequality, we obtain that $k(t) = 0$ for all t. Hence $\tilde{x}_1^+(y,\eta) = 0$, therefore the trajectory $(x(t,y,\eta), \xi(t,y,\eta))$ equals $\phi(t,0)\circ\tilde{\mathcal{F}}_1^+(0,\xi)$ for $\xi = \xi^+(y,\eta)$. □

This ends the proof of Theorem 1.9.6 except for the regularity properties of $(Y_1(t,\xi), E_1(t,\xi))$. They will follow from the estimates (1.9.20) proven in the following proposition.

Proposition 1.9.12
(i) Assume that

$$\partial_\xi^\beta \tilde{Y}_1(t,\xi) \in O(t), \quad |\beta| = 1.$$

(For the moment, we do not assume that the differential equation (1.9.21) is satisfied.) Then there exists T_1 such that, for every $T_1 \leq t_1 \leq t_2 \leq \infty$, there exists a trajectory

$$s \mapsto (Y_1(s,t_1,t_2,\xi), E_1(s,t_1,t_2,\xi))$$

such that

$$\begin{cases} Y_1(t_1,t_1,t_2,\xi) = \tilde{Y}_1(t_1, E_1(t_1,t_1,t_2,\xi)), \\ E_1(t_2,t_1,t_2,\xi) = \xi. \end{cases}$$

It satisfies, uniformly for $T_1 \leq t_1 \leq s \leq t_2$, the estimates

$$\partial_\xi^\beta (Y_1(s,t_1,t_2,\xi) - (s-t_1)\xi - \tilde{Y}_1(t_1,\xi)) \in o(s), \quad |\beta| \leq 1. \tag{1.9.20}$$

(ii) Assume, additionally,

$$\partial_t \tilde{Y}_1(t,\xi) = \xi - \nabla_\xi \tilde{Y}_1(t,\xi) F(t, \tilde{Y}_1(t,\xi)). \tag{1.9.21}$$

Let $Y_1(t,\xi)$ be the family of trajectories constructed in Theorem 1.9.6. Then for any s,

$$\tilde{Y}_1(t,\xi) = Y_1(t,s,t,\xi),$$

$$Y_1(t,\xi) = Y_1(t,s,\infty,\xi).$$

Proof. Using the notation of Theorem 1.5.1, we can write

$$(Y_1(s,t_1,t_2,\xi), E_1(s,t_1,t_2,\xi)) = (\tilde{y}(s,t_1,t_2,x,\xi), \tilde{\eta}(s,t_1,t_2,x,\xi)),$$

where x is a function of t_1, t_2, ξ, which satisfies

$$x = \tilde{Y}_1(t_1, \tilde{\eta}(t_1, t_1, t_2, x, \xi)). \tag{1.9.22}$$

The equation (1.9.22) can be written as

$$x = \mathcal{P}(x),$$

where $\mathcal{P}$ is a map on X that depends on the parameters t_1, t_2 and ξ.

Let us show that, for big enough t_1, the map $\mathcal{P}$ is a contraction. In fact,

$$\nabla_x \mathcal{P}(x) = \nabla_{\tilde{\eta}} \tilde{Y}_1(t, \tilde{\eta}(t_1, t_1, t_2, x, \xi)) \nabla_x \tilde{\eta}(t_1, t_1, t_2, x, \xi).$$

By (1.5.5) and (1.9.13), this is $O(t_1) o(t_1^{-1}) = o(t_1^0)$. Hence, for big enough T_1 and $T_1 \le t_1$,

$$\|\nabla_x \mathcal{P}(x)\| \le \frac{1}{2},$$

which implies that the map $\mathcal{P}$ is a contraction and the equation (1.9.22) has a unique solution. This ends the proof of the existence of the family of trajectories $(Y_1(s, t_1, t_2, \xi), E_1(s, t_1, t_2, \xi))$.

We note that

$$\begin{aligned} \nabla_\xi \mathcal{P}(x) &= \nabla_{\tilde{\eta}} \tilde{Y}_1(t_1, \tilde{\eta}(t_1, t_1, t_2, x, \xi)) \nabla_\xi \tilde{\eta}(t_1, t_1, t_2, x, \xi) \\ &\in O(t_1) o(t_1^0) = o(t_1). \end{aligned}$$

Therefore, from

$$\nabla_\xi x = (1 - \nabla_x \mathcal{P}(x))^{-1} \nabla_\xi \mathcal{P}(x)$$

we conclude that

$$\nabla_\xi x \in o(t_1). \tag{1.9.23}$$

Now let us prove the estimates (1.9.20) . We write

$$\begin{aligned} &Y_1(s, t_1, t_2, \xi) - (s - t_1)\xi - \tilde{Y}_1(t_1, \xi) \\ &= \tilde{y}(s, t_1, t_2, x, \xi) - (s - t_1)\xi - x \\ &+ \tilde{Y}_1(t_1, \tilde{\eta}(t_1, t_1, t_2, x, \xi)) - \tilde{Y}_1(t_1, \xi) = I_1(s) + I_2(s), \end{aligned} \tag{1.9.24}$$

where x is the solution of (1.9.22). Now if we use (1.5.3), (1.5.4) and (1.9.23), then we see that

$$\partial_\xi^\beta I_1(s) \in o(s), \quad |\beta| = 0, 1.$$

If we use (1.9.13) (1.5.5) and (1.9.22), then we get

$$\partial_\xi^\beta I_2(s) \in o(t_1), \quad |\beta| = 0, 1.$$

This implies the estimates (1.9.20) and ends the proof of *(i)*.

Let us show *(ii)*. We have

$$Y_1(t,t,\infty,\xi) = \tilde{Y}_1(t, E_1(t,t,\infty,\xi)), \quad E_1(\infty,t,\infty,\xi) = \xi.$$

Hence, by the uniqueness part of Theorem 1.9.6, we have

$$(Y_1(t,\xi), E_1(t,\xi)) = (Y_1(t,t,\infty,\xi), E_1(t,t,\infty,\xi)).$$

Therefore, for any s,

$$(Y_1(s,\xi), E_1(s,\xi)) = (Y_1(s,t,\infty,\xi), E_1(s,t,\infty,\xi)). \tag{1.9.25}$$

Next we note the following obvious identity valid for any t, s:

$$Y_1(t,s,\infty,\xi) = Y_1(t,s,t,E_1(t,s,\infty,\xi)).$$

Using (1.9.25), we obtain

$$Y_1(t,\xi) = Y_1(t,s,t,E_1(t,\xi)).$$

But we know that

$$Y_1(t,\xi) = \tilde{Y}_1(t, E_1(t,\xi)).$$

and, for sufficiently big t, the map $\xi \mapsto E_1(t,\xi)$ is bijective. This shows that, for any η,

$$Y_1(t,s,t,\eta) = \tilde{Y}_1(t,\eta).$$

□

Let us now describe a version of Proposition 1.9.12 valid for conservative potentials.

Proposition 1.9.13
(i) Assume that $F(t,x)$ is conservative. Assume that a function $S_1(t,\xi)$ satisfies

$$\partial_\xi^\beta S_1(t,\xi) \in O(t), \quad |\beta| = 2.$$

Then, for big enough T_1 and $T_1 \le t_1 \le t_2$, there exists a unique $C^{1,1}(X')$ function $S_1(t_1,t_2,\xi)$ such that

$$\begin{cases} \partial_{t_2} S_1(t_1,t_2,\xi) = \frac{1}{2}\xi^2 + V(t_2, \nabla_\xi S_1(t_1,t_2,\xi)), \\ S_1(t_1,t_1,\xi) = S_1(t_1,\xi). \end{cases}$$

It satisfies, uniformly for $T_1 \le t_1 \le t_2$,

$$\partial_\xi^\beta \left(S_1(t_1,t_2,\xi) - \frac{1}{2}(t_2 - t_1)\xi^2 - S_1(t_1,\xi) \right) \in o(t_2), \quad |\beta| = 1,2.$$

(ii) Assume, in addition, that

$$\partial_t S_1(t,\xi) = \frac{1}{2}\xi^2 + V(t, \nabla_\xi S_1(t,\xi)).$$

Then for any t, t_1, we have

$$S_1(t,\xi) = S_1(t_1,t,\xi).$$

(iii) There exists the limit

$$\lim_{t\to\infty}(S_1(t,\xi) - S(t,\xi)) =: \sigma^+(\xi). \tag{1.9.26}$$

Moreover,

$$\nabla_\xi \sigma^+(\xi) = f^+(\xi).$$

Proof. *(i)* and *(ii)* follow from Proposition 1.9.12 and the standard theory presented in Appendix A.3 if we set

$$\begin{aligned} S_1(t_1,t_2,\xi) &:= S_1(t_1,E_1(t_1)) \\ &+ \textstyle\int_{t_1}^{t_2}(\frac{1}{2}E_1^2(s) + V(s,Y_1(s)) - \langle Y_1(s), \nabla_x V(s,Y_1(s))\rangle)\mathrm{d}s, \end{aligned}$$

where $(Y_1(s), E_1(s)) = (Y_1(s,t_1,t_2,\xi,x), E_1(s,t_1,t_2,x,\xi))$.

To prove *(iii)*, we note that by the Hamilton-Jacobi equation

$$\partial_t(S(t,\xi) - S_1(t,\xi)) = V(t,\nabla_\xi S(t,\xi)) - V(t,\nabla_\xi S_1(t,\xi)). \tag{1.9.27}$$

But, by Theorem 1.9.6,

$$\nabla_\xi S(t,\xi) - \nabla_\xi S_1(t,\xi) = \tilde{Y}(t,\xi) - \tilde{Y}_1(t,\xi)$$

is bounded. Therefore the right-hand side of (1.9.27) is integrable. This implies the existence of the limit (1.9.26). □

Note that under the assumptions of Proposition 1.9.13, the modified free flow $\phi_1(t)$ is symplectic and $S_1(t,\xi) + \langle x,\xi\rangle$ is its generating function. Hence the modified wave transformation $\mathcal{F}_1^+$ is also symplectic. The generating function of $\phi(s,0)\circ\mathcal{F}_1^+$ is given by

$$\lim_{t\to\infty}(S(s,t,x,\xi) - S_1(t,\xi)) =: \Phi_1^+(s,x,\xi).$$

Just as $\Phi^+(s,x,\xi)$, the function $\Phi_1^+(s,x,\xi)$ also solves the eikonal equation

$$\begin{cases} -\partial_s\Phi_1^+(s,x,\xi) = \frac{1}{2}(\nabla_x\Phi_1^+(s,x,\xi))^2 + V(s,x), \\ \lim_{s\to\infty}\nabla_x\Phi_1^+(s,x,\xi) = \xi. \end{cases} \tag{1.9.28}$$

We also have

$$\Phi^+(s,x,\xi) - \Phi_1^+(s,x,\xi) = \sigma^+(\xi).$$

1.10 Smoothness of Trajectories

In order to prove the differentiability of trajectories with respect to parameters, one needs to make certain assumptions on the differentiability of the forces. Let us state the conditions that we will use in this section:

$$\int_0^\infty \|\partial_x^\alpha F(t,\cdot)\|_\infty \langle t\rangle^{|\alpha|+\mu} \mathrm{d}t < \infty, \quad \alpha \in \mathbb{N}^n. \tag{1.10.1}$$

We will always assume that $\mu \geq 0$ in (1.10.1). The most important cases of (1.10.1) will be $\mu = 0$, which can be called the *slow-decaying smooth condition*, and $\mu = 1$, which can be named the *fast-decaying smooth condition*.

The condition (1.10.1) is akin to the conditions used by specialists in pseudo-differential operators to define "semi-classical symbols" with t^{-1} playing the role of Planck's constant.

We would like to describe some bounds on the derivatives of the solutions $(\tilde{y}(s,t_1,t_2,x,\xi), \tilde{\eta}(s,t_1,t_2,x,\xi))$, which were constructed in Theorem 1.5.1.

Theorem 1.10.1
Assume (1.10.1) with $\mu \geq 0$. Then one has, uniformly for $T \leq t_1 \leq s \leq t_2 \leq \infty$, $(x,\xi) \in X \times X'$, the following estimate:

$$\partial_\xi^\beta (\tilde{y}(s,t_1,t_2,x,\xi) - x - (s-t_1)\xi) \in o(s^0)\langle t_1\rangle^{-\mu}|s-t_1|. \tag{1.10.2}$$

Moreover, for some functions f_β, $f_{\alpha,\beta} \in L^1(\mathrm{d}u)$, one has

$$\begin{aligned} &|\partial_\xi^\beta(\tilde{y}(s,t_1,t_2,x,\xi) - x - (s-t_1)\xi)| \\ &\quad \leq \textstyle\int_{t_1}^\infty f_\beta(u)\langle u\rangle^{1-\mu}\mathrm{d}u, \quad \mu \geq 1, \end{aligned} \tag{1.10.3}$$

$$\begin{aligned} &|\partial_x^\alpha\partial_\xi^\beta(\tilde{y}(s,t_1,t_2,x,\xi) - x - (s-t_1)\xi)| \\ &\quad \leq \langle t_1\rangle^{1-|\alpha|} \textstyle\int_{t_1}^\infty f_{\alpha,\beta}(u)\langle u\rangle^{-\mu}\mathrm{d}u, \quad |\alpha| \geq 1, \end{aligned} \tag{1.10.4}$$

$$|\partial_\xi^\beta(\tilde{\eta}(s,t_1,t_2,x,\xi) - \xi)| \leq \textstyle\int_s^\infty f_\beta(u)\langle u\rangle^{-\mu}\mathrm{d}u, \tag{1.10.5}$$

$$\begin{aligned} |\partial_x^\alpha\partial_\xi^\beta(\tilde{\eta}(s,t_1,t_2,x,\xi) - \xi)| \leq \langle t_1\rangle^{1-|\alpha|} \textstyle\int_s^\infty f_{\alpha,\beta}(u)\langle u\rangle^{-1-\mu}\mathrm{d}u,& \\ |\alpha| \geq 1.& \end{aligned} \tag{1.10.6}$$

Note that the somewhat uncommon right-hand sides of the above estimates will be convenient in applications. To see what they mean, assume that $f(u) \in L^1(\mathrm{d}u)$ and set

$$g(s) := \int_s^\infty f(u)\langle u\rangle^{-\nu}\mathrm{d}u.$$

Then it is easy to see that

$$g(s) \in o(s^{-\nu}) \text{ if } \nu \geq 0,$$

$$g(s) \in \langle s\rangle^{1-\nu} L^1(\mathrm{d}s) \text{ if } \nu \geq 1.$$

Proof. The proof of this theorem is a natural continuation of that of Theorem 1.5.1, where we already proved all the estimates with $|\alpha| + |\beta| \leq 1$ under the assumption (1.5.1). We will use the notation of the proof of Theorem 1.5.1.

We use Faa di Bruno's formula to compute $\partial_x^\alpha \partial_\xi^\beta \tilde{z}(s)$. We obtain

$$\begin{aligned} &\partial_x^\alpha \partial_\xi^\beta \tilde{z}(s) \\ &= -\sum \textstyle\int_{t_1}^{t_2} \zeta_{t_1,s}(u) \nabla_{\tilde{y}}^q F(u, \tilde{y}(u)) \partial_x^{\gamma_1} \partial_\xi^{\delta_1} \tilde{y}(u) \cdots \partial_x^{\gamma_q} \partial_\xi^{\delta_q} \tilde{y}(u) \mathrm{d}u. \end{aligned} \tag{1.10.7}$$

or, equivalently,

$$\begin{aligned} &\partial_x^\alpha \partial_\xi^\beta \tilde{z}(s) + \textstyle\int_{t_1}^{t_2} \zeta_{t_1,s}(u) \nabla_{\tilde{y}} F(u, \tilde{y}(u)) \partial_x^\alpha \partial_\xi^\beta \tilde{z}(u) \mathrm{d}u \\ &= - \textstyle\int_{t_1}^{t_2} \zeta_{t_1,s}(u) \nabla_{\tilde{y}} F(u, \tilde{y}(u)) \partial_x^\alpha \partial_\xi^\beta (x + (u - t_1)\xi) \mathrm{d}u \\ &- \sum_{q \neq 1} \textstyle\int_{t_1}^{t_2} \zeta_{t_1,s}(u) \nabla_{\tilde{y}}^q F(u, \tilde{y}(u)) \partial_x^{\gamma_1} \partial_\xi^{\delta_1} \tilde{y}(u) \cdots \partial_x^{\gamma_q} \partial_\xi^{\delta_q} \tilde{y}(u) \mathrm{d}u. \end{aligned} \tag{1.10.8}$$

We rewrite (1.10.8) as

$$(1 - \nabla_{\tilde{z}} \mathcal{P}(\tilde{z})) \partial_x^\alpha \partial_\xi^\beta \tilde{z} = h_{\alpha,\beta}, \tag{1.10.9}$$

where the map $\mathcal{P}$ was introduced in the proof of Theorem 1.5.1.

Now to prove estimate (1.10.2) for $|\beta| \geq 1$ we use the induction with respect to $|\beta|$. The induction hypothesis $H(n)$ is

$$\partial_\xi^\beta \tilde{z}(s) \in |s - t_1| o(s^0) \langle t_1\rangle^{-\mu}, \quad 1 \leq |\beta| \leq n. \tag{1.10.10}$$

$H(0)$ is empty, hence true.

We assume that $H(n-1)$ is true. Consider β with $|\beta| = n$. The induction assumption $H(n-1)$ implies

$$\partial_\xi^\delta \tilde{y}(u) \in O(u), \qquad 1 \leq |\delta| \leq n - 1. \tag{1.10.11}$$

Using (1.10.11), we easily see that in that case

$$|s - t_1|^{-1} h_\beta(s) \in o(s^0) \langle t_1\rangle^{-\mu}. \tag{1.10.12}$$

We recall from the proof of Theorem 1.5.1 that $(1 - \nabla_{\tilde{z}} \mathcal{P}(\tilde{z}))$ is invertible on $Z_{t_1,\infty}^1$ for $t_1 \geq T$. Therefore, we can use the identity

$$\partial_\xi^\beta \tilde{z} = (1 - \nabla_{\tilde{z}} \mathcal{P}(\tilde{z}))^{-1} h_\beta \tag{1.10.13}$$

to show that

$$|s - t_1|^{-1} \partial_\xi^\beta \tilde{z}(s) \in o(s^0) \langle t_1\rangle^{-\mu}.$$

We want now to show by induction the following estimate, which we call $H'(n)$

$$\partial_x^\alpha \partial_\xi^\beta \tilde{y}(s) \in O(t_1^{1-|\alpha|}), \quad |\alpha| \geq 1, \quad |\alpha| + |\beta| \leq n. \tag{1.10.14}$$

Assume that we know that $H'(n-1)$ is true. Let $|\alpha| + |\beta| = n$ and $|\alpha| \geq 1$. The induction assumption $H'(n-1)$ says that

$$\partial_x^\gamma \partial_\xi^\delta \tilde{y}(u) \in O(t_1^{1-|\gamma|}), \quad |\gamma| \geq 1, \quad |\gamma| + |\delta| \leq n-1. \tag{1.10.15}$$

Using (1.10.11) and (1.10.15), we easily see that

$$h_{\alpha,\beta}(s) \in o(t_1^{1-|\alpha|-\mu}). \tag{1.10.16}$$

We know from the proof of Theorem 1.5.1 that $(1 - \nabla_{\tilde{z}} \mathcal{P}(\tilde{z}))$ is invertible on $Z_{t_1}^0$ for $t_1 \geq T$. Therefore, we can use the identity (1.10.9) to show that

$$\partial_x^\alpha \partial_\xi^\beta \tilde{z}(s) \in o(t_1^{1-|\alpha|-\mu}).$$

This implies (1.10.14).

Now (1.10.3) follows immediately from (1.10.7) and (1.10.14).

To see (1.10.4), let us consider one of the terms on the right-hand side of (1.10.7). Let q_1 be the number of the derivatives of $\tilde{y}$ with $|\gamma_i| = 0$. Then by (1.10.14) and (1.10.15)

$$\begin{aligned} &\left| \int_{t_1}^{t_2} \zeta_{t_1,s}(u) \nabla_y^q F(u, \tilde{y}(u)) \partial_x^{\gamma_1} \partial_\xi^{\delta_1} \tilde{y}(u) \cdots \partial_x^{\gamma_q} \partial_\xi^{\delta_q} \tilde{y}(u) \mathrm{d}u \right| \\ &\leq C \langle t_1 \rangle^{q-q_1-|\alpha|} \int_{t_1}^\infty \|\nabla_y^q F(u, \cdot)\|_\infty \langle u \rangle^{q_1+1} \mathrm{d}u. \end{aligned} \tag{1.10.17}$$

Now (1.10.4) follows if we take into account that $|\alpha| \geq 1$ implies $q_1 + 1 \leq q$.

To get the estimates on the derivatives of $\dot{\tilde{z}}$, we differentiate the identity (1.5.17). Using again Faa di Bruno's formula, we obtain

$$\begin{aligned} &\partial_x^\alpha \partial_\xi^\beta \dot{\tilde{z}}(s) \\ &= -\sum \int_s^{t_2} \nabla_y^q F(u, \tilde{y}(u)) \partial_x^{\gamma_1} \partial_\xi^{\delta_1} \tilde{y}(u) \cdots \partial_x^{\gamma_q} \partial_\xi^{\delta_q} \tilde{y}(u) \mathrm{d}u. \end{aligned} \tag{1.10.18}$$

Then we use (1.10.14) and (1.10.15). □

The remaining estimates of this section are essentially corollaries of the above theorem.

Proposition 1.10.2
Assume the fast-decaying smooth condition, that is, (1.10.1) with $\mu = 1$. Then

$$\partial_x^\alpha \partial_\xi^\beta (z_{\mathrm{fd}}^+(s, x, \xi) - x + s\xi) \in o(s^{-|\alpha|}),$$

$$\partial_x^\alpha \partial_\xi^\beta (z_{\mathrm{fd}}^+(s, x, \xi) - x + s\xi) \in \langle s \rangle^{1-|\alpha|} L^1(\mathrm{d}s), \quad |\alpha| \geq 1,$$

$$\partial_x^\alpha \partial_\xi^\beta (\zeta^+(s, x, \xi) - \xi) \in o(s^{-1-|\alpha|}),$$

$$\partial_x^\alpha \partial_\xi^\beta (\zeta^+(s, x, \xi) - \xi) \in \langle s \rangle^{-|\alpha|} L^1(\mathrm{d}s).$$

Proof. We use Theorem 1.10.1. For the first estimate, we use, in addition, (1.7.1).

□

Proposition 1.10.3
Assume the slow-decaying smooth condition, that is, (1.10.1) with $\mu = 0$. Then

$$z_{\rm sd}^{+}(s,x,\xi) - x + Y(s,\xi) \in o(s^0)|x - Y(s,\xi)|,$$

$$\partial_\xi^\beta(z_{\rm sd}^{+}(s,x,\xi) - x + Y(s,\xi)) \in o(s^0)(|x - Y(s,\xi)| + \langle s\rangle),$$

$$\partial_x^\alpha\partial_\xi^\beta(z_{\rm sd}^{+}(s,x,\xi) - x + Y(s,\xi)) \in o(s^{1-|\alpha|}), \quad |\alpha| \geq 1,$$

$$\partial_x^\alpha\partial_\xi^\beta(\zeta^{+}(s,x,\xi) - \xi) \in o(s^{-|\alpha|}),$$

$$\partial_x^\alpha\partial_\xi^\beta(\zeta^{+}(s,x,\xi) - \xi) \in \langle s\rangle^{1-|\alpha|}L^1(\mathrm{d}s), \quad |\alpha| \geq 1.$$

Proof. We use Theorem 1.10.1. For the first two estimates, we use, in addition, (1.7.3). □

Next we consider various functions defined the conservative case, which we describe in the following proposition.

The following proposition follows immediately from Theorem 1.10.1.

Proposition 1.10.4
Assume (1.10.1) with $\mu \geq 0$, and that the force is conservative. Suppose that $|\alpha|+|\beta| \geq 1$. Then the function $S(t_1,t_2,x,\xi)$ satisfies, uniformly for $T \leq t_1 \leq t_2$, $(x,\xi) \in X \times X'$, the following estimates:

$$\partial_\xi^\beta\left(S(t_1,t_2,x,\xi) - \langle x,\xi\rangle - \tfrac{1}{2}(t_2-t_1)\xi^2\right) \in |t_2 - t_1| o(t_2^0)\langle t_1\rangle^{-\mu}, \quad |\beta| \geq 1,$$

$$\partial_x^\alpha\partial_\xi^\beta\left(S(t_1,t_2,x,\xi) - \langle x,\xi\rangle - \tfrac{1}{2}(t_2-t_1)\xi^2\right) \in o(t_1^{1-|\alpha|-\mu}), \quad |\alpha| + \mu \geq 1,$$

$$\partial_x^\alpha\partial_\xi^\beta\left(S(t_1,t_2,x,\xi) - \langle x,\xi\rangle - \tfrac{1}{2}(t_2-t_1)\xi^2\right) \in \langle t_1\rangle^{2-|\alpha|-\mu}L^1(\mathrm{d}t_1), \quad |\alpha| + \mu \geq 2.$$

The following two propositions describe the regularity of the solutions of the eikonal equation associated with the fast- and slow-decaying case. They follow immediately from Propositions 1.10.2 and 1.10.3.

Proposition 1.10.5
Assume that the force is conservative and (1.10.1) with $\mu = 1$ holds. Then

$$\partial_x^\alpha\partial_\xi^\beta(\Phi_{\rm fd}^{+}(s,x,\xi) - \langle x,\xi\rangle + \tfrac{1}{2}s\xi^2) \in o(s^{-|\alpha|}), \quad |\alpha| + |\beta| \geq 1,$$

$$\partial_x^\alpha\partial_\xi^\beta(\Phi_{\rm fd}^{+}(s,x,\xi) - \langle x,\xi\rangle + \tfrac{1}{2}s\xi^2) \in \langle s\rangle^{1-|\alpha|}L^1(\mathrm{d}s), \quad |\alpha| \geq 1.$$

Proposition 1.10.6
Assume that the force is conservative and (1.10.1) with $\mu = 0$ holds. Then

$$\partial_\xi^\beta(\Phi_{\rm sd}^+(s,x,\xi) - \langle x,\xi\rangle + S(s,\xi)) \in o(s^0)|x - Y(s,\xi)|, \quad |\beta| = 1,$$

$$\partial_\xi^\beta(\Phi_{\rm sd}^+(s,x,\xi) - \langle x,\xi\rangle + S(s,\xi)) \in o(s^0)(|x - Y(s,\xi)| + \langle s\rangle), \quad |\beta| \geq 1,$$

$$\partial_x^\alpha\partial_\xi^\beta(\Phi_{\rm sd}^+(s,x,\xi) - \langle x,\xi\rangle + S(s,\xi)) \in o(s^{1-|\alpha|}), \quad |\alpha| \geq 1,$$

$$\partial_x^\alpha\partial_\xi^\beta(\Phi_{\rm sd}^+(s,x,\xi) - \langle x,\xi\rangle + S(s,\xi)) \in \langle s\rangle^{2-|\alpha|}L^1({\rm d}s), \quad |\alpha| \geq 2.$$

The proof of the following proposition is an obvious extension of the proof of Proposition 1.9.13.

Proposition 1.10.7
Assume the hypotheses of Proposition 1.9.13 (i) and (ii). Suppose also that, for some t_1,

$$|\partial_\xi^\beta S_1(t_1,\xi)| \leq C_\beta, \quad |\beta| \geq 2.$$

Then the function $S_1(t,\xi)$ constructed in Proposition 1.9.13 satisfies, uniformly for $t_1 \leq t$, $\xi \in X'$,

$$\partial_\xi^\beta\left(S_1(t,\xi) - \frac{1}{2}(t-t_1)\xi^2 - S_1(t_1,\xi)\right) \in o(t), \quad |\beta| \geq 1.$$

1.11 Comparison of Two Dynamics

In this section we compare two different classical dynamics. In the first theorem we give conditions when trajectories of two slow-decaying systems are asymptotic to each other.

Theorem 1.11.1
Suppose that the forces $F_1(t,x)$, $F_2(t,x)$ satisfy

$$\int_0^\infty \|\partial_x^\alpha F_i(t,\cdot)\|_\infty \langle t\rangle^{|\alpha|}{\rm d}t < \infty, \quad |\alpha| = 0,1, \quad i = 1,2, \tag{1.11.1}$$

$$\int_0^\infty \|F_1(t,\cdot) - F_2(t,\cdot)\|_\infty \langle t\rangle {\rm d}t < \infty. \tag{1.11.2}$$

Let the time T be chosen such that the conditions of Theorem 1.5.1 are satisfied for both $F_1(t,x)$ and $F_2(t,x)$. Let $Y_i(t,\xi)$ and $\tilde{Y}_i(t,\xi)$ be defined as in Sects. 1.5 and 1.6, using the force $F_i(t,x)$. Then the following limits exist and are equal:

$$\lim_{t\to\infty}(Y_1(t,\xi) - Y_2(t,\xi)) = \lim_{t\to\infty}(\tilde{Y}_1(t,\xi) - \tilde{Y}_2(t,\xi)).$$

Proof. Let $\tilde{y}_i(s, t_1, t_2, x, \xi)$ be the solution of the boundary value problem constructed in Theorem 1.5.1 for the force $F_i(t, x)$. Let us first show that, for $T \le t_1 \le s \le t_2 \le \infty$, there exists a uniform bound

$$|\tilde{y}_1(s, t_1, t_2, x, \xi) - \tilde{y}_2(s, t_1, t_2, x, \xi)| \le C. \tag{1.11.3}$$

In fact, we have the following identity:

$$\begin{aligned}\tilde{y}_1(s) - \tilde{y}_2(s) &= \int_{t_1}^{t_2} \zeta_{t_1,s}(u)(F_1(u, \tilde{y}_1(u)) - F_2(u, \tilde{y}_2(u)))du \\ &= \int_{t_1}^{t_2} \zeta_{t_1,s}(u)(F_1(u, \tilde{y}_1(u)) - F_2(u, \tilde{y}_1(u)))du \\ &+ \int_{t_1}^{t_2} \zeta_{t_1,s}(u)(F_2(u, \tilde{y}_1(u)) - F_2(u, \tilde{y}_2(u)))du.\end{aligned}$$

Hence

$$\begin{aligned}|\tilde{y}_1(s) - \tilde{y}_2(s)| &\le \int_{t_1}^{\infty} |u - t_1| \|F_1(u, \cdot) - F_2(u, \cdot)\|_\infty du \\ &+ \int_{t_1}^{\infty} |u - t_1| \|\nabla_{\tilde{y}} F_2(u, \cdot)\|_\infty |\tilde{y}_1(u) - \tilde{y}_2(u)| du.\end{aligned}$$

Therefore, the bound (1.11.3) follows by the Gronwall inequality.

Next recall that

$$Y_i(t, \xi) = \tilde{y}_i(t, T, \infty, 0, \xi), \qquad \tilde{Y}_i(t, \xi) = \tilde{y}_i(t, T, t, 0, \xi).$$

Hence,

$$\begin{aligned}&Y_1(t) - Y_2(t) \\ &= \int_T^\infty \zeta_{T,t}(u)(F_1(u, \tilde{y}_1(u, T, \infty, 0, \xi)) - F_2(u, \tilde{y}_2(u, T, \infty, 0, \xi)))du,\end{aligned} \tag{1.11.4}$$

$$\begin{aligned}&\tilde{Y}_1(t) - \tilde{Y}_2(t) \\ &= \int_T^t (u - T)(F_1(u, \tilde{y}_1(u, T, t, 0, \xi)) - F_2(u, \tilde{y}_2(u, T, t, 0, \xi)))du.\end{aligned} \tag{1.11.5}$$

Using the bound (1.11.3) and the assumptions on the potentials (1.11.1) and (1.11.2), we can estimate the integrands in (1.11.4) and (1.11.5) by an integrable function. Thus, by Lebesgue's theorem, both (1.11.4) and (1.11.5) converge to

$$\int_T^\infty (u - T)(F_1(u, \tilde{y}_1(u, T, \infty, 0, \xi)) - F_2(u, \tilde{y}_2(u, T, \infty, 0, \xi)))du. \tag{1.11.6}$$

□

In quantum scattering, it is useful to know when the difference of two solutions of the Hamilton-Jacobi equation converges.

Theorem 1.11.2
Suppose that the forces $F_i(t, x)$ are conservative and $F_i(t, x) = -\nabla_x V_i(t, x)$. Let $S_i(t, \xi)$ be the solution of the Hamilton-Jacobi equation

$$\begin{cases} \partial_t S_i(t,\xi) = \frac{1}{2}\xi^2 + V_i(t, \nabla_\xi S_i(t,\xi)), \\ S_i(T,\xi) = 0. \end{cases} \tag{1.11.7}$$

Suppose that

$$\int_0^\infty \|V_1(t,\cdot) - V_2(t,\cdot)\|_\infty \mathrm{d}t < \infty. \tag{1.11.8}$$

Assume, in addition, either one of the following hypotheses:
(i)

$$\begin{aligned} &\textstyle\int_0^\infty \|\partial_x^\alpha F_i(t,\cdot)\|_\infty \langle t\rangle^{|\alpha|}\mathrm{d}t < \infty, \quad |\alpha| = 0,1, \quad i = 1,2, \\ &\textstyle\int_0^\infty \|F_1(t,\cdot) - F_2(t,\cdot)\|_\infty \langle t\rangle \mathrm{d}t < \infty; \end{aligned}$$

or
(ii)

$$\begin{aligned} &\textstyle\int_0^\infty \|F_i(t,\cdot)\|_\infty \langle t\rangle^{1/2}\mathrm{d}t < \infty, \quad i = 1,2, \\ &\textstyle\int_0^\infty \|\partial_x^\alpha F_i(t,\cdot)\|_\infty \langle t\rangle \mathrm{d}t < \infty, \quad |\alpha| = 1, \quad i = 1,2. \end{aligned}$$

Then there exists the uniform limit

$$\lim_{t\to\infty} (S_1(t,\xi) - S_2(t,\xi)). \tag{1.11.9}$$

Proof. We have

$$\partial_t(S_1(t,\xi) - S_2(t,\xi)) = V_1(t, \nabla_\xi S_1(t,\xi)) - V_2(t, \nabla_\xi S_2(t,\xi)) \tag{1.11.10}$$

Hence

$$\begin{aligned} |\partial_t(S_1(t,\xi) - S_2(t,\xi))| \; &\le \|V_1(t,\cdot) - V_2(t,\cdot)\|_\infty \\ &\quad + \|\nabla_x V_2(t,\cdot)\|_\infty |\tilde{Y}_1(t,\xi) - \tilde{Y}_2(t,\xi)|. \end{aligned} \tag{1.11.11}$$

In the case *i)*, (1.11.11) is integrable, because by Theorem 1.11.1 $|\tilde{Y}_1(t,\xi)) - \tilde{Y}_2(t,\xi)|$ is bounded.

In the case *(ii)*, we first see that

$$|\tilde{Y}_i(t,\xi) - t\xi| \le C\langle t\rangle^{1/2}, \quad i = 1,2.$$

Hence

$$|\tilde{Y}_1(t,\xi) - \tilde{Y}_2(t,\xi)| \le C\langle t\rangle^{1/2}. \tag{1.11.12}$$

Therefore, (1.11.11) is integrable also in this case. □

1.12 More Examples of Modified Free Dynamics

So far, in order to define modified free dynamics, we used the modifiers $\tilde{Y}(t, \xi)$ or, more generally, modifiers $\tilde{Y}_1(t, \xi)$. They have quite special properties, which are expressed in the equations (1.9.13). In the literature on long-range scattering theory, one can find other choices of modified free dynamics of the form

$$\phi_W(t)(x, \xi) := (x + W(t, \xi), \xi),$$

where $W(t, \xi)$ is an appropriately chosen function. Such dynamics $\phi_W(t)$ conserve the momentum and are (t, ξ)-dependent translations in the x variable. This is why they are called "free". The main requirement on $W(t, \xi)$ is the existence of the limit

$$\lim_{t\to\infty} \phi(0, t)\phi_W(t). \tag{1.12.1}$$

Moreover, in the conservative case, one also wants $\phi_W(t)$ to be symplectic.

For a given wave transformation $\mathcal{F}_1^+$, there exist many functions $W(t, \xi)$ that satisfy these requirements. One of them is naturally distinguished: it is the function $\tilde{Y}_1(t, \xi)$ in the notation of Sect. 1.9. One can argue that $\phi_{\tilde{Y}_1}(t)$ is in a sense "the best" modified free dynamics for a given modified wave transformation $\mathcal{F}_1^+$. Nevertheless, in practice, in order to calculate it, one has to solve the equations of motion, which is usually hard. The modified free flows that can be found in the literature are often much easier to calculate.

Clearly, the existence of (1.12.1) is guaranteed by the existence of the limit

$$\lim_{t\to\infty} \phi_{\rm sd}^{-1}(t)\phi_W(t), \tag{1.12.2}$$

or, equivalently, by the existence of

$$\lim_{t\to\infty} (\tilde{Y}(t, \xi) - W(t, \xi)). \tag{1.12.3}$$

In the proposition below, we will give a family of examples of $W(t, \xi)$ that have been considered by various authors and we will give the conditions for (1.12.3) to exist.

Proposition 1.12.1
Assume that, for some $N \geq 1$, the force $F(t, x)$ satisfies

$$\begin{aligned} &\textstyle\int_0^\infty \langle t\rangle^{|\alpha|+\frac{1}{N}} \|\partial_x^\alpha F(t,\cdot)\|_\infty \mathrm{d}t < \infty, \quad |\alpha| \leq N-1, \\ &\textstyle\int_0^\infty \langle t\rangle^{N} \|\partial_x^\alpha F(t,\cdot)\|_\infty \mathrm{d}t < \infty, \quad |\alpha| = N. \end{aligned} \tag{1.12.4}$$

For a function $[0, \infty[\times X' \ni (t, \xi) \mapsto W(t, \xi) \in X$, we set (at least formally)

$$\mathcal{P}(W)(t, \xi) := t\xi - \int_0^t \nabla_\xi W(u, \xi) F(u, W(u, \xi)) \mathrm{d}u. \tag{1.12.5}$$

Moreover, we introduce

$$\begin{aligned} W_1(t,\xi) &:= t\xi, \\ W_n(t,\xi) &:= \mathcal{P}(W_{n-1})(t,\xi). \end{aligned} \tag{1.12.6}$$

Then for $1 \leq n \leq N$, the functions $W_n(t,\xi)$ are well defined and there exists the uniform limit

$$\lim_{t\to\infty}(\tilde{Y}(t,\xi) - W_N(t,\xi)). \tag{1.12.7}$$

The dynamics $\phi_{W_1}(t)$ is the ordinary free dynamics ϕ_0, which is used in the fast-decaying case.

The dynamics $\phi_{W_2}(t)$ is usually called the *Dollard modified dynamics* and goes back to [Do1] in the quantum case. One has

$$W_2(t,\xi) = t\xi - \int_0^t uF(u,u\xi)\mathrm{d}u. \tag{1.12.8}$$

The dynamics $\phi_{W_N}(t)$ was first introduced also in the quantum case by Buslaev and Matveev [BuMa]. Its construction for the classical case can be found in [He].

Proof of Proposition 1.12.1. We will first show by induction that, for $n = 1, \ldots, N-1$,

$$\partial_\xi^\beta(W_n(t,\xi) - \tilde{Y}(t,\xi)) \in o(t^{1-n\frac{1}{N}}), \quad |\beta| \leq N-n. \tag{1.12.9}$$

Note first that, by arguing exactly as in Sect. 1.10, we obtain that that $\tilde{Y}(t,\xi)$ satisfies the following estimates:

$$\partial_\xi^\beta \tilde{Y}(t,\xi) \in O(t), \quad |\beta| \leq N. \tag{1.12.10}$$

Using (1.12.10) and the equation

$$\partial_t\tilde{Y}(t,\xi) = \xi - \nabla_\xi\tilde{Y}(t,\xi)F(t,\tilde{Y}(t,\xi)), \tag{1.12.11}$$

we easily obtain

$$\partial_\xi^\beta(\tilde{Y}(t,\xi) - t\xi) \in o(t^{1-\frac{1}{N}}), \quad |\beta| \leq N-1,$$

which is the estimate (1.12.9) for $n = 1$.

Let us prove the induction step for (1.12.9). First note that the functions $W_n(t,\xi)$ satisfy the following substitute for (1.12.11):

$$\partial_t W_n(t,\xi) = \xi - \nabla_\xi W_{n-1}(t,\xi)F(t,W_{n-1}(t,\xi)). \tag{1.12.12}$$

If we subtract (1.12.11) from (1.12.12), then we obtain

$$\begin{aligned} &\tfrac{\mathrm{d}}{\mathrm{d}t}(\tilde{Y}(t,\xi) - W_n(t,\xi)) \\ &= \nabla_\xi\tilde{Y}(t,\xi)F(t,\tilde{Y}(t,\xi)) - \nabla_\xi W_{n-1}(t,\xi)F(t,W_{n-1}(t,\xi)), \end{aligned}$$

and

$$\begin{aligned}
&\tfrac{\mathrm{d}}{\mathrm{d}t}\partial_\xi^\beta(\tilde{Y}(t,\xi)-W_n(t,\xi))\\
&=\textstyle\sum_{\beta_1+\beta_2=\beta}C_{\beta_1,\beta_2}(\partial_\xi^{\beta_1}\nabla_\xi\tilde{Y}(t,\xi)-\partial_\xi^{\beta_1}\nabla W_{n-1}(t,\xi))\partial_\xi^{\beta_2}F(t,\tilde{Y}(t,\xi))\\
&+\textstyle\sum_{\beta_1+\beta_2=\beta}C_{\beta_1,\beta_2}\partial_\xi^{\beta_1}\nabla_\xi\tilde{Y}(t,\xi)(\partial_\xi^{\beta_2}F(t,\tilde{Y}(t,\xi))-\partial_\xi^{\beta_2}F(t,W_{n-1}(t,\xi))).
\end{aligned}$$

We will use now the induction assumption (1.12.9), which gives, for $n = 2,\ldots,N-1$,

$$\begin{aligned}
&\tfrac{\mathrm{d}}{\mathrm{d}t}\partial_\xi^\beta(\tilde{Y}(t,\xi)-W_n(t,\xi))\\
&\in\textstyle\sum_{j=0}^{|\beta|+1}o(t^{1+j-\frac{n-1}{N}})\|\nabla_x^jF(t,\cdot)\|_\infty,\ \ |\beta|\le N-n.
\end{aligned}\tag{1.12.13}$$

By integrating (1.12.13), we obtain (1.12.9). If $n = N$, $\beta = 0$, then (1.12.13) becomes

$$\frac{\mathrm{d}}{\mathrm{d}t}(\tilde{Y}(t,\xi)-W_N(t,\xi))\in o(t^{\frac{1}{N}})\|F(t,\cdot)\|_\infty+o(t^{1+\frac{1}{N}})\|\nabla_xF(t,\cdot)\|_\infty,$$

which is integrable and proves the existence of the limit (1.12.7). □

If the force is conservative, then the modified dynamics from the above propositions are symplectic. Let us describe how one can construct their generating functions and let us compare them with $S(t,\xi)$, which is a generating function of $\phi_{\mathrm{sd}}(t)$.

Proposition 1.12.2
Assume that the force is conservative and, for some $N\ge 1$, satisfies

$$\begin{aligned}
&\textstyle\int_0^\infty\langle t\rangle^{|\alpha|+\frac{1}{N}}\|\partial_x^\alpha F(t,\cdot)\|_\infty\mathrm{d}t<\infty,\ \ 0\le|\alpha|\le N-2,\\
&\textstyle\int_0^\infty\langle t\rangle^{N-1}\|\partial_x^\alpha F(t,\cdot)\|_\infty\mathrm{d}t<\infty,\ \ |\alpha|=N-1.
\end{aligned}\tag{1.12.14}$$

If $[0,\infty[\times X'\ni(t,\xi)\mapsto Z(t,\xi)\in\mathbb{R}$, then we set

$$\tilde{\mathcal{P}}(Z)(t,\xi):=\frac{1}{2}t\xi^2+\int_0^t V(u,\nabla_\xi Z(u,\xi))\mathrm{d}u,\tag{1.12.15}$$

Furthermore, we define

$$\begin{aligned}
Z_1(t,\xi)&:=\tfrac{1}{2}t\xi^2,\\
Z_n(t,\xi)&:=\tilde{\mathcal{P}}(Z_{n-1})(t,\xi).
\end{aligned}\tag{1.12.16}$$

Then for $1\le n\le N$, the functions $Z_n(t,\xi)$ are well defined, $\nabla_\xi Z_n(t,\xi) = W_n(t,\xi)$ and there exists the uniform limit

$$\lim_{t\to\infty}(S(t,\xi)-Z_N(t,\xi)).\tag{1.12.17}$$

Again, for $N = 2$, we obtain

$$Z_2(t,\xi) = \frac{1}{2}t\xi^2 + \int_0^t V(u, u\xi)du.$$

Proof. Note first the following formal identity:

$$\nabla_\xi \tilde{\mathcal{P}}(Z) = \mathcal{P}(\nabla_\xi Z), \tag{1.12.18}$$

where $\mathcal{P}$ was defined in (1.12.5). Using (1.12.18), we obtain by induction on n that $\nabla_\xi Z_n(t,\xi) = W_n(t,\xi)$. To prove the existence of the limit (1.12.17), we compute

$$\partial_t(S(t,\xi) - Z_N(t,\xi)) = V(t, \nabla_\xi S(t,\xi)) - V(t, W_{N-1}(t,\xi)). \tag{1.12.19}$$

Looking at the proof of Proposition 1.12.1, we see that, under the assumptions (1.12.14), one has

$$W_{N-1}(t,\xi) - \tilde{Y}(t,\xi) = W_{N-1}(t,\xi) - \nabla_\xi S(t,\xi) \in o(t^{\frac{1}{N}}),$$

which shows that (1.12.19) is integrable in time and that the limit (1.12.17) exists. □

2. Classical 2-Body Hamiltonians

2.0 Introduction

In this chapter we study scattering theory for time-independent Hamiltonians of the form

$$H(x,\xi) = \frac{1}{2}\xi^2 + V(x). \tag{2.0.1}$$

The previous chapter, devoted to time-decaying forces, can be viewed as a kind of an introduction to the present one. Many concepts and properties of scattering theory are more transparent and easier to describe in the time-decaying case. Therefore, it is in the previous chapter where we tried to explain them as thoroughly as possible. In the present chapter we will often use facts proven in the time-decaying framework.

Throughout this chapter we assume that the force is conservative and we use the Hamiltonian formalism, since it allows for some minor simplifications.

Let us briefly describe the content of this chapter.

In Sect. 2.1 we give a couple of definitions and facts about unbounded trajectories in rather general dynamical systems.

Beginning from the second section, we restrict our attention to Hamiltonians of the form (2.0.1). Typical assumptions on the potentials that we have in mind are the following: for some $\mu > 0$,

$$|\partial_x^\alpha V(x)| \leq C\langle x\rangle^{-\mu-|\alpha|}, \qquad |\alpha| = 1, 2. \tag{2.0.2}$$

If $\mu > 1$ then we say that the system is short-range; if $1 \geq \mu > 0$ then we say that it is long-range.

In Sect. 2.2 we prove among other things the following bound on the trajectories with zero energy:

$$|x(t,y,\eta)| \leq C\langle t\rangle^{2/(2+\mu)}.$$

This bound implies in particular that for zero energy trajectories we have

$$\lim_{t\to\pm\infty} \frac{x(t)}{t} = 0. \tag{2.0.3}$$

Trajectories with this property that are unbounded for $\pm t > 0$ we call *almost-bounded trajectories* for $t \to \pm\infty$. These trajectories are not well behaved from the point of view of scattering theory.

Section 2.3 is probably the most important of the whole chapter. In this section we concentrated most new concepts that we did not introduce in the previous chapter.

First of all, if one studies time-independent systems one needs to make clear which trajectories are likely to have good properties from the point of view of scattering theory. The right condition turns out to be the existence of $C_0 > 0$ and T such that

$$|x(t, y, \eta)| \geq C_0(\pm t - T), \quad \pm t \geq 0.$$

The trajectories satisfying this condition we call *scattering trajectories* for $t \to \pm\infty$.

The basic tool in the study of scattering trajectories is the classical counterpart of the so-called *Mourre estimate.* Roughly speaking, this estimate says that a certain observable $a_Q(x, \xi)$, which is equal for large x to $\langle x, \xi \rangle$, increases along the trajectories with a positive energy. Using the classical Mourre estimate, one can show that the set of scattering trajectories is equal to the set of unbounded trajectories with a positive energy.

Thus, all the trajectories fall into three disjoint categories: bounded, almost-bounded and scattering trajectories.

Non-trapping energies are those energies for which all the trajectories escape to infinity. Some basic properties of the dynamics for non-trapping energies are described in Sect. 2.4.

The momentum always has a limit as $t \to \infty$ along scattering trajectories. Nevertheless, for example along bounded trajectories, in general, it does not have a limit. The quantity that has a limit along all trajectories is $x(t, y, \eta)/t$. We call this limit the *asymptotic velocity* and denote it by $\xi^+(y, \eta)$. It is a substitute for the asymptotic momentum of the previous chapter. It turns out that the scattering trajectories are exactly the trajectories with a non-zero asymptotic velocity.

The construction of the asymptotic velocity is contained in Sect. 2.5. Section 2.6 is devoted to the short-range case. These two sections are parallel to Sects. 1.3 and 1.4 about time-decaying forces. The main difference consists in the fact that the wave transformation is not well defined for the zero momentum.

Suppose that we want to study trajectories with the absolute value of the asymptotic velocity greater than $C_0 > 0$. One way to do this is to choose a function $J \in C^\infty(X)$ such that $0 \notin \mathrm{supp} J$ and $J = 1$ on a neighborhood of $\{\xi \mid |\xi| \geq C_0\}$ and to introduce the "effective time-dependent force"

$$F_J(t, x) := -J\left(\frac{x}{t}\right) \nabla_x V(x).$$

It is easy to see that this time-dependent force belongs to the category of forces considered in the previous chapter. Moreover, on any scattering trajectory with $|\xi^+| > C_0$, this time-dependent force coincides with $-\nabla_x V(x)$ for large enough time. Therefore, many statements on time-independent systems, especially from Sects. 2.5 and 2.7, follow easily from Chap. 1 with help of this trick.

In Sect. 2.7 we develop the long-range case. It is parallel to Sects. 1.5, 1.6 and 1.8 about slow-decaying forces. The main new difficulty is due to the fact that the boundary value problem considered in Sect. 1.5, in general, does not have a global solution in the time-independent case. But if we restrict ourselves to an appropriate outgoing region, then we can solve this boundary problem.

Similarly, in general, there does not exists a function $S(t, \xi)$ that solves the Hamilton-Jacobi equation

$$\partial_t S(t, \xi) = \frac{1}{2}\xi^2 + V(\nabla_\xi S(t, \xi)), \tag{2.0.4}$$

for all $(t, \xi) \in \mathbb{R}^+ \times X'$. Because of this difficulty, it is in general not possible to introduce a modified free dynamics for time-independent systems as cleanly as it was done in the previous chapter. Therefore, we content ourselves with a more complicated and less natural definition of the modified free flow in the time-independent case. The main input into this definition is a function $S(t, \xi)$ that, for any $\epsilon > 0$, solves the Hamilton-Jacobi equation (2.0.4) for $|\xi| > \epsilon$ and $t > T_\epsilon$.

Section 2.9 describes some bounds on the derivatives with respect to parameters of various objects that we constructed. In these bounds, we restrict ourselves to the so-called outgoing region, that is, roughly speaking, to the subset of the phase space in which the momentum is bounded away from zero and the angle between the momentum ξ and the position x is less than π. For the initial or scattering data in an appropriately chosen outgoing region, time-independent systems behave like time-decaying systems considered in the previous chapter. Therefore, we can use the results of Sect. 1.10.

Sections 2.4 and 2.9 can be skipped by a reader who is interested just in the basic material.

We already described a part of the literature on classical scattering theory in the introduction to the previous chapter. Here we are going to comment just on the work of various authors about the concepts that we introduce in this chapter and that are not direct analogs of concepts that we discussed in Chap. 1.

Proposition 2.1.2 about trapping energies can be found in [GeSj]. Proposition 2.1.4 about the zero measure of trapped but not bounded trajectories is due to Siegel [Sie].

As we mentioned earlier, most of the ideas of this chapter that were not introduced in the previous one are contained in Sect. 2.3. It is in this section where we introduce the classical Mourre estimate. Apparently, it appeared first in [GeMa1]. The Mourre estimate was invented by E. Mourre [Mo1, Mo2]. Other papers devoted to the Mourre estimate include [PSS, FH1]. The original Mourre estimate was devoted to quantum N-body systems.

In our version of the classical Mourre estimate, we also use some of the ideas of G.M. Graf [Gr] that suggest how to modify the observable we are constructing so that its Poisson bracket with the Hamiltonian is everywhere positive.

The fact that, for non-zero energies, all the trajectories are either bounded or escape to infinity as $|x(t)| \geq C_0(t-T)$ for some $C_0 > 0$, which we prove in Theorem 2.3.3, was probably first proven in [Hu2].

The name almost-bounded trajectories was introduced in [De7]. In celestial mechanics, they have been usually called parabolic trajectories. The study of almost-bounded trajectories presented in [De7] was a by-product of the proof of asymptotic completeness for N-body long-range systems contained in [De8]. In particular, Example 2.2.4 and the a priori bound on almost-bounded trajectories of Lemma 2.2.1 come from [De7].

The idea that trajectories with outgoing initial conditions have good properties from the point of view of scattering theory is probably as old as scattering theory itself, nevertheless it seems to have been first exploited successfully on a larger scale in the quantum problem by V. Enss [E1].

The construction of a function that solves the Hamilton-Jacobi equation in a domain that is large enough for applications in scattering theory, which we give in Theorem 2.7.5, is due to L. Hörmander [Hö2].

2.1 General Facts About Dynamical Systems

In this section we will describe some general results about dynamical systems on a non-compact manifold.

Suppose that M is a non-compact manifold. Let $\phi(t)$ be a continuous flow on M, that is, a continuous map $\phi(t) : \mathbb{R} \times M \to M$ such that

$$\phi(t) \circ \phi(s) = \phi(t+s), \quad \phi(0) = 1\!\!1.$$

We assume that $H : M \to \mathbb{R}$ is a continuous function invariant with respect to ϕ and $\mathrm{d}\mu$ is a Borel measure on M invariant with respect to ϕ. (We can think of M as of a symplectic manifold with the symplectic measure $\mathrm{d}\mu$ and of $\phi(t)$ as of the flow generated by a Hamiltonian H).

We can now introduce the following definition:

Definition 2.1.1
A point $\rho \in M$ belongs to a trajectory bounded at $\pm\infty$ *if $\phi(t)(\rho)$ stays in a compact set for $t \in \mathbb{R}^{\pm}$. We will denote the set of such points by $\mathcal{B}^+$ (respectively $\mathcal{B}^-$). We denote by $\mathcal{B} = \mathcal{B}^+ \cap \mathcal{B}^-$ the union of all* bounded trajectories *and by $\mathcal{R}^{\pm}$ the set $M \backslash \mathcal{B}^{\pm}$, which is the union of all* trajectories unbounded at $\pm\infty$.

Proposition 2.1.2
The following subsets of $\mathbb{R}$ are equal:

$$H(\mathcal{B}^+) = H(\mathcal{B}^-) = H(\mathcal{B}).$$

Proof. It is enough to show that $H(\mathcal{B}^+)$ is contained in $H(\mathcal{B})$. Assume that $\lambda \in H(\mathcal{B}^+)$, and pick $\rho \in \mathcal{B}^+ \cap H^{-1}(\{\lambda\})$. Then $\{\phi(t)(\rho) \mid t \geq 0\}$ is contained in a compact set K. Therefore, there exist $\rho_\infty \in H^{-1}(\{\lambda\})$ and a sequence $t_n \to \infty$ such that $\rho_n := \phi(t_n)(\rho) \to \rho_\infty$.

We claim that $\rho_\infty \in \mathcal{B}$. Indeed, given an arbitrary $T \in \mathbb{R}$ and $\epsilon > 0$, the continuity of $\phi(T)$ implies that there exist $n = n(\epsilon, T)$ such that $\mathrm{dist}(\phi(T)(\rho_\infty), \phi(T)(\rho_n)) \leq \epsilon$. We can, moreover, pick n large enough such that $T + t_n \geq 0$, and hence $\phi(T)(\rho_n) = \phi(T + t_n)(\rho) \in K$. Since ϵ is arbitrary, we obtain that $\phi(T)(\rho_\infty) \in K$. This proves that $\rho_\infty \in \mathcal{B}$. □

Definition 2.1.3
The set $H(\mathcal{B}^+)$ described in Proposition 2.1.2 will be called the set of trapping energy levels *and it will be denoted by σ.*

The second abstract result is due to Siegel [Sie].

Proposition 2.1.4
The sets $\mathcal{B}^+ \backslash \mathcal{B}$ and $\mathcal{B}^- \backslash \mathcal{B}$ are of measure zero.

Proof. If K is a compact set in M, then denote by $\mathcal{B}_K^\pm$ the set of $\rho \in M$ such that $\phi(t)(\rho) \in K$ for $t \in \mathbb{R}^\pm$. We set $\mathcal{B}_K = \mathcal{B}_K^+ \cap \mathcal{B}_K^-$. Using the group property of $\phi(t)$, one sees easily that

$$\bigcap_{n \in \mathbb{N}} \phi(n)(\mathcal{B}_K^+) = \mathcal{B}_K.$$

One also has

$$\phi(n+1)(\mathcal{B}_K^+) \subset \phi(n)(\mathcal{B}_K^+).$$

Using these two facts, we have

$$\begin{aligned} \mathcal{B}_K^+ \backslash \mathcal{B}_K &= \cup_{n=0}^{\infty} (\phi(n)(\mathcal{B}_K^+) \backslash \phi(n+1)(\mathcal{B}_K^+)) \\ &= \cup_{n=0}^{\infty} \phi(n)(\mathcal{B}_K^+ \backslash \phi(1)(\mathcal{B}_K^+)). \end{aligned}$$

A little attention shows that the above union is disjoint. Hence, using the invariance of the measure $\mathrm{d}\mu$, we get

$$\mu(\mathcal{B}_K^+ \backslash \mathcal{B}_K) = \sum_{n=0}^{\infty} \mu(\mathcal{B}_K^+ \backslash \phi(1)(\mathcal{B}_K^+)). \tag{2.1.1}$$

So $\mu(\mathcal{B}_K^+ \backslash \mathcal{B}_K)$ is either equal to 0 or to ∞. But since K is compact, the measure of $\mathcal{B}_K^+ \backslash \mathcal{B}_K$ is finite, hence equal to 0. If we take a sequence K_n of compact sets converging to M, then we have

$$\mathcal{B}^+ \backslash \mathcal{B} = \bigcup_{n=1}^{\infty} \mathcal{B}_{K_n}^+ \backslash \bigcup_{n=1}^{\infty} \mathcal{B}_{K_n} \subset \bigcup_{n=1}^{\infty} \mathcal{B}_{K_n}^+ \backslash \mathcal{B}_{K_n}.$$

This clearly implies the desired result. □

2.2 Upper Bounds on Trajectories

From now on we will assume that M is the symplectic manifold $X \times X'$ and that H is a 2-body Hamiltonian of the form

$$H(x, \xi) = \frac{1}{2}\xi^2 + V(x),$$

where for the moment we only assume that the potential $V(x)$ is bounded and that the force

$$F(x) := -\nabla_x V(x)$$

belongs to $C^{0,1}(X)$. The free Hamiltonian will be, as usual,

$$H_0(x, \xi) := \frac{1}{2}\xi^2.$$

We will consider the equations of motion generated by the Hamiltonian $H(x, \xi)$, i.e.

$$\begin{cases} \dot{x}(t) = \xi(t), \\ \dot{\xi}(t) = -\nabla_x V(x(t)). \end{cases} \tag{2.2.1}$$

We will denote by $(x(t, y, \eta), \xi(t, y, \eta))$ the solutions of (2.2.1) with the initial conditions

$$\begin{cases} x(0, y, \eta) = y, \\ \xi(0, y, \eta) = \eta. \end{cases}$$

We will also use the notation $\phi(t)(y, \eta) = (x(t, y, \eta), \xi(t, y, \eta))$ and $\phi_0(t)(x, \xi) = (x + t\xi, \xi)$. Sometimes we will also use the notation

$$(x(t, s, y, \eta), \xi(t, s, y, \eta)) := (x(t-s, y, \eta), \xi(t-s, y, \eta)).$$

Note that in the case of our Hamiltonian the local Lipschitz condition of the force together with the boundedness of the potential guarantees not only the local existence and uniqueness of the trajectories, but also the completeness of the flow, that means, no trajectory will escape to infinity in a finite time. In fact, since $V(x)$ is bounded and the energy is conserved, for any initial conditions (y, η),

$$|\xi(t)| \le C.$$

Therefore,

$$|x(t) - y| \le C|t|.$$

Hence, within a finite interval of time, the solution $(x(t), \xi(t))$ stays inside a bounded set, which means that the flow is complete.

Let us remark that, thanks to the conservation of energy, a point $(y, \eta) \in X \times X'$ belongs to $\mathcal{B}^+$ if and only if $x(t, y, \eta)$ stays bounded for all $t \in \mathbb{R}^+$.

Next we would like to give a certain useful a priori upper bound on trajectories with a given value of energy. Of course, it will not restrict the generality if we assume that the energy is zero, because we may add a constant to the potential without changing the equations of motion.

Let

$$G(r) := \inf_{|x|=r} V(x).$$

If there exists a sequence of $r_n \to \infty$ such that $G(r_n) \geq 0$, then all the trajectories with zero energy are bounded, because they cannot cross the spheres $\{|x| = r_n\}$.

We will say that $G(r)$ is negative at infinity if there exists R_0 such that, for $r \geq R_0$, we have $G(r) < 0$. In such a case, we define

$$R = \inf\{r \mid G(s) < 0, \quad s \geq r\},$$

fix $r_0 > R$ and set

$$T := \int_R^{r_0} (-2G(r_1))^{\frac{1}{2}} \mathrm{d}r_1.$$

We denote by $[-T, \infty[\ni t \mapsto \omega(t)$ the unique solution of

$$\begin{cases} \frac{1}{2}(\dot{\omega}(t))^2 + G(\omega(t)) = 0, \\ \omega(0) = r_0, \quad \dot{\omega}(0) > 0. \end{cases}$$

The solution $\omega(t)$ can be computed as follows. For $r > r_0$, we set

$$K(r) = \int_{r_0}^{r} (-2G(r_1))^{-\frac{1}{2}} \mathrm{d}r_1.$$

Then $\omega(t)$ is the inverse function of

$$[r_0, \infty[\ni r \mapsto K(r).$$

We have the following result.

Proposition 2.2.1
Let $G(r)$ be negative at infinity. Then for every $(y, \eta) \in H^{-1}(\{0\})$, there exists t_0 such that, for all $t \geq 0$,

$$|x(t, y, \eta)| \leq \omega(t - t_0).$$

Proof. The trajectory $x(t)$ cannot cross the sphere $\{x \mid |x| = R\}$. Therefore, it is enough to consider the case when $|x(t)| > R$ for all times. We have

$$\frac{\mathrm{d}|x(t)|}{\mathrm{d}t} \leq \left|\frac{\mathrm{d}x(t)}{\mathrm{d}t}\right| \leq (-2V(x(t)))^{\frac{1}{2}} \leq (-2G(|x(t)|))^{\frac{1}{2}}.$$

Therefore,

$$\frac{\mathrm{d}K(|x(t)|)}{\mathrm{d}t} \leq 1.$$

Hence, for $t_0 := -K(|x(0)|)$, we obtain

$$K(|x(t)|) \leq t - t_0,$$

from which we deduce directly that

$$|x(t)| \leq \omega(t - t_0).$$

□

Remark. If the potential $V(x)$ is spherically symmetric and $G(r)$ is negative at infinity, then for any unit vector v, there are zero-energy trajectories of the form

$$x(t) = \omega(t - t_0)v.$$

Therefore, the bound of Proposition 2.2.1 is optimal for such potentials.

From now on, will assume that

$$\lim_{|x|\to\infty} V(x) = 0. \tag{2.2.2}$$

Under this assumption, zero-energy trajectories have special properties that we describe below.

Proposition 2.2.2
Assume (2.2.2). Let $(y, \eta) \in H^{-1}(\{0\})$. *Then*

$$\lim_{t\to\pm\infty} \frac{x(t, y, \eta)}{t} = 0. \tag{2.2.3}$$

Proof. $\lim_{r\to\infty} G(r) = 0$ implies $\lim_{r\to\infty} K(r)/r = \infty$. Therefore, $\lim_{t\to\infty} \omega(t)/t = 0$. Hence (2.2.3) follows from Proposition 2.2.1. □

Motivated by the above proposition we introduce the following definition.

Definition 2.2.3
The trajectory $x(t)$ *is called* almost-bounded at ∞ *if and only if*

$$\lim_{t\to\infty} \frac{x(t)}{t} = 0$$

and $x(t)$ *is not bounded for* $t \geq 0$.

Analogously, we define trajectories almost-bounded at $-\infty$. Proposition 2.2.2 says that all the zero-energy trajectories are either bounded or almost-bounded at $\pm\infty$. Now let us give an example of almost-bounded trajectories.

Example 2.2.4 If $\mu > 0$ and $V(x) = C_0|x|^{-\mu}$ is a one-dimensional potential, then there are trajectories in $H^{-1}(\{0\})$ of the form $x(t) = C_1 t^\delta$, for $\delta = 2/(2+\mu)$ and $C_1 = (\frac{C_0}{2}(2+\mu)^2)^{1/(2+\mu)}$. More generally, in any dimension, if $\mu > 0$ and $|V(x)| \leq C\langle x\rangle^{-\mu}$, then we have the bound

$$|\omega(t)| \leq C\langle t\rangle^\delta.$$

with the same δ.

2.3 The Mourre Estimate and Scattering Trajectories

In this section we strengthen our assumptions on the potentials. In addition to (2.2.2), we will assume that

$$\lim_{|x|\to\infty} x\nabla_x V(x)\rangle = 0. \tag{2.3.1}$$

The following class of trajectories will be the main object of investigations throughout this whole chapter.

Definition 2.3.1
The trajectory $x(t, y, \eta)$ is called a scattering trajectory for $t \to \infty$ *if there exists some T and $C_0 > 0$ such that, for $t \geq 0$,*

$$|x(t, y, \eta)| \geq C_0(t - T).$$

Scattering trajectories for $t \to -\infty$ are defined in an analogous way.

Note that almost-bounded trajectories and scattering trajectories are two disjoint categories of unbounded trajectories. As we will see below, if we assume (2.2.2) and (2.3.1), then every point in $\mathcal{R}^+$ belongs either to an almost-bounded trajectory (if the energy is zero) or to a scattering trajectory (if the energy is positive).

Along a scattering trajectory, we can translate a spatial decay of the force $-\nabla_x V(x)$ into its time decay along a scattering trajectory. Therefore, as we will see in the following sections, scattering trajectories are better behaved than general unbounded trajectories from the point of view of scattering theory.

Let us now introduce some special observables that are useful in the study of scattering trajectories.

Let $f \in C^\infty(\mathbb{R}^+)$, $f' \geq 0$, $f = 0$ on $]0, 1/2]$ and $f = 1$ on $[1, \infty[$. For $r > 0$, we set

$$F(r) := 1 + \int_1^r f(r_1)\mathrm{d}r_1.$$

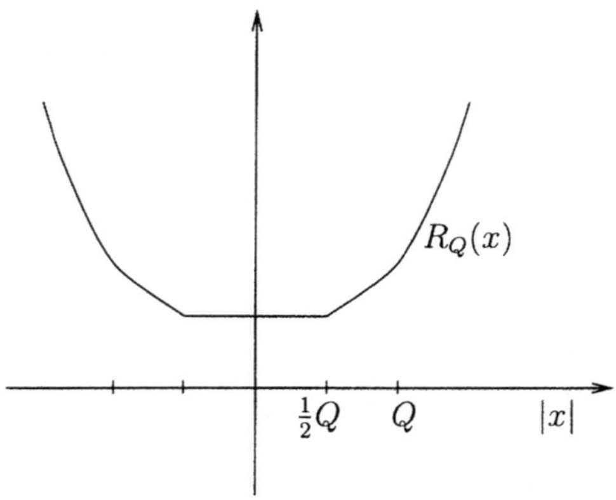

Fig. 2.1. Graph of the function R_Q

For any $Q > 0$, we define

$$R_Q(x) := \frac{1}{2} Q^2 F\left(\frac{x^2}{Q^2}\right),$$

$$a_Q(x, \xi) := \langle \xi, x \rangle f\left(\frac{x^2}{Q^2}\right) = \frac{1}{2}\left\{\xi^2, R_Q\right\},$$

where $\{\cdot, \cdot\}$ denotes the Poisson bracket.

Note that $|x| \geq Q$ if and only if $R_Q(x) \geq \frac{1}{2}Q^2$, and in that case

$$R_Q(x) = \frac{1}{2}x^2.$$

The following proposition can be considered as a classical version of the celebrated *Mourre estimate* for the quantum problem.

Proposition 2.3.2
We assume that the potential $V(x)$ satisfies (2.2.2) and (2.3.1). Then, for any $\gamma > 0$, there exists Q such that

$$\{H, a_Q\}(x, \xi) \geq 0 \quad \text{if} \quad 2H(x, \xi) \geq \gamma, \tag{2.3.2}$$

$$\{H, a_Q\}(x, \xi) \geq 2H(x, \xi) - \gamma \quad \text{if} \quad |x| \geq Q. \tag{2.3.3}$$

Proof. We compute:

$$\{H, a_Q\} = \frac{\langle \xi, x \rangle^2}{Q^2} f'\left(\frac{x^2}{Q^2}\right) + \xi^2 f\left(\frac{x^2}{Q^2}\right) - \langle x, \nabla_x V(x) \rangle f\left(\frac{x^2}{Q^2}\right)$$

$$\geq (2H(x,\xi) - 2V(x) - \langle x, \nabla_x V(x)\rangle) f\left(\frac{x^2}{Q^2}\right).$$

Since $V(x)$ and $\langle x, \nabla_x V(x)\rangle$ go to 0 when $|x|$ goes to ∞, this proves the proposition. □

It will be useful to fix notation for some special subsets of the phase space $X \times X'$. Let $R \geq 0$, $\epsilon \geq 0$ and $-1 \leq \sigma \leq 1$. Then we define

$$\Gamma^+_{\epsilon,\sigma} := \{(x,\xi) \in X \times X' \mid |\xi| \geq \epsilon,\ \langle x, \xi\rangle \geq \sigma |x||\xi|\},$$

$$\Gamma^+_{R,\epsilon,\sigma} := \Gamma^+_{\epsilon,\sigma} \backslash \{(x,\xi) \in X \times X' \mid |x| \leq R.\}$$

An easy geometric argument shows that if $\epsilon > 0$ and $\sigma > -1$, then there exists $C_0 > 0$ such that, for any $(x,\xi) \in \Gamma^+_{\epsilon,\sigma}$, we have

$$|x + s\xi| \geq C_0(s + |x|),\ s \geq 0.$$

As we will see below, a similar estimate is true for the full flow if $\epsilon > 0$, $\sigma > -1$, R is big enough and the initial conditions belong to $\Gamma^+_{R,\epsilon,\sigma}$. One usually says that $\Gamma^+_{\epsilon,\sigma}$ is *outgoing* for the free flow $\phi_0(t)$ and $\Gamma^+_{R,\epsilon,\sigma}$ is *outgoing* for the full flow $\phi(t)$.

The most important consequence of Proposition 2.3.2 is the following theorem.

Theorem 2.3.3
Assume (2.2.2) and (2.3.1). Then the following statements are true:
(i) If $(y,\eta) \in H^{-1}(]0,\infty[) \cap \mathcal{R}^+$, then (y,η) belongs to a scattering trajectory.
(ii) The set $H^{-1}(]0,\infty[) \cap \mathcal{R}^+$ is open.
(iii) If $\epsilon > 0$ and $-1 < \sigma$, then there exists $R > 0$ such that $\Gamma^+_{R,\epsilon,\sigma} \subset H^{-1}(]0,\infty[) \cap \mathcal{R}^+$; moreover, there exists $C_0 > 0$ such that, for any $(y,\eta) \in \Gamma^+_{R,\epsilon,\sigma}$ and $t > 0$,

$$|x(t,y,\eta)| \geq C_0(t + |y|).$$

(iv) If K is a compact subset of $H^{-1}(]0,\infty[) \cap \mathcal{R}^+$, then there exist $C_0 > 0$ and T such that, for any $(y,\eta) \in K$ and $t > T$,

$$|x(t,y,\eta)| \geq C_0(t - T).$$

(v) For any $\lambda_0 > 0$, there exists Q such that if $(y,\eta) \in \mathcal{B}^+ \cap H^{-1}(]\lambda_0,\infty[)$, then

$$\limsup_{t\to\infty} |x(t,y,\eta)| \leq Q.$$

Proof. We will first prove *(i)*.

Fix $2\lambda_0 > \gamma > 0$. Choose Q such that a_Q satisfies (2.3.2) and (2.3.3) for this γ.

Let $(y,\eta) \in \mathcal{R}^+ \cap H^{-1}(]\lambda_0,\infty[)$ and let $(x(t),\xi(t))$ be the trajectory starting at (y,η). Set

$$Q(t) := a_Q(x(t), \xi(t)), \qquad R_Q(t) := R_Q(x(t)).$$

Note the following properties of the functions $a_Q(t)$ and $R_Q(t)$, which we will use in our proof.

$$\begin{cases} \frac{\mathrm{d}}{\mathrm{d}t} R_Q(t) = a_Q(t), \\ \frac{\mathrm{d}}{\mathrm{d}t} a_Q(t) \geq 0, \\ \text{if} \;\; R_Q(t) \geq \frac{1}{2}Q^2, \;\; \text{then} \;\; \frac{\mathrm{d}}{\mathrm{d}t} a_Q(t) \geq 2\lambda_0 - \gamma. \end{cases}$$

The first equality follows from the equations of motion, the second and third are consequences of (2.3.2) and (2.3.3).

Since $a_Q(t)$ is an increasing function, there exists $a_Q^+ := \lim_{t\to\infty} a_Q(t)$. Consider two cases.

Case (1) $a_Q^+ > 0$. Then for sufficiently big t,

$$a_Q(t) = \frac{\mathrm{d}}{\mathrm{d}t} R_Q(t) \geq c > 0.$$

Hence, for $t > t_0$ where t_0 is sufficiently big, we have $R_Q(t) \geq \frac{1}{2}Q^2$. Therefore, for $t \geq t_0$,

$$\frac{1}{2}|x(t)|^2 = R_Q(t) \geq a + bt + \frac{t^2}{2}(2\lambda_0 - \gamma). \tag{2.3.4}$$

Therefore (y, η) belongs to a scattering trajectory.

Case (2) $a_Q^+ \leq 0$. Then for all t,

$$\frac{\mathrm{d}}{\mathrm{d}t} R_Q(t) \leq 0. \tag{2.3.5}$$

Thus, clearly, $R_Q(t)$ is bounded and hence (y, η) belongs to a bounded trajectory. This ends the proof of *(i)*.

Clearly, in *Case (2)* there exists $\lim_{t\to\infty} R_Q(t)$. If this limit was greater than $\frac{1}{2}Q^2$, then (2.3.4) would be true for $t \geq t_0$ with t_0 sufficiently big and we would be back in *Case (1)*. Therefore, this limit is less than $\frac{1}{2}Q^2$. Therefore, $|x(t)| \leq Q$ for t sufficiently big. This proves *(v)*.

Next note that $(y, \eta) \mapsto a_Q^+(y, \eta)$ is the limit of an increasing family of continuous functions $a_Q(t)$ on the open set $H^{-1}(]\lambda_0, \infty[)$. Hence a_Q^+ is a lower semicontinuous function. Thus the set $\{(y, \eta) \mid a_Q^+(y, \eta) > 0, \;\; H(y, \eta) > \lambda_0\}$ is open. But *(i)* says that this set equals $H^{-1}(]\lambda_0, \infty[) \cap \mathcal{R}^+$. This implies *(ii)*.

Now let us prove *(iii)*. Let us first fix some constants $c_0 > 0$, $\epsilon_0 > 0$ such that

$$1 - 2c_0 - |\sigma_-| > \epsilon_0,$$

where $\sigma_- = \min\{\sigma, 0\}$. By Proposition 2.3.2, we can choose Q such that, for $|x| > Q$,

$$\{a_Q, h\}(x, \xi) \geq 2H(x, \xi)(1 - c_0) \geq \xi^2(1 - 2c_0). \tag{2.3.6}$$

Let R satisfy

$$R^2(1 - |\sigma_-|) > Q^2,$$

and let $(y, \eta) \in \Gamma^+_{R,\epsilon,\sigma}$. Since $R > Q$, we have, by (2.3.6), for small enough time t,

$$\begin{aligned} \tfrac{d^2}{dt^2} R_Q(t) &\geq 2H(x(t), \xi(t))(1 - c_0) \\ &\geq 2H(y, \eta)(1 - c_0) \geq \eta^2(1 - 2c_0). \end{aligned}$$

Moreover,

$$\begin{aligned} R_Q(0) &= \tfrac{1}{2}y^2, \\ a_Q(0) &= \langle y, \eta\rangle \geq -|y||\eta||\sigma_-|. \end{aligned}$$

Hence as long as $|x(t)| > Q$ we have

$$\begin{aligned} R_Q(t) &= \tfrac{1}{2}y^2 + t\langle y, \eta\rangle + \tfrac{1}{2}t^2\eta^2(1 - 2c_0) \\ &\geq \tfrac{1}{2}y^2 - t|y||\eta||\sigma_-| + \tfrac{1}{2}t^2\eta^2(1 - 2c_0) \\ &\geq \tfrac{1}{2}y^2(1 - |\sigma_-|) + \tfrac{1}{2}t^2\eta^2(1 - 2c_0 - |\sigma_-|) + \tfrac{1}{2}|\sigma_-|(|y| - t|\eta|)^2 \\ &\geq \tfrac{1}{2}R^2(1 - |\sigma_-|) + \tfrac{1}{2}t^2\epsilon_0\epsilon^2. \end{aligned} \tag{2.3.7}$$

Since $R^2(1 - |\sigma_-|) > Q^2$, we deduce from (2.3.7) that $|x(t)|$ is greater than Q for all times and, consequently, (2.3.7) holds for all times, which proves *(iii)*.

Let us now prove *(iv)*. Fix any R, ϵ, σ are such as in *(iii)*. From the proof of *(i)* we see that if $(y_0, \eta_0) \in K \subset \mathcal{R}^+ \cap H^{-1}(]\lambda_0, \infty[)$, then we will find T_0 such that

$$(x(T_0, y_0, \eta_0), \xi(T_0, y_0, \eta_0)) \in \Gamma^+_{R,\epsilon,\sigma}.$$

It follows by the continuity of the flow that we can find an open neighborhood U of (y_0, η_0) such that if $(y, \eta) \in U$ then $(x(T_0, y, \eta), \xi(T_0, y, \eta)) \in \Gamma^+_{R,\epsilon,\sigma}$. If we now apply *(iii)*, then we see that

$$|x(t, y, \eta)| \geq C_0(t - T),$$

for some $C_0 > 0$ and T, uniformly for $(y, \eta) \in U$. To extend this onto the whole compact set K, we use the standard covering argument. This completes the proof of the theorem. □

2.4 Non-trapping Energies

In this section we would like to study the trajectories for non-trapping energies. Results of this section will not be used in this chapter.

We start with the following simple proposition about the set of trapping energies introduced in Definition 2.1.2.

Proposition 2.4.1
The set $]0,\infty[\backslash\sigma$ *is open.*

Proof. Let $0<\lambda_0<\lambda$. Let $\lambda_n \in]\lambda_0,\infty[\cap H(\mathcal{B}^+)$ such that $\lambda_n \to \lambda$. Then, by Theorem 2.3.3 *(v)*, there exist

$$(x_n,\xi_n)\in H^{-1}(\{\lambda_n\})\cap\mathcal{B}^+\cap\{(x,\xi)|\ |x|\le Q\}.$$

Next we use the compactness argument, and we see that, by taking a subsequence, we can guarantee that there exists

$$\lim_{n\to\infty}(x_n,\xi_n)=:(x_0,\xi_0).$$

Now since we know by Theorem 2.3.3 *(ii)* that the set $H^{-1}(]\lambda_0,\infty[)\cap\mathcal{R}^+$ is open, $(x_0,\xi_0)\in\mathcal{B}^+$. Therefore, $\lambda\in H(\mathcal{B}^+)$, which implies the closedness of $H(\mathcal{B}^+)$ in $\mathbb{R}^+$. □

For non-trapping energies, one can strengthen the classical Mourre estimate. By modifying the observable a_Q, one can make its Poisson bracket bounded below by a positive constant in any interval of non-trapping energies.

Proposition 2.4.2
Assume that the potential satisfies (2.2.2) and (2.3.1). If I *is a compact subset of* $\mathbb{R}^+\backslash\sigma$ *and* $\gamma>0$, *then there exists a function* $\tilde{a}(\cdot,\cdot)$ *on* $X\times X'$ *such that, for* $|x|$ *large enough,*

$$\tilde{a}(x,\xi)=\langle x,\xi\rangle$$

and, for all $(x,\xi)\in H^{-1}(I)$,

$$\{H,\tilde{a}\}(x,\xi)\ge 2H(x,\xi)-\gamma. \tag{2.4.1}$$

Proof. Let Q and a_Q satisfy the Mourre estimate as in Proposition 2.3.2 with γ replaced with $\gamma/2$. Let $G\in C_0^\infty(X)$ such that $G\ge 0$ and $G=1$ for $|x|<Q$. Set

$$r(y,\eta):=-\int_0^\infty G(x(t,y,\eta))\mathrm{d}t.$$

Let $I_1\subset\mathbb{R}^+\backslash\sigma$ be a compact set that contains I in its interior. By the compactness of $H^{-1}(I_1)\cap\{x|\ |x|<Q\}$ and Theorem 2.3.3 *(iv)*, we see that r is a bounded function on $H^{-1}(I_1)$. Moreover, $\{H,r\}=G(x)$. Let $g\in C_0^\infty(\mathbb{R})$ be such that $g\ge 0$, $g=0$ outside I_1 and $g(\lambda)=\lambda$ on I. We put

$$r_R(x,\xi)=G\left(R^{-1}x\right)r(x,\xi)g(H(x,\xi)).$$

We have

$$\begin{aligned}\{H,r_R\}(x,\xi)&=R^{-1}\langle\xi,\nabla G\left(R^{-1}x\right)\rangle r(x,\xi)g(H(x,\xi))\\&\quad+G\left(R^{-1}x\right)G(x)g(H(x,\xi)).\end{aligned}$$

By picking R large enough, we can make the first term smaller than $\gamma/2$ and we can guarantee that $G(R^{-1}x)G(x) = G(x)$. For such R, we set

$$\tilde{a}(x,\xi) := a_Q(x,\xi) + r_R(x,\xi).$$

We obtain

$$\{H,\tilde{a}\}(x,\xi) \geq \{H,a_Q\}(x,\xi) + G(x)g(H(x,\xi)) - \frac{\gamma}{2},$$

from which the estimate (2.4.1) follows at once. □

The variant of the classical Mourre estimate contained in Proposition 2.4.2 allows one to estimate the time that is needed for a trajectory with a non-trapping energy to become outgoing.

Theorem 2.4.3
Assume that the potential satisfies (2.2.2) and (2.3.1). Let I be a compact subset of $\mathbb{R}^+\backslash\sigma$. Then there exists $C_0 > 0$ with the following property. For any $(y,\eta) \in H^{-1}(I)$, we will find T such that

$$\langle y \rangle \geq C_0 T,$$

and if $t \in \mathbb{R}$, then

$$\langle x(t,y,\eta)\rangle \geq C_0|t-T|. \tag{2.4.2}$$

Proof. We will use the observable $\tilde{a}(x,\xi)$ constructed in Proposition 2.4.2 that, for $(x,\xi) \in H^{-1}(I)$, satisfies

$$\{H,\tilde{a}\}(x,\xi) \geq C_1 > 0. \tag{2.4.3}$$

It is easy to see that, for $(x,\xi) \in H^{-1}(I)$,

$$C_2\langle x \rangle \geq |\tilde{a}(x,\xi)|. \tag{2.4.4}$$

Set $\tilde{a}(t) := \tilde{a}(x(t,y,\eta),\xi(t,y,\eta))$. By (2.4.3), for any $(y,\eta) \in H^{-1}(I)$,

$$\frac{\mathrm{d}}{\mathrm{d}t}\tilde{a}(t) \geq C_1. \tag{2.4.5}$$

Let T be defined by the equality $\tilde{a}(T) = 0$. Then, by integrating (2.4.5) and using (2.4.4), we obtain

$$C_2\langle y \rangle \geq |\tilde{a}(0)| = \left|\textstyle\int_0^T \frac{\mathrm{d}}{\mathrm{d}s}\tilde{a}(s)\mathrm{d}s\right| \geq C_1 T,$$
$$C_2\langle x(t,y,\eta)\rangle \geq |\tilde{a}(t)| = \left|\textstyle\int_T^t \frac{\mathrm{d}}{\mathrm{d}s}\tilde{a}(s)\mathrm{d}s\right| \geq C_1|t-T|.$$

□

2.5 Asymptotic Velocity

The aim of this section is to introduce the basic asymptotic quantity – the *asymptotic velocity*. It will be the analog of the asymptotic momentum constructed in Theorem 1.3.1 in the time-decaying case.

We will need the following additional condition on the force:

$$\int_0^\infty \sup_{|x|\geq R} |F(x)| \mathrm{d}R < \infty. \tag{2.5.1}$$

Actually, as we will see from the proposition below, this condition implies (2.3.1) and "almost implies" (2.2.2), which are the assumptions that we used before.

Proposition 2.5.1
Assume (2.5.1).
(i) If the dimension of X is greater than 1 than there exists

$$\lim_{|x|\to\infty} V(x). \tag{2.5.2}$$

If the dimension is equal to 1 *then there exist both* $\lim_{x\to\pm\infty} V(x)$.
(ii)

$$\lim_{|x|\to\infty} |x||F(x)| = 0.$$

Proof. *(ii)* follows from Lemma A.1.3. Let us show *(i)*.

Clearly, for any unit vector v, there exists

$$\lim_{t\to\infty} V(tv).$$

If the dimension is bigger than 1, then we can connect two unit vectors $v(0)$ and $v(1)$ with a curve $[0,1] \ni \tau \mapsto v(\tau)$ such that $|v(\tau)| = 1$. Now,

$$\lim_{t\to\infty} |V(tv(0)) - V(tv(1))| \leq C \lim_{t\to\infty} \left(\sup_{\tau\in[0,1]} t|F(tv(\tau))| \right) = 0,$$

where in the last step we used *(ii)*. □

Now we can show the existence of the asymptotic velocity. This result is the first step of the classification of all the scattering trajectories, which is the main goal of this chapter.

Theorem 2.5.2
Assume the hypotheses (2.2.2) and (2.5.1). Then for any $(y,\eta) \in X \times X'$, the following limit exists:

$$\lim_{t\to\infty} \frac{x(t,y,\eta)}{t} =: \xi^+(y,\eta). \tag{2.5.3}$$

The function $\xi^+(\cdot,\cdot)$ *has the following properties.*
(i) If $\xi^+(y,\eta) \neq 0$, *then*

$$\lim_{t\to\infty} \xi(t,y,\eta) = \xi^+(y,\eta). \tag{2.5.4}$$

(ii) The set $(\xi^+)^{-1}(X'\backslash\{0\})$ *is open and is equal to*

$$H^{-1}(]0,\infty[) \cap \mathcal{R}^+$$

and also to the union of all scattering trajectories.
(iii) The map

$$(\xi^+)^{-1}(X'\backslash\{0\}) \ni (y,\eta) \mapsto \xi^+(y,\eta) \in X'$$

is continuous.
(iv) If $\epsilon > 0$ *and* $-1 < \sigma$, *then there exists* R *such that on* $\Gamma^+_{R,\epsilon,\sigma}$

$$\xi^+(y,\eta) - \eta \in o(|y|^0). \tag{2.5.5}$$

Proof. If $(y,\eta) \in \mathcal{B}^+$, then, obviously,

$$\lim_{t\to\infty} \frac{x(t,y,\eta)}{t} = 0.$$

Using Proposition 2.2.2, we get that if $(y,\eta) \in H^{-1}(\{0\})$, then we also have

$$\lim_{t\to\infty} \frac{x(t,y,\eta)}{t} = 0.$$

Clearly, $H^{-1}(\lambda)$ is a compact set if $\lambda < 0$. Hence $H^{-1}(]-\infty,0[) \subset \mathcal{B}$. Therefore it remains to consider the set $\mathcal{R}^+ \cap H^{-1}(]0,\infty[)$.

Let K be a compact set contained in $\mathcal{R}^+ \cap H^{-1}(]0,\infty[)$. Then, by Theorem 2.3.3 *(iv)*, for all $(y,\eta) \in K$, we will find T and $C_0 > 0$ such that

$$|x(t,y,\eta)| \geq C_0(t-T).$$

Now choose $J \in C^\infty(X)$ such that $0 \notin \text{supp} J$ and $J = 1$ on a neighborhood of $\{x \mid |x| > C_0\}$. Note that all the trajectories starting in K are, for $t > T_1$, also trajectories for the following time-dependent force:

$$F_J(t,x) := J\left(\frac{x}{t}\right) F(x). \tag{2.5.6}$$

The force (2.5.6) satisfies the assumptions of Theorem 1.3.1. Hence the existence of (2.5.3) for scattering trajectories as well as the statements *(i)*, *(ii)* and *(iii)* follow from Theorem 1.3.1. Clearly,

$$\xi^+(y,\eta) - \eta = -\int_0^\infty F(x(u,y,\eta))\mathrm{d}u. \tag{2.5.7}$$

Now *(iv)* follows from (2.5.7) and the estimates on $x(t,y,\eta)$ of Theorem 2.3.3 *(iii)*. This completes the proof of the theorem. □

We will use the trick of replacing the time-independent force $F(x)$ with a time-dependent one $J\left(\frac{x}{t}\right)F(x)$ many times. We will refer to it in the sequel as "introducing an effective time-dependent force".

The function $\xi^+(\cdot,\cdot)$ will be called the *asymptotic velocity.* (In the previous chapter the analogous function was called the asymptotic momentum, but now the name asymptotic velocity seems more justified, since, for bounded trajectories, it is the limit (2.5.3) that always exists, and (2.5.4) does not have to be true). We will sometimes write $\xi(\infty,y,\eta)$ instead of $\xi^+(y,\eta)$.

Example 2.5.3 below shows that in general the asymptotic velocity is not continuous on $(\xi^+)^{-1}(\{0\})$.

Example 2.5.3
Consider a one-dimensional potential $V(x)$ that satisfies the assumptions of Theorem 2.5.2, has a global maximum at $x=0$ and goes to zero as $|x|\to\infty$. Then $(x(t),\xi(t)) = (0,0)$ is a bounded trajectory, hence $\xi^+(0,0)=0$. On the other hand, if $\eta = \pm\sqrt{2(V(0)-V(y))}$ and $\pm y>0$, then $\xi^+(y,\eta) = \pm\sqrt{2V(0)}$.

2.6 Short-Range Case

We are now going to study scattering theory for short-range time-independent potentials. The short-range case means roughly that the force $F(x)$ decays like $\langle x\rangle^{-1-\mu}$ for some $\mu>1$.

The first result is analogous to Theorem 1.4.1. It says that one can define the asymptotic position exactly as in Chap. 1. Essentially, the only difference is that now one has to assume that the asymptotic momentum is non-zero.

Theorem 2.6.1
Assume that

$$\int_0^\infty \sup_{|x|\geq R} |F(x)|\langle R\rangle \mathrm{d}R < \infty. \tag{2.6.1}$$

Let $\xi^+(y,\eta)\neq 0$. *Then there exists*

$$\lim_{t\to\infty}(x(t,y,\eta) - t\xi^+(y,\eta)) =: x_{\mathrm{sr}}^+(y,\eta). \tag{2.6.2}$$

Moreover, the following statements are true.
(i) The function

$$(\xi^+)^{-1}(X'\backslash\{0\}) \ni (y,\eta) \mapsto x_{\mathrm{sr}}^+(y,\eta) \in X$$

is continuous.

(ii) If $\epsilon > 0$ *and* $-1 < \sigma$*, then there exists* R *such that on* $\Gamma^{+}_{R,\epsilon,\sigma}$

$$x^{+}_{\rm sr}(y,\eta) - y \in o(|y|^{0}).$$

Proof. We introduce an effective time-dependent force as in the proof of Theorem 2.5.2, and then apply Theorem 1.4.1. By the equality

$$x^{+}_{\rm sr}(y,\eta) - y = \int_{0}^{\infty} tF(x(t,y,\eta))\mathrm{d}t,$$

the property *(ii)* follows from the estimates on $x(t,y,\eta)$ of Theorem 2.3.3 *(iii)*. □

We will call $x^{+}_{\rm sr}(y,\eta)$ the *asymptotic position* .

As in Sect. 1.4, for $t \in [0,\infty[$ and $(x,\xi) \in X \times X'$, we denote by $[0,t] \ni s \mapsto (y_{\rm sr}(s,t,x,\xi), \eta_{\rm sr}(s,t,x,\xi))$ the solution of

$$\begin{cases} \partial_s y_{\rm sr}(s,t,x,\xi) = \eta_{\rm sr}(s,t,x,\xi), \\ \partial_s \eta_{\rm sr}(s,t,x,\xi) = F(y_{\rm sr}(s,t,x,\xi)), \\ y_{\rm sr}(t,t,x,\xi) = x + t\xi, \qquad \eta_{\rm sr}(t,t,x,\xi) = \xi. \end{cases} \tag{2.6.3}$$

Note that, for every $r \in \mathbb{R}$,

$$\begin{aligned} &(y_{\rm sr}(s,t,x,\xi), \eta_{\rm sr}(s,t,x,\xi)) \\ &= (y_{\rm sr}(s-r,t-r,x+r\xi,\xi), \eta_{\rm sr}(s-r,t-r,x+r\xi,\xi)). \end{aligned} \tag{2.6.4}$$

The following theorem is an analog of Theorem 1.4.2 of the previous chapter. The main difference is that, this time, one has to restrict oneself to the case $\xi \neq 0$.

Theorem 2.6.2
Assume that the force satisfies (2.6.1) and, in addition,

$$\int_{0}^{\infty} \sup_{|x|\geq R} |\partial_x^{\alpha} F(x)| \langle R \rangle \mathrm{d}R < \infty, \ |\alpha| = 1. \tag{2.6.5}$$

Then the trajectory $(y_{\rm sr}(s,t,x,\xi), \eta_{\rm sr}(s,t,x,\xi))$ *converges as* $t \to \infty$ *uniformly for* (s,x,ξ) *in compact sets of* $\mathbb{R} \times X \times (X' \backslash \{0\})$ *to a trajectory*

$$s \mapsto (y_{\rm sr}(s,\infty,x,\xi), \eta_{\rm sr}(s,\infty,x,\xi)),$$

which satisfies

$$\begin{cases} \lim_{s\to\infty} (y_{\rm sr}(s,\infty,x,\xi) - x - s\xi) = 0, \\ \lim_{s\to\infty} (\eta_{\rm sr}(s,\infty,x,\xi) - \xi) = 0. \end{cases} \tag{2.6.6}$$

Moreover, the following facts are true:
(i) The trajectory $(y_{\rm sr}(s,\infty,x,\xi), \eta_{\rm sr}(s,\infty,x,\xi))$ *is the only one that satisfies (2.6.6).*

(ii) The mapping

$$[0,\infty] \times X \times (X'\backslash\{0\}) \ni (t,x,\xi)$$
$$\mapsto (y_{\rm sr}(s,t,x,\xi) - x - s\xi, \ \ \eta_{\rm sr}(s,t,x,\xi) - \xi) \in C_\infty(\mathbb{R}_s, X \times X')$$

is continuous.

Proof. Let Θ be a compact subset of $X \times (X'\backslash\{0\})$. We can find $\epsilon > 0$, $-1 < \sigma$ and r such that if $(x,\xi) \in \Theta$, then $(x + r\xi, \xi) \in \Gamma^+_{\epsilon,\sigma}$. Therefore, using (2.6.4), we see that it is enough to prove the theorem for $(x,\xi) \in \Gamma^+_{\epsilon,\sigma}$.

Clearly, there exists $C_0 > 0$ such that if $(x,\xi) \in \Gamma^+_{\epsilon,\sigma}$ and $s \geq 0$, then

$$|x + s\xi| \geq C_0 s. \tag{2.6.7}$$

Introduce a cut-off function $J(x)$ and a time-dependent effective force $F_J(t,x)$ as we did in the proof of Theorem 2.5.2. Clearly, the force $F_J(t,x)$ satisfies the assumptions of Theorem 1.4.2.

Let $y_{{\rm fd},J}(s,t,x,\xi)$ be the trajectory constructed in Theorem 1.4.2 with the force $F_J(t,x)$. We know from this theorem that it satisfies

$$y_{{\rm fd},J}(s,t,x,\xi) - x - s\xi \in o(s^0). \tag{2.6.8}$$

uniformly in (t,x,ξ). Therefore, if we put together (2.6.7) and (2.6.8), we see that if T_1 is large enough, then this trajectory is also a solution of (2.2.1) for $s > T_1$.

The fact that the trajectory $(y_{\rm sr}(s,\infty,x,\xi), \eta_{\rm sr}(s,\infty,x,\xi))$ constructed in this way is the only solution of (2.6.6) can be seen by the following argument. If there exist two solutions $(y_{{\rm sr},1}(s), \eta_{{\rm sr},1}(s))$ and $(y_{{\rm sr},2}(s), \eta_{{\rm sr},2}(s))$ that satisfy (2.6.6) then, for s big enough, they both are solutions of the problem considered in Theorem 1.4.2 with the time-dependent force $F_J(t,x)$. By the uniqueness of the problem considered in Theorem 1.4.2, they coincide for $s > T$, hence for all s. □

The following identities are true:

$$\phi(s-t)\phi_0(t)(x,\xi) = (y_{\rm sr}(s,t,x,\xi), \eta_{\rm sr}(s,t,x,\xi)),$$

$$\phi_0(-t)\phi(t)(y,\eta) = (x(t,y,\eta) - t\xi(t,y,\eta), \ \xi(t,y,\eta)).$$

As in Sect. 1.4, we will now summarize the results obtained so far by introducing the *wave transformations.*

Theorem 2.6.3
(i) Assume (2.6.1). Then there exists the limit

$$\lim_{t\to\infty} \phi_0^{-1}(t)\phi(t) \tag{2.6.9}$$

uniformly on compact sets in $(\xi^+)^{-1}(X'\backslash\{0\})$. *The limit is a continuous map from* $(\xi^+)^{-1}(X'\backslash\{0\})$ *into* $X \times (X'\backslash\{0\})$.
(ii) Assume, in addition, that (2.6.5). Then there exists the limit

$$\lim_{t\to\infty} \phi(-t)\phi_0(t) =: \mathcal{F}^+_{\rm sr} \tag{2.6.10}$$

uniformly on compact sets in $X \times (X'\backslash\{0\})$. *The map*

$$\mathcal{F}^+_{\rm sr} : X \times (X'\backslash\{0\}) \to (\xi^+)^{-1}(X'\backslash\{0\})$$

defined by (2.6.10) is continuous and bijective. Moreover, (2.6.9) is equal to $(\mathcal{F}^+_{\rm sr})^{-1}$.
(iii) If $(y, \eta) = \mathcal{F}^+_{\rm sr}(x, \xi)$, *one has*

$$\lim_{t\to\infty} (\phi(t)(y, \eta) - \phi_0(t)(x, \xi)) = 0. \tag{2.6.11}$$

(iv) The mapping $\mathcal{F}^+_{\rm sr}$ *is symplectic.*
(v) The wave transformation intertwines the full and the free dynamics:

$$H \circ \mathcal{F}^+_{\rm sr} = H_0,$$

$$\phi(t) \circ \mathcal{F}^+_{\rm sr} = \mathcal{F}^+_{\rm sr} \circ \phi_0(t).$$

Note also that

$$(\mathcal{F}^+_{\rm sr})^{-1}(y, \eta) = (x^+_{\rm sr}(y, \eta), \xi^+(y, \eta)).$$

2.7 Long-Range Case

In this section we consider the case of long-range time-independent potentials. This means roughly that $F(x)$ decays at infinity like $\langle x\rangle^{-1-\mu}$ for some $0 < \mu \le 1$. In particular, the physically important *Coulomb potential* is of the long-range type. Our main assumption on the potentials in the long-range case will be

$$\int_0^\infty \sup_{|x|\ge R} |\partial_x^\alpha F(x)| \langle R\rangle^{|\alpha|} \mathrm{d}R < \infty, \qquad |\alpha| \le 1. \tag{2.7.1}$$

As in Sect. 1.5, we start by a study of a mixed problem where boundary conditions are the initial position and the final momentum.

Theorem 2.7.1
Assume (2.7.1). Then for any $\epsilon > 0$, $\sigma > -1$, *there exist* R *and* ϵ_0 *such that, for any* $t \ge 0$, $(x, \xi) \in \Gamma^+_{R,\epsilon,\sigma}$, *there exists a unique trajectory*

$$[0, t] \ni s \mapsto (\tilde{y}(s, t, x, \xi), \tilde{\eta}(s, t, x, \xi))$$

such that

$$\begin{cases} \partial_s \tilde{y}(s,t,x,\xi) = \tilde{\eta}(s,t,x,\xi), \\ \partial_s \tilde{\eta}(s,t,x,\xi) = F(\tilde{y}(s,t,x,\xi)), \\ \tilde{y}(0,t,x,\xi) = x, \quad \tilde{\eta}(t,t,x,\xi) = \xi, \end{cases} \tag{2.7.2}$$

and

$$|\tilde{y}(s,t,x,\xi) - x - s\xi| \le \epsilon_0 s. \tag{2.7.3}$$

Moreover, the following estimates hold uniformly for $0 \le s \le t \le \infty$, $(x,\xi) \in \Gamma^+_{R,\epsilon,\sigma}$*:*

$$\partial_\xi^\beta(\tilde{y}(s,t,x,\xi) - x - s\xi) \in o((\langle x\rangle + \langle s\rangle)^0)s, \quad |\beta| \le 1,$$
$$\partial_x^\alpha(\tilde{y}(s,t,x,\xi) - x - s\xi) \in o(\langle x\rangle^0), \quad |\alpha| = 1,$$
$$\partial_x^\alpha\partial_\xi^\beta(\tilde{\eta}(s,t,x,\xi) - \xi) \in o((\langle x\rangle + \langle s\rangle)^{-|\alpha|}), \quad |\alpha| + |\beta| \le 1,$$
$$\sup_{\{(s,t,x,\xi)\ |0\le s\le t,\ r\le s+|x|\ (x,\xi)\in\Gamma^+_{R,\epsilon,\sigma}\}} |\partial_x^\alpha(\tilde{\eta}(s,t,x,\xi) - \xi)| \in L^1(\mathrm{d}r), \quad |\alpha| = 1.$$

Proof. We will reduce ourselves to the proof of Theorem 1.5.1 by introducing an effective time-dependent force. For ϵ, σ as in the theorem, there exists $C_0 > 0$ such that if $(x,\xi) \in \Gamma^+_{R,\epsilon,\sigma}$, then

$$|x + (s - t_1)\xi| \ge C_0(|s - t_1| + |x|). \tag{2.7.4}$$

We fix $0 < \epsilon_0 < C_0$ and we introduce a cut-off function $J \in C^\infty(X)$ such that $0 \notin \mathrm{supp} J$ and $J = 1$ on a neighborhood of $\{x \mid |x| > C_0 - \epsilon_0\}$. Using J, we define the effective time-dependent force $F_J(t,x)$ as in the proof of Theorem 2.5.2. It follows from (2.7.1) that $F_J(t,x)$ satisfies the hypotheses of Theorem 1.5.1. Therefore, we can find T such that the boundary value problem considered in Theorem 1.5.1 possesses a unique solution for any $T \le t_1 \le t_2$ and any x, ξ. Let us denote it by $(\tilde{y}_J(s,t_1,t_2,x,\xi), \tilde{\eta}_J(s,t_1,t_2,x,\xi))$. By enlarging T if needed, we can guarantee that

$$|\tilde{y}_J(s,t_1,t_2,x,\xi) - x - (s - t_1)\xi| \le \epsilon_0|s - t_1|. \tag{2.7.5}$$

From (2.7.4) and (2.7.5) we see that we see that

$$|\tilde{y}_J(s,t_1,t_2,x,\xi)| \ge (C_0 - \epsilon_0)|s - t_1| + C_0|x|. \tag{2.7.6}$$

We claim that if $R = T(C_0 - \epsilon_0)/C_0$ and $(x,\xi) \in \Gamma^+_{R,\epsilon,\sigma}$, then we can solve our boundary value problem by putting

$$\begin{aligned} &(\tilde{y}(s,t,x,\xi), \tilde{\eta}(s,t,x,\xi)) \\ &:= (\tilde{y}_J(s+r,r,t+r,x,\xi), \tilde{\eta}_J(s+r,r,t+r,x,\xi)), \end{aligned} \tag{2.7.7}$$

where $r = |x|C_0/(C_0 - \epsilon_0)$. In fact, obviously we have

$$(\tilde{y}(s,t,x,\xi), \tilde{\eta}(s,t,x,\xi)) := (\tilde{y}(s+r,r,t+r,x,\xi), \tilde{\eta}(s+r,r,t+r,x,\xi)).$$

Moreover, by (2.7.6) we have

$$|\tilde{y}_J(s+r,r,t+r,x,\xi)| \geq (C_0 - \epsilon_0)|s+r|.$$

Hence,

$$F_J(s+r,\tilde{y}_J(s+r,r,t+r,x,\xi)) = F(\tilde{y}_J(s+r,r,t+r,x,\xi)).$$

Therefore the function (2.7.7) solves the boundary value problem (2.7.2) with the initial time-independent force.

The uniqueness of the solution comes from the fact that any solution of (2.7.2) with $|\tilde{y}(s) - x - s\xi| \leq \epsilon_0 s$ is also a solution of the problem considered in Theorem 1.5.1 for the force $F_J(t,x)$ if time is large enough. Finally, the estimates on $(\tilde{y}(s,t,x,\xi), \tilde{\eta}(s,t,x,\xi))$ are obtained directly from those of Theorem 1.5.1 using the identity (2.7.7) and replacing s, t_1, t_2 there by $s + \langle x\rangle, \langle x\rangle, t + \langle x\rangle$. □

It is easy to see that in general there is no *global* uniqueness for the solution of (2.7.2).

We will now study scattering trajectories in a way that is parallel to that of the previous chapter. We start with a discussion of the comparison of trajectories, which is an obvious analog of Theorem 1.5.2.

Theorem 2.7.2
Assume that the potential $V(x)$ satisfies the estimates (2.7.1).
(i) Let $(y_1(s), \eta_1(s))$ and $(y_2(s), \eta_2(s))$ be two trajectories such that

$$\lim_{s\to\infty} \frac{y_1(s)}{s} = \lim_{s\to\infty} \frac{y_2(s)}{s} \neq 0.$$

Then there exists

$$\lim_{s\to\infty} (y_1(s) - y_2(s)).$$

(ii) Let $(y_1(s), \eta_1(s))$ be a trajectory such that

$$\lim_{s\to\infty} \frac{y_1(s)}{s} \neq 0,$$

and let $x \in X$. Then there exists a unique trajectory $(y_2(s), \eta_2(s))$ such that

$$\lim_{s\to\infty} \eta_1(s) = \lim_{s\to\infty} \eta_2(s),$$
$$\lim_{s\to\infty} (y_1(s) - y_2(s)) = x.$$

Proof. The proof is reduced to the one of Theorem 1.5.2 by introducing an effective time-dependent force. □

The above theorem provides a complete classification of the union of scattering trajectories. All the points with a given non-zero asymptotic momentum are labeled with elements of the affine space X.

Next we would like to discuss the Hamilton-Jacobi equation in the time-independent case. Our main aim will be to construct a certain solution of this equation that will be used to define the modified wave transformation. This solution will also be useful later on, when we will consider the quantum 2-body case.

The problem of finding solutions to the Hamilton-Jacobi equation is more difficult in the time-independent case than in the time-dependent case considered in the previous chapter. This difficulty stems from the fact that, in general, the boundary value problem considered in Theorem 2.7.1 does not possess a global solution. What is possible though is to solve the Hamilton-Jacobi equation in an appropriate outgoing region. As we will see, this will be enough for the purposes of scattering theory.

The following proposition is an immediate consequence of Theorems A.3.3 and 2.7.1.

Proposition 2.7.3
Assume (2.7.1). Let ϵ, σ, R be as in Theorem 2.7.1. For $(t,x,\xi)\in[0,\infty[\times\Gamma^+_{R,\epsilon,\sigma}$ we set

$$S(t,x,\xi)=\langle\xi,\tilde y(t,t,x,\xi)\rangle-\textstyle\int_0^t\left(\frac{1}{2}\tilde\eta^2(s,t,x,\xi)-V(\tilde y(s,t,x,\xi))\right)\mathrm{d}s.$$

(i) The function $S(t,x,\xi)$ is the only $C^{1,1}(X')$ solution on $[0,\infty[\times\Gamma^+_{R,\epsilon,\sigma}$ of the problem

$$\begin{cases}\partial_t S(t,x,\xi) &= \frac{1}{2}\xi^2+V(\nabla_\xi S(t,x,\xi)),\\ S(0,x,\xi)=\langle x,\xi\rangle.\end{cases}$$

(ii) The function $S(t,x,\xi)$ is the only $C^{1,1}(X)$ solution on $[0,\infty[\times\Gamma^+_{R,\epsilon,\sigma}$ of the problem

$$\begin{cases}\partial_t S(t,x,\xi) &= \frac{1}{2}(\nabla_x S(t,x,\xi))^2+V(x),\\ S(0,x,\xi)=\langle x,\xi\rangle.\end{cases}$$

(iii) $\nabla_x S(t,x,\xi)=\tilde\eta(0,t,x,\xi)$ and $\nabla_\xi S(t,x,\xi)=\tilde y(t,t,x,\xi)$.
(iv) The following estimates are true uniformly for $0\le t$, $(x,\xi)\in\Gamma^+_{R,\epsilon,\sigma}$:

$$\partial_\xi^\beta\left(S(t,x,\xi)-\langle x,\xi\rangle-\frac{1}{2}t\xi^2\right)\in o((t+|x|)^0)|t|,\quad |\beta|\le 2;$$

$$\partial_x^\alpha\partial_\xi^\beta\left(S(t,x,\xi)-\langle x,\xi\rangle-\frac{1}{2}t\xi^2\right)\in o(|x|^{1-|\alpha|}),\quad |\alpha|\ge 1,\ |\alpha|+|\beta|\le 2;$$

$$\sup_{(t,x,\xi)\in\mathbb{R}^+\times\Gamma^+_{r,\epsilon,\sigma}}\left|\partial_x^\alpha\left(S(t,x,\xi)-\langle x,\xi\rangle-\frac{1}{2}t\xi^2\right)\right|\in L^1(\mathrm{d}r),\qquad |\alpha|=2.$$

Next we would like to give a simple condition for the solvability of the Hamilton-Jacobi equation with given initial conditions. This criterion can be viewed as a time-independent analog of Proposition 1.9.13, from which it easily follows.

Proposition 2.7.4
(i) Assume (2.7.1). Suppose that a function $[0,\infty[\times X' \ni (t,\eta) \mapsto S(t,\eta)$ *satisfies the following condition: for any* $\epsilon > 0$*, uniformly for* $|\eta| > \epsilon$*, we have*

$$\partial_\eta^\beta \left(S_1(t,\eta) - \frac{1}{2}t\eta^2 \right) \in o(t), \quad |\beta| \le 2.$$

Then for any $\epsilon_1 > 0$*, there exists* T_1 *such that, for* $T_1 \le t_1 < t_2 \le \infty$ *and* $|\xi| > \epsilon_1$*, there exists a unique family of trajectories depending continuously on* ξ

$$s \mapsto (Y_1(s,t_1,t_2,\xi), E_1(s,t_1,t_2,\xi))$$

satisfying the following conditions:

$$Y_1(t_1,t_1,t_2,\xi) = \nabla_\eta S_1(t_1, E_1(t_1,t_1,t_2,\xi)), \qquad E_1(t_2,t_1,t_2,\xi) = \xi.$$

They satisfy, uniformly for $T_1 \le t_1 \le s \le t_2$ *and* $|\xi| > \epsilon_1$*, the following estimates:*

$$\partial_\xi^\beta (Y_1(s,t_1,t_2,\xi) - s\xi) \in o(s), \quad |\beta| \le 1.$$

If $\epsilon \ge \epsilon_1$*, then let us denote*

$$\Omega_{\epsilon,T} := \{(s, E_1(s,T,\infty,\xi)) \mid s > T, \quad |\xi| > \epsilon\},$$
$$\Xi_{\epsilon,T} := [T,\infty[\times\{\xi \mid |\xi| > \epsilon\}.$$

If $0 < \epsilon_0 < \epsilon_1 < \epsilon_2$*, then by choosing* t_1 *big enough we can guarantee that*

$$\Xi_{\epsilon_2,t_1} \subset \Omega_{\epsilon_1,t_1} \subset \Xi_{\epsilon_0,t_1}.$$

(ii) Suppose that ϵ_1*,* t_1 *are chosen as in (i). Then, on* Ω_{ϵ_1,t_1}*, there exists a unique* $C^{1,1}(X')$ *solution of the Hamilton-Jacobi equation*

$$\begin{cases} \partial_t S_1(t,t_1,\xi) = \frac{1}{2}\xi^2 + V(\nabla_\xi S_1(t,t_1,\xi)), \\ S_1(t_1,t_1,\xi) = S_1(t_1,\xi). \end{cases} \tag{2.7.8}$$

This solution satisfies

$$\partial_\xi^\beta \left(S_1(t,t_1,\xi) - \frac{1}{2}(t-t_1)\xi^2 - S_1(t_1,\xi) \right) \in o(t), \quad |\beta| \le 2. \tag{2.7.9}$$

(iii) If, on Ω_{ϵ_1,t_1}*,*

$$\partial_t S_1(t,\xi) = \frac{1}{2}\xi^2 + V(\nabla_\xi S_1(t,\xi)),$$

then, for $(t, \xi) \in \Omega_{\epsilon_1, t_1}$, *we have*

$$S_1(t, t_1, \xi) = S_1(t, \xi).$$

Proof. Suppose that $\epsilon_0 < \epsilon_1$. We will find T_1 and $C_0 > 0$ such that, for $t \geq T_1$ and $|\eta| \geq \epsilon_0$,

$$|x(t, \nabla_\eta S_1(t, \eta), \eta)| \geq C_0 t.$$

We choose a cut-off J and define the effective force $F_J(t, x)$ as we did in the proof of Theorem 2.5.2. By Proposition 1.9.12, for all $t_1 \leq t_2$ big enough and $|\xi| > \epsilon$, we will find families of trajectories

$$s \mapsto (Y_{1,J}(s, t_1, t_2, \xi), E_{1,J}(s, t_1, t_2, \xi))$$

satisfying the conditions

$$Y_{1,J}(t_1, t_1, t_2, \xi) = \nabla_\eta S_1(t_1, E_{1,J}(t_1, t_1, t_2, \xi)), \quad E_{1,J}(t_2, t_1, t_2, \xi)) = \xi.$$

They satisfy the estimates

$$\partial_\xi^\beta (Y_{1,J}(s, t_1, t_2, \xi) - (s - t_1)\xi - \nabla_\xi S_1(t_1, \xi)) \in o(s), \quad |\beta| \leq 1. \qquad (2.7.10)$$

If t_1 is big enough, then $F(x)$ coincides with $F_J(t, x)$ along those trajectories. Therefore, we can write $(Y_1(s), E_1(s))$ instead of $(Y_{1,J}(s), E_{1,J}(s))$. This ends the proof of *(i)*.

To show *(ii)*, we define

$$\begin{aligned} \tilde{S}_1(t, t_1, \xi) &:= S_1(t_1, E_1(t_1))) \\ &\quad + \int_{t_1}^{t} (\tfrac{1}{2} E_1^2(s) + V(Y_1(s)) - \langle Y_1(s), \nabla V(Y_1(s)) \rangle) ds, \end{aligned}$$

where $(Y_1(s), E_1(s)) = (Y_1(s, t_1, t_2, \xi), E_1(s, t_1, t_2, \xi)$.

It follows from (2.7.10) that (2.7.9) is true for $|\beta| = 1, 2$. Next we note that

$$S_1(t, t_1, \xi) - \frac{1}{2} t \xi^2 = S_1(t_1, \xi) + \int_{t_1}^{t} V(\nabla_\xi S_1(s, t_1, \xi)) ds.$$

But by (2.2.2)

$$V(\nabla_\xi S_1(s, t_1, \xi)) \in o(s^0).$$

Therefore, (2.7.9) is true also for $|\beta| = 0$.

(iii) follows from the uniqueness of the solution of (2.7.8). □

In the time-decaying case, in order to construct a global solution of the Hamilton-Jacobi equation, it was enough to fix an origin in position and in time for time large enough. In the time-independent case, one has to make a more complicated construction. Moreover, in general, we cannot demand that the function $S(t,\xi)$ be a solution of the Hamilton-Jacobi equation for all $t \geq T$, uniformly in ξ.

Proposition 2.7.4 can be used to construct solutions of the Hamilton-Jacobi equation for the momenta outside of an arbitrarily small neighborhood of zero. Below we construct a function $S(t,\xi)$ that solves the Hamilton-Jacobi equation only for large enough time and non uniformly in the momentum, but at least is defined everywhere.

We will also construct the family $Y(t,\xi)$ of trajectories with the asymptotic momentum ξ related to the function $S(t,\xi)$ similarly as in the previous chapter. Both $Y(t,\xi)$ and $S(t,\xi)$ will be used afterwards to define the *modified free flow* and the *modified wave transformations.*

Theorem 2.7.5
Under the hypothesis (2.7.1), there exists a function $S(t,\xi)$ that has the following property: for any $\epsilon > 0$, there exists T_ϵ such that

$$\partial_t S(t,\xi) = \frac{1}{2}\xi^2 + V(\nabla_\xi S(t,\xi)), \text{ for } |\xi| \geq \epsilon,\ t \geq T_\epsilon.$$

The function $S(t,\xi)$ satisfies, uniformly for $|\xi| \geq \epsilon$,

$$\partial_\xi^\beta \left(S(t,\xi) - \frac{1}{2}t\xi^2 \right) \in o(t), \quad |\beta| \leq 2.$$

Moreover, there exists a family of trajectories $(Y(t,\xi), E(t,\xi))$ for $\xi \neq 0$ such that

$$\lim_{t\to\infty} E(t,\xi) = \xi,$$
$$\partial_\xi^\beta (Y(t,\xi) - t\xi) \in o(t), \quad |\beta| \leq 1,$$

and that is related to $S(t,\xi)$ by the following property: for any $\epsilon > 0$, there exists T_ϵ such that if $t \geq T_\epsilon$ and $|\xi| \geq \epsilon$, then

$$\nabla_\eta S(t, E(t,\xi)) = Y(t,\xi).$$

Proof. Our construction of the function $S(t,\xi)$ is inspired by a similar construction of Hörmander [Hö2, Thm 30.3.3].

The trajectories $(Y(t,\xi), E(t,\xi))$ will be constructed consecutively for $|\xi| > 2^{-n}$. For any $\epsilon > 0$, T it will be useful to have the notation

$$\Omega_{\epsilon,T} := \{(t, E(t,\xi)) \mid t > T,\ |\xi| > \epsilon\}.$$

We will also use the sets $\Xi_{\epsilon,T}$, which were defined in Proposition 2.7.4.

Apart from trajectories, we will define an increasing sequence of times $t_n \in \mathbb{R}^+$ and functions $[0,\infty[\times X' \ni (t,\eta) \mapsto S_n(t,\eta)$ that satisfy the following conditions:

$$(i) \quad \partial_\xi^\beta \left(S_n(t,\xi) - \tfrac{1}{2}t\xi^2 \right) \in o(t), \quad |\beta| \le 2,$$

$$(ii) \quad \partial_t S_n(t,\xi) = \tfrac{1}{2}\xi^2 + V(\nabla_\xi S_n(t,\xi)), \quad n \ge 1, \quad (t,\xi) \in \Omega_{2^{-n},t_n},$$

$$(iii) \quad S_n(t,\xi) = S_{n-1}(t,\xi), \quad n \ge 1, \quad t \le t_n,$$

$$(iv) \quad Y(t,\xi) = \nabla_\eta S_n(t, E(t,\xi)), \quad n \ge 1, \quad (t,\xi) \in \Xi_{2^{-n},t_n}.$$

We start the induction by setting

$$S_0(t,\xi) := \frac{1}{2}t\xi^2.$$

Suppose that we defined $S_{n-1}(t,\xi)$. Besides, suppose that we defined the trajectories $(Y(t,\xi), E(t,\xi))$ for $|\xi| > 2^{-n}$. By Proposition 2.7.4 *(i)*, we will find t_n such that there exists a family of trajectories $(Y_n(t,\xi), E_n(t,\xi))$ for $|\xi| > 2^{-n-3}$ satisfying

$$Y_n(t_n,\xi) = \nabla_\eta S_{n-1}(t_n, E_n(t_n,\xi)), \qquad E_n(\infty,\xi) = \xi.$$

For $\epsilon > 2^{-n-3}$, we set

$$\Omega^n_{\epsilon,T} := \{(t, E_n(t,\xi)) \mid t > T, \ |\xi| > \epsilon\}.$$

By enlarging t_n if needed, we can guarantee that

$$\Omega^n_{2^{-n},t_n} \subset \Xi_{2^{-n-1},t_n}, \qquad \Xi_{2^{-n-2},t_n} \subset \Omega^n_{2^{-n-3},t_n}.$$

Clearly, for $|\xi| > 2^{-n+1}$, the new trajectories $(Y_n(t,\xi), E_n(t,\xi))$ coincide with $(Y(t,\xi), E(t,\xi))$. For $2^{-n} \le |\xi| < 2^{-n+1}$, we set

$$((Y(t,\xi), E(t,\xi)) := (Y_n(t,\xi), E_n(t,\xi)).$$

From Proposition 2.7.4 *(ii)* we obtain the existence on $\Omega^n_{2^{-n-3},t_n}$ of the solution of the problem

$$\begin{cases} \partial_t \tilde{S}_n(t,\xi) = \tfrac{1}{2}\xi^2 + V(\nabla_\xi \tilde{S}_n(t,\xi)), \\ \tilde{S}_n(t_n,\xi) = S_{n-1}(t_n,\xi). \end{cases}$$

Note that, on $\Omega^n_{2^{-n+1},t_n} = \Omega_{2^{-n+1},t_n}$, we have

$$S_{n-1}(t,\xi) = \tilde{S}_n(t,\xi).$$

Then we set

$$S_n(t,\xi) := \begin{cases} S_{n-1}(t,\xi), & \text{for } t \le t_n, \\ \tilde{S}_n(t,\xi)(1-\chi_n(\xi)) + \tfrac{1}{2}t\xi^2\chi_n(\xi) & \text{otherwise,} \end{cases}$$

where $\chi_n \in C_0^\infty(X')$ such that $\chi_n(\xi) = 1$ for $|\xi| < 2^{-n-2}$ and $\chi_n(\xi) = 0$ for $|\xi| > 2^{-n-3}$. Eventually, we set

$$S(t,\xi) := \lim_{n\to\infty} S_n(t,\xi).$$

The above limit exists trivially: for any t, ξ, the number $S_n(t,\xi)$ does not depend on n for big enough n. This completes the proof of the theorem. □

Now we are prepared to define the asymptotic position and the modified wave operator in the long-range case. As in Sects. 1.5, 1.6 and 1.8, we have the choice of using, in these definitions, either the family of trajectories $Y(t,\xi)$ or the generating function $S(t,\xi)$ (both introduced in Theorem 2.7.5).

We start with an analog of Theorem 1.5.3. It follows e.g. by introducing an effective time-dependent force.

Theorem 2.7.6
Assume (2.7.1). Let $\xi^+(y,\eta) \neq 0$. Then there exists

$$\lim_{t\to\infty}(x(t,y,\eta) - Y(t,\xi^+(y,\eta))) =: x_{\mathrm{lr}}^+(y,\eta). \tag{2.7.11}$$

Moreover, the following statements are true:
(i) The function

$$(\xi^+)^{-1}(X'\backslash\{0\}) \ni (y,\eta) \mapsto x_{\mathrm{lr}}^+(y,\eta) \in X$$

is continuous.
(ii) If $\epsilon > 0$ and $-1 < \sigma$, then there exists R such that, on $\Gamma_{R,\epsilon,\sigma}^+$, we have

$$x_{\mathrm{lr}}^+(y,\eta) - y \in o(\langle y\rangle).$$

(iii) The convergence of (2.7.11) is uniform on compact subsets of $\xi^{-1}(X'\backslash\{0\})$.
(iv) The map

$$\xi^{-1}(X'\backslash\{0\}) \ni (y,\eta) \mapsto (x_{\mathrm{lr}}^+(y,\eta), \xi^+(y,\eta)) \in X \times (X'\backslash\{0\}) \tag{2.7.12}$$

is bijective.

Definition 2.7.7
We denote the inverse of (2.7.12) by $\mathcal{F}_{\mathrm{lr}}^+$ and call it the modified wave transformation.

This completes the construction of the modified wave transformation that is parallel to that of Sect. 1.5.

Next we would like to present a construction of $\mathcal{F}_{\mathrm{lr}}^+$ that is parallel to that of Sect. 1.6. We start with an analog of Theorem 1.6.1.

Theorem 2.7.8
Assume (2.7.1). Let $\xi^+(y,\eta) \neq 0$. Then

$$x_{\mathrm{lr}}^{+}(y,\eta) = \lim_{t\to\infty}(x(t,y,\eta) - \nabla_\xi S(t,\xi(t,y,\eta))).$$

The convergence of (2.7.11) is uniform on compact subsets of $\xi^{-1}(X'\backslash\{0\})$.

The next result is similar to Theorem 1.6.2.

For any $(x,\xi) \in X \times (X'\backslash\{0\})$, $t \in [0,\infty[$, we will denote by $s \mapsto (y_{\mathrm{lr}}(s,t,x,\xi), \eta_{\mathrm{lr}}(s,t,x,\xi))$ the unique trajectory such that

$$\begin{cases} y_{\mathrm{lr}}(t,t,x,\xi) = x + \nabla_\xi S(t,\xi)(t,\xi), \\ \eta_{\mathrm{lr}}(t,t,x,\xi) = \xi. \end{cases}$$

Theorem 2.7.9
Assume (2.7.1). Then the trajectory

$$\mathbb{R}^+ \ni s \mapsto (y_{\mathrm{lr}}(s,t,x,\xi), \eta_{\mathrm{lr}}(s,t,x,\xi))$$

converges uniformly on compact sets of $\mathbb{R} \times X \times (X'\backslash\{0\})$ *as* $t \to \infty$ *to a trajectory*

$$\mathbb{R}^+ \ni s \mapsto (y_{\mathrm{lr}}(s,\infty,x,\xi), \eta_{\mathrm{lr}}(s,\infty,x,\xi)),$$

which satisfies

$$\begin{cases} \lim_{s\to\infty}(y_{\mathrm{lr}}(s,\infty,x,\xi) - x - \nabla_\xi S(s,\xi)) = 0, \\ \lim_{s\to\infty}(\eta_{\mathrm{lr}}(s,\infty,x,\xi) - \xi) = 0. \end{cases} \tag{2.7.13}$$

Moreover, the following statements are true:
(i) The trajectory $(y_{\mathrm{lr}}(s,\infty,x,\xi), \eta_{\mathrm{lr}}(s,\infty,x,\xi))$ *is the only one that satisfies (2.7.13).*
(ii) The map

$$\begin{aligned} &[0,\infty] \times X \times (X'\backslash\{0\}) \ni (t,x,\xi) \\ &\mapsto (y_{\mathrm{lr}}(s,t,x,\xi) - x - y_{\mathrm{lr}}(s,t,0,\xi),\quad \eta_{\mathrm{lr}}(s,\infty,x,\xi) - \xi) \in C_\infty(\mathbb{R}_s, X \times X') \end{aligned}$$

is continuous.

Proof. Fix a compact set $K \subset X \times (X'\backslash\{0\})$. If $(x,\xi) \in K$, there exist $C_0 > 0$ and T_0 such that, for $s \geq T_0$,

$$|x + s\xi| \geq C_0 s. \tag{2.7.14}$$

Let us introduce a cutoff function J and a time-dependent force $F_J(t,x)$ as in the proof of Theorem 2.5.2. We claim that we can find a unique trajectory

$$(y_{\mathrm{sd},J}(s,t,x,\xi), \eta_{\mathrm{sd},J}(s,t,x,\xi))$$

for the force $F_J(t,x)$ such that

$$\begin{cases} y_{\mathrm{sd},J}(t,t,x,\xi) = x + \nabla_\xi S(t,\xi), \\ \eta_{\mathrm{sd},J}(t,t,x,\xi) = \xi, \end{cases} \tag{2.7.15}$$

and this trajectory converges as $t \to \infty$ to a trajectory

$$s \mapsto (y_{\mathrm{sd},J}(s,\infty,x,\xi), \eta_{\mathrm{sd},J}(s,\infty,x,\xi)).$$

In fact, the force $F_J(t,x)$ satisfies the conditions of Theorem 1.6.2. Moreover, for $t \geq T_1$ and $|\xi| > \epsilon$, there exists a family of trajectories $(Y(t,\xi), E(t,\xi))$ that satisfies

$$\nabla_\eta S(t, E(t,\xi)) = Y(t,\xi).$$

This family of trajectories clearly solves (2.7.15) with $x = 0$. Now the existence of $(y_{\mathrm{sd},J}(s,t,x,\xi), \eta_{\mathrm{sd},J}(s,t,x,\xi))$ follows from the arguments of Theorem 1.6.2.

To be exact, the trajectories $Y(t,\xi)$ and the modifier $\nabla_\xi S(t,\xi)$ that we are now using differ from those used in Theorem 1.6.2, where they were fixed by the condition $Y(T,\xi) = 0$. The reader will easily convince himself that, as long as $|\xi| \geq \epsilon$, we can use the arguments of Theorem 1.6.2.

We also note the following estimate:

$$y_{\mathrm{sd},J}(s,t,x,\xi) - x - s\xi \in o(s^0)\langle x\rangle. \tag{2.7.16}$$

Therefore, from (2.7.14) and (2.7.16) we infer that there exists T_2 such that, for $s > T_2$,

$$F_J(s, y_{\mathrm{lr},J}(s,t,x,\xi)) = F(y_{\mathrm{lr},J}(s,t,x,\xi)).$$

Hence, for $s \geq T_2$, the trajectory $y_{\mathrm{lr},J}(s,t,x,\xi)$ is also a trajectory for the original force $F(x)$. By setting $y_{\mathrm{lr}}(s) := y_{\mathrm{sd},J}(s)$ for $s \geq T_1$ and extending it for $s \leq T_1$ by the flow, we obtain the trajectory $y_{\mathrm{lr}}(s)$ satisfying (2.7.13). The statements *(i)* and *(ii)* of the theorem are direct consequences of Theorem 1.6.2. □

We can now formulate the results obtained so far in the now familiar language of *wave transformations.*

Definition 2.7.10
We define the modified free flow *by*

$$X \times X' \ni (x,\xi) \mapsto \phi_{\mathrm{lr}}(t)(x,\xi) := (x + \nabla_\xi S(t,\xi), \xi) \in X \times X'.$$

One has the following corollary of Theorems 2.7.8 and 2.7.9.

Theorem 2.7.11
(i) Assume that the hypothesis (2.7.1) holds. Then the following limit exists uniformly on compact sets $X \times (X' \backslash \{0\})$:

$$\mathcal{F}^+_{\mathrm{lr}} := \lim_{t\to\infty} \phi(-t)\phi_{\mathrm{lr}}(t) \tag{2.7.17}$$

and the following limit exists uniformly on compact sets of $(\xi^+)^{-1}(X'\backslash\{0\})$*:*

$$\lim_{t\to\infty} \phi_{\rm lr}^{-1}(t)\phi(t). \tag{2.7.18}$$

The limit in (2.7.18) is equal to $(\mathcal{F}_{\rm lr}^+)^{-1}$.
(ii) If $(y,\eta) = \mathcal{F}_{\rm lr}^+(x,\xi)$*, then one has*

$$\lim_{t\to\infty} (\phi(t,0)(y,\eta) - (x + Y(t,\xi),\xi)) = 0.$$

(iii) The mapping $\mathcal{F}_{\rm lr}^+$ *is symplectic.*
(iv) The modified wave transformation intertwines the full and the free dynamics:

$$\begin{aligned} H \circ \mathcal{F}_{\rm lr}^+ &= H_0, \\ \phi(t) \circ \mathcal{F}_{\rm lr}^+ &= \mathcal{F}_{\rm lr}^+ \circ \phi_0(t). \end{aligned}$$

2.8 The Eikonal Equation

The eikonal equation is especially interesting in the time-independent case, because it does not involve the time variable.

First let us describe the solution of the eikonal equation that is natural in the short-range case.

Proposition 2.8.1
Assume (2.6.1) and (2.6.5). Let R, ϵ, σ *be as in Theorem 2.7.1. Then, for* $(x,\xi) \in \Gamma_{R,\epsilon,\sigma}^+$*, the following limit exists:*

$$\lim_{t\to\infty} \left(S(t,x,\xi) - \frac{1}{2}t\xi^2 \right) =: \Phi_{\rm sr}^+(x,\xi).$$

It satisfies the eikonal equation

$$\frac{1}{2}\xi^2 = \frac{1}{2}(\nabla_x \Phi_{\rm sr}^+(x,\xi))^2 + V(x).$$

It is a generating function of $\mathcal{F}_{\rm sr}^+$*, that is*

$$(x, \nabla_x \Phi_{\rm sr}^+(x,\xi)) = \mathcal{F}_{\rm sr}^+(\nabla_\xi \Phi_{\rm sr}^+(x,\xi),\xi).$$

We have the identities

$$\nabla_x \Phi_{\rm sr}^+(x,\xi) = \tilde{\eta}(0,\infty,x,\xi),$$

$$\begin{aligned} \nabla_\xi \Phi_{\rm sr}^+(x,\xi) &= \lim_{t\to\infty} (\tilde{y}(t,\infty,x,\xi) - t\xi) \\ &= \lim_{t\to\infty} (\tilde{y}(t,t,x,\xi) - t\xi). \end{aligned}$$

Next we describe the solution of the eikonal equation that, in the long-range case, is naturally associated to the function $S(t,\xi)$.

Proposition 2.8.2
Assume (2.7.1). Let R, ϵ, σ be as in Theorem 2.7.1. Then, for $(x,\xi)\in\Gamma^+_{R,\epsilon,\sigma}$, there exists

$$\lim_{t\to\infty}(S(t,x,\xi)-S(t,\xi)) =: \Phi^+_{\mathrm{lr}}(x,\xi).$$

It solves the eikonal equation

$$\frac{1}{2}\xi^2=\frac{1}{2}(\nabla_x\Phi^+_{\mathrm{lr}}(x,\xi))^2+V(x).$$

It is a generating function of $\mathcal{F}^+_{\mathrm{sr}}$, that is

$$(x,\nabla_x\Phi^+_{\mathrm{lr}}(x,\xi))=\mathcal{F}^+_{\mathrm{lr}}(\nabla_\xi\Phi^+_{\mathrm{lr}}(x,\xi),\xi).$$

We have the identities

$$\nabla_x\Phi^+_{\mathrm{lr}}(x,\xi)=\tilde\eta(0,\infty,x,\xi), \tag{2.8.1}$$

$$\begin{aligned}\nabla_\xi\Phi^+_{\mathrm{lr}}(x,\xi)&=\lim_{t\to\infty}(\tilde y(t,\infty,x,\xi)-Y(t,\xi))\\&=\lim_{t\to\infty}(\tilde y(t,t,x,\xi)-\tilde Y(t,\xi)).\end{aligned}\tag{2.8.2}$$

The following estimates hold uniformly for $(x,\xi)\in\Gamma^+_{R,\epsilon,\sigma}$:

$$\partial_x^\alpha\partial_\xi^\beta(\Phi^+_{\mathrm{lr}}(x,\xi)-\langle x,\xi\rangle)\in o(\langle x\rangle^{1-|\alpha|}),\qquad |\alpha|+|\beta|\le 2, \tag{2.8.3}$$

$$\sup_{(x,\xi)\in\Gamma^+_{r,\epsilon,\sigma}}|\partial_x^\alpha(\Phi^+_{\mathrm{lr}}(x,\xi)-\langle x,\xi\rangle)|\in L^1(\mathrm{d}r),\quad |\alpha|=2.$$

Proof. The proof is analogous to that of Proposition 1.8.3 and left to the reader. The estimate (2.8.3) for $|\alpha|=0$ is proven as in the proof of Proposition 2.9.5. □

2.9 Smoothness of Trajectories

Now we are going to study the smoothness with respect to parameters of various functions constructed in the previous section. We will use the following assumptions on the forces:

$$\int_0^\infty \sup_{|x|\ge R}|\partial_x^\alpha F(x)|\langle R\rangle^{|\alpha|+\mu}\mathrm{d}R<\infty,\quad \alpha\in\mathbb{N}^n. \tag{2.9.1}$$

We will always assume that $\mu\ge 0$ in (2.9.1). The case $\mu=0$ will be called the "smooth long-range condition", and the case $\mu=1$ will go under the name of the "smooth short-range condition".

Note that (2.9.1) is akin to the conditions used in the theory of pseudo-differential operators to define "symbols of order $-\mu$".

We begin with some estimates on the derivatives of the solutions $(\tilde{y}(s,t,x,\xi), \tilde{\eta}(s,t,x,\xi))$ to the boundary value problem considered in Theorem 2.7.1. It will be an analog of Theorem 1.10.1.

Theorem 2.9.1
Assume that the force $F(x)$ satisfies (2.9.1) with $\mu \geq 0$. Let R, $\epsilon > 0$ and σ be as in Theorem 2.7.1. Then one has, uniformly for $(x,\xi) \in \Gamma^+_{R,\epsilon,\sigma}$, $0 \leq s \leq t$,

$$\partial_\xi^\beta(\tilde{y}(s,t,x,\xi) - x - s\xi) \in o((\langle x\rangle + s)^0)\langle x\rangle^{-\mu}|s|. \tag{2.9.2}$$

Moreover, for some $f_\beta, f_{\alpha,\beta} \in L^1(\mathrm{d}u)$, we have

$$|\partial_\xi^\beta(\tilde{y}(s,t,x,\xi) - x - s\xi)| \leq \int_{\langle x\rangle}^\infty f_\beta(u)\langle u\rangle^{1-\mu}\mathrm{d}u, \quad \mu \geq 1, \tag{2.9.3}$$

$$|\partial_x^\alpha\partial_\xi^\beta(\tilde{y}(s,t,x,\xi) - x - s\xi)| \leq \langle x\rangle^{1-|\alpha|}\int_{\langle x\rangle}^\infty f_{\alpha,\beta}(u)\langle u\rangle^{-\mu}\mathrm{d}u, \quad |\alpha| \geq 1, \tag{2.9.4}$$

$$|\partial_\xi^\beta(\tilde{\eta}(s,t,x,\xi) - \xi)| \leq \int_{\langle x\rangle+s}^\infty f_\beta(u)\langle u\rangle^{-\mu}\mathrm{d}u, \tag{2.9.5}$$

$$|\partial_x^\alpha\partial_\xi^\beta(\tilde{\eta}(s,t,x,\xi) - \xi)| \leq \langle x\rangle^{1-|\alpha|}\int_{\langle x\rangle+s}^\infty f_{\alpha,\beta}(u)\langle u\rangle^{-1-\mu}\mathrm{d}u, \quad \text{for } |\alpha| \geq 1. \tag{2.9.6}$$

Proof. Applying the trick used in the proof of Theorem 2.7.1, we see that these estimates are obtained from those of Theorem 1.10.1 by replacing (s,t_1,t_2) by $(s+\langle x\rangle, \langle x\rangle, t+\langle x\rangle)$. □

Let us now state some estimates on solutions of the Hamilton-Jacobi and eikonal equations.

Proposition 2.9.2
Assume (2.9.1) with $\mu \geq 0$. Let R, ϵ and σ be as in Theorem 2.7.1. Then the function $S(t,x,\xi)$ satisfies, uniformly for $(t,x,\xi) \in [0,\infty] \times \Gamma^+_{R,\epsilon,\sigma}$, the following estimates:

$$\partial_\xi^\beta\left(S(t,x,\xi) - \langle x,\xi\rangle - \frac{1}{2}t\xi^2\right) \in o((\langle x\rangle + t)^0)\langle x\rangle^{-\mu}|t|,$$

$$\partial_x^\alpha\partial_\xi^\beta\left(S(t,x,\xi) - \langle x,\xi\rangle - \frac{1}{2}t\xi^2\right) \in o(\langle x\rangle^{1-|\alpha|-\mu}), \quad |\alpha| + \mu \geq 1,$$

$$\sup_{(t,x,\xi)\in\mathbb{R}^+\times\Gamma_{r,\epsilon,\sigma}} \left|\partial_x^\alpha\partial_\xi^\beta\left(S(t,x,\xi)-\langle x,\xi\rangle-\frac{1}{2}t\xi^2\right)\right| \in r^{2-|\alpha|-\mu}L^1(\mathrm{d}r), \quad |\alpha|+\mu\geq 2.$$

Next we consider the smoothness properties of the solution of the eikonal equation associated with the short-range wave transformation.

Proposition 2.9.3
Assume (2.9.1) with $\mu \geq 1$. Let R, ϵ and σ be such as in Theorem 2.7.1. Then uniformly for $(x,\xi)\in\Gamma^+_{R,\epsilon,\sigma}$, we have

$$|\partial_x^\alpha\partial_\xi^\beta(\Phi^+_{\mathrm{sr}}(x,\xi)-\langle x,\xi\rangle)| \in o(\langle x\rangle^{-|\alpha|}),$$

$$\sup_{(x,\xi)\in\Gamma^+_{r,\epsilon,\sigma}} |\partial_x^\alpha\partial_\xi^\beta(\Phi^+_{\mathrm{sr}}(x,\xi)-\langle x,\xi\rangle)| \in r^{1-|\alpha|}L^1(\mathrm{d}r), \quad |\alpha|\geq 1.$$

Next we study the regularity of functions that we defined in the long-range case.

Proposition 2.9.4
Assume (2.9.1) with $\mu = 0$. Then the function $S(t,\xi)$ and the trajectories $Y(t,\xi)$ constructed in Theorem 2.7.5 for $\epsilon > 0$, uniformly for $|\xi|\geq\epsilon$, satisfy the estimates

$$\partial_\xi^\beta\left(S(t,\xi)-\tfrac{1}{2}t\xi^2\right)\in o(t),$$

$$\partial_\xi^\beta(Y(t,\xi)-t\xi)\in o(t).$$

Proposition 2.9.5
Assume (2.9.1) with $\mu = 0$. Let R, ϵ and σ be as in Theorem 2.7.1. Then, uniformly for $(x,\xi)\in\Gamma^+_{R,\epsilon,\sigma}$, we have

$$\partial_x^\alpha\partial_\xi^\beta(\Phi^+_{\mathrm{lr}}(x,\xi)-\langle x,\xi\rangle)\in o(\langle x\rangle^{1-|\alpha|}),$$

$$\sup_{(x,\xi)\in\Gamma^+_{r,\epsilon,\sigma}} |\partial_x^\alpha\partial_\xi^\beta(\Phi^+_{\mathrm{lr}}(x,\xi)-\langle x,\xi\rangle)| \in r^{2-|\alpha|}L^1(\mathrm{d}r), \quad |\alpha|\geq 2.$$

Proof. The case $|\alpha|\geq 1$ follows immediately from (2.8.1) and Theorem 2.9.1.

Let us show the estimates for $|\alpha| = 0$, $|\beta|\geq 1$. We use the first identity of (2.8.2):

$$\begin{aligned}\nabla_\xi\Phi^+_{\mathrm{lr}}(x_0,\xi) &= \lim_{t\to\infty}(\tilde{y}(t,\infty,x_0,\xi)-Y(t,\xi))\\ &= \lim_{t\to\infty}(\tilde{y}(t,\infty,\tilde{y}(s,\infty,x_0,\xi),\xi)-\tilde{y}(t,\infty,Y(s,\xi),\xi))\\ &= \tilde{y}(s,\infty,x_0,\xi)-Y(s,\xi)\\ &+\textstyle\int_0^\infty(\tilde{\eta}(t,\infty,\tilde{y}(s,\infty,x_0,\xi),\xi)-\tilde{\eta}(t,\infty,Y(s,\xi),\xi))\mathrm{d}t.\end{aligned}$$

where we chose s big enough so that $(Y(s,\xi),\xi) \in \Gamma^+_{R,\epsilon,\sigma}$ and $(\tilde{y}(s,\infty,x_0,\xi),\xi) \in \Gamma^+_{R,\epsilon,\sigma}$ for appropriate R,ϵ,σ. Then we can write

$$\begin{aligned}&\tilde{\eta}(t,\infty,\tilde{y}(s,\infty,x_0,\xi),\xi) - \tilde{\eta}(t,\infty,Y(s,\xi),\xi)\\&= \textstyle\int_0^1 \nabla_x\tilde{\eta}(t,\infty,\tau\tilde{y}(s,\infty,x_0,\xi) + (1-\tau)Y(s,\xi),\xi)(\tilde{y}(s,\infty,x_0,\xi) - Y(s,\xi))\mathrm{d}\tau.\end{aligned}$$

and use the estimates of Theorem 2.9.1.

The case $|\alpha| = |\beta| = 0$ will not be used later on and is left to the reader. □

3. Quantum Time-Decaying Hamiltonians

3.0 Introduction

Our presentation of classical 2-body scattering was divided into two chapters. In the first chapter we studied scattering in the presence of forces that decay in time. In the second chapter we investigated potentials that are time-independent but decay in space. In the quantum case, we will also consider separately two analogous classes of 2-body systems.

In this chapter we will treat time-dependent Hamiltonians of the form

$$H(t) = \tfrac{1}{2}D^2 + V(t,x). \tag{3.0.1}$$

We will make assumptions on the temporal decay of $\partial_x^\alpha V(t,x)$ that are uniform in x. We will study various objects that describe the asymptotics of the evolution defined by (3.0.1) for $t \to \infty$.

In the literature, scattering theory for time-dependent Hamiltonians of the form (3.0.1) is rarely studied as the end in itself. They appear usually as auxiliary objects useful in the study of time-independent Hamiltonians

$$H = \tfrac{1}{2}D^2 + V(x) \tag{3.0.2}$$

with a potential that decays in space. As a matter of fact, results obtained in this chapter will be used in the next chapter devoted to scattering theory for Hamiltonians of the form (3.0.2). Nevertheless, we think that time-dependent Hamiltonians deserve our attention. As we will see throughout this chapter, under suitable conditions on $V(t,x)$, scattering theory for such Hamiltonians has very good mathematical properties and can serve as an excellent "training ground" to learn some of the concepts of scattering theory for time-independent Hamiltonians.

Let us briefly describe the contents of this chapter.

In Sect. 3.1 we fix the notation and discuss the relation between the evolution $U(t,s)$ and the time-dependent Hamiltonian $H(t)$ that generates this evolution. This question is studied in an abstract setting in Appendix B.3.

In Sect. 3.2 we introduce the asymptotic momentum, that is, the self-adjoint operator defined by

$$D^+ := \lim_{t\to\infty} U(0,t)DU(t,0), \tag{3.0.3}$$

where the limit (3.0.3) has to be understood in a special sense described in Appendix B.2. The existence of (3.0.3) is true under quite general assumptions on the potential, e.g. if $V(t,x) = V_0(t,x) + V_1(t,x)$, where

$$|V_0(t,x)| \leq C\langle t\rangle^{-\mu_0}, \qquad \mu_0 > 1,$$
$$|\nabla_x V_1(t,x)| \leq C\langle t\rangle^{-1-\mu_1}, \qquad \mu_1 > 0.$$

Another, equivalent definition of D^+ is possible:

$$D^+ = \lim_{t\to\infty} U(0,t)\frac{x}{t}U(t,0). \tag{3.0.4}$$

A large part of Sect. 3.1 is devoted to a proof of (3.0.4). This part of Sect. 3.1 essentially will not be used in this chapter.

Sect. 3.3 is devoted to the fast-decaying scattering theory. In the fast-decaying case, one can compare the dynamics with the evolution generated by the free Hamiltonian

$$H_0 := \frac{1}{2}D^2.$$

We prove that if we assume, for instance, that

$$|V(t,x)| \leq C\langle t\rangle^{-\mu}, \qquad \mu > 1,$$

then the wave operator for the fast-decaying case

$$\text{s-}\lim_{t\to\infty} U(0,t)\mathrm{e}^{-itH_0} =: \Omega_{\text{fd}}^+ \tag{3.0.5}$$

exists and is unitary. The unitarity of the wave operator goes under the name of *asymptotic completeness.*

The wave operator implements the unitary equivalence of D^+ and D:

$$D^+ = \Omega_{\text{fd}}^+ D \Omega_{\text{fd}}^{+*}. \tag{3.0.6}$$

In the slow-decaying case, the limit (3.0.5), in general, does not exist. We need to replace in (3.0.5) the free dynamics e^{-itH_0} with a modified one $\mathrm{e}^{-iS(t,D)}$. It turns out that if we chose appropriately the function $S(t,\xi)$, then the modified wave operator

$$\text{s-}\lim_{t\to\infty} U(0,t)\mathrm{e}^{-iS(t,D)} =: \Omega_{\text{sd}}^+ \tag{3.0.7}$$

exists and is unitary. It satisfies

$$D^+ = \Omega_{\text{sd}}^+ D \Omega_{\text{sd}}^{+*}. \tag{3.0.8}$$

The function $S(t,\xi)$ is not uniquely defined. The choice that we usually make is a solution of the Hamilton-Jacobi equation with a certain potential that is close to the potential $V(t,x)$.

The construction of the modified wave operator and the proof of its unitarity are described in Sects. 3.4 and 3.5. In Sect. 3.4 we impose very weak conditions on the potential, roughly speaking, we demand that

$$|\partial_x^\alpha V(t,x)| \leq C\langle t\rangle^{-\mu-|\alpha|}, \quad \mu > 0, \quad |\alpha| = 1, 2. \tag{3.0.9}$$

In Sect. 3.5 we develop slow-decaying scattering theory under more restrictive hypotheses

$$|\partial_x^\alpha V(t,x)| \leq C_\alpha\langle t\rangle^{-\mu-|\alpha|}, \quad \mu > 0, \quad |\alpha| \geq 1. \tag{3.0.10}$$

The proofs in Sect. 3.4 are quite technical. Therefore we decided for the convenience of the reader to give an independent treatment of this subject in Sect. 3.5. We recommend the reader to skip Sect. 3.4 on the first reading (the results of this section are not used in the remaining part of this chapter).

In practice, it is useful to know how to construct a modified free dynamics that can be put in (3.0.7) without solving the Hamilton-Jacobi equation. If the potential satisfies

$$|\partial_x^\alpha V(t,x)| \leq C\langle t\rangle^{-\mu-|\alpha|}, \quad \mu > \tfrac{1}{2}, \quad |\alpha| = 1,$$

then one can define wave operators using the so-called Dollard dynamics. Beside its simplicity, the Dollard dynamics has the advantage of being applicable if the system has some "internal degrees of freedom". Dollard wave operators are described in Sect. 3.6.

In Sect. 3.7 we present a construction of the modified wave operator that is an adaptation to the time-dependent case of a construction of Isozaki-Kitada [IK1]. This construction uses a Fourier integral operator

$$J_{\mathrm{sd}}^+(s)\phi(x) := (2\pi)^{-n}\int\int \mathrm{e}^{i\Phi_{\mathrm{sd}}^+(s,x,\xi)-i\langle y,\xi\rangle}\phi(y)\mathrm{d}y\mathrm{d}\xi, \tag{3.0.11}$$

where $\Phi_{\mathrm{sd}}^+(s,x,\xi)$ is a solution of the eikonal equation. We show that

$$\Omega_{\mathrm{sd}}^+ = \lim_{s\to\infty} U(0,s)J_{\mathrm{sd}}^+(s). \tag{3.0.12}$$

Note that (3.0.7) was the strong limit, whereas (3.0.12), under the hypotheses on potentials that we use, is the norm limit. If the choices of $S(t,\xi)$ and $\Phi_{\mathrm{sd}}^+(s,x,\xi)$ are related to one another in the way described in Chap. 1, then the modified wave operators defined by (3.0.7) and (3.0.12) are equal.

In Sect. 3.8 we describe examples of time-dependent potentials for which the asymptotic velocity D^+ and the wave operator Ω_{fd}^+ exist but asymptotic completeness fails, that is, $\mathrm{Ran}\Omega_{\mathrm{fd}}^+ \neq L^2(X)$. Moreover, D^+ has some pure point spectrum, therefore D^+ cannot be unitarily equivalent to D. The first class of examples, due to Yafaev [Yaf1], is given in Subsect. 3.8.2. It is based on the adiabatic approximation described in Subsect. 3.8.1. In these counterexamples, $\partial_x^\alpha V(t,x)$, $|\alpha| = 2$, decays a little bit slower than $O(t^{-2})$. In Subsect. 3.8.3 we give sharper counterexamples, with $\partial_x^\alpha V(t,x) \in O(t^{-2})$, $|\alpha| = 2$, which is the borderline for asymptotic completeness.

Similarly as in the classical case, if we assume the following verion of the fast-decaying condition:

$$|\partial_x^\alpha V(t,x)| \leq C_\alpha\langle t\rangle^{-\mu-|\alpha|}, \quad \mu > 1, \quad |\alpha| \geq 1, \tag{3.0.13}$$

then wave operators have especially good properties. It turns out that (3.0.13) implies that Ω_{fd}^{+} is a bounded pseudo-differential operator in the following sense: there exists a function $a^{+}(x,\xi)$ such that

$$\begin{aligned} &\Omega_{\mathrm{fd}}^{+}\phi(x) = (2\pi)^{-n}\int\int a^{+}(x,\xi)\mathrm{e}^{i(x-y)\xi}\phi(y)\mathrm{d}y\mathrm{d}\xi, \\ &|\partial_x^{\alpha}\partial_{\xi}^{\beta}a^{+}(x,\xi)| \leq C_{\alpha,\beta}. \end{aligned} \tag{3.0.14}$$

These properties of fast-decaying scattering theory are proven in Sect. 3.9.

It turns out that the condition (3.0.10) does not imply that Ω_{sd}^{+} is a pseudo-differential operator in the sense of (3.0.14). Instead, we show in Sect. 3.0.15 that the time-translated modified wave operator is a Fourier integral operator in the following sense: for s big enough, there exists an amplitude $a^{+}(s,x,\xi)$ such that

$$\begin{aligned} &U(s,0)\Omega_{\mathrm{sd}}^{+}\phi(x) = (2\pi)^{-n}\int\int a^{+}(s,x,\xi)\mathrm{e}^{i\Phi_{\mathrm{sd}}^{+}(s,x,\xi)-i\langle y,\xi\rangle}\phi(y)\mathrm{d}y\mathrm{d}\xi, \\ &\partial_x^{\alpha}\partial_{\xi}^{\beta}(a^{+}(s,x,\xi)-1) \in o(s^{-|\alpha|}). \end{aligned} \tag{3.0.15}$$

A reader who needs just a short introduction to the basic construction of modified wave operators can restrict himself to the first part of Sect. 3.2 and Sect.3.5. On the first reading, it is also a good idea to learn the alternative constructions of modified wave operators given in Sects. 3.6 and 3.7.

The existence of the asymptotic momentum and fast-decaying scattering theory are very easy in the case of Hamiltonians considered in this chapter. The most difficult subject of this chapter is slow-decaying scattering theory. The results about scattering in the slow-decaying case will be used in the next chapter, where we will study the long-range problem. In fact, a desire to give a clear exposition of the main technical difficulties of the long-range problem led us to write a separate chapter on time-dependent Hamiltonians.

Let us sketch the history of long-range scattering theory. The definition of a modified wave operator in the case of the Coulomb potential was first given by Dollard in [Do1]. It was extended to a much larger class of potentials in [BuMa]. Other early papers on the subject include [AlKa, AMM]. In [Hö1], Hörmander introduced modified free dynamics defined by exact solutions of the Hamilton-Jacobi equation.

Asymptotic completeness for long-range two-body systems was first proven by Saito [Sai1, Sai2] and Kitada [Ki1, Ki2, Ki3]. Their proofs used the stationary approach.

A fully time-dependent proof was first given in [E5] and later in [Pe1] in the case $\mu > 1/2$ (see (3.0.10)). A time-dependent proof for potentials with a slower decay was given in [KiYa1, KiYa2]. Let us note that this proof allowed for time-dependent potentials.

A different construction of modified wave operators, which uses a Fourier integral modifier whose phase is a solution of the eikonal equation, was given in [Kako, IK1].

It is possible to give other constructions of modified wave operators. One of them, which instead of the modified free dynamics $\mathrm{e}^{-iS(t,D)}$ uses a dynamics of the form

$$I_{\Psi}(t)\phi(x) := t^{-\frac{n}{2}} e^{i\Psi(t,x)} \phi(\tfrac{x}{t}),$$

where $\Psi(t,x)$ is an appropriate function, was used by Yafaev in [Yaf3] where the existence of this type of modified wave operators was shown. Asymptotic completeness of such wave operators under the same hypotheses as used in this chapter was proved in [DeGe2].

A time-independent proof of asymptotic completeness using very weak assumptions on the decay of the potential was given in [Hö2, vol IV]. This proof used (mostly unpublished) ideas of Agmon.

A very appealing time-dependent proof of asymptotic completeness was given by Sigal in [Sig2].

The proof given in this chapter in Sect. 3.5 follows to a great extent that of [Sig2] in its slightly simplified form contained in [De6]. In Sect. 3.4 we present an improved version of this proof that works under much less restrictive hypotheses on the potential. In this proof, some of the ideas of [Hö2, vol. IV] are incorporated into the method of Sigal.

Regularity properties of wave operators were considered in [I3, Ag1] and, recently, in [JN, HeSk1].

Among numerous other papers on long-range scattering theory and related subjects, let us mention [Ar, BC, Com1, Com2, Geo, GGNT, E6, Ike1, Ike2, IkeI1, I1, I2, IK2, IK3, IK4].

3.1 Time-Dependent Schrödinger Hamiltonians

Let us start with describing notation and facts concerning time-dependent Schrödinger Hamiltonians. The basic notation concerning Hilbert spaces is given in Appendix B.1.

Most of the time, we will work with the Hilbert space $L^2(X)$ where $X = \mathbb{R}^n$. X is equipped with a scalar product. $|x|$ will denote the length of $x \in X$.

$D := i^{-1}\nabla$ will denote the momentum operator (which is a vector of commuting self-adjoint operators). The free Hamiltonian is defined as

$$H_0 := \frac{1}{2}D^2 = -\frac{1}{2}\Delta.$$

We will sometimes use the scale of Sobolev spaces

$$H^m(X) := \{\phi \in \mathcal{D}'(X) \mid (1-\Delta)^{\frac{m}{2}}\phi \in L^2(X)\}.$$

A time-dependent Schrödinger operator is a function $\mathbb{R}^+ \mapsto H(t)$ with values in self-adjoint operators of the form

$$H(t) = \frac{1}{2}D^2 + V(t,x), \tag{3.1.1}$$

where $\mathbb{R}^+ \times X \ni (t,x) \mapsto V(t,x) \in \mathbb{R}$ is a measurable function satisfying certain conditions that permit to define a unitary dynamics $U(t,s)$ generated by $H(t)$.

There are many different conditions that one can impose on $V(t,x)$ for this purpose. One set of conditions that we can use in this chapter, which will be sufficient for applications in the following chapter, is as follows:

$$\begin{aligned} &t \mapsto \|V(t,\cdot)\|_\infty \quad \text{belongs to} \quad L^1_{\text{loc}}(\mathrm{d}t), \\ &t \mapsto \|(1+D^2)^{-\frac{1}{2}}[D^2, V(t,x)](1+D^2)^{-\frac{1}{2}}\| \quad \text{belongs to} \quad L^1_{\text{loc}}(\mathrm{d}t). \end{aligned} \tag{3.1.2}$$

Then we define $U(t,s)$ by the following convergent expansion:

$$U(t,s) = \sum_{n=0}^{\infty} \quad \int_{t\geq u_n \geq \cdots \geq u_1 \geq s} \cdots \int \quad \mathrm{e}^{i(t-u_n)H_0} V(u_n,x) \cdots V(u_1,x) \mathrm{e}^{i(u_1-s)H_0} \mathrm{d}u_n \cdots \mathrm{d}u_1.$$

By Proposition B.3.6, it follows from Hypothesis (3.1.2) that the unitary dynamics $U(t,s)$ is D^2-regularly generated by $H(t)$ in the sense of Definition B.3.2. Note also that $U(t,s)$ preserves $H^1(X)$.

The hypothesis (3.1.2) has the following disadvantage. It implies that, for almost all times, the potential $V(t,x)$ has to be bounded. This is somewhat disappointing, because in most of this chapter the boundedness of $V(t,x)$ does not play a role, it is the boundedness of $\nabla_x V(t,x)$ that is important. Alternatively, instead of (3.1.2), we can assume a much more general hypothesis, which, unfortunately, is not explicit. We can just suppose that $U(t,s)$ is a unitary dynamics in the sense of Definition B.3.1 that is B-regularly generated by $H(t)$ in the sense of Definition B.3.2 with $B = D^2 + x^2$.

We will denote by $\nabla_x V(t,x)$ the distributional derivative of $V(t,x)$, which is equal to the (possibly unbounded) operator $[D, iV(t,x)]$. Note that $[D, iV(x)]$ is bounded iff the distributional derivative $\nabla_x V(x)$ is in $L^\infty(X)$, and then $[D, iV(x)] = \nabla_x V(x)$.

We define the Heisenberg derivative associated with $H(t)$:

$$\mathbf{D} := \frac{\mathrm{d}}{\mathrm{d}t} + i[H(t), \cdot].$$

3.2 Asymptotic Momentum

As in the classical case, we will start our exposition of scattering theory for time-dependent potentials with a construction of the asymptotic momentum, which is a basic asymptotic quantity common to the fast- and slow-decaying cases. Our first result will be the quantum analog of Theorem 1.3.1.

Theorem 3.2.1
Suppose that

$$\int_0^\infty \|[(1+D^2)^{-1}, V(t,x)]\| \mathrm{d}t < \infty. \tag{3.2.1}$$

Then there exists the limit

$$\text{s}-C_\infty-\lim_{t\to\infty} U(0,t)DU(t,0) =: D^+. \tag{3.2.2}$$

D^+ *is a vector of commuting self-adjoint operators. If we assume, in addition, that*

$$V(t,x) = V_0(t,x) + V_1(t,x)$$

such that

$$\int_0^\infty \|V_0(t,x)\|\mathrm{d}t < \infty, \tag{3.2.3}$$

$$\lim_{\epsilon\to 0}\int_0^\infty \|[(1+\epsilon D^2)^{-1}, V_1(t,x)]\|\mathrm{d}t = 0, \tag{3.2.4}$$

then D^+ *is densely defined.*

Remark. Let us note that the conditions (3.2.3) and (3.2.4) imply (3.2.1). Moreover, (3.2.4) follows from the following condition: for some $\sigma < 1/2$,

$$\int_0^\infty \|\langle D\rangle^{-\sigma}\partial_x^\alpha V_1(t,x)\langle D\rangle^{-\sigma}\|dt < \infty, \quad |\alpha| = 1. \tag{3.2.5}$$

Proof of Theorem 3.2.1. Let us first prove the existence of the limit in (3.2.2). By a density argument, it is enough to show the existence of

$$\text{s}-\lim_{t\to\infty} U(0,t)g(D)U(t,0) \tag{3.2.6}$$

for $g \in C_0^\infty(X)$. Now,

$$\tfrac{\mathrm{d}}{\mathrm{d}t}U(0,t)g(D)U(t,0) = U(0,t)[iV(t,x), g(D)]U(t,0).$$

By Lemma C.1.2,

$$\|[iV(t,x), g(D)]\| \le C\|[(1+D^2)^{-1}, V(t,x)]\|.$$

This is integrable. So the existence of (3.2.6) follows by integration.

Next let us prove that D^+ has a dense domain. Let $U_1(t,s)$ denote the dynamics generated by the Hamiltonian

$$H_1(t) := \frac{1}{2}D^2 + V_1(t,x).$$

Then it is very easy to see that there exists

$$\lim_{t\to\infty} U(0,t)U_1(t,0).$$

Now,

$$\begin{aligned}&\text{s}-\lim_{t\to\infty} U_1(0,t)\left(1+\epsilon D^2\right)^{-1} U_1(t,0) - \left(1+\epsilon D^2\right)^{-1} \\ &= \textstyle\int_0^\infty U_1(0,t)[\left(1+\epsilon D^2\right)^{-1}, V_1(t,x)]U_1(t,0)\mathrm{d}t.\end{aligned} \tag{3.2.7}$$

This converges to zero as $\epsilon \to 0$. Therefore

$$\begin{aligned} &\text{s-}\lim_{\epsilon\to 0}\left(1+\epsilon(D^+)^2\right)^{-1} \\ &= \text{s-}\lim_{\epsilon\to 0}\left(\text{s-}\lim_{t\to\infty} U_1(0,t)\left(1+\epsilon D^2\right)^{-1} U_1(t,0)\right) = 1 \end{aligned}$$

By (B.2.2), this implies that the self-adjoint operator D^+ has a dense domain.

□

The observable D^+ classifies the states in $L^2(X)$ according to their asymptotic behavior in momentum space. It turns out that there exists an alternative method of constructing D^+ using the operator $\frac{x}{t}$ instead of D. Consequently, D^+ describes also the asymptotic behavior of states in position space.

Theorem 3.2.2
Assume (3.2.3) and (3.2.4). Then

$$D^+ = \text{s-}C_\infty\text{-}\lim_{t\to\infty} U(0,t)\frac{x}{t}U(t,0). \tag{3.2.8}$$

To prove this theorem, we will need some additional techniques, which will be further developed in a somewhat different situation in the next chapter. Note that the remaining part of this section will not be used in this chapter except for Sect. 3.8.

Proposition 3.2.3
Assume (3.2.1). Suppose that $j, g \in C_0^\infty(X)$ and $\mathrm{supp}j \cap \mathrm{supp}g = \emptyset$. *Then*

$$\int_1^\infty \left\| j\left(\frac{x}{t}\right) g(D)U(t,0)\phi\right\|^2 \frac{\mathrm{d}t}{t} \le C\|\phi\|^2,\ \phi\in L^2(X). \tag{3.2.9}$$

If, moreover, $J \in C_0^\infty(X)$ such that $J = 1$ on a neighborhood of $\mathrm{supp}g$ *then*

$$\text{s-}\lim_{t\to\infty} U(0,t)J\left(\frac{x}{t}\right) g(D)U(t,0) = g(D^+). \tag{3.2.10}$$

Proof. We will prove the proposition by constructing a suitable propagation observable and applying Lemma B.4.1 of the Appendix B.4. By a covering argument, we may assume that the support of g and j are very close respectively to $\xi_0 \in X'$ and $x_0 \in X$, with $\xi_0 \neq x_0$. We can then find $v \in X'$ and $\theta_1 < \theta_2$ such that

$$\mathrm{supp}g \subset \{x \mid \langle v, x\rangle > \theta_1\}, \qquad \mathrm{supp}j \subset \{x \mid \langle v, x\rangle < \theta_2\}.$$

Choose a function $\tilde{J} \in C^\infty(\mathbb{R})$ such that $\tilde{J}' \in C_0^\infty(\mathbb{R})$ and

$$\nabla_x \tilde{J}(\langle v, x\rangle) \ge j^2(x).$$

Set $J(x) := \tilde{J}(\langle v, x\rangle)$.

We consider the following propagation observable:

$$\Phi(t) := g(D)J\left(\frac{x}{t}\right)g(D),$$

which is uniformly bounded in t. We compute its Heisenberg derivative. We obtain

$$\begin{aligned}\mathbf{D}\Phi(t) &= [V(t,x), ig(D)]J(\tfrac{x}{t})g(D) + \mathrm{hc}\\ &+\tfrac{1}{2t}g(D)(D-\tfrac{x}{t})\nabla_x J(\tfrac{x}{t})g(D) + \mathrm{hc}.\end{aligned} \tag{3.2.11}$$

We claim that, for some $C_0 > 0$, the second term on the right-hand side of (3.2.11) is greater than or equal to

$$C_0\frac{1}{t}g(D)j^2\left(\frac{x}{t}\right)g(D) + O(t^{-2}). \tag{3.2.12}$$

One way to prove (3.2.12) is to use the pseudo-differential calculus. In fact, by Proposition D.5.3 we can rewrite the left-hand side of (3.2.12) as

$$\frac{1}{t}r^{\mathrm{w}}(t,x,D) + O(t^{-2}), \quad \text{where} \quad r(t,x,\xi) = g^2(\xi)\left(\xi - \frac{x}{t}\right)\nabla_x J\left(\frac{x}{t}\right).$$

The right-hand side of (3.2.12) can be rewritten as

$$\frac{1}{t}p^{\mathrm{w}}(t,x,D) + O(t^{-2}), \quad \text{where} \quad p(t,x,\xi) = g^2(\xi)j^2\left(\frac{x}{t}\right).$$

Both $r(t,x,\xi)$ and $p(t,x,\xi)$ are symbols of the class $S(1, g_0(t))$ (see Appendix D.5). We clearly have

$$r(t,x,\xi) \geq (\theta_2 - \theta_1)p(t,x,\xi). \tag{3.2.13}$$

Using (3.2.13) and the sharp Gårding inequality (see Proposition D.5.4), we get

$$\frac{1}{t}r^{\mathrm{w}}(t,x,D) \geq \frac{1}{t}(\theta_2-\theta_1)p^{\mathrm{w}}(t,x,D) + O(t^{-2}),$$

which proves (3.2.12).

Applying Lemma B.4.1, we see that (3.2.12) implies (3.2.9). Let us now consider $J \in C^\infty(X)$ such that $\nabla_x J \in C_0^\infty(X)$ and $\mathrm{supp}\nabla_x J \cap \mathrm{supp}g = \emptyset$. We will prove now that there exists

$$\mathrm{s-}\lim_{t\to\infty} U(0,t)J\left(\frac{x}{t}\right)g(D)U(t,0). \tag{3.2.14}$$

In fact, take $\tilde{j}, \tilde{g} \in C_0^\infty(X)$ such that $\tilde{j}\nabla_x J = \nabla_x J$, $\tilde{g}g = g$ and $\mathrm{supp}\tilde{j} \cap \mathrm{supp}g = \emptyset$. Then we can estimate

$$\left|\left(\phi\middle|\left(\mathbf{D}J\left(\frac{x}{t}\right)g(D)\right)\phi\right)\right| \leq Ct^{-1}\left\|\tilde{j}\left(\frac{x}{t}\right)\tilde{g}(D)\phi\right\|^2 + O(t^{-2}). \tag{3.2.15}$$

This is integrable along the evolution by (3.2.9). Hence (3.2.14) exists.

If we assume, in addition, that $J \in C_0^\infty(X)$ and $\mathrm{supp} J \cap \mathrm{supp} g = \emptyset$, then we know by (3.2.9) that

$$\int_1^\infty \left\| J\left(\frac{x}{t}\right) g(D)U(t,0)\phi \right\|^2 \frac{\mathrm{d}t}{t} < \infty. \tag{3.2.16}$$

Clearly, if a function $f(t)$ satisfies

$$\lim_{t\to\infty} f(t) \text{ exists}, \qquad \int_1^\infty f^2(t)\frac{\mathrm{d}t}{t} < \infty,$$

then $\lim_{t\to\infty} f(t) = 0$. Hence (3.2.16) and the existence of (3.2.14) imply that (3.2.14) is zero if $J \in C_0^\infty(X), \mathrm{supp} J \cap \mathrm{supp} g = \emptyset$.

Unfortunately, this is not the end of the proof of (3.2.10), since we need to show that there is no propagation for large $|x|/t$. To this end, take functions $F \in C^\infty(\mathbb{R})$, $f \in C_0^\infty(\mathbb{R})$ such that $F = 0$ on a neighborhood of 0, $F = 1$ on a neighborhood of ∞, and $F' = f^2$. Set

$$\Phi_R(t) := g(D)F\left(\frac{|x|}{Rt}\right) g(D).$$

$$\begin{aligned}
-\mathbf{D}\Phi_R(t) &= [V(t,x), g(D)]F\left(\tfrac{|x|}{Rt}\right) g(D) + \mathrm{hc} \\
&\quad + \tfrac{1}{t} g(D) f^2\left(\tfrac{|x|}{Rt}\right) \tfrac{|x|}{Rt} g(D) \\
&\quad + \tfrac{1}{2tR} g(D) D\tfrac{x}{|x|} f^2\left(\tfrac{|x|}{Rt}\right) g(D) + \mathrm{hc}.
\end{aligned} \tag{3.2.17}$$

Let $\tilde{g} \in C_0^\infty(X)$ such that $\tilde{g}g = g$. The third term on the right-hand side of (3.2.17) equals

$$\begin{aligned}
&\tfrac{1}{tR} g(D) f\left(\tfrac{|x|}{Rt}\right) \left(\tilde{g}(D) D\tfrac{x}{|x|} + \mathrm{hc}\right) f\left(\tfrac{|x|}{Rt}\right) g(D) + O(t^{-2}R^{-2}) \\
&\geq -\tfrac{C_0}{tR} g(D) f\left(\tfrac{|x|}{Rt}\right)^2 g(D) + O(t^{-2}R^{-2}).
\end{aligned}$$

Hence, for $R \geq C_0$,

$$-\mathbf{D}\Phi_R(t) \geq -C\|[V(t,x), g(D)]\| + O(t^{-2}R^{-2}). \tag{3.2.18}$$

Therefore, for $R > C_0$ and any $t_0 \geq 0$,

$$\begin{aligned}
\mathrm{s-}\lim_{t\to\infty} U(0,t)\Phi_R(t)U(t,0) &\leq U(0,t_0)\Phi_R(t_0)U(t_0,0) \\
&\quad + C\textstyle\int_{t_0}^\infty \|[V(t,x), g(D)]\|\mathrm{d}t + O(t_0^{-1}R^{-2}).
\end{aligned} \tag{3.2.19}$$

By choosing t_0 big enough, we can make the integral on the right-hand side of (3.2.19) as small as we wish. For a fixed t_0, the first and third terms on the right-hand side of (3.2.19) go to zero as $R \to \infty$. Hence

$$\mathrm{s-}\lim_{R\to\infty}\left(\mathrm{s-}\lim_{t\to\infty} U(0,t)\Phi_R(t)g(D)U(t,0)\right) = 0. \tag{3.2.20}$$

But we already know that, for big enough R_1, R_2,

$$\text{s-}\lim_{t\to\infty} U(0,t)\left(\Phi_{R_2}(t)-\Phi_{R_1}(t)\right)U(t,0)=0,$$

using the fact that the function

$$F\left(\frac{|x|}{R_2}\right)-F\left(\frac{|x|}{R_1}\right)$$

has a compact support. Therefore, for big R,

$$\text{s-}\lim_{t\to\infty} U(0,t)\Phi_R(t)U(t,0)=0.$$

This ends the proof of (3.2.10). □

Proposition 3.2.4
Assume (3.2.1). Suppose that $g, J\in C_0^\infty(X)$ *and* $\text{supp} g\cap\text{supp}\nabla J=\emptyset$. *Then*

$$\int_1^\infty \left\|\left|\frac{x}{t}-D\right| J\left(\frac{x}{t}\right) g(D)U(t,0)\phi\right\|^2 \frac{\mathrm{d}t}{t}\leq C\|\phi\|^2,\ \phi\in L^2(X), \tag{3.2.21}$$

$$\text{s-}\lim_{t\to\infty}\left|\frac{x}{t}-D\right| J\left(\frac{x}{t}\right) g(D)U(t,0)=0. \tag{3.2.22}$$

Proof. We consider the following propagation observable

$$\Phi(t) := -g(D)J\left(\frac{x}{t}\right)\left(\frac{x}{t}-D\right)^2 J\left(\frac{x}{t}\right)g(D),$$

which is bounded uniformly in t, and compute its Heisenberg derivative. We obtain

$$\begin{aligned}\mathbf{D}\Phi(t) &= -\tfrac{1}{t}g(D)(\mathbf{D}J(\tfrac{x}{t}))(\tfrac{x}{t}-D)^2J(\tfrac{x}{t})g(D)+\text{hc}\\ &-[V(t,x),ig(D)]J(\tfrac{x}{t})(\tfrac{x}{t}-D)^2J(\tfrac{x}{t})g(D)+\text{hc}\\ &+\tfrac{1}{t}g(D)J(\tfrac{x}{t})(\tfrac{x}{t}-D)^2J(\tfrac{x}{t})g(D)\\ &+g(D)J(\tfrac{x}{t})(\tfrac{x}{t}-D)[V(t,x),iD]J(\tfrac{x}{t})g(D)+\text{hc}.\end{aligned} \tag{3.2.23}$$

Using Lemma C.1.2, we see that the terms in the second and fourth line of the right-hand side of (3.2.23) are integrable in norm. Let $\tilde{j}\in C_0^\infty(X)$ such that $\tilde{j}=1$ on $\text{supp}\nabla_x J$ and $\text{supp} g\cap\text{supp}\tilde{j}=\emptyset$. Then the first line of the right-hand side of (3.2.23) can be written as

$$\frac{1}{2t}g(D)\nabla J(\frac{x}{t})\left(\frac{x}{t}-D\right)^3\tilde{j}\left(\frac{x}{t}\right)g(D)+\text{hc}+O(t^{-3}).$$

This is integrable along the evolution by Proposition 3.2.3. Now by Lemma B.4.1 we obtain (3.2.21).

To prove (3.2.22), we observe that there exists

$$\text{s-}\lim_{t\to\infty} U(0,t)\Phi(t)U(t,0), \tag{3.2.24}$$

because the Heisenberg derivative of $\Phi(t)$ is integrable by Proposition 3.2.3. Moreover, $\Phi(t) \le 0$ and, by (3.2.21), we have

$$\int_1^\infty (\phi, U(0,t)\Phi(t)U(t,0)\phi)\frac{dt}{t} < \infty. \tag{3.2.25}$$

Therefore,

$$\lim_{t\to\infty} U(0,t)\Phi(t)U(t,0) = 0,$$

which proves (3.2.22). □

Proof of Theorem 3.2.2. Let $f, g \in C_0^\infty(X)$ and $\phi \in L^2(X)$. Since the domain of D^+ is dense, it suffices to show that

$$\lim_{t\to\infty} U(0,t)f\left(\frac{x}{t}\right)U(t,0)g(D^+)\phi. = f(D^+)g(D^+)\phi \tag{3.2.26}$$

Choose $J \in C_0^\infty(X)$ such that $J = 1$ on a neighborhood of supp$g$. Then, by (3.2.10), the difference between the two sides of (3.2.26) is equal to

$$\lim_{t\to\infty} U(0,t)\left(f\left(\tfrac{x}{t}\right) - f(D)\right)J\left(\tfrac{x}{t}\right)g(D)U(t,0)\phi. \tag{3.2.27}$$

By the Baker-Campbell-Hausdorff formula, we have

$$\begin{aligned} f(D) - f\left(\tfrac{x}{t}\right) &= \int_0^1 \nabla f\left(\tau D + (1-\tau)\tfrac{x}{t}\right)\left(D - \tfrac{x}{t}\right)d\tau \\ &\quad + \tfrac{i}{2t}\int_0^1 \Delta f\left(\tau D + (1-\tau)\tfrac{x}{t}\right)d\tau. \end{aligned} \tag{3.2.28}$$

Therefore (3.2.27) equals

$$\lim_{t\to\infty} B(t)\left(\tfrac{x}{t} - D\right)J\left(\tfrac{x}{t}\right)g(D)U(t,0)\phi + O(t^{-1}),$$

where $B(t)$ is bounded. This equals zero by (3.2.22). □

3.3 Fast-Decaying Case

The asymptotic momentum constructed in Theorem 3.2.1 gives a classification of the states in $L^2(X)$ according to their behavior under the dynamics $U(t,0)$. We would like to know whether the asymptotic momentum D^+ is unitarily equivalent to the momentum. The answer is positive only if we assume some additional conditions on the potential. In the fast-decaying case, one can construct a unitary operator that intertwines the momentum and the asymptotic momentum in a particularly simple way.

Theorem 3.3.1
Suppose that the potential can be written as

$$V(t,x) = V_0(t,x) + V_1(t,x),$$

such that

$$\int_0^\infty \|V_0(t,\cdot)\|_\infty \mathrm{d}t < \infty, \tag{3.3.1}$$

$$\int_0^\infty \|\nabla_x V_1(t,x\cdot\|_\infty \langle t\rangle \mathrm{d}t < \infty, \tag{3.3.2}$$

and $V_1(t,0) \in L^1_{\mathrm{loc}}(\mathrm{d}t)$. *Set*

$$\theta(t) := \int_0^t V_1(s,0)\mathrm{d}s.$$

Then there exist

$$\mathrm{s-}\lim_{t\to\infty} U(0,t)\mathrm{e}^{-itH_0 - i\theta(t)}, \tag{3.3.3}$$

$$\mathrm{s-}\lim_{t\to\infty} e^{itH_0 + i\theta(t)} U(t,0). \tag{3.3.4}$$

If we denote (3.3.3) by $\Omega^+_{\mathrm{fd},\theta}$ *then (3.3.4) equals* $\Omega^{+*}_{\mathrm{fd},\theta}$. *Moreover,* $\Omega^+_{\mathrm{fd},\theta}$ *is unitary.*
The conditions of Theorem 3.2.1 are satisfied. Hence D^+ *exists and*

$$D^+ = \Omega^+_{\mathrm{fd},\theta} D \Omega^{+*}_{\mathrm{fd},\theta}. \tag{3.3.5}$$

Proof. It is enough to assume that $V_1(t,0) = 0$ and $\theta(t) = 0$. Let us prove the existence of (3.3.4), the case of (3.3.3) being simpler.

We introduce an auxiliary dynamics $U_1(s,t)$ generated by

$$H_1(t) := \frac{1}{2}D^2 + V_1(t,x).$$

First we see that there exists

$$\lim_{t\to\infty} U_1(0,t)U(t,0). \tag{3.3.6}$$

Next we note that the following identity is true:

$$U_1(0,t)xU(t,0) = x + tD + \int_0^t U_1(0,s)\nabla_x V_1(s,x)U_1(s,0)(t-s)\mathrm{d}s. \tag{3.3.7}$$

Hence

$$\|xU_1(t,0)\langle D\rangle^{-1}\langle x\rangle^{-1}\| \le C\langle t\rangle. \tag{3.3.8}$$

Let us now prove the existence of

$$\text{s-}\lim_{t\to\infty} \mathrm{e}^{itH_0} U_1(t,0). \tag{3.3.9}$$

Consider a vector $\phi = \langle D\rangle^{-1}\langle x\rangle^{-1}\psi$. Then

$$\begin{aligned}&\tfrac{\mathrm{d}}{\mathrm{d}t}\mathrm{e}^{itH_0}U_1(t,0)\phi\\ &= \mathrm{e}^{itH_0}\textstyle\int_0^1 \nabla V_1(t,\tau x)x\mathrm{d}\tau U_1(t,0)\langle D\rangle^{-1}\langle x\rangle^{-1}\psi,\end{aligned}$$

which is integrable by (3.3.2) and (3.3.8). Hence (3.3.9) exists.

Now the existence of (3.3.6) and (3.3.9) imply the existence of (3.3.4). □

3.4 Slow-Decaying Case – Hörmander Potentials

In this section we begin our study of scattering theory in the slow-decaying case. In this case, the asymptotic momentum is well defined, although the usual wave operators, characteristic of the fast-decaying case, in general do not exist. Nevertheless, for a very large class of potentials, one can show the existence and completeness of modified wave operators that have almost the same properties as the usual wave operators. They are defined using a modified free evolution – a unitary evolution that conserves the momentum and resembles the free evolution but takes into account the shape of the slow-decaying potential. Modified wave operators intertwine the momentum and the asymptotic momentum.

In this section we prove that modified wave operators exist and are unitary under quite general assumptions on the potential. This proof is unfortunately rather involved. The reader who prefers an easier exposition of the slow-decaying case under more restrictive hypotheses should go directly to Sect. 3.5.

For slow-decaying potentials satisfying only the hypotheses of Sect. 3.2, the asymptotic momentum D^+ can have a spectral measure that is different from the one of D. In fact, we will give in Sect. 3.8 examples of a time-dependent slow-decaying potential for which $\mathbb{1}_{\{0\}}(D^+)$ is an infinite dimensional projection.

The main result of this section is the following theorem.

Theorem 3.4.1
Assume that

$$V(t,x) = V_{\mathrm{s}}(t,x) + V_{\mathrm{l}}(t,x)$$

such that

$$\int_0^\infty \|V_{\mathrm{s}}(t,\cdot)\|_\infty \mathrm{d}t < \infty,$$

$$\int_0^\infty \langle t\rangle^{|\alpha|-1}\|\partial_x^\alpha V_{\mathrm{l}}(t,\cdot)\|_\infty \mathrm{d}t < \infty, \quad |\alpha| = 1,2. \tag{3.4.1}$$

Then there exists a C^∞ function $S(t,\xi)$ such that

$$\text{s-}\lim_{t\to\infty} U(0,t)\mathrm{e}^{-iS(t,D)}, \tag{3.4.2}$$

$$\text{s-}\lim_{t\to\infty} e^{iS(t,D)}U(t,0). \tag{3.4.3}$$

exist. If we denote (3.4.2) by $\Omega^+_{\rm sd}$, then (3.4.3) equals $\Omega^{+*}_{\rm sd}$. Moreover, $\Omega^{+*}_{\rm sd}$ is unitary and

$$D^+ = \Omega^+_{\rm sd} D \Omega^{+*}_{\rm sd}. \tag{3.4.4}$$

Remark. Note that by Lemma 3.4.5 *(i)* if

$$\int_0^\infty \langle t\rangle^{1/2} \|\partial_x^\alpha V_1(t,\cdot)\|_\infty \mathrm{d}t < \infty, \quad |\alpha| = 1,$$

then the hypotheses of Theorem 3.4.1 are satisfied. Under the above conditions, one can also use the Dollard construction of modified wave operators (see Sect. 3.6).

One might want to know what is the relationship of the function $S(t,\xi)$ and the potentials that appear in the statement of the theorem. It is natural to ask whether, as this function, we can take a solution of the Hamilton-Jacobi equation with the potential $V_1(t,x)$. It turns out that this is possible if we strengthen the assumptions of the theorem, as we describe in the following proposition.

Proposition 3.4.2
If instead of (3.4.1) the potential $V_1(t,x)$ satisfies one of the following two hypotheses:

$$\int_0^\infty \langle t\rangle^{|\alpha|-1} \|\partial_x^\alpha V_1(t,\cdot)\|_\infty \mathrm{d}t < \infty, \quad |\alpha| = 1,2,3, \tag{3.4.5}$$

or

$$\begin{aligned} &\textstyle\int_0^\infty \langle t\rangle^{1/2} \|\partial_x^\alpha V_1(t,\cdot)\|_\infty \mathrm{d}t < \infty, \quad |\alpha| = 1, \\ &\textstyle\int_0^\infty \langle t\rangle \|\partial_x^\alpha V_1(t,\cdot)\|_\infty \mathrm{d}t < \infty, \quad |\alpha| = 2, \end{aligned} \tag{3.4.6}$$

then, as the function $S(t,\xi)$ in Theorem 3.4.1, we can take the solution of the Hamilton-Jacobi equation

$$\begin{cases} \partial_t S(t,\xi) = \frac{1}{2}\xi^2 + V_1(t, \nabla_\xi S(t,\xi)), \\ S(T,\xi) = 0, \end{cases} \tag{3.4.7}$$

which exists for large enough T.

Proof. The proposition follows from Lemma 3.4.5 *(iii)*, Theorems 3.4.1 and 1.11.2. □

The proof of Theorem 3.4.1 will be divided into a series of lemmas. First we need some additional analysis of classical scattering that was not contained in Sect. 1.10.

Recall that, under the assumption

$$\int_0^\infty \|\partial_x^\alpha F(t,\cdot)\|_\infty \langle t\rangle^{|\alpha|} \mathrm{d}t < \infty, \quad |\alpha| = 0, 1, \tag{3.4.8}$$

for $T \leq t_1 \leq s \leq t_2$, we constructed in Theorem 1.5.1 the solutions $\tilde{y}(s, t_1, t_2, x, \xi)$ of the classical boundary problem, where we fixed the initial position and the final momentum. In Theorem 1.10.1, assuming the so-called smooth slow-decaying condition, we showed some estimates on the derivatives of these solutions. Unfortunately, in this section, we will deal with a much wider class of potentials and we need to generalize a part of Theorem 1.10.1.

Note that, in the following proposition, we do not assume the force to be conservative.

Proposition 3.4.3
Suppose that, for $n = 0, 1, \ldots$, we fix positive numbers $\kappa(n)$ that satisfy

$$\kappa(n) + \kappa(m) \leq \kappa(n+m). \tag{3.4.9}$$

(Note that this implies $\kappa(0) = 0$). Assume that

$$\int_0^\infty \|\partial_x^\alpha F(t,\cdot)\|_\infty \langle t\rangle^{|\alpha|-\kappa(|\alpha|-1)} \mathrm{d}t < \infty, \quad |\alpha| \geq 1. \tag{3.4.10}$$

Then, uniformly for $T \leq t_1 \leq s \leq t_2 < \infty$, we have the estimate

$$\partial_\xi^\beta(\tilde{y}(s, t_1, t_2, x, \xi) - x - (s - t_1)\xi) \in o(t_1^0)|s - t_1|\langle t_2\rangle^{\kappa(|\beta|-1)}. \tag{3.4.11}$$

Proof. Recall from the proof of Theorem 1.5.1 that

$$\tilde{z}(s) := \tilde{y}(s) - x - (s - t_1)\xi$$

satisfies

$$\tilde{z}(s) = -\int_{t_1}^{t_2} \zeta_{t_1,s}(u) F(u, \tilde{y}(u)) \mathrm{d}u. \tag{3.4.12}$$

We will prove our proposition by induction with respect to $|\beta|$. The induction hypothesis $H(n)$ will be

$$\partial_\xi^\beta \tilde{z}(s) \in o(t_1^0)|s - t_1|\langle t_2\rangle^{\kappa(|\beta|-1)}, \quad 1 \leq |\beta| \leq n. \tag{3.4.13}$$

Let us assume that $H(n-1)$ is true. Consider β such that $|\beta| = n$. We use the Faa di Bruno formula to compute $\partial_\xi^\beta \tilde{z}(s)$, and we obtain

$$\begin{aligned}
&\partial_\xi^\beta \tilde{z}(s) + \textstyle\int_{t_1}^{t_2} \zeta_{t_1,s}(u) \nabla_{\tilde{y}} F(u, \tilde{y}(u)) \partial_\xi^\beta \tilde{z}(u) \mathrm{d}u \\
&= \textstyle\int_{t_1}^{t_2} \zeta_{t_1,s}(u) \nabla_{\tilde{y}} F(u, \tilde{y}(u)) \partial_\xi^\beta (x + (u - t_1)\xi) \mathrm{d}u \\
&- \textstyle\sum_{q \neq 1} \int_{t_1}^{t_2} \zeta_{t_1,s}(u) \nabla_{\tilde{y}}^q F(u, \tilde{y}(u)) \partial_\xi^{\delta_1} \tilde{y}(u) \cdots \partial_\xi^{\delta_q} \tilde{y}(u) \mathrm{d}u.
\end{aligned} \tag{3.4.14}$$

(3.4.14) can be rewritten as

$$\partial_\xi^\beta \tilde{z} - \nabla_{\tilde{z}} \mathcal{P}(\tilde{z}) \partial_\xi^\beta \tilde{z} = \sum g_\delta, \tag{3.4.15}$$

where the map $\mathcal{P}$ was introduced in the proof of Theorem 1.5.1. The induction hypothesis $H(n-1)$ implies

$$|\partial_\xi^\delta \tilde{y}(u)| \leq C|u - t_1|\langle t_2\rangle^{\kappa(|\delta|-1)}, \quad 1 \leq |\delta| \leq n-1.$$

Therefore,

$$\begin{aligned} |g_\delta(s)| &\leq C\langle t_2\rangle^{\kappa(|\delta_1|-1)+\dots+\kappa(|\delta_q|-1)} \int_{t_1}^{t_2} \zeta_{t_1,s}(u) \|\nabla_{\tilde{y}}^q F(u,\cdot)\|_\infty \langle u\rangle^q \mathrm{d}u \\ &\leq C\langle t_2\rangle^{\kappa(q-1)+\kappa(|\delta_1|-1)+\dots+\kappa(|\delta_q|-1)} |s-t_1| \int_{t_1}^{t_2} \|\nabla_{\tilde{y}}^q F(u,\cdot)\|_\infty \langle u\rangle^{q-\kappa(q-1)} \mathrm{d}u \\ &\in C\langle t_2\rangle^{\kappa(|\beta|-1)} |s-t_1| o({t_1}^0). \end{aligned}$$

Moreover, we know that $(1 - \nabla_{\tilde{z}}\mathcal{P}(\tilde{z}))$ is uniformly invertible on $Z_{t_1}^1$ for $T \leq t_1 \leq t_2 \leq \infty$. Therefore, we can use the identity

$$\partial_\xi^\beta \tilde{z} = (1 - \nabla_{\tilde{z}}\mathcal{P}(\tilde{z}))^{-1} \sum g_\delta$$

to show that (3.4.13) is true. □

From now on we assume that the force is conservative and $F(t,x) = -\nabla_x V(t,x)$. Recall that the functions $\tilde{y}(s,t_1,t_2,x,\xi)$ are used to define the function $S(t,\xi)$, which is the solution of the problem

$$\begin{cases} \partial_t S(t,\xi) = \frac{1}{2}\xi^2 + V(t, \nabla_\xi S(t,\xi)), \\ S(T,\xi) = 0. \end{cases} \tag{3.4.16}$$

Below we give estimates on $S(t,\xi)$ that follow from Proposition 3.4.3.

Corollary 3.4.4
Suppose that

$$\begin{aligned} &\textstyle\int_0^\infty \|\partial_x^\alpha V(t,\cdot)\|_\infty \langle t\rangle^{|\alpha|-1} \mathrm{d}t < \infty, \quad |\alpha| = 1,2, \\ &\textstyle\int_0^\infty \|\partial_x^\alpha V(t,\cdot)\|_\infty \langle t\rangle^{|\alpha|-1-\kappa(|\alpha|-2)} \mathrm{d}t < \infty, \quad |\alpha| \geq 2. \end{aligned} \tag{3.4.17}$$

Then

$$\begin{aligned} &\partial_\xi^\beta \left(S(t,\xi) - \tfrac{1}{2}t\xi^2\right) \in o(t), \quad |\beta| = 1,2, \\ &\partial_\xi^\beta \left(S(t,\xi) - \tfrac{1}{2}t\xi^2\right) \in o(t^{1+\kappa(|\beta|-2)}), \quad |\beta| \geq 2. \end{aligned} \tag{3.4.18}$$

Below we will show how we can change the splitting of the potential into a slow-decaying and a fast-decaying part such that the results of Proposition 3.4.3 will be applicable to $V_1(t,x)$.

Lemma 3.4.5
(i) Suppose that $V_1(t,x)$ satisfies

$$\int_0^\infty \langle t\rangle^{1/2}\|\partial_x^\alpha V_1(t,\cdot)\|_\infty \mathrm{d}t < \infty,\ |\alpha| = 1.$$

Then there exists a splitting

$$V_1(t,x) = \tilde{V}_s(t,x) + \tilde{V}_1(t,x)$$

such that

$$\textstyle\int_0^\infty \|\tilde{V}_s(t,\cdot)\|_\infty \mathrm{d}t < \infty,$$

$$\textstyle\int_0^\infty \langle t\rangle^{\frac{1}{2}|\alpha|}\|\partial_x^\alpha \tilde{V}_1(t,\cdot)\|_\infty \mathrm{d}t < \infty,\quad |\alpha| \geq 1.$$

(ii) Suppose that $V_1(t,x)$ satisfies

$$\int_0^\infty \langle t\rangle^{|\alpha|-1}\|\partial_x^\alpha V_1(t,\cdot)\|_\infty \mathrm{d}t < \infty,\ |\alpha| = 1,2.$$

Then there exists a splitting

$$V_1(t,x) = \tilde{V}_s(t,x) + \tilde{V}_1(t,x)$$

such that

$$\int_0^\infty \|\tilde{V}_s(t,\cdot)\|_\infty \mathrm{d}t < \infty, \tag{3.4.19}$$

$$\int_0^\infty \|\partial_x^\alpha \tilde{V}_1(t,\cdot)\|_\infty \langle t\rangle^{|\alpha|-1}\mathrm{d}t < \infty,\quad |\alpha| = 1,2 \tag{3.4.20}$$

$$\int_0^\infty \|\partial_x^\alpha \tilde{V}_1(t,\cdot)\|_\infty \langle t\rangle^{\frac{1}{2}|\alpha|}\mathrm{d}t < \infty,\quad |\alpha| \geq 2. \tag{3.4.21}$$

(iii) Suppose that $V_1(t,x)$ satisfies

$$\int_0^\infty \langle t\rangle^{|\alpha|-1}\|\partial_x^\alpha V_1(t,\cdot)\|_\infty \mathrm{d}t < \infty,\ |\alpha| = 1,2,3.$$

Then, in addition to (3.4.19), the potential $\tilde{V}_s(t,x)$ satisfies

$$\int_0^\infty \langle t\rangle\|\partial_x^\alpha \tilde{V}_s(t,\cdot)\|_\infty \mathrm{d}t < \infty,\quad |\alpha| = 1. \tag{3.4.22}$$

Proof. Consider first *(i)*. Choose $\jmath \in C_0^\infty(X)$ such that $\int \jmath(x)\mathrm{d}x = 1$. Set

$$\begin{aligned}\tilde{V}_1(t,x) &:= \textstyle\int V_1(t,x+t^{\frac{1}{2}}y)\jmath(y)\mathrm{d}y,\\ \tilde{V}_s(t,x) &:= V_1(t,x) - \tilde{V}_1(t,x).\end{aligned}$$

Now,

$$\begin{aligned}\tilde{V}_s(t,x) &= \textstyle\int (V_1(t,x) - V_1(t,x+t^{\frac{1}{2}}y))\mathrm{d}y\\ &= \textstyle\int\int_0^1 \nabla V_1(t,x+t^{\frac{1}{2}}y)t^{\frac{1}{2}}y\jmath(y)\mathrm{d}\tau\mathrm{d}y.\end{aligned}$$

Moreover,

$$\nabla_x^k \tilde{V}_l(t,x) = (-1)^{k-1} t^{-\frac{1}{2}(k-1)} \int \nabla V_l(t, x + t^{\frac{1}{2}} y) \nabla_y^{k-1} j(y) \mathrm{d}y.$$

Let us now prove *(ii)*. This time we assume additionally that $\int j(x) x \mathrm{d}x = 0$. We define $\tilde{V}_s(t,x)$ and $\tilde{V}_l(t,x)$ as above. Now,

$$\tilde{V}_s(t,x) = \int \nabla_x V_l(t,x) t^{\frac{1}{2}} y j(y) \mathrm{d}y + O(t^{-1}) \|\nabla_x^2 V_l(t,.)\|_\infty, \tag{3.4.23}$$

The first term on the right-hand side of (3.4.23) is zero. Hence (3.4.19) is true.
Moreover,

$$\nabla_x^k \tilde{V}_l(t,x) = (-1)^{k-2} t^{-\frac{1}{2}(k-2)} \int \nabla^2 V_l(t, x + t^{\frac{1}{2}} y) \nabla_y^{k-2} j(y) \mathrm{d}y.$$

This implies (3.4.21).
The proof of *(iii)* is similar. □

Corollary 3.4.6
Suppose that we are given a potential $V(t,x)$ satisfying the assumptions of Theorem 3.4.1. Then we can introduce a new splitting

$$V(t,x) = V_s(t,x) + V_l(t,x)$$

such that

$$\int_0^\infty \|V_s(t,\cdot)\|_\infty \mathrm{d}t < \infty,$$
$$\int_0^\infty \|\partial_x^\alpha V_l(t,\cdot)\|_\infty \langle t\rangle^{|\alpha|-1} \mathrm{d}t < \infty, \quad |\alpha| = 1,2$$
$$\int_0^\infty \|\partial_x^\alpha V_l(t,\cdot)\|_\infty \langle t\rangle^{\frac{1}{2}|\alpha|} \mathrm{d}t < \infty, \quad |\alpha| \geq 2.$$

For T big enough, let $S(t,\xi)$ be the solution of the Hamilton-Jacobi equation (3.4.16) with this new $V_l(t,x)$. Then

$$\begin{aligned} &\partial_\xi^\beta \left(S(t,\xi) - \tfrac{1}{2} t\xi^2 \right) \in o(t), \quad |\beta| = 1,2, \\ &\partial_\xi^\beta \left(S(t,\xi) - \tfrac{1}{2} t\xi^2 \right) \in o(t^{\frac{1}{2}|\beta|}), \quad |\beta| \geq 2. \end{aligned} \tag{3.4.24}$$

Finally, if we set

$$g(t,x,\xi) := \int_0^1 F_l(t, \tau x + (1-\tau)\nabla_\xi S(t,\xi)) \mathrm{d}\tau,$$

where $F_l(t,x) = -\nabla_x V_l(t,x)$, then

$$\begin{aligned} &\int_0^\infty \|g(t,\cdot,\cdot)\|_\infty \mathrm{d}t < \infty, \\ &\int_0^\infty \|\partial_x^\alpha \partial_\xi^\beta g(t,\cdot,\cdot)\|_\infty \langle t\rangle^{\frac{1}{2}(1+|\alpha|-|\beta|)} \mathrm{d}t < \infty, \quad |\alpha| + |\beta| \geq 1. \end{aligned} \tag{3.4.25}$$

Proof. Lemma 3.4.5 *(ii)* implies immediately that we can change the splitting of $V(t,x)$.

The estimates on $S(t,\xi)$ follow immediately from Corollary 3.4.4 with $\kappa(n) = n/2$.

It remains to show the estimates on $g(t,x,\xi)$. By the Faa di Bruno formula,

$$\begin{aligned}
&\partial_x^\alpha \partial_\xi^\beta F_1(t, \tau x + (1-\tau)\nabla_\xi S(t,\xi)) \\
&= \sum \partial_x^\alpha \nabla_x^q F_1(t, \tau x + (1-\tau)\nabla_\xi S(t,\xi))\partial_\xi^{\delta_1}\nabla_\xi S(t,\xi)\cdots\partial_\xi^{\delta_q}\nabla_\xi S(t,\xi) \\
&= \sum f_\delta(t,x,\xi).
\end{aligned}$$

Now,

$$\begin{aligned}
|f_\delta(t,x,\xi)| &\le C\|\nabla_x^{|\alpha|+q+1} V_1(t,\cdot)\|_\infty \langle t\rangle^{\frac{1}{2}(|\delta_1|+1)+\cdots+\frac{1}{2}(|\delta_q|+1)} \\
&= C\|\nabla_x^{|\alpha|+q+1} V_1(t,\cdot)\|_\infty \langle t\rangle^{\frac{1}{2}(|\beta|+q)}.
\end{aligned}$$

This implies (3.4.25). □

Now let

$$r(t,x,\xi) := \mathrm{div}_\xi g(t,x,\xi).$$

Set

$$G(t) := g(t,x,D), \qquad R(t) := r(t,x,D).$$

Lemma 3.4.7
One has

$$V_1(t,x) - V_1(t,\nabla_\xi S(t,D)) = G(t)(x - \nabla_\xi S(t,D)) + R(t). \tag{3.4.26}$$

Moreover

$$\|G(t)\|,\ \|[x,G(t)]\|,\ \|[\nabla_\xi S(t,D),G(t)]\|,\ \|R(t)\| \in L^1(\mathrm{d}t).$$

Proof. First note that

$$\begin{aligned}
V_1(t,x) - V_1(t,\nabla_\xi S(t,D)) &= xG(t) - G(t)\nabla_\xi S(t,D) \\
&= G(t)(x - \nabla_\xi S(t,D)) + \textstyle\sum_{i=1}^n [x_i, G_i(t)].
\end{aligned}$$

This implies (3.4.26).

Then we note that

$$g(t,x,\xi), \nabla_\xi g(t,x,\xi) \in L^1(\mathrm{d}t, S(\langle t\rangle^{-1}\mathrm{d}x^2 + \langle t\rangle \mathrm{d}\xi^2)),$$

where we used the notations of Appendix D.1. Hence

$$G(t), R(t), [x,G(t)] \in L^1(\mathrm{d}t, \Psi(\langle t\rangle^{-1}\mathrm{d}x^2 + \langle t\rangle \mathrm{d}\xi^2)).$$

Finally, note that

$$\nabla_x g(t,x,\xi) \in L^1(\langle t\rangle \mathrm{d}t, S(\langle t\rangle^{-1}\mathrm{d}x^2 + \langle t\rangle \mathrm{d}\xi^2)),$$

$$\nabla_\xi^2 S(t,\xi) \in S(\langle t\rangle, \langle t\rangle^{-1}\mathrm{d}x^2 + \langle t\rangle \mathrm{d}\xi^2).$$

Hence, by Proposition D.5.2,

$$[\nabla_\xi S(t,D), G(t)] \in L^1(\mathrm{d}t, \Psi(\langle t\rangle^{-1}\mathrm{d}x^2 + \langle t\rangle \mathrm{d}\xi^2)).$$

□

We define

$$\begin{aligned} H_1(t) &:= \tfrac{1}{2}D^2 + V_1(t,x), \\ \tilde{H}_1(t) &:= \tfrac{1}{2}D^2 + V_1(t,x) + R(t) \\ &= \tfrac{1}{2}D^2 + V_1(t, \nabla_\xi S(t,D)) + G(t)(x - \nabla_\xi S(t,D)), \\ \tilde{\mathbf{D}}_1 &:= \tfrac{\mathrm{d}}{\mathrm{d}t} + i[\tilde{H}_1(t), \cdot]. \end{aligned}$$

We also define $U_1(t,s)$ to be the dynamics generated by $H_1(t)$.

Note that the operator $\tilde{H}_1(t)$ is not, in general, self-adjoint.

Lemma 3.4.8
There exists a two-parameter family of uniformly bounded operators $\tilde{U}_1(t,s)$ defined by

$$\begin{cases} i\partial_t \tilde{U}_1(t,s) = \tilde{H}_1(t)\tilde{U}_1(t,s) \\ \tilde{U}_1(t,t) = 1. \end{cases}$$

We also have

$$i\partial_s \tilde{U}_1(t,s) = \tilde{U}_1(t,s)\tilde{H}_1(s).$$

Proof. Let

$$W(t,s) := U_1(s,t)\tilde{U}_1(t,s).$$

It satisfies

$$\begin{cases} i\partial_t W(t,s) = Z(t,s)W(t,s), \\ W(t,t) = 1, \end{cases}$$

where $Z(t,s) := U_1(s,t)R(t)U_1(t,s)$. We observe that $\|Z(\cdot,s)\| \in L^1(\mathrm{d}t)$, which by Proposition B.3.6 implies the lemma. □

Lemma 3.4.9

$$(x - \nabla_\xi S(t,D))\tilde{U}_1(t,T)\langle x\rangle^{-1} \tag{3.4.27}$$

is uniformly bounded.

Proof. We compute:

$$\begin{aligned}\tilde{\mathbf{D}}_1(x-\nabla_\xi S(t,D)) &= D+\nabla_x V_1(t,\nabla_\xi S(t,D))\nabla_\xi^2 S(t,D)-\partial_t\nabla_\xi S(t,D)\\ &\quad +i[G(t),(x-\nabla_\xi S(t,D)](x-\nabla_\xi S(t,D))\\ &= i[G(t),(x-\nabla_\xi S(t,D)](x-\nabla_\xi S(t,D)).\end{aligned}$$

Let

$$f(t) := \|\tilde{U}_1(T,t)(x-\nabla_\xi S(t,D))\tilde{U}_1(t,T)\langle x\rangle^{-1}\|.$$

Then

$$\begin{aligned}\tfrac{\mathrm{d}}{\mathrm{d}t}f(t) &\le \|\tfrac{\mathrm{d}}{\mathrm{d}t}\tilde{U}_1(T,t)(x-\nabla_\xi S(t,D))\tilde{U}_1(t,T)\langle x\rangle^{-1}\|\\ &= \|\tilde{U}_1(T,t)(\tilde{\mathbf{D}}_1(x-\nabla_\xi S(t,D)))\tilde{U}_1(t,T)\langle x\rangle^{-1}\| \le g(t)f(t),\end{aligned}$$

where

$$g(t) := \|\tilde{U}_1(T,t)[(x-\nabla_\xi S(t,D),G(t)]\tilde{U}_1(t,T)\|$$

is integrable by Lemma 3.4.7. Therefore, by the Gronwall inequality

$$f(t) \le Cf(T).$$

But $f(T) = \|x\langle x\rangle^{-1}\|$ is bounded. This completes the proof of the lemma. □

Proof of Theorem 3.4.1. We observe that, by the arguments of Theorem 3.3.1, the norm limit

$$\lim_{t\to\infty} U_1(T,t)U(t,T)$$

exists. Moreover,

$$\|H_1(t)-\tilde{H}_1(t)\| = \|R(t)\| \in L^1(\mathrm{d}t),$$

which implies that the norm limit

$$\lim_{t\to\infty} U_1(T,t)\tilde{U}_1(t,T)$$

exist. By the chain rule of wave operators, it suffices to prove that the limits

$$\text{s-}\lim_{t\to\infty} \tilde{U}_1(T,t)\mathrm{e}^{-iS(t,D)}, \tag{3.4.28}$$

$$\text{s-}\lim_{t\to\infty} \mathrm{e}^{iS(t,D)}\tilde{U}_1(t,T) \tag{3.4.29}$$

exist.

Next we observe that

$$\mathrm{e}^{iS(t,D)}(x-\nabla_\xi S(t,D))\mathrm{e}^{-iS(t,D)} = x,$$

which implies

$$\|(x-\nabla_\xi S(t,D))\mathrm{e}^{-iS(t,D)}\langle x\rangle^{-1}\| \le C, \quad t \ge T. \tag{3.4.30}$$

We have, for $\phi \in \mathcal{D}(\langle x\rangle)$,

$$\frac{\mathrm{d}}{\mathrm{d}t}\tilde{U}_{\mathrm{l}}(T,t)\mathrm{e}^{-iS(t,D)}\phi = \tilde{U}_{\mathrm{l}}(T,t)G(t)(x - \nabla_\xi S(t,D))\mathrm{e}^{-iS(t,D)}\phi$$

This is integrable by Lemmas 3.4.8 and 3.4.7, and by (3.4.30), which proves that the limit (3.4.28) exists.

Similarly, for $\phi \in \mathcal{D}(\langle x\rangle)$ by Lemmas 3.4.8, 3.4.7 and 3.4.9

$$\tfrac{\mathrm{d}}{\mathrm{d}t}\mathrm{e}^{iS(t,D)}\tilde{U}_{\mathrm{l}}(t,T)\phi = -\mathrm{e}^{iS(t,D)}G(t)(x - \nabla_\xi S(t,D))\tilde{U}_{\mathrm{l}}(t,T)\phi$$

is integrable, which proves the existence of (3.4.29).

The proof of (3.4.4) is immediate. This completes the proof of the theorem. □

3.5 Slow-Decaying Case – Smooth Potentials

In this section we give an independent treatment of the topics discussed in Sect. 3.4 if the slow-decaying part of the potentials satisfies the smooth slow-decaying condition. For potentials of this class, one can avoid some of the technicalities of the previous section and give a simpler proof of the existence and completeness of wave operators. The main result of this section is the following analog of Theorem 3.4.1.

Theorem 3.5.1
Assume that

$$V(t,x) = V_{\mathrm{s}}(t,x) + V_{\mathrm{l}}(t,x)$$

such that

$$\begin{aligned} &\textstyle\int_0^\infty \|V_{\mathrm{s}}(t,\cdot)\|_\infty \mathrm{d}t < \infty, \\ &\textstyle\int_0^\infty \langle t\rangle^{|\alpha|-1}\|\partial_x^\alpha V_{\mathrm{l}}(t,\cdot)\|_\infty \mathrm{d}t < \infty, \quad |\alpha| \geq 1. \end{aligned} \tag{3.5.1}$$

For T big enough, let $S(t,\xi)$ be the solution of the problem

$$\begin{cases} \partial_t S(t,\xi) = \frac{1}{2}\xi^2 + V_{\mathrm{l}}(t, \nabla_\xi S(t,\xi)), \\ S(T,\xi) = 0. \end{cases} \tag{3.5.2}$$

Then the limits

$$\mathrm{s-}\lim_{t\to\infty} U(0,t)\mathrm{e}^{-iS(t,D)}, \tag{3.5.3}$$

$$\mathrm{s-}\lim_{t\to\infty} \mathrm{e}^{iS(t,D)}U(t,0) \tag{3.5.4}$$

exist. If we denote (3.4.2) by Ω_{sd}^+, then (3.4.3) equals $\Omega_{\mathrm{sd}}^{+*}$. Moreover, Ω_{sd}^+ is unitary and

$$D^+ = \Omega_{\mathrm{sd}}^+ D \Omega_{\mathrm{sd}}^{+*}. \tag{3.5.5}$$

We will use the notation introduced in Appendix D.1. Note that the force $F_l(t,x) = -\nabla_x V_l(t,x)$ can be regarded both as an element of $L^1(\mathrm{d}t, S(g_0(t)))$ and as an element of $L^1(\mathrm{d}t, \Psi(g_0(t)))$.

We set

$$g(t,x,\xi) := \int_0^1 F_l(t, \tau x + (1-\tau)\nabla_\xi S(t,\xi))\mathrm{d}\tau,$$

$$r(t,x,\xi) := \mathrm{div}_\xi g(t,x,\xi).$$

We define

$$G(t) := g(t,x,D), \qquad R(t) := r(t,x,D).$$

Lemma 3.5.2
One has

$$V_l(t,x) - V_l(t, \nabla_\xi S(t,D)) = G(t)(x - \nabla_\xi S(t,D)) + R(t). \tag{3.5.6}$$

Moreover, $G(t)$ *and* $R(t)$ *belong to* $L^1(\mathrm{d}t, \Psi(g_0(t)))$.

Proof. First note that

$$\begin{aligned} V_l(t,x) - V_l(t, \nabla_\xi S(t,D)) &= xG(t) - G(t)\nabla_\xi S(t,D) \\ &= G(t)(x - \nabla_\xi S(t,D)) + \textstyle\sum_{i=1}^n [x_i, G_i(t)]. \end{aligned}$$

Let us show the following estimate on $g(t,x,\xi)$:

$$\int_0^\infty \|\partial_x^\alpha \partial_\xi^\beta g(t,\cdot,\cdot)\|_\infty \langle t\rangle^{|\alpha|}\mathrm{d}t < \infty, \ \alpha, \beta \in \mathbb{N}^n. \tag{3.5.7}$$

Let us recall that, in Proposition 1.10.7, we proved that

$$|\partial_\xi^\beta S(t,\xi)| \leq C_\beta \langle t\rangle, \ \ |\beta| \geq 1. \tag{3.5.8}$$

By the Faa di Bruno formula,

$$\begin{aligned} &\partial_x^\alpha \partial_\xi^\beta F_l(t, \tau x + (1-\tau)\nabla_\xi S(t,\xi)) \\ &= \textstyle\sum \partial_x^\alpha \nabla_x^q F_l(t, \tau x + (1-\tau)\nabla_\xi S(t,\xi)) \partial_\xi^{\delta_1}\nabla_\xi S(t,\xi) \cdots \partial_\xi^{\delta_q}\nabla_\xi S(t,\xi) \\ &= \textstyle\sum f_\delta(t,x,\xi). \end{aligned}$$

We have

$$|f_\delta(t,x,\xi)| \leq C\|\nabla_x^{|\alpha|+q+1} V_l(t,\cdot)\|_\infty \langle t\rangle^q.$$

Therefore, $g(t,x,\xi)$ and $r(t,x,\xi)$ belong to $L^1(\mathrm{d}t, S(g_0(t)))$, which implies that $G(t)$ and $R(t)$ belong to $L^1(\mathrm{d}t, \Psi(g_0(t)))$. □

We define

$$H_l(t) := \tfrac{1}{2}D^2 + V_l(t,x),$$

$$\mathbf{D}_l := \partial_t + i[H_l(t), \cdot].$$

Let $U_l(t,s)$ be the unitary propagator associated with $H_l(t)$.

Lemma 3.5.3

$$(x - \nabla_\xi S(t, D))U_1(t, T)\langle x\rangle^{-1} \tag{3.5.9}$$

is uniformly bounded.

Proof. We compute:

$$\begin{aligned}\mathbf{D}_1(x - \nabla_\xi S(t, D)) &= D + [V_1(t, x), \nabla_\xi S(t, D)] - \partial_t \nabla_\xi S(t, D)\\ &= i[(V_1(t, x) - V_1(t, \nabla_\xi S(t, D)), (x - \nabla_\xi S(t, D)]\\ &= i[G(t), (x - \nabla_\xi S(t, D)](x - \nabla_\xi S(t, D))\\ &+i[R(t), (x - \nabla_\xi S(t, D)].\end{aligned}$$

Let

$$f(t) := \|U_1(T, t)(x - \nabla_\xi S(t, D))U_1(t, T)\langle x\rangle^{-1}\|$$

Then

$$\begin{aligned}\tfrac{\mathrm{d}}{\mathrm{d}t} f(t) &\leq \|\tfrac{\mathrm{d}}{\mathrm{d}t} U_1(T, t)(x - \nabla_\xi S(t, D))U_1(t, T)\langle x\rangle^{-1}\|\\ &= \|U_1(T, t)(\mathbf{D}_1(x - \nabla_\xi S(t, D)))U_1(t, T)\langle x\rangle^{-1}\|\\ &\leq g(t)f(t) + h(t),\end{aligned}$$

where the functions

$$g(t) := \|U_1(T, t)[(x - \nabla_\xi S(t, D), G(t)]U_1(t, T)\|$$

$$h(t) := \|U_1(T, t)[(x - \nabla_\xi S(t, D), R(t)]U_1(t, T)\langle x\rangle^{-1}\|$$

are integrable by Lemma 3.5.2. Therefore, by the Gronwall inequality,

$$f(t) \leq C(1 + f(T))$$

is bounded. □

Proof of Theorem 3.5.1. We start with the proof of the existence of (3.5.3) and (3.5.4). We first observe that, by the arguments of Theorem 3.3.1, the norm limit

$$\lim_{t\to\infty} U_1(T, t)U(t, 0)$$

exists. So, by the chain rule of wave operators, it suffices to prove that the limits

$$\mathrm{s-}\lim_{t\to\infty} U_1(T, t)\mathrm{e}^{-iS(t,D)}, \tag{3.5.10}$$

$$\mathrm{s-}\lim_{t\to\infty} \mathrm{e}^{iS(t,D)}U_1(t, T) \tag{3.5.11}$$

exist.

We first observe that

$$\mathrm{e}^{iS(t,D)}(x - \nabla_\xi S(t, D))\mathrm{e}^{-iS(t,D)} = x,$$

which implies

$$\|(x - \nabla_\xi S(t, D))\mathrm{e}^{-iS(t,D)}\langle x\rangle^{-1}\| \le C, \quad t \ge T. \tag{3.5.12}$$

Let $\phi \in \mathcal{D}(\langle x\rangle)$. By (3.5.12) and Lemma 3.5.2,

$$\begin{aligned}\tfrac{\mathrm{d}}{\mathrm{d}t}U_1(T,t)\mathrm{e}^{-iS(t,D)}\phi &= U_1(T,t)(V_1(t,x) - V_1(t, \nabla_\xi S(t,D)))\mathrm{e}^{-iS(t,D)}\phi \\ &= U_1(T,t)G(t)(x - \nabla_\xi S(t,D))\mathrm{e}^{-iS(t,D)}\phi \\ &+U_1(T,t)R(t)\mathrm{e}^{-iS(t,D)}\phi\end{aligned}$$

is integrable, which proves that the limit (3.5.10) exists.

Similarly, for $\phi \in \mathcal{D}(\langle x\rangle)$, by Lemmas 3.5.3 and 3.5.2,

$$\begin{aligned}\tfrac{\mathrm{d}}{\mathrm{d}t}\mathrm{e}^{iS(t,D)}U_1(t,T)\phi &= -\mathrm{e}^{iS(t,D)}G(t)(x - \nabla_\xi S(t,D))U_1(t,T)\phi \\ &-\mathrm{e}^{iS(t,D)}R(t)U_1(t,T)\phi\end{aligned}$$

is integrable, which proves the existence of (3.5.11).

The proof of (3.5.5) is immediate. This completes the proof of the theorem. □

3.6 Dollard Wave Operators

In this section we would like to describe the construction of wave operators using the so-called Dollard dynamics. To use this construction, roughly speaking, one has to demand that the potentials decay as $|\nabla_x V(t,x)| \le C\langle t\rangle^{-1-\mu}$ for $\mu > 1/2$, which is a severe restriction as compared with the results of Sects. 3.4 and 3.5, where we used solutions of the Hamilton-Jacobi equation to define modified free dynamics. Nevertheless, Dollard wave operators have big advantages. First of all, their construction is very simple. Secondly, they easily allow to handle the case of additional degrees of freedom.

Assume that our system possesses some "internal degrees of freedom" (e.g. spin). More precisely, let us assume that the Hilbert space of our system is $L^2(X) \otimes \mathcal{H}_1$ where $\mathcal{H}_1$ is a certain auxiliary Hilbert space. Suppose that the time-dependent Hamiltonian has the form

$$\begin{aligned}&H(t) := \tfrac{1}{2}D^2 \otimes 1_{\mathcal{H}_1} + V(t,x), \\ \text{where} \quad &V^*(t,x) = V(t,x) \in B(\mathcal{H}_1)\end{aligned}$$

and the Hamiltonian $H(t)$ generates a flow $U(t,s)$ in the sense described in Sect. B.3.

Scattering theory for such $H(t)$ in the fast-decaying case is completely analogous to what we described in Sect. 3.3. In the slow-decaying case, however, if the potential couples the internal degrees of freedom in a nontrivial way, the constructions of Sects. 3.4 and 3.5 do not go through. In fact, we cannot even write the Hamilton-Jacobi equation. Nevertheless, it turns out that the so-called Dollard modified wave operators, which we are going to present in the following theorem, work in the case of internal degrees of freedom.

We will assume that the time-dependent potential $V(t,x)$ is equal to

$$V(t,x) = V_s(t,x) + V_l(t,x),$$

where, for almost all (t,x), the operators $V_s(t,x)$ and $V_l(t,x)$ are self-adjoint on $B(\mathcal{H}_1)$,

$$\begin{aligned} &\int_0^\infty \sup_{x\in X} \|V_s(t,x)\|_{B(\mathcal{H}_1)} \mathrm{d}t < \infty, \\ &\int_0^\infty \langle t\rangle^{\frac{1}{2}} \sup_{x\in X} \|\partial_x^\alpha V_l(t,x)\|_{B(\mathcal{H}_1)} \mathrm{d}t < \infty, \quad |\alpha| = 1. \end{aligned} \tag{3.6.1}$$

By Lemma 3.4.5 *(i)*, we can change the splitting of $V(t,x)$ so that

$$\begin{aligned} &\int_0^\infty \sup_{x\in X} \|V_s(t,x)\|_{B(\mathcal{H}_1)} \mathrm{d}t < \infty, \\ &\int_0^\infty \langle t\rangle^{\frac{1}{2}} \sup_{x\in X} \|\partial_x^\alpha V_l(t,x)\|_{B(\mathcal{H}_1)} \mathrm{d}t < \infty, \quad |\alpha| = 1, \\ &\int_0^\infty \langle t\rangle \sup_{x\in X} \|\partial_x^\alpha V_l(t,x)\|_{B(\mathcal{H}_1)} \mathrm{d}t < \infty, \quad |\alpha| = 2. \end{aligned} \tag{3.6.2}$$

Now we can introduce the Dollard modified dynamics. For $\xi \in X'$, we denote by

$$T\left(\mathrm{e}^{-i\int_0^t V_l(s,s\xi)\mathrm{d}s}\right)$$

the unitary propagator on $\mathcal{H}_1$ for the time-dependent Hamiltonian $V(t,t\xi)$. The symbol T stands for "time-ordering" and is defined in Definition B.3.5.

Definition 3.6.1
We define the Dollard modified dynamics $U_\mathrm{D}(t)$ by

$$U_\mathrm{D}(t) := \mathrm{e}^{-i\frac{1}{2}tD^2} T\left(\mathrm{e}^{-i\int_0^t V_l(s,sD)\mathrm{d}s}\right).$$

The modified free dynamics $U_\mathrm{D}(t)$ was essentially first introduced by Dollard [Do1].

We have the following result:

Theorem 3.6.2
Under the above conditions, the limits

$$\mathrm{s-}\lim_{t\to\infty} U(0,t)U_\mathrm{D}(t) =: \Omega_\mathrm{D}^+, \tag{3.6.3}$$

$$\mathrm{s-}\lim_{t\to\infty} U_\mathrm{D}(t)^* U(t,0) \tag{3.6.4}$$

exist and the limit in (3.6.4) is equal to Ω_D^{+*}. Moreover, Ω_D^+ is unitary and

$$D^+ = \Omega_{\rm D}^+ D \Omega_{\rm D}^{+*}.$$

Proof. By the chain rule of wave operators and the arguments used in the proof of Theorem 3.3.1, we may replace in $H(t)$ the potential $V_{\rm s}(t,x)$ by 0, and assume that $U(t,0) = U_{\rm l}(t,0)$.

We first claim that

$$(x - tD)U(t,0)\langle x\rangle^{-1} \in O(t^{\frac{1}{2}}), \tag{3.6.5}$$

$$(x - tD)U_{\rm D}(t)\langle x\rangle^{-1} \in O(t^{\frac{1}{2}}). \tag{3.6.6}$$

Indeed, we compute

$$\begin{aligned} &\tfrac{\rm d}{{\rm d}t}U(0,t)(x-tD)U(t,0)\langle x\rangle^{-1} \\ &= U(0,t)t\nabla_x V(t,x)U(t,0)\langle x\rangle^{-1} \in \langle t\rangle^{\frac{1}{2}}L^1({\rm d}t), \end{aligned}$$

which proves (3.6.5) by integration from 0 to t. Similarly

$$\begin{aligned} &\tfrac{\rm d}{{\rm d}t}U_{\rm D}(t)^*(x-tD)U_{\rm D}(t)\langle x\rangle^{-1} \\ &= U_{\rm D}(t)^* t\nabla_x V(t,tD)U_{\rm D}(t)\langle x\rangle^{-1} \in \langle t\rangle^{\frac{1}{2}}L^1({\rm d}t), \end{aligned}$$

which proves (3.6.6).

Let us now prove the existence of the limit (3.6.4). For $\phi \in \mathcal{D}(\langle x\rangle)$, we compute

$$\tfrac{\rm d}{{\rm d}t}U_{\rm D}(t)^*U(t,0)\phi = iU_{\rm D}(t)^*(V_{\rm l}(t,tD) - V_{\rm l}(t,x))U(t,0)\phi. \tag{3.6.7}$$

Now let $\phi \in \mathcal{D}(\langle x\rangle)$. Using the Baker-Campbell-Hausdorff formula (3.2.28) applied to $V(t,x)$ and the estimates (3.6.2), (3.6.5), we obtain that

$$\begin{aligned} &\|U_{\rm D}(t)^*(V_{\rm l}(t,tD) - V_{\rm l}(t,x))U(t,0)\phi\| \\ &\leq C\|\nabla_x V_{\rm l}(t,\cdot)\|_\infty \|(x-tD)U(t,0)\langle x\rangle^{-1}\| \|\langle x\rangle\phi\| + Ct\|\Delta_x V_{\rm l}(t,\cdot)\|_\infty\|\phi\|, \end{aligned}$$

is integrable, which proves the existence of the limit (3.6.4). The proof of the existence of the limit (3.6.3) is analogous except that we use (3.6.6) instead of (3.6.5). □

3.7 Isozaki-Kitada Construction

In this section we introduce another construction of wave operators in the slow-decaying case – a time-dependent version of the one introduced in the time-independent case by Isozaki-Kitada [IK1]. We will prove that, for smooth slow-decaying time-dependent potentials, the two wave operators coincide.

We will assume that the potential satisfies the so-called smooth slow-decaying condition, that is

$$\int_0^\infty \langle t\rangle^{|\alpha|-1}\|\partial_x^\alpha V(t,\cdot)\|_\infty \mathrm{d}t < \infty, \quad |\alpha| \geq 1. \tag{3.7.1}$$

First let us recall some facts from Chap. 1. In Proposition 1.8.3 we constructed a function $\Phi_{\mathrm{sd}}^+(s,x,\xi)$ that, for $s \geq T$, solves the eikonal equation

$$-\partial_s \Phi_{\mathrm{sd}}^+(s,x,\xi) = \frac{1}{2}(\nabla_\xi \Phi_{\mathrm{sd}}^+(s,x,\xi))^2 + V(s,x).$$

By Proposition 1.10.6, the function $\Phi_{\mathrm{sd}}^+(s,x,\xi)$ satisfies the estimates

$$\begin{aligned}
&\partial_x^\alpha \partial_\xi^\beta (\Phi_{\mathrm{sd}}^+(s,x,\xi) - \langle x,\xi\rangle + \tfrac{1}{2}s\xi^2) \in o(s^{1-|\alpha|}), \quad |\alpha| \geq 1, \\
&\partial_x^\alpha \partial_\xi^\beta (\Phi_{\mathrm{sd}}^+(s,x,\xi) - \langle x,\xi\rangle + \tfrac{1}{2}s\xi^2) \in \langle s\rangle^{2-|\alpha|}L^1(\mathrm{d}s), \quad |\alpha| \geq 2.
\end{aligned} \tag{3.7.2}$$

We define next, for $s \geq T$, the following operator:

$$J_{\mathrm{sd}}^+(s)\phi(x) := (2\pi)^{-n}\int\int \mathrm{e}^{i\Phi_{\mathrm{sd}}^+(s,x,\xi)-i\langle y,\xi\rangle}\phi(y)\mathrm{d}y\mathrm{d}\xi. \tag{3.7.3}$$

By Theorem D.13.2 and (3.7.2), the operator $J_{\mathrm{sd}}^+(s)$ is uniformly bounded on $L^2(X)$ for s sufficiently big.

The main result of this section is the following theorem:

Theorem 3.7.1
Assume that the potential $V(t,x)$ satisfies the estimates (3.7.1). Then the following results hold:
(i) the norm limit

$$\lim_{s\to\infty} U(0,s)J_{\mathrm{sd}}^+(s) \tag{3.7.4}$$

exists.
(ii) (3.7.4) is equal to Ω_{sd}^+ defined in Theorem 3.5.1.

To prove part *(ii)* of Theorem 3.7.1, for large s,t, we will construct an approximation of $U(s,t)$ as a Fourier integral operator. A more refined construction will be given in Sect. 3.10. Similar constructions first appeared in [Ki5], where they were used to prove the existence and completeness of wave operators for smooth time-dependent potentials.

In order to construct this approximation, let us recall some constructions of the classical case. In Chap. 1 we introduced the function $S(s,t,x,\xi)$, which is the solution of the Hamilton-Jacobi equation

$$\begin{cases} -\partial_s S(s,t,x,\xi) = \frac{1}{2}(\nabla_x S(s,t,x,\xi))^2 + V(s,x), \\ S(t,t,x,\xi) = \langle x,\xi\rangle. \end{cases}$$

By Proposition 1.10.4, the function $S(s,t,x,\xi)$ satisfies, for $s \leq t$, the estimates

$$
\begin{aligned}
&\partial_x^\alpha\partial_\xi^\beta(S(s,t,x,\xi)-\langle x,\xi\rangle)\in o(s^{1-|\alpha|}),\quad |\alpha|\geq 1,\\
&\textstyle\int_T^\infty \sup_{\{(t,x,\xi)\ |\ s\leq t\}} |\partial_x^\alpha\partial_\xi^\beta(S(s,t,x,\xi)-\langle x,\xi\rangle)|s^{|\alpha|-2}\mathrm{d}s<\infty.\ |\alpha|\geq 2.
\end{aligned}
\tag{3.7.5}
$$

For $T\leq s\leq t<\infty$, we denote by $I(s,t)$ the operator

$$
I(s,t)\phi(x) \ := \ (2\pi)^{-n}\int \mathrm{e}^{iS(s,t,x,\xi)-i\langle y,\xi\rangle}\phi(y)\mathrm{d}y\mathrm{d}\xi. \tag{3.7.6}
$$

By Theorem D.13.2 and (3.7.6), the operator $I(s,t)$ is uniformly bounded on $L^2(X)$ for $t\geq s$ sufficiently big.

Recall from Proposition 1.8.3 that the functions $S(s,\xi)$ and $\Phi_{\mathrm{sd}}^+(s,x,\xi)$ are related by the identity

$$
\Phi_{\mathrm{sd}}^+(s,x,\xi)=\lim_{t\to\infty}(S(s,t,x,\xi)-S(t,\xi)). \tag{3.7.7}
$$

Proposition 3.7.2
Assume that the potential satisfies (3.7.1). Then

$$
\sup_{s\leq t}\|U(t,s)-I(t,s)\|\in o(s^0). \tag{3.7.8}
$$

Proof. We compute

$$
\begin{aligned}
\tfrac{\mathrm{d}}{\mathrm{d}s}U(t,s)I(s,t) &= U(t,s)(iH(s)+\partial_s)I(s,t)\\
&= U(t,s)P(s,t),
\end{aligned}
\tag{3.7.9}
$$

where $P(t,s)$ is the operator defined as

$$
P(t,s)\phi(x):=(2\pi)^{-n}\int\int \mathrm{e}^{iS(s,t,x,\xi)-i\langle y,\xi\rangle}\Delta_x S(s,t,x,\xi)\phi(y)\mathrm{d}\xi\mathrm{d}y. \tag{3.7.10}
$$

Note that

$$
I(t,t)=U(t,t)=1.
$$

Therefore

$$
I(s,t)-U(s,t)=-\int_s^t U(s,u)P(u,t)\mathrm{d}u,
$$

and

$$
\sup_{\{t\ |\ s\leq t\}}\|I(s,t)-U(s,t)\|\leq\int_s^\infty \sup_{\{t'\ |\ u\leq t'\leq\infty\}}\|P(u,t')\|\mathrm{d}u. \tag{3.7.11}
$$

Now (3.7.5) implies that, for $T_0\leq s\leq t$,

$$
\begin{aligned}
&\textstyle\int_T^\infty \sup_{\{t\ |\ s\leq t\leq\infty\}}\|\partial_x^\alpha\partial_\xi^\beta\Delta_x S(s,t,\cdot,\cdot)\|_\infty\mathrm{d}s<\infty,\\
&|\partial_x^\alpha\partial_\xi^\beta S(s,t,x,\xi)|\leq C_{\alpha,\beta},\ |\alpha|\geq 1,\\
&|\nabla_x\nabla_\xi S(s,t,x,\xi)-1|\leq \tfrac{1}{2}.
\end{aligned}
\tag{3.7.12}
$$

From the estimates (3.7.12) and Theorem D.13.2, we infer that the right-hand side of (3.7.11) is $o(s^0)$. □

Proof of Theorem 3.7.1. Let us now prove *(i)*. We compute the derivative

$$\frac{\mathrm{d}}{\mathrm{d}s}U(0,s)J_{\mathrm{sd}}^{+}(s) = U(0,s)Q_{\mathrm{sd}}^{+}(s), \tag{3.7.13}$$

where

$$Q_{\mathrm{sd}}^{+}(s)\phi(y) := (2\pi)^{-n}\int\int \Delta_x \Phi_{\mathrm{sd}}^{+}(s,x,\xi)\mathrm{e}^{i\Phi_{\mathrm{sd}}^{+}(s,x,\xi)-i\langle y,\xi\rangle}\phi(y)\mathrm{d}\xi\mathrm{d}y. \tag{3.7.14}$$

We have the following estimates for $s \geq T_0$:

$$\begin{aligned}
&\partial_x^\alpha \partial_\xi^\beta \Delta_x \Phi_{\mathrm{sd}}^{+} \in L^1(\mathrm{d}s),\\
&|\partial_x^\alpha \partial_\xi^\beta \Phi_{\mathrm{sd}}^{+}(s,x,\xi)| \leq C_{\alpha,\beta},\ |\alpha| \geq 1,\\
&|\nabla_x \nabla_\xi \Phi_{\mathrm{sd}}^{+}(s,x,\xi) - 1| \leq \tfrac{1}{2}.
\end{aligned} \tag{3.7.15}$$

Therefore, by Theorem D.13.2 $\|Q_{\mathrm{sd}}^{+}(s)\|$ is integrable, which proves that the norm limit (3.7.4) exists.

Let us now prove *(ii)*. We first claim that

$$J_{\mathrm{sd}}^{+}(s)^* = \mathrm{s-}\lim_{t\to\infty} \mathrm{e}^{iS(t,D)}I(s,t)^*. \tag{3.7.16}$$

To see this, it is enough to prove that if $\chi \in C_0^\infty(X)$, then

$$\lim_{t\to\infty}(\mathrm{e}^{iS(t,D)}I(s,t)^* - J_{\mathrm{sd}}^{+}(s)^*)\chi(x) = 0. \tag{3.7.17}$$

But

$$\begin{aligned}
&(\mathrm{e}^{iS(t,D)}I(s,t)^* - J_{\mathrm{sd}}^{+}(s)^*)\chi(x)\phi(x)\\
&= (2\pi)^{-n}\textstyle\int\int \mathrm{e}^{-i\Phi_{\mathrm{sd}}^{+}(s,x,\xi)+i\langle\xi,y\rangle}b(s,t,x,\xi)\phi(x)\mathrm{d}\xi\mathrm{d}x,
\end{aligned} \tag{3.7.18}$$

where

$$b(s,t,x,\xi) = \left(\mathrm{e}^{-iS(s,t,x,\xi)+iS(t,\xi)+i\Phi_{\mathrm{sd}}^{+}(s,x,\xi)} - 1\right)\chi(x).$$

By (3.7.7), the amplitude $b(s,t,x,\xi)$ goes uniformly to zero as $t \to \infty$ together with all its derivatives. Therefore (3.7.17) is true.

Let us now fix a vector $\phi \in L^2(X)$. Using first (3.7.16), and then Proposition 3.7.2, we see that, for $s \leq t$, one has

$$\begin{aligned}
J_{\mathrm{sd}}^{+}(s)^*U(s,0)\phi &= \mathrm{e}^{iS(t,D)}I(s,t)^*U(s,0)\phi + o(t^0)\\
&= \mathrm{e}^{iS(t,D)}U(t,s)U(s,0)\phi + o(s^0) + o(t^0).
\end{aligned} \tag{3.7.19}$$

Therefore, if we let $t \to \infty$ in (3.7.19), we obtain

$$J_{\mathrm{sd}}^{+}(s)^*U(s,0)\phi = \Omega_{\mathrm{sd}}^{+*}\phi + o(s^0). \tag{3.7.20}$$

Letting then s go to ∞, we obtain

$$\lim_{s\to\infty} J_{\mathrm{sd}}^{+}(s)^*U(s,0)\phi = \Omega_{\mathrm{sd}}^{+*}\phi,$$

which is the desired result. □

3.8 Counterexamples to Asymptotic Completeness

There exists a large class of time-dependent potentials for which the asymptotic momentum D^+ and the wave operator $\Omega_{\rm fd}^+$ are well defined but asymptotic completeness breaks down, that is, $\mathrm{Ran}\Omega_{\rm fd}^+ \neq L^2(X)$. Moreover, it is possible to construct potentials such that D^+ has some pure point spectrum, hence it is not unitarily equivalent to D. For such potentials, asymptotic completeness fails even if we try to use modified wave operators.

In this section, we construct some potentials with such a property. Such examples were first found by Yafaev [Yaf1]. A related construction giving a sharper class of counterexamples was given by Yajima [Ya1]. The breakdown of asymptotic completeness is related to the *adiabatic approximation*. We will start this section with a rather general discussion about the adiabatic approximation. These considerations lead easily to the counterexamples in [Yaf1], which we give in Subsect. 3.8.2. The sharper counterexample in [Ya1] uses the fact that certain quadratic time-dependent Hamiltonians are exactly solvable and will be given in Subsect. 3.8.3.

3.8.1 Adiabatic Evolution

Let $\mathcal{H}$ be a Hilbert space and $\mathbb{R} \ni t \mapsto H(t)$ a family of self-adjoint operators with a fixed domain $\mathcal{D}$ that is C^1 in norm-resolvent sense. Let $U(t,s)$ be the unitary evolution generated by $H(t)$. Let $t \mapsto P(t)$ be the spectral projection for one eigenvalue $\lambda(t)$ of $H(t)$, and let us assume that $P(t)$ and $\lambda(t)$ are C^1. Note that we will often drop (t) from $H(t)$, $P(t)$, etc.

Definition 3.8.1
The adiabatic evolution *is the unitary evolution $U_{\rm ad}(t,s)$ generated by the time-dependent Hamiltonian*

$$H_{\rm ad}(t) := H(t) + [\dot{P}(t), iP(t)].$$

Proposition 3.8.2
The adiabatic evolution satisfies

$$U_{\rm ad}(t,s)P(s) = P(t)U_{\rm ad}(t,s).$$

Proof. Differentiating the identity $P^2 = P$, we obtain

$$\dot{P}P + P\dot{P} = \dot{P},$$

which implies that

$$P\dot{P}P = (1-P)\dot{P}(1-P) = 0.$$

Thus

$$
\begin{aligned}
(\tfrac{\mathrm{d}}{\mathrm{d}t} + i[H_{\mathrm{ad}}, \cdot])P &= \dot{P} - [[\dot{P}, P], P] \\
&= (1 - P)\dot{P}(1 - P) + P\dot{P}P = 0.
\end{aligned}
$$

□

We would like to know if the adiabatic evolution approximates the exact evolution for large times. To this end, we have to investigate the existence of the limit

$$\lim_{t\to\infty} U(0,t)U_{\mathrm{ad}}(t,0) =: \Omega_{\mathrm{ad}}^{+}. \tag{3.8.1}$$

The simplest method of proving the existence of such limit is to prove that $[P, i\dot{P}] \in L^1(\mathrm{d}t)$. It turns out that this will never hold for the Hamiltonians that we would like to study. Typically we have only $[P, i\dot{P}] \in O(t^{-1})$. It turns out, however, that the following criterion is useful.

Proposition 3.8.3
Set

$$K(t) := i[(H(t) - \lambda(t))^{-1}(1 - P(t)), \dot{P}(t)].$$

Assume that

$$K(t) = o(t^0), \qquad \dot{K}(t) \in L^1(\mathrm{d}t), \qquad K(t)[P(t), \dot{P}(t)] \in L^1(\mathrm{d}t). \tag{3.8.2}$$

Then there exists the limit (3.8.1).

Proof. We have

$$\mathrm{s-}\lim_{t\to\infty} U(0,t)U_{\mathrm{ad}}(t,0) = \mathrm{s-}\lim_{t\to\infty} U(0,t)(1 + K(t))U_{\mathrm{ad}}(t,0).$$

Moreover,

$$[H, K] = [\dot{P}, iP]. \tag{3.8.3}$$

Therefore,

$$\tfrac{\mathrm{d}}{\mathrm{d}t} U(0,t)(1 + K)U_{\mathrm{ad}}(t,0) = U(0,t)(\dot{K} - K[P, \dot{P}])U_{\mathrm{ad}}(t,0)$$

is integrable. □

3.8.2 Counterexample Based on the Adiabatic Approximation

Let $W(x) \in C_0^\infty(X)$. Let $g(t)$ be a C^∞ function such that $g(0) = 1$. We set

$$
\begin{aligned}
H(t) &:= \tfrac{1}{2}D^2 + V(t,x), \\
\text{where} \quad V(t,x) &= g^{-2}(t)W(\tfrac{x}{g(t)}).
\end{aligned}
\tag{3.8.4}
$$

Suppose that $\lambda(0) \in \sigma_{\mathrm{disc}}(H(0))$, where

$$H(0) = \frac{1}{2}D^2 + W(x).$$

Let $P(0)$ be the corresponding projection. Note that if A is the generator of dilations, then we have

$$\begin{aligned} H(t) &= g^{-2}(t)g^{-iA}(t)H(0)g^{iA}(t), \\ P(t) &= g^{-iA}(t)P(0)g^{iA}(t), \\ \lambda(t) &= g^{-2}(t)\lambda(0). \end{aligned}$$

Note that $[A, (z - H(0))^{-1}]$ is bounded for $z \notin \sigma(H(0))$, therefore $[A, P(0)]$ is bounded too.

Now we fix $g(t) = \sqrt{\langle t \rangle}(1 + \log\langle t \rangle)^{-\epsilon}$ for $\epsilon > 1/2$. Then we have the following estimates on $V(t, x)$:

$$\begin{aligned} \partial_x^\alpha V(t,x) &\in O(t^{-\frac{3}{2}}(\log t)^{3\epsilon}), \quad |\alpha| = 1, \\ \partial_x^\alpha V(t,x) &\in O(t^{-2}(\log t)^{4\epsilon}), \quad |\alpha| = 2. \end{aligned}$$

As we will see from the theorem below, the Hamiltonian defined in (3.8.4) does not satisfy asymptotic completeness.

Theorem 3.8.4
(i) The asymptotic momentum D^+ for $H(t)$ exists.
(ii) The usual wave operator

$$\text{s-}\lim_{t\to\infty} U(0,t)\mathrm{e}^{-itH_0} =: \Omega_{\mathrm{fd}}^+$$

exists, $D^+\Omega_{\mathrm{fd}}^+ = \Omega_{\mathrm{fd}}^+ D$, and

$$\mathrm{Ran}\Omega_{\mathrm{fd}}^+ = \mathrm{Ran}\mathbb{1}_{X\setminus\{0\}}(D^+) = \mathcal{H}_{\mathrm{c}}(D^+).$$

(iii) The norm limit

$$\lim_{t\to\infty} U(0,t)U_{\mathrm{ad}}(t,0) =: \Omega_{\mathrm{ad}}^+$$

exists and is unitary. Moreover,

$$\mathrm{Ran}\Omega_{\mathrm{ad}}^+ P(0) \subset \mathrm{Ran}\mathbb{1}_{\{0\}}(D^+) = \mathcal{H}_{\mathrm{pp}}(D^+). \tag{3.8.5}$$

Hence $\mathrm{Ran}\Omega_{\mathrm{fd}}^+ \neq L^2(X)$.

Proof. The potential $V(t,x)$ satisfies the conditions of Theorem 3.2.1, hence *(i)* is true.

Let us show *(ii)*. Let $g, J \in C_0^\infty(X)$ such that $0 \notin \mathrm{supp}J$ and $J = 1$ on a neighborhood of $\mathrm{supp}g$. By the methods of Sect. 3.2, we show that

$$\begin{aligned} &\text{s-}\lim_{t\to\infty} U_{\mathrm{ad}}(0,t)\mathrm{e}^{-itH_0}g^2(D) \\ &= \text{s-}\lim_{t\to\infty} U_{\mathrm{ad}}(0,t)g(D)J\left(\tfrac{x}{t}\right)g(D)\mathrm{e}^{-itH_0} \end{aligned} \tag{3.8.6}$$

exists. Then we use the fact that vectors of the form $g(D)\phi$ with $g \in C_0^\infty(X)$ with $0 \notin \mathrm{supp} g$ are dense in $L^2(X)$.

Let us show *(iii)*. We will use Proposition 3.8.3. We have

$$\begin{aligned}
\dot{P}(t) &= -\dot{g}g^{-1}g^{-iA}[A, iP(0)]g^{iA} \\
&\in O(\dot{g}(t)g^{-1}(t)), \\
K(t) &= -ig\dot{g}g^{-iA}[(H(0)-\lambda(0))^{-1}(1-P(0)), [A, P(0)]]g^{iA} \\
&\in O(g(t)\dot{g}(t)), \\
\dot{K}(t) &= -\dot{g}^2 g^{-iA}\left[A, [(H(0)-\lambda(0))^{-1}(1-P(0)), [A, P(0)]]\right] g^{iA} \\
&\quad -(\dot{g}^2 + g\ddot{g})g^{-iA}[(H(0)-\lambda(0))^{-1}(1-P(0)), [A, P(0)]]g^{iA} \\
&\in O(g(t)\ddot{g}(t)) + O(\dot{g}^2(t)), \\
K(t)[P(t), \dot{P}(t)] &\in O(\dot{g}^2(t)).
\end{aligned}$$

But we have

$$\begin{aligned}
O(\dot{g}(t)g(t)) &= O(\langle \log t\rangle^{-2\epsilon}), \\
O(\ddot{g}(t)g(t)) = O(\dot{g}^2(t)) &= O(\langle t\rangle^{-1}\langle \log t\rangle^{-2\epsilon}),
\end{aligned}$$

hence the assumptions of Proposition 3.8.3 are satisfied.

Let us show (3.8.5). Let $J \in C_0^\infty(X)$ with $0 \notin \mathrm{supp} J$. Then

$$\begin{aligned}
J(D^+)\Omega_{\mathrm{ad}}^+ P(0) &= \mathrm{s-}\lim_{t\to\infty} U(0,t)J(\tfrac{x}{t})U(t,0)\Omega_{\mathrm{ad}}^+ P(0) \\
&= \mathrm{s-}\lim_{t\to\infty} U(0,t)J(\tfrac{x}{t})U_{\mathrm{ad}}(t,0)P(0).
\end{aligned}$$

But

$$\begin{aligned}
J(\tfrac{x}{t})U_{\mathrm{ad}}(t,0)P(0) &= J(\tfrac{x}{t})P(t)U_{\mathrm{ad}}(t,0) \\
&= g^{-iA}(t)J(\tfrac{xg(t)}{t})P(0)g^{iA}(t)U_{\mathrm{ad}}(t,0)
\end{aligned}$$

converges to zero in norm. □

3.8.3 A Sharper Counterexample

In this subsection we give a sharper counterexample using the special properties of certain quadratic Hamiltonians.

Let us fix a certain cutoff function $\chi \in C_0^\infty(\mathbb{R})$ equal to 1 near the origin and $C > 1/8$. Define

$$\begin{aligned}
H(t) &= \tfrac{1}{2}D^2 + V(t,x), \\
\text{where}\quad V(t,x) &:= \tfrac{Cx^2}{t^2}\chi\left(\tfrac{|x|}{\sqrt{t|\log t|}}\right).
\end{aligned} \tag{3.8.7}$$

We denote by $U(t,s)$ the unitary evolution generated by the Hamiltonian $H(t)$. Note that one has

$$\partial_x^\alpha V(t,x) \in O(t^{-3/2}(\log t)^{\frac{1}{2}}), \quad |\alpha| = 1,$$
$$\partial_x^\alpha V(t,x) \in O(t^{-2}), \quad |\alpha| = 2,$$

so $V(t,x)$ almost (but not quite) satisfies the conditions of the existence and completeness of modified wave operators for general slow-decaying potentials in Theorem 3.4.1 or of the existence and completeness of Dollard wave operators in Theorem 3.6.2.

It will be useful to introduce another Hamiltonian

$$H_{\mathrm{Y}}(t) := \frac{1}{2}D^2 + \frac{Cx^2}{t^2}.$$

The evolution $U_{\mathrm{Y}}(t,s)$ generated by $H_{\mathrm{Y}}(t)$ can be computed explicitly. Indeed, if we put

$$T_t\phi(x) := t^{-n/4}\phi(\tfrac{x}{t^{\frac{1}{2}}}),$$
$$U_0\phi(x) := \mathrm{e}^{-ix^2/4}\phi(x),$$
$$H_{\mathrm{Y}} := \tfrac{1}{2}D^2 + (C - \tfrac{1}{8})x^2,$$

then it is an easy computation to check that

$$U_{\mathrm{Y}}(t,1)\phi = T_t U_0 \mathrm{e}^{-i\log t H_{\mathrm{Y}}}\phi.$$

As we will show below, the Hamiltonian $H(t)$ violates asymptotic completeness.

Theorem 3.8.5
(i) There exists the asymptotic momentum for $H(t)$

$$D^+ := \mathrm{s}\text{-}C_\infty\text{-}\lim_{t\to\infty} U(1,t)DU(t,1).$$

(ii) The usual wave operator

$$\mathrm{s}\text{-}\lim_{t\to\infty} U(1,t)\mathrm{e}^{-i(t-1)H_0} =: \Omega_{\mathrm{fd}}^+$$

exists, satisfies $D^+\Omega_{\mathrm{fd}}^+ = \Omega_{\mathrm{fd}}^+ D$, and

$$\mathrm{Ran}\Omega_{\mathrm{fd}}^+ = \mathrm{Ran}\mathbb{1}_{X\setminus\{0\}}(D^+) = \mathcal{H}_{\mathrm{c}}(D^+).$$

(iii) There exists another wave operator

$$\lim_{t\to\infty} U(1,t)U_{\mathrm{Y}}(t,1) =: \Omega_{\mathrm{Y}}^+.$$

Moreover,

$$\mathrm{Ran}\Omega_{\mathrm{Y}}^+ \subset \mathrm{Ran}\mathbb{1}_{\{0\}}(D^+) = \mathcal{H}_{\mathrm{pp}}(D^+). \tag{3.8.8}$$

Hence $(\mathrm{Ran}\Omega_{\mathrm{fd}}^{+})^{\perp}$ *is infinite dimensional.*

Proof. *(i)* and *(ii)* are proven as in Theorem 3.8.4.

Let us show *(iii)*. Suppose that ψ is an eigenfunction of H_Y for the eigenvalue λ. Then

$$U_{\mathrm{Y}}(t,1)\psi(x) = t^{-n/4}\mathrm{e}^{-i\lambda\log t + ix^2/4t}\psi(\tfrac{x}{t^{\frac{1}{2}}}), \tag{3.8.9}$$

Moreover,

$$\tfrac{\mathrm{d}}{\mathrm{d}t}U(1,t)U_{\mathrm{Y}}(t,1)\psi = U(1,t)\tfrac{Cx^2}{t^2}\left(1-\chi\left(\tfrac{|x|}{\sqrt{t|\log t|}}\right)\right)U_{\mathrm{Y}}(t,1)\psi \tag{3.8.10}$$

If we use (3.8.9), (3.8.10) and the exponential decay of ψ (which is an eigenfunction of a harmonic oscillator), then we see that

$$\|\tfrac{\mathrm{d}}{\mathrm{d}t}U(1,t)U_{\mathrm{Y}}(t,1)\psi\| \le C_1 t^{-1}\exp\left(-C_2\log t\right),$$

which is integrable. Hence Ω_{Y}^{+} exists. □

3.9 Smoothness of Wave Operators in the Fast-Decaying Case

In this section we will study scattering theory for potentials that satisfy the so-called smooth fast-decaying condition. More precisely, we will assume that the potential satisfies

$$\int_0^\infty \langle t\rangle^{|\alpha|}\|\partial_x^\alpha V(t,\cdot)\|_\infty \mathrm{d}t < \infty, \quad |\alpha|\ge 1. \tag{3.9.1}$$

To simplify, we will also suppose that

$$\int_0^\infty |V(t,0)|\mathrm{d}t < \infty, \tag{3.9.2}$$

which implies that we do not need to renormalize the free dynamics in order to define wave operators.

In the classical case, under these assumptions, the wave transformation is smooth and all its derivatives are bounded. In the quantum case, there is an analog of this property, which can be expressed using an appropriate class of pseudo-differential operators

The main result of this section can be formulated in the following theorem (see Appendix D.4 for the notation concerning pseudo-differential operators).

Theorem 3.9.1
Assume (3.9.1) and (3.9.2). Then

$$\Omega_{\mathrm{fd}}^{+} \in \Psi(1, g_0).$$

In other words, we can write

$$\Omega_{\rm fd}^{+}\phi(x)=(2\pi)^{-n}\int\int a^{+}(x,\xi)\mathrm{e}^{i(x-y)\xi}\phi(y)\mathrm{d}y\mathrm{d}\xi,$$

where $a\in S(1,g_0)$.

Remark. Set

$$\phi_\xi(x):=\mathrm{e}^{ix\xi},\quad \phi_\xi^{+}(x):=a^{+}(x,\xi)\mathrm{e}^{ix\xi}.$$

Note that ϕ_ξ is a generalized eigenvector of D with the eigenvalue ξ. Similarly, ϕ_ξ^{+} is a generalized eigenvector of D^{+} with the eigenvalue ξ. It follows from Theorem 3.9.1 and Lemma D.6.1 that the wave operator $\Omega_{\rm fd}^{+}$ is bounded on all the weighted spaces $\langle x\rangle^m L^2(X)$. Therefore, the following identity makes sense if we treat ϕ_ξ,ϕ_ξ^{+} as elements of $\langle x\rangle^m L^2(X)$ for $m>n/2$:

$$\Omega_{\rm fd}^{+}\phi_\xi=\phi_\xi^{+}.$$

The class of operators $\Psi(1,g_0)$ is associated in a natural way with the problem we are looking at in this chapter. Unfortunately, in this class we do not have a "semi-classical parameter", and hence no symbolic calculus is available. A natural semi-classical parameter s^{-1} appears if we allow our quantities to depend on the initial time s.

Let us define

$$\Omega_{\rm fd}(t,s):=U(s,t)U_0(t-s).$$

We extend the definition of $\Omega_{\rm fd}(t,s)$ to $t=\infty$ in the obvious way. Clearly, $\Omega_{\rm fd}(\infty,0)=\Omega_{\rm fd}^{+}$. The following theorem is an extension of Theorem 3.9.1.

Theorem 3.9.2
Assume (3.9.1) and (3.9.2). Then, uniformly for $s\le t\le\infty$, *we have*

$$\Omega_{\rm fd}(t,s)-1\in\Psi(o(s^0),g_0(s)),\tag{3.9.3}$$

In other words, there exist $a(t,s,x,\xi)$ *such that*

$$\partial_x^\alpha\partial_\xi^\beta(a(t,s,x,\xi)-1)\in o(\langle s\rangle^{-|\alpha|}),$$

$$\Omega_{\rm fd}(t,s)\phi(x)=(2\pi)^{-n}\int\int a^{+}(t,s,x,\xi)\mathrm{e}^{i(x-y)\xi}\phi(y)\mathrm{d}y\mathrm{d}\xi.$$

We first prove the following lemma.

Lemma 3.9.3
Let

$$\int_0^\infty t^{|\beta|}\|\mathrm{ad}_x^\alpha \mathrm{ad}_D^\beta P(t)\|\mathrm{d}t < \infty,$$

or, using the notation of Sect. D.5, $P(t) \in L^1(\mathrm{d}t, \Psi(g_0(t)))$. *Let* $W(t,s)$ *be the unique solution of*

$$\begin{cases} \partial_t W(t,s) = W(t,s)P(t), \\ W(s,s) = 1. \end{cases}$$

Then

$$W(t,s) - 1 \in \begin{cases} \Psi(o(s^0), g_0(s)), & s \le t, \\ \Psi(o(t^0), g_0(t)), & t \le s. \end{cases}$$

Proof. Clearly,

$$W(t,s) - 1 \in \begin{cases} o(s^0), & s \le t, \\ o(t^0), & t \le s. \end{cases}$$

Let us now prove by induction on $|\alpha| + |\beta|$ that

$$\mathrm{ad}_x^\alpha \mathrm{ad}_D^\beta (W(t,s) - 1) \in \begin{cases} o(s^{-|\beta|}), & s \le t, \\ o(t^{-|\beta|}), & t \le s. \end{cases} \tag{3.9.4}$$

Assume that (3.9.4) holds for $|\alpha| + |\beta| \le n-1$. Using the Leibniz rule, we obtain

$$\partial_t \mathrm{ad}_x^\alpha \mathrm{ad}_D^\beta W(t,s) = \sum_{(\gamma_1,\delta_1)+(\gamma_2,\delta_2)=(\alpha,\beta)} C_{\gamma,\delta} \mathrm{ad}_x^{\gamma_1} \mathrm{ad}_D^{\delta_1} W(t,s) \mathrm{ad}_x^{\gamma_2} \mathrm{ad}_D^{\delta_2} P(t).$$

We can rewrite this as

$$\partial_t \mathrm{ad}_x^\alpha \mathrm{ad}_D^\beta W(t,s) + (\mathrm{ad}_x^\alpha \mathrm{ad}_D^\beta W(t,s))P(t) = R_{\alpha,\beta}(t,s).$$

By the induction assumption,

$$R_{\alpha,\beta}(t,s) \in \begin{cases} o(s^{-|\beta|})L^1(\mathrm{d}t), & s \le t, \\ o(t^{-|\beta|})L^1(\mathrm{d}t), & t \le s. \end{cases}$$

This implies (3.9.4) for $|\alpha| + \beta| = n$ by the Gronwall lemma and proves the desired result. □

Proof of Theorem 3.9.2. One has

$$\begin{cases} \partial_t \Omega_{\mathrm{fd}}(t,s) = -\mathrm{i}\Omega_{\mathrm{fd}}(t,s) V^{\mathrm{w}}(t, x + (t-s)D), \\ \Omega_{\mathrm{fd}}(s,s) = 1. \end{cases} \tag{3.9.5}$$

It follows from (3.9.1) that, uniformly for $0 \le s \le t$,

$$V^{\mathrm{w}}(t, x + (t-s)D) \in L^1(\mathrm{d}t, \Psi(g_0(t))),$$

Applying Lemma 3.9.3 to $\Omega_{\mathrm{fd}}(t,s)$ gives (3.9.3). □

3.10 Smoothness of Wave Operators in the Slow-Decaying Case

In this section we will show that in the smooth slow-decaying case case the time translated wave operator

$$\Omega^+_{\rm sd}(s) \;:=\; U(s,0)\Omega^+_{\rm sd},$$

and the evolution $U(s,t)$ are Fourier integral operators associated with the canonical transformation $\phi(s,0)\circ\mathcal{F}^+_{\rm sd}$ and $\phi(s,t)$ respectively in the sense described by the following theorem.

Theorem 3.10.1
Assume (3.7.1). Then, for $T_0 \le s \le t$, there exist functions $a(s,t,x,\xi)$ and $a^+(s,x,\xi)$ such that

$$\begin{aligned} &a^+(s,x,\xi) \;:=\; \lim_{t\to\infty} a(s,t,x,\xi),\\ &\partial_x^\alpha\partial_\xi^\beta(a(s,t,x,\xi)-1)\in o(s^{-|\alpha|}),\quad T_0\le s\le t<\infty,\\ &\partial_x^\alpha\partial_\xi^\beta(a^+(s,x,\xi)-1)\in o(s^{-|\alpha|}),\quad T_0\le s<\infty, \end{aligned} \tag{3.10.1}$$

$$U(s,t)\phi(x) = (2\pi)^{-n}\int\!\!\int e^{iS(s,t,x,\xi)-i\langle y,\xi\rangle}a(s,t,x,\xi)\phi(y)\mathrm{d}y\mathrm{d}\xi, \tag{3.10.2}$$

$$\Omega^+_{\rm sd}(s)\phi(x) = (2\pi)^{-n}\int\!\!\int e^{i\Phi^+_{\rm sd}(s,x,\xi)-i\langle y,\xi\rangle}a^+(s,x,\xi)\phi(y)\mathrm{d}y\mathrm{d}\xi. \tag{3.10.3}$$

Proof. Note that it follows from Proposition D.14.1 that

$$I(s,t)I^*(s,t)-1\in\Psi(o(s^0),g_0(s)).$$

Hence, for $s\ge T_0$,

$$\|I(s,t)I^*(s,t)-1\|\le C_0<1,$$

and so $I(s,t)I^*(s,t)$ is invertible. Using the Neumann series and the Beals criterion (see Theorem D.4.1), we obtain

$$(I(s,t)I^*(s,t))^{-1}-1\in\Psi(o(s^0),g_0(s)). \tag{3.10.4}$$

Set

$$W(s,t) := U(t,s)I(s,t).$$

Then it follows from (3.7.9) that, for $T_0\le s\le t$,

$$\begin{cases} \frac{\mathrm{d}}{\mathrm{d}s}W(s,t) = W(s,t)\tilde{P}(s,t),\\ W(t,t)=1, \end{cases} \tag{3.10.5}$$

where

$$\tilde{P}(s,t) := I^{-1}(s,t)P(s,t)$$
$$= I^*(s,t)(I(s,t)I^*(s,t))^{-1}P(s,t).$$

and $P(t,s)$ was defined in (3.7.10).

By (3.10.4) and Propositions D.14.2 and D.14.1, we have

$$\tilde{P}(s,t) \in L^1(\mathrm{d}s, \Psi(g_0(s))).$$

Therefore by Lemma 3.9.3 (with the role of s and t reversed), we get

$$W(s,t) - 1 \in \Psi(o(s^0), g_0(s)). \tag{3.10.6}$$

We know that

$$U(s,t) = (I(s,t)I^*(s,t))^{-1}I(s,t)W^*(s,t). \tag{3.10.7}$$

Now (3.10.4), (3.10.6), (3.10.7) and Proposition D.14.2 imply the properties of $U(s,t)$ stated in our theorem.

If we know that $U(s,t)$ can be written as a Fourier integral operator (3.10.2), then it is easy to see that $\Omega_{\mathrm{sd}}^+(s)$ can be written as a Fourier integral operator (3.10.3). In fact, we note that

$$\begin{aligned} &\mathrm{e}^{iS(t,D)}U(s,t)^*\phi(y) \\ &= (2\pi)^{-n} \int\int \bar{a}(s,t,x,\xi)\mathrm{e}^{i\langle y,\xi\rangle - iS(s,t,x,\xi) + iS(t,\xi)}\phi(x)\mathrm{d}\xi\mathrm{d}x. \end{aligned} \tag{3.10.8}$$

Using similar arguments as in the proof of (3.7.16), we see that (3.10.8) goes to

$$\Omega_{\mathrm{sd}}^{+*}(s)\phi(y) = (2\pi)^{-n} \int\int \bar{a}^+(s,x,\xi)\mathrm{e}^{i\langle y,\xi\rangle - i\Phi_{\mathrm{sd}}^+(s,x,\xi)}\phi(x)\mathrm{d}\xi\mathrm{d}x.$$

when t goes to ∞. □

4. Quantum 2-Body Hamiltonians

4.0 Introduction

In this chapter we study scattering theory for quantum time-independent 2-particle systems. They are described by Hamiltonians on $L^2(X)$ of the form

$$H = \frac{1}{2}D^2 + V(x), \tag{4.0.1}$$

where $V(x)$ decays in space in all directions. The previous chapter was devoted to Hamiltonians with time-decaying potentials and should be viewed as an introduction to the present one. In particular, various objects related to long-range scattering theory were discussed in Chap. 3. Therefore we advise the reader to first become familiar with the previous chapter, in particular with Sects. 3.2, 3.3 and 3.5.

In the mathematical physics literature, the name scattering theory is usually reserved for Hamiltonians of the form (4.0.1). Classical systems, studied in Chaps. 1 and 2, and time-decaying quantum systems described in Chap. 3 are seldom studied for their own sake; they are usually considered only as tools that are used to study (4.0.1).

The most obvious classification of states in the case of a time-independent 2-body Hamiltonian is

$$L^2(X) = \mathcal{H}_{\mathrm{pp}}(H) \oplus \mathcal{H}_{\mathrm{c}}(H),$$

where $\mathcal{H}_{\mathrm{pp}}(H)$ is the space of bound states of H and $\mathcal{H}_{\mathrm{c}}(H)$ the continuous spectral subspace. It is not a priori clear that any state in $\mathcal{H}_{\mathrm{c}}(H)$ behaves asymptotically for large times as a free particle. Actually, we saw in Chap. 2 that in the classical case there are unbounded trajectories with zero energy that behave at infinity very differently from free trajectories. In the quantum case, it turns out that under rather weak conditions on the interaction, any state in $\mathcal{H}_{\mathrm{c}}(H)$ behaves asymptotically like a free particle. In more precise mathematical terms, $\mathcal{H}_{\mathrm{c}}(H)$ is equal to the range of the wave operator. This fact is called the asymptotic completeness of 2-body scattering and will be the central result proven in this chapter.

Let us describe the main strategy of our proof of asymptotic completeness. The first step is contained in Chap. 3, where we consider time-decaying potentials

$V(t,x)$ and the dynamics $U(t,s)$ they generate. We show the existence of wave operators for this dynamics, which (in the long-range case) are defined as

$$\mathrm{s}-\lim_{t\to\infty} U(0,t)\mathrm{e}^{-iS(t,D)}. \tag{4.0.2}$$

Along with the existence of (4.0.2), we show the existence of

$$\mathrm{s}-\lim_{t\to\infty} \mathrm{e}^{iS(t,D)}U(t,0), \tag{4.0.3}$$

which implies the unitarity of the wave operator for the dynamics $U(t,s)$.

Note that the existence of (4.0.3) is straightforward in the fast-decaying case; it is much less easy in the slow-decaying case.

The second step consists of a number of propagation estimates for the dynamics e^{-itH}. Their aim is to show that if $\chi \in C_0^\infty(\mathbb{R})$ is supported away of 0 and eigenvalues of H, then any state $\mathrm{e}^{-itH}\chi(H)\phi$ moves away from the origin in a controllable way. This enables us to replace the time-independent potential by an effective time-decaying one

$$V(t,x) := J(\tfrac{x}{t})V(x),$$

and the dynamics e^{-itH} by $U(t,0)$. More precisely, we are able to show the existence of the limit

$$\mathrm{s}-\lim_{t\to\infty} U(0,t)\mathrm{e}^{-itH}\chi(H). \tag{4.0.4}$$

Now (4.0.3), (4.0.4) and the chain rule yield the existence of

$$\mathrm{s}-\lim_{t\to\infty} \mathrm{e}^{iS(t,D)}\mathrm{e}^{-itH}\mathbb{1}^{\mathrm{c}}(H), \tag{4.0.5}$$

which implies asymptotic completeness.

The concept of the wave operator was introduced by Møller in [Mø]. Therefore, wave operators are sometimes called Møller operators.

The literature on mathematical aspects of quantum scattering theory is very rich and contains numerous techniques. The reader will find a bibliographical review of this subject up to the early eighties in [RS, vol III].

Among the main approaches to the problem of asymptotic completeness, one can distinguish the abstract Kato-Birman approach, the stationary approach and the time-dependent approach.

The Kato-Birman approach is a time-dependent method which has two basic ingredients. The first one is the existence and completeness of wave operators for a pair (H_0, H) of Hamiltonians whose difference is of trace class. The second one is the invariance of wave operators under the replacement of H_0 and H with $f(H_0)$ and $f(H)$ for suitable functions f. It applies only to a rather narrow class of fast decaying short-range potentials. The reader can find its description in [RS, vol III] and [Yaf4].

The stationary method is usually more efficient. It is based on the study of the boundary values of the resolvent $(z-H)^{-1}$. A modern exposition can be found in [Hö2, vol II] (the short-range case) and [Hö2, vol IV] (the long-range case), see also [Ag1] and [RS, vol III and IV].

We never use the stationary method in our monograph. The main objective of our exposition is to prove various properties related to asymptotic completeness. We think that studying the resolvent $(z - H)^{-1}$ is a kind of a detour for our purposes and, in our opinion, leads to unnecessary complication. But the most convincing argument against the stationary method is the following: if the potential is time-dependent, then the time-dependent method works quite well, while the stationary method is much less useful (see, however, [How, Yaf5] for the short-range case).

On the other hand, the stationary method gives some important additional information that seems not to follow easily by the time-dependent method. By the stationary method, one obtains some properties of the boundary values of the resolvent, which are especially useful in the study of the scattering operator and eigenfunction expansions. But these topics are not covered by our monograph.

In the time-dependent approach, the main object under study is the dynamics e^{-itH}. It was used in the first proof of the existence of wave operators in [Co1]. On the other hand, in the problem of asymptotic completeness time-dependent techniques entered the literature quite late with papers of V. Enss [E1, E2]. Ideas of Enss aroused a considerable interest and inspired a number of papers by other authors, among them: [Dav, KiYa1, KiYa2, Pe1, Sim2].

The time-dependent approach to scattering theory proved to be the most successful. It led to a series of remarkable results about N-body systems beginning with [E5, SS1], which we discuss later on. There exist various techniques related to the time-dependent approach. In particular, one of the techniques that proved to be important is the so-called method of positive commutators (or, more exactly, the method of positive Heisenberg derivatives) – see Appendix B.4 for references.

Even the time-dependent approach comes in a number of different varieties. The original argument due to V. Enss [E1, E2], who was first to use the time-dependent method to show asymptotic completeness, was based on the so-called RAGE theorem. We do not use this idea, because it seems not to work for more than 3 particles. The approach that we use is based on positive commutators and weak propagation estimates. Positive commutators have a relatively long history, culminating in the work of E. Mourre [Mo1]. Weak propagation estimates showed their extraordinary efficiency in the work of Sigal and Soffer [SS1], who proved asymptotic completeness in the N-body case. In our presentation of the propagation estimates of this chapter, we also use ideas of Graf [Gr] and Yafaev [Yaf1].

Let us now briefly describe the content of this chapter.

Section 4.1 is devoted to the problem of the self-adjointness of Schrödinger operators and some of their general properties. We refer the reader to [RS, vol I and II] for more details.

In Sect. 4.2 we prove so-called weak large velocity estimates. This is a simple example of a *weak propagation estimate.* Their typical form is as follows:

$$\int_0^\infty \|B(t)\mathrm{e}^{-itH}\phi\|^2\frac{\mathrm{d}t}{t} < \infty,$$

where $B(t)$ is a certain uniformly bounded observable. The weak maximal velocity estimates say roughly that the velocity of a particle is bounded by the square root of its energy. These estimates were first found by Sigal and Soffer [SS1, SS3], although our presentation follows that of Graf [Gr].

One says that a Hamiltonian H satisfies the Mourre estimate on the energy interval Δ with conjugate operator A if

$$\mathbb{1}_\Delta(H)[H,iA]\mathbb{1}_\Delta(H) \geq C_0\mathbb{1}_\Delta(H) + K, \tag{4.0.6}$$

where $C_0 > 0$ and K is a compact operator. Hamiltonians of the form (4.0.1), under certain assumptions on $V(x)$, satisfy the Mourre estimate on any positive energy interval with a conjugate operator equal to the generator of dilations

$$A := \frac{1}{2}(\langle x, D\rangle + \langle D, x\rangle).$$

The existence of an estimate as (4.0.6) has deep consequences on the spectral theory of H on the interval Δ. It was used in a fundamental paper by Mourre [Mo1] to prove the absence of singular continuous spectrum for H in Δ. His ideas were related to some earlier work by Lavine [La3]. The abstract Mourre commutator method has been extended and refined and applied to a wide variety of problems (see, among others, [Mo2, PSS, Yaf2, JMP, ABG, BG]).

The Mourre estimate is shown for 2-body systems in Sect. 4.3, where we also describe a number of its consequences. In particular, we will show that the point spectrum can accumulate only at 0.

In our presentation of scattering theory, we made an attempt to describe its basic steps in a form that stresses various natural objects and we attach less importance to technical estimates that involve arbitrary cutoff functions. Besides, we prefer arguments that generalize easily to the N-body case. An example of such a natural construction is the *asymptotic velocity,* which we introduce in Sect. 4.4. The asymptotic velocity is the self-adjoint operator defined by

$$P^+ := \mathrm{s}{-}C_\infty{-}\lim_{t\to\infty} \mathrm{e}^{itH}\frac{x}{t}\mathrm{e}^{-itH}, \tag{4.0.7}$$

Another equivalent definition of P^+ is

$$P^+ := \mathrm{s}{-}C_\infty{-}\lim_{t\to\infty} \mathrm{e}^{itH}D\mathrm{e}^{-itH}\mathbb{1}^{\mathrm{c}}(H). \tag{4.0.8}$$

(The limits (4.0.7) (4.0.8) are strong C_∞ limits defined in Appendix B.2.) The existence of (4.0.7) holds under very weak assumptions on the potential, e.g. $V(x)$ has to be $\Delta-$compact and, roughly speaking,

$$|\nabla_x V(x)| \leq C\langle x\rangle^{-1-\mu}, \quad \mu > 0.$$

The existence of the observable P^+ in the 2-body case follows easily from the existence of (modified) wave operators. Therefore in the literature it was usually not considered for its own sake. It becomes much less trivial in the N-body case, where its first explicit construction was given in [De6]. (Nevertheless, it appeared implicitly in the earlier work on the subject, especially in [Gr]).

We think that, even in the 2-body case, introducing the asymptotic velocity is a useful idea. First of all, it helps to organize the proof of asymptotic completeness. Secondly, it can be shown to exist under rather weak assumptions. In fact, as we will see in Sect. 4.10, there are 2-body Hamiltonians for which the asymptotic velocity exists, and thus one can argue that some kind of a scattering theory for such systems is available, but for which the wave operators fail to be complete.

The proof of the existence of (4.0.7) relies on a number of weak propagation estimates. The idea of using this type of estimates in scattering theory is due to Sigal and Soffer [SS1]. In our presentation, we follow a very elegant approach due to Graf [Gr] with some modifications inspired by Yafaev [Yaf1].

We also show that

$$\mathbb{1}_{\{0\}}(P^+) = \mathbb{1}^{\mathrm{pp}}(H), \tag{4.0.9}$$

which means that the states of zero asymptotic velocity coincide with the bound states of H. As we saw in Chap. 2, an analog of this property is false in classical mechanics. This property follows from the so-called *minimal velocity estimate* [Gr], whose proof is based on the Mourre estimate. The property (4.0.9) plays an important role in the proof of asymptotic completeness.

Section 4.4 is probably the central section of the whole chapter. The result of this section enable us to reduce the study of time-independent potentials to the framework of Chap. 3.

In Sect. 4.5 we describe the joint spectrum of the asymptotic velocity and the energy. This result can be viewed as a weak form of asymptotic completeness. It will not be used in the proof of the asymptotic completeness of wave operators.

Section 4.6 is devoted to short-range scattering theory. Roughly speaking we assume that

$$|V(x)| \leq C\langle x\rangle^{-\mu}, \quad \mu > 1.$$

Using Sect. 4.3 and the results of Chap. 3, one proves the existence of the short-range wave operators

$$\mathrm{s-}\lim_{t\to\infty} \mathrm{e}^{itH}\mathrm{e}^{-itH_0} =: \Omega_{\mathrm{sr}}^+,$$

and the fact that

$$\mathrm{Ran}\,\Omega_{\mathrm{sr}}^+ = \mathcal{H}_{\mathrm{c}}(H).$$

This property of the wave operator goes under the name of *asymptotic completeness.* The wave operator implements the unitary equivalence of P^+ and D on $\mathcal{H}_{\mathrm{c}}(H)$, meaning that

$$P^+ = \Omega_{\mathrm{sr}}^+ D \Omega_{\mathrm{sr}}^{+*}.$$

Section 4.7 is devoted to long-range scattering theory. In this section we treat in a parallel way potentials with the long-range part $V_{\rm l}$ satisfying roughly

$$|\partial_x^\alpha V_{\rm l}(x)| \leq C_\alpha \langle x\rangle^{-|\alpha|-\mu}, \quad \mu > 0, \quad |\alpha| = 1, 2, \tag{4.0.10}$$

or the stronger condition

$$|\partial_x^\alpha V_{\rm l}(x)| \leq C_\alpha \langle x\rangle^{-|\alpha|-\mu}, \quad \mu > 0, \quad |\alpha| \geq 1. \tag{4.0.11}$$

The second condition allows for some simplifications, whereas the first condition is essentially optimal for asymptotic completeness. Asymptotic completeness under condition (4.0.10) is due to Hörmander, who used a slightly stronger hypothesis in [Hö2] to show asymptotic completeness by the time-independent method. The results of Sect. 4.3 allow us to reduce ourselves quite easily to the case of long-range time-dependent potentials treated in Sect. 3.4. In this way, we can prove the asymptotic completeness of the modified wave operators

$$\Omega_{\rm lr}^+ := \text{s-}\lim_{t\to\infty} \mathrm{e}^{itH}\mathrm{e}^{-iS(t,D)}, \tag{4.0.12}$$

where $S(t,\xi)$ is a solution of an appropriate Hamilton-Jacobi equation.

If $\mu > 1/2$, then it is possible to construct wave operators in a simpler way, using the so-called Dollard modifiers, instead of solutions of the Hamilton-Jacobi equation. Besides, one can include Hamiltonians with internal degrees of freedom. This is the subject of Sect. 4.8.

In Sect. 4.9 we consider another construction of long-range wave operators due to Isozaki and Kitada [IK1]. We assume in this section that the potentials satisfy the *smooth long-range condition*, roughly, (4.0.11). The Isozaki-Kitada construction is based on a time-independent modifier, which is a Fourier integral operator $J_{\rm lr}^+$ defined by

$$J_{\rm lr}^+\phi(x) := (2\pi)^{-n}\int \mathrm{e}^{i\Phi_{\rm lr}^+(x,\xi)-i\langle y,\xi\rangle} q^+(x,\xi)\phi(y)\mathrm{d}y\mathrm{d}\xi, \tag{4.0.13}$$

associated with a solution $\Phi_{\rm lr}^+(x,\xi)$ of the eikonal equation, where $q^+(x,\xi)$ is a cutoff equal to $\chi(\frac{1}{2}\xi^2)$ in an appropriate outgoing region. It turns out that the limit

$$\text{s-}\lim_{t\to\infty} \mathrm{e}^{itH} J_{\rm lr}^+ \mathrm{e}^{-itH_0} \tag{4.0.14}$$

exists and is equal to the usual long-range wave operator $\Omega_{\rm lr}^+$ multiplied by an energy cut-off $\chi_0(H_0)$. This type of a construction proved useful to study various properties of the scattering matrix (see for example [IK3, GeMa2]), because it works well within the stationary approach to scattering theory.

This ends the main part of this chapter devoted to the existence and completeness of wave operators. The remaining sections of this chapter describe more special topics.

In Sect. 4.10 we describe a Schrödinger operator for which the short-range wave operator and the asymptotic velocity exist but the wave operator is not complete. We prove this by constructing an additional wave operator whose range is orthogonal both to the bound states and to the usual scattering states. Such an example was first given by Yafaev [Yaf2] and we essentially follow his construction. Its main ingredient can be viewed as a certain version of the Born-Oppenheimer approximation. The variables of the configuration space are divided into two components, x and y. If we choose the potential $V(x,y)$ in an appropriate way, then there exists a nontrivial channel that describes a particle moving away from the origin as $C_0 t$ for $C_0 > 0$ and spreading in the direction of the y coordinate at the rate of $C\sqrt{t}$.

Weak propagation estimates, which we used so far in this chapter, give very weak information on the decay of $\|B(t)\mathrm{e}^{-itH}\phi\|$ for some observables $B(t)$. Their advantage consists in very weak assumptions on the potentials and the fact that they are valid for all ϕ in the Hilbert space. One can also study the so-called *strong propagation estimates.* They describe a faster decay of observables, typically $B(t)\mathrm{e}^{-itH}\phi \in O(t^{-N})$, but they are valid only for ϕ in a certain dense subspace of the Hilbert space. Usually, they also require stronger assumptions on the potentials.

It seems that weak propagation estimates are more important than the strong ones. In particular, they are sufficient for proving the existence and completeness of wave operators. Nevertheless, in order to prove certain detailed results on wave operators for smooth potentials we need some strong propagation estimates, notably the strong low velocity estimate.

The next three sections are devoted to strong propagation estimates. They are due to Sigal and Soffer [SS3]. The abstract method used to obtain these estimates consists in finding a positive (unbounded) observable with a negative Heisenberg derivative. A similar method is also used in partial differential equations to prove the propagation of the wave front set along bicharacteristics (see [Hö3]).

In Sect. 4.11 we prove the strong large velocity estimate. It says roughly that, for $\chi \in C_0^\infty(\mathrm{I\!R})$ and θ large enough,

$$\mathbb{1}_{[\theta,\infty[}\left(\tfrac{|x|}{t}\right)\chi(H)\mathrm{e}^{-itH}\langle x\rangle^{-N} \in O(t^{-N}). \tag{4.0.15}$$

This estimate is relatively easy to show and it requires very weak assumptions on the potentials.

The estimate that is much more useful is the so-called strong minimal velocity estimate. It says, more or less, that, for $\chi \in C_0^\infty(\mathrm{I\!R}\backslash\{0\})$ and θ small enough,

$$\mathbb{1}_{[0,\theta]}\left(\tfrac{|x|}{t}\right)\chi(H)\mathrm{e}^{-itH}\langle x\rangle^{-N} \in O(t^{-N}). \tag{4.0.16}$$

Its proof is more difficult than the proof of (4.0.15) and it requires essentially that the potentials are smooth. In order to show this estimate, one has first to prove a strong propagation estimate for the generator of dilations A, roughly:

$$\mathbb{1}_{[-\infty,\theta^2]}\left(\tfrac{A}{t}\right)\chi(H)\mathrm{e}^{-itH}\langle A\rangle^{-N} \in O(t^{-N}). \tag{4.0.17}$$

The estimate (4.0.17) is shown in Sect. 4.12 and the estimate (4.0.16) is shown in Sect. 4.13.

Note that there exists an alternative approach to strong propagation estimates based on the study of boundary values of the resolvent, which is due to Jensen, Mourre and Perry [JMP, Jen].

Scattering theory, especially in the long-range case, is intimately connected with the theory of pseudo-differential and Fourier integral operators associated with the metric $\langle x\rangle^{-2}\mathrm{d}x^2+\mathrm{d}\xi^2$. In the remaining part of this chapter, we would like to explore these relationships. Roughly speaking, we assume that $V(x)$ is a symbol of the class $S(\langle x\rangle^{-\mu}, \langle x\rangle^{-2}\mathrm{d}x^2)$, where $\mu>0$ in the long-range case and $\mu>1$ in the short-range case. Under such a condition, if $\chi\in C_0^\infty(\mathbb{R})$, then $\chi(H)$ itself is a pseudo-differential operator. This property and other simple properties of functions of H are described in Sect. 4.14.

In Sect. 4.15 we present a construction of a Fourier integral operator I^+ that has the same phase as the Isozaki-Kitada modifier (4.0.13) but its amplitude in the outgoing region solves asymptotically the appropriate transport equation. We call it an improved Isozaki-Kitada modifier.

In Sect. 4.16 we show a number of strong propagation estimates that use microlocal cutoffs. There are various possible ways to show these estimates. Originally these estimates were obtained in [IK4] using similar estimates on the resolvent $(H-\lambda)^{-1}$ and the Fourier transform. In our approach, we first show these estimates for the free evolution e^{-itH_0}, using the non-stationary phase method. Then we use the improved Isozaki-Kitada modifiers and the Duhamel formula to obtain similar estimates for the full evolution e^{-itH}. The crucial step of this proof is an application of the strong minimal velocity estimates of Sect. 4.13.

If we multiply the wave operator by a pseudo-differential cutoff supported in an outgoing region, then it equals an improved Isozaki-Kitada modifier modulo terms in $\Psi(\langle x\rangle^{-\infty})$. In particular, this means that the wave operator with a cutoff in an outgoing region is a pseudo-differential operator in the short-range case and a Fourier integral operator in the long-range case. These facts follow from the microlocal propagation estimates of Sect. 4.16 and are proven in Sect. 4.17.

One can also ask about regularity properties of wave operators without a microlocal cutoff. Such properties are the subject of Sect. 4.18. We show that, for $\chi\in C_0^\infty(\mathbb{R}\backslash\{0\})$ and $s'<s$,

$$\langle x\rangle^{-s}\Omega^+\chi(H_0)\langle x\rangle^{s'} \text{ is bounded.} \tag{4.0.18}$$

This result was obtained by Jensen-Nakamura [JN] and Herbst-Skibsted [HeSk1].

In the case of a positive s, one can strengthen this property and show that (4.0.18) is true for $0<s=s'$. This result is due to Isozaki [I3] and requires a somewhat different proof.

4.1 Schrödinger Hamiltonians

In this section we will describe notation and facts concerning Schrödinger Hamiltonians. The basic notation concerning Hilbert spaces is given in Appendix B.1.

Most of the time, we will work with the Hilbert space $L^2(X)$ where $X = \mathbb{R}^n$. Basic notation concerning this Hilbert space was given in Sect. 3.1.

We will also use the notation $\langle x \rangle := \sqrt{1+x^2}$ and

$$|s|_- := \begin{cases} |s|, s < 0, \\ 0, \ s \geq 0, \end{cases} \qquad |s|_+ := \begin{cases} 0, \ s \leq 0, \\ |s|, s > 0, \end{cases}$$

$$\langle s \rangle_- := \begin{cases} \langle s \rangle, s < 0, \\ 1, \ s \geq 0, \end{cases} \qquad \langle s \rangle_+ := \begin{cases} 1, \ s \leq 0, \\ \langle s \rangle, s > 0. \end{cases}$$

A Schrödinger operator is an operator of the form

$$H = \frac{1}{2}D^2 + V(x), \tag{4.1.1}$$

where $V(x)$ is a real-valued function on X satisfying appropriate conditions that make it possible to define (4.1.1) as a self-adjoint operator.

Let us list conditions that are useful in defining H:

Definition 4.1.1
(1a) $V(x)$ is H_0-bounded with the H_0-bound a_1 if

$$\lim_{\lambda\to\infty} \|(\lambda + H_0)^{-1}V(x)\| = a_1.$$

(1b) $V(x)$ is H_0-compact if

$$V(x)(1 + H_0)^{-1} \text{ is compact.} \tag{4.1.2}$$

(2a) $V(x)$ is H_0-form bounded with the H_0-bound a_2 if

$$\lim_{\lambda\to\infty} \|(\lambda + H_0)^{-\frac{1}{2}}V(x)(\lambda + H_0)^{-\frac{1}{2}}\| = a_2.$$

(2b) $V(x)$ is H_0-form compact if

$$(1 + H_0)^{-\frac{1}{2}}V(x)(1 + H_0)^{-\frac{1}{2}} \text{ is compact.} \tag{4.1.3}$$

Let us note the following implications.

Proposition 4.1.2
(i) $V(x)$ is H_0-bounded with H_0-bound $a_1 \Rightarrow V(x)$ is H_0-form bounded with H_0-form bound $\leq a_1$.
(ii) $V(x)$ is H_0-compact $\Rightarrow V(x)$ is H_0-bounded with H_0-bound 0.
(iii) $V(x)$ is H_0-form compact $\Rightarrow V(x)$ is H_0-form bounded with H_0-form bound 0.
(iv) $V(x)$ is H_0-compact $\Rightarrow V(x)$ is H_0-form compact.

The following proposition gives certain sufficient conditions for the self-adjointness of H.

Proposition 4.1.3
(i) If $V(x)$ is H_0-bounded with H_0-bound less than 1, then H is a self-adjoint operator with the domain

$$\mathcal{D}(H) = \mathcal{D}(H_0) = H^2(X),$$

(ii) If $V(x)$ is H_0-form bounded with H_0-bound less than 1, then H is a self-adjoint operator with the form domain

$$\mathcal{Q}(H) = \mathcal{Q}(H_0) = H^1(X).$$

The H_0-form compactness of $V(x)$ means that $V(x)$ decays (in some mean sense) in all directions. This is typical for 2-body interactions and this assumption will be satisfied most of the time in this chapter.

We will usually assume a stronger assumption, namely, the H_0-compactness of $V(x)$, because it is technically somewhat easier than the H_0-form compactness.

Let us state the so-called Weyl theorem for Schrödinger operators.

Proposition 4.1.4
Assume (4.1.3). Then for any $\chi \in C_0^\infty(\mathbb{R})$, the operator $\chi(H)-\chi(H_0)$ is compact and

$$\sigma_{\mathrm{ess}}(H) = [0, \infty[. \tag{4.1.4}$$

Our main object of interest will be the one-parameter group of unitary operators e^{-itH} generated by H. If $\phi \in L^2(X)$ we will sometimes use the notation

$$\phi_t := \mathrm{e}^{-itH}\phi.$$

If $A(t)$ is an operator-valued function, then the *Heisenberg derivative* of $A(t)$ is defined as

$$\mathbf{D}A(t) := \tfrac{\mathrm{d}}{\mathrm{d}t}A(t) + i[H, A(t)].$$

Note that

$$\tfrac{\mathrm{d}}{\mathrm{d}t}\mathrm{e}^{-itH}A(t)\mathrm{e}^{-itH} = \mathrm{e}^{-itH}\left(\mathbf{D}A(t)\right)\mathrm{e}^{-itH}.$$

4.2 Weak Large Velocity Estimates

In this and the next section we will assume that (4.1.3) holds. Note that, compared to the assumptions used in the remainder of this chapter, (4.1.3) is a very weak condition.

The main result of this section gives a rigorous meaning to the idea that, for large time, the probability of finding the particle in the region $x^2 > 2Ht^2$ goes to zero in a certain weak sense. This result gives us very little control on the rate at which this probability goes to zero. Therefore it is called a *weak* large velocity estimate. The weak large velocity estimate will be used very often in this chapter in the proof of the existence and completeness of wave operators.

Proposition 4.2.1
(i) If $\chi \in C_0^\infty(\mathbb{R})$, $\theta_1 < \theta_2$ *and* $\mathrm{supp}\chi \subset]-\infty, \frac{1}{2}\theta_1^2[$, *then*

$$\int_1^\infty \left\| 1\!\!1_{[\theta_1,\theta_2]}\left(\frac{|x|}{t}\right)\chi(H)\phi_t\right\|^2 \frac{\mathrm{d}t}{t} \leq C\|\phi\|^2.$$

(ii) Let $\chi \in C_0^\infty(\mathbb{R})$, $\mathrm{supp}\chi \subset]-\infty, \frac{1}{2}\theta_1^2[$, $F \in C^\infty(\mathbb{R})$ *with* $F' \in C_0^\infty(\mathbb{R})$ *and* $\mathrm{supp}F \subset]\theta_1, \infty[$. *Then*

$$\mathrm{s-}\lim_{t\to\infty} \mathrm{e}^{itH} F\left(\frac{|x|}{t}\right)\chi(H)\mathrm{e}^{-itH} = 0.$$

Before the proof of this proposition let us state a simple lemma.

Lemma 4.2.2
(i) Let $\chi \in C_0^\infty(\mathbb{R})$ *and* $f \in C_0^\infty(X)$. *Then*

$$\left\|\left[\chi(H), f\left(\frac{x}{t}\right)\right]\right\| \in O(t^{-1}).$$

(ii) If, moreover, $0 \leq \chi \leq 1$, $\mathrm{supp}\chi \subset [-\infty, \frac{1}{2}\theta^2]$ *and* $0 \leq f \leq 1$, $0 \notin \mathrm{supp}f$, *one has*

$$\left\|\chi(H)D\frac{x}{|x|}f\left(\frac{|x|}{t}\right)\right\| \leq \theta + o(t^0). \tag{4.2.1}$$

Proof. *(i)* follows from the bound

$$\|(H+i)^{-1}[H,x](H+i)^{-1}\| < C$$

and the methods of Appendix C.1.

Let us now prove *(ii)*. We first claim that

$$\chi(H)f\left(\frac{|x|}{t}\right)V(x)\chi(H) = o(t^0). \tag{4.2.2}$$

Indeed, by (4.1.3) $\chi(H)V(x)\chi(H)$ is compact, moreover, $\text{s}-\lim_{t\to\infty} f\left(\frac{|x|}{t}\right) = 0$. Therefore,

$$\lim_{t\to\infty} f\left(\frac{|x|}{t}\right)\chi(H)V(x)\chi(H) = 0,$$

which, using *(i)* proves (4.2.2). Using (4.2.2), we compute

$$\begin{aligned}
&\chi(H)D\tfrac{x}{|x|}f^2\left(\tfrac{|x|}{t}\right)\tfrac{x}{|x|}D\chi(H) \\
&\leq \chi(H)\left(Df^2\left(\tfrac{|x|}{t}\right)D + 2f^2\left(\tfrac{|x|}{t}\right)V(x)\right)\chi(H) + o(t^0) \\
&= 2\chi(H)f^2\left(\tfrac{|x|}{t}\right)H\chi(H) + o(t^0).
\end{aligned} \tag{4.2.3}$$

This clearly has the norm less than $\theta^2 + o(t^0)$. □

Proof of Proposition 4.2.1. We will prove the proposition by constructing a suitable propagation observable. Let $\theta_{-2} < \theta_{-1} < \theta_0 < \theta_1 < \theta_2$ such that $\text{supp}\chi \subset [-\infty, \frac{1}{2}\theta_{-2}^2[$. Choose $f \in C_0^\infty(\mathbb{R})$ so that $\text{supp}f \subset [\theta_{-1}, \infty]$, $f = 1$ on $[\theta_0, \theta_1]$. Define

$$F(s) := \int_{-\infty}^{s} f^2(s_1)\text{d}s_1.$$

Our propagation observable will be

$$\Phi(t) := \chi(H)F\left(\frac{|x|}{t}\right)\chi(H).$$

We compute:

$$\begin{aligned}
-\mathbf{D}\Phi(t) &= t^{-1}\chi(H)f^2\left(\tfrac{|x|}{t}\right)\tfrac{|x|}{t}\chi(H) \\
&\quad -\tfrac{1}{2}t^{-1}\chi(H)\left(D\tfrac{x}{|x|}f^2\left(\tfrac{|x|}{t}\right) + \text{hc}\right)\chi(H).
\end{aligned} \tag{4.2.4}$$

Choose $\tilde{\chi} \in C_0^\infty(\mathbb{R})$, $\tilde{f} \in C_0^\infty(\mathbb{R})$ such that $\chi\tilde{\chi} = \chi$, $f\tilde{f} = f$, $0 \leq \tilde{\chi} \leq 1$, $0 \leq \tilde{f} \leq 1$, $\text{supp}\tilde{\chi} \subset [-\infty, \frac{1}{2}\theta_{-2}^2]$ and $0 \notin \text{supp}\tilde{f}$. Using Lemma 4.2.2 *(i)*, we see that the second term on the right of (4.2.4) equals

$$\frac{1}{t}\chi(H)f\left(\frac{|x|}{t}\right)\tilde{\chi}(H)\left(D\frac{x}{|x|}\tilde{f}\left(\frac{|x|}{t}\right) + \text{hc}\right)\chi(H) + O(t^{-2}). \tag{4.2.5}$$

Using Lemma 4.2.2 *(ii)* to estimate (4.2.5), we see that (4.2.4) can be estimated from below by

$$C_0t^{-1}\chi(H)f^2\left(\frac{|x|}{t}\right)\chi(H), \tag{4.2.6}$$

where $C_0 := \theta_0 - \theta_{-1} > 0$. By Lemma B.4.1, this implies

$$\int_1^\infty \left\| f\left(\frac{|x|}{t}\right) \chi(H)\phi_t \right\|^2 \frac{\mathrm{d}t}{t} \leq C\|\phi\|^2. \tag{4.2.7}$$

and ends the proof of *(i)*.

Let us prove *(ii)*. Let the function F satisfy the conditions described in *(ii)*. Clearly, we can assume additionally that $F \geq 0$ and $F(s) = 1$ for $s \geq R_0$. Choose $\tilde{f} \in C_0^\infty(X)$ such that $\tilde{f} = 1$ on $\mathrm{supp}F'$ and $\mathrm{supp}\tilde{f} \subset [\theta_1, \infty[$. Then

$$\mathbf{D}\Phi(t) = t^{-1}\chi(H)\tilde{f}\left(\frac{|x|}{t}\right) B(t)\tilde{f}\left(\frac{|x|}{t}\right)\chi(H) + O(t^{-2}),$$

where $B(t)$ is uniformly bounded. Therefore, by *(i)*, there exists the limit

$$\mathrm{s-}\lim_{t\to\infty} \mathrm{e}^{itH}\Phi(t)\mathrm{e}^{-itH}. \tag{4.2.8}$$

If, in addition, F has a compact support, then, by *(i)*, we have

$$\textstyle\int_1^\infty (\phi_t|\Phi(t)\phi_t)\frac{\mathrm{d}t}{t} \leq C\|\phi\|^2. \tag{4.2.9}$$

Thus, if F satisfies the conditions in *(ii)* and has a compact support, the limit (4.2.8) is zero.

Let us now take functions $F_1 \in C^\infty(\mathbb{R})$, $f \in C_0^\infty(\mathbb{R})$ such that $\mathrm{supp}F_1 \subset [\theta_0, \infty[$, $F_1 = 1$ on a neighborhood of ∞, and $F_1' = f^2$. Set

$$\Phi_R(t) := \chi(H)F_1\left(\frac{|x|}{Rt}\right)\chi(H).$$

By the previous discussion, we know that, for $R \geq 1$, the limit

$$\mathrm{s-}\lim_{t\to\infty} \mathrm{e}^{itH}\Phi_R(t)\mathrm{e}^{-itH}$$

exists. Repeating the calculations of the proof of *(i)* and keeping track of R we obtain

$$\begin{aligned} -\mathbf{D}\Phi_R(t) &= \tfrac{1}{t}\chi(H)f^2\left(\tfrac{|x|}{Rt}\right)\tfrac{|x|}{Rt}\chi(H) - \tfrac{1}{2tR}\chi(H)D\tfrac{x}{|x|}f^2\left(\tfrac{|x|}{Rt}\right)\chi(H) + \mathrm{hc} \\ &\geq \tfrac{1}{t}\left(1 - \tfrac{C_1}{R}\right)\chi(H)f^2\left(\tfrac{|x|}{Rt}\right)\tfrac{|x|}{Rt}\chi(H) + O(t^{-2}R^{-2}). \end{aligned} \tag{4.2.10}$$

Hence, for $R \geq C_1$,

$$-\mathbf{D}\Phi_R(t) \geq O(t^{-2}R^{-2}). \tag{4.2.11}$$

Therefore, for $t_0 \geq 0$, we have

$$\begin{aligned} \mathrm{s-}\lim_{t\to\infty} \mathrm{e}^{itH}\Phi_R(t)\mathrm{e}^{itH} &= \mathrm{e}^{it_0H}\Phi_R(t_0)\mathrm{e}^{-it_0H} \\ &\quad + \textstyle\int_{t_0}^\infty \mathrm{e}^{isH}\left(\mathbf{D}\Phi(s)\right)\mathrm{e}^{-isH}\mathrm{d}s \\ &\leq \mathrm{e}^{it_0H}\Phi_R(t_0)\mathrm{e}^{-it_0H} + O(t_0^{-1}R^{-2}). \end{aligned} \tag{4.2.12}$$

For a fixed t_0, the terms on the right-hand side of (4.2.12) go strongly to zero as $R \to \infty$. Hence

$$\text{s-}\lim_{R\to\infty}\left(\text{s-}\lim_{t\to\infty} \mathrm{e}^{itH}\Phi_R(t)\chi(H)\mathrm{e}^{-itH}\right) = 0. \tag{4.2.13}$$

We remark now that, for $R \geq 1$, the function $F_1(|x|) - F_1\left(\frac{|x|}{R}\right)$ has a compact support included in $]\theta_0, \infty[$. So,

$$\text{s-}\lim_{t\to\infty} \mathrm{e}^{itH}\left(\Phi_1(t) - \Phi_R(t)\right)\chi(H)\mathrm{e}^{-itH} = 0. \tag{4.2.14}$$

Letting R tend to infinity in (4.2.14) and using (4.2.13), we obtain

$$\text{s-}\lim_{t\to\infty} \mathrm{e}^{itH}\Phi_1(t)\mathrm{e}^{-itH} = 0.$$

This ends the proof of *(ii)*. □

4.3 The Mourre Estimate and Its Consequences

We first define the self-adjoint operator

$$A := \tfrac{1}{2}(\langle x, D\rangle + \langle D, x\rangle)$$

called the *generator of dilations.* A is defined as the infinitesimal generator of the unitary group T_t defined by

$$T_t\phi(x) := \mathrm{e}^{-\frac{nt}{2}}\phi(\mathrm{e}^{-t}x),\ \phi \in L^2(X).$$

It is easy to verify that

$$\mathcal{D}(A) = \{\phi \in L^2(X) \mid A\phi \in L^2(X)\}.$$

For an H_0−bounded potential $V(x)$, we will denote by $x\nabla_x V(x)$ the (possibly unbounded) operator $[A, iV(x)]$, defined as a form on $\mathcal{D}(A) \cap \mathcal{D}(H_0)$.

The following theorem describes the Mourre estimate for 2-body Hamiltonians.

Theorem 4.3.1
Suppose that (4.1.2) holds and

$$(1-\Delta)^{-1}x\nabla_x V(x)(1-\Delta)^{-1} \quad \textit{is compact.} \tag{4.3.1}$$

Then, for any $\lambda_1 < \lambda_2$, there exists a compact operator K such that

$$\mathbb{1}_{[\lambda_1,\lambda_2]}(H)[H, iA]\mathbb{1}_{[\lambda_1,\lambda_2]}(H) = 2H\mathbb{1}_{[\lambda_1,\lambda_2]}(H) + K. \tag{4.3.2}$$

Moreover, for any $\lambda \notin \sigma_{\mathrm{pp}}(H)$ and $\delta > 0$, we can find an open neighborhood Δ containing λ such that

$$\mathbb{1}_{\Delta}(H)[H, iA]\mathbb{1}_{\Delta}(H) \geq 2(\lambda - \delta)\mathbb{1}_{\Delta}(H). \tag{4.3.3}$$

Proof. As a form on $\mathcal{D}(A) \cap \mathcal{D}(H)$, we have

$$[H, iA] = D^2 - x\nabla_x V = 2H - 2V(x) - x\nabla_x V(x).$$

Using (4.1.2) and (4.3.1), we see that

$$(1 - \Delta)^{-1}(2V(x) + x\nabla_x V(x))(1 - \Delta)^{-1}$$

is compact and

$$\mathbb{1}_{[\lambda_1, \lambda_2]}(H)(1 - \Delta)$$

is bounded. This implies (4.3.2).

To prove (4.3.3), we use the fact that if $\lambda \notin \sigma_{\mathrm{pp}}(H)$, we have

$$\text{s-}\lim_{\kappa \to 0} \mathbb{1}_{[\lambda-\kappa, \lambda+\kappa]}(H) = \mathbb{1}_{\{\lambda\}}(H) = 0.$$

Since K is compact, this implies that $K\mathbb{1}_{[\lambda-\kappa, \lambda+\kappa]}(H)$ tends to 0 in norm when κ tends to 0, which proves (4.3.3). □

Theorem 4.3.1 is analogous to the classical Mourre estimate proven in Proposition 2.3.2 in Chap. 2.

The next theorem is known as the *virial theorem* and has been proven by various authors [Wei, Kal, Mo1, PSS]. Our proof follows [PSS].

Theorem 4.3.2
Assume that H is self-adjoint with domain $H^2(X)$ and

$$(H + i)^{-1} x\nabla_x V(x)(H + i)^{-1} \textit{ is bounded.}$$

Then for any $\lambda \in \mathbb{R}$,

$$\mathbb{1}_{\{\lambda\}}(H)[H, iA]\mathbb{1}_{\{\lambda\}}(H) = 0.$$

Proof. Let $H_0 = \frac{1}{2}D^2$. As maps on $\mathcal{S}(X)$, we have

$$H_0(1 + i\epsilon A) = (1 + 2\epsilon + i\epsilon A)H_0,$$

which gives

$$H_0(1 + i\epsilon A)^{-1} = (1 + 2\epsilon + i\epsilon A)^{-1}H_0,$$

and hence

$$\begin{aligned}&(H_0+i)(1+i\epsilon A)^{-1}(H_0+i)^{-1}\\&=(1+2\epsilon+i\epsilon A)^{-1}H_0(H_0+i)^{-1}+i(1+i\epsilon A)^{-1}(H_0+i)^{-1}.\end{aligned}\tag{4.3.4}$$

This implies that $(1+i\epsilon A)^{-1}$ is bounded on $H^2(X)$. We also deduce from (4.3.4) that

$$\text{s}-\lim_{\epsilon\to 0}(H_0+i)(1+i\epsilon A)^{-1}(H_0+i)^{-1}=1.\tag{4.3.5}$$

If we define

$$A_\epsilon := A(1+i\epsilon A)^{-1}=i\epsilon^{-1}(1+i\epsilon A)^{-1}-i\epsilon^{-1},$$

we claim that

$$\begin{aligned}&\text{w}-\lim_{\epsilon\to 0}(H_0+i)^{-1}[H,iA_\epsilon](H_0+i)^{-1}\\&=(H_0+i)^{-1}[H,iA](H_0+i)^{-1}.\end{aligned}\tag{4.3.6}$$

Indeed, we have

$$\begin{aligned}&(H_0+i)^{-1}[H,iA_\epsilon](H_0+i)^{-1}\\&=-\epsilon^{-1}(H_0+i)^{-1}[H,(1+i\epsilon A)^{-1}](H_0+i)^{-1}\\&=(H_0+i)^{-1}(1+i\epsilon A)^{-1}[H,iA](1+i\epsilon A)^{-1}(H_0+i)^{-1}.\end{aligned}$$

This implies (4.3.6), using (4.3.5) and the fact that $(H_0+i)^{-1}[H,iA](H_0+i)^{-1}$ is bounded.

Let now ψ_i, $i=1,2$, be eigenvectors for H with the energy λ. Since $\psi_i\in H^2(X)$ and $H\psi_i=\lambda\psi_i$, we have, by (4.3.6),

$$\begin{aligned}(\psi_1|[H,iA]\psi_2)&=\lim_{\epsilon\to 0}(\psi_1|[H,iA_\epsilon]\psi_2)\\&=\lim_{\epsilon\to 0}(H\psi_1,iA_\epsilon\psi_2)-(iA_\epsilon\psi_1,H\psi_2)=0,\end{aligned}$$

which proves the proposition. □

The next result is due to Mourre [Mo1].

Theorem 4.3.3
Assume the hypotheses of Theorem 4.3.1. Then for any $\lambda_1\le\lambda_2$ such that $0\notin[\lambda_1,\lambda_2]$,

$$\dim \mathbb{1}^{\text{pp}}(H)\mathbb{1}_{[\lambda_1,\lambda_2]}(H)<\infty.$$

Proof. If $\lambda_2<0$, the result follows immediately from (4.1.4).

Assume now that $\lambda_1>0$. By (4.3.2), for some compact operator K, we have

$$\mathbb{1}_{[\lambda_1,\lambda_2]}(H)[H,iA]\mathbb{1}_{[\lambda_1,\lambda_2]}(H)\ge 2\lambda_1\mathbb{1}_{[\lambda_1,\lambda_2]}(H)+K.\tag{4.3.7}$$

Let ψ_n, $n\in\mathbb{N}$, be orthonormal eigenfunctions with eigenvalues in $[\lambda_1,\lambda_2]$. Using (4.3.7) and Proposition 4.3.2, we obtain

$$0 = (\psi_n, \mathbb{1}_{[\lambda_1,\lambda_2]}(H)[H, iA]\mathbb{1}_{[\lambda_1,\lambda_2]}(H)\psi_n) \geq 2\lambda_1\|\psi_n\|^2 + (\psi_n, K\psi_n).$$

If the set $\{\psi_n\}_{n\in\mathbb{N}}$ is infinite, the sequence ψ_n tends weakly to zero, so

$$\lim_{n\to\infty}(\psi_n, K\psi_n) = 0,$$

since K is compact. But this is a contradiction, since $\lambda_1 > 0$. This completes the proof of the theorem. □

Note that Theorem 4.3.3 implies that the only accumulation point of $\sigma_{\mathrm{pp}}(H)$ can be at 0. Therefore

$$\mathbb{1}^{\mathrm{pp}}(H) = \mathbb{1}_{\overline{\sigma_{\mathrm{pp}}(H)}}(H). \tag{4.3.8}$$

Bound states of a 2-body Hamiltonian fall into three categories. Bound states with a negative energy are the most physical ones. Sometimes a Hamiltonian may possess a zero energy bound state. Finally, a priori one should not rule out positive energy bound states.

One might think that positive energy bound states can be an obstacle in developing scattering theory. It turns out that they are not. We know by Theorem 4.3.3 that they are discrete in $]0, \infty[$. Therefore, for instance, in proofs of asymptotic completeness we just localize in energy outside of the pure point spectrum. Nevertheless, it is good to know that under quite general assumptions on the potentials there are no positive bound states whatsoever. This result is due to Froese-Herbst [FH2]. It will follow from the more general Theorems 6.5.1 and 6.5.4 valid in the N-body case.

Theorem 4.3.4
(i) Assume the hypotheses of Theorem 4.3.1. Let $\psi \in H^2(X)$ satisfy $H\psi = E\psi$ with $E > 0$. Then for any θ, we have

$$\mathrm{e}^{\theta|x|}\psi \in H^2(X).$$

(ii) If $E < 0$, then

$$\mathrm{e}^{(|-E|^{1/2}-\epsilon)|x|}\psi \in H^2(X), \quad \epsilon > 0.$$

(iii) If, moreover,

$$\lim_{\lambda\to\infty}\|(\lambda - \tfrac{1}{2}\Delta)^{-\frac{1}{2}}x\nabla_x V(x)(\lambda - \tfrac{1}{2}\Delta))^{-\frac{1}{2}}\| < 1,$$

then H has no positive eigenvalues.

4.4 Asymptotic Velocity

In this section we construct the fundamental asymptotic observable for time-independent 2-body Hamiltonians, namely, the *asymptotic velocity* and describe its basic properties. It will be the analog of the asymptotic momentum constructed in Theorem 3.2.1 in the time-decaying case. Beside its independent interest, this observable will be useful in our proof of the asymptotic completeness of wave operators for the Hamiltonians we study in this chapter.

The main results of this section are stated in the following theorem.

Theorem 4.4.1
Assume that (4.1.2) holds and

$$\int_1^\infty \left\|(1-\Delta)^{-1}\nabla_x V(x)\mathbb{1}_{[1,\infty[}(\tfrac{|x|}{R})(1-\Delta)^{-1}\right\| \mathrm{d}R < \infty. \tag{4.4.1}$$

Then (4.1.3) holds, and hence the conclusion of Theorem 4.2.1 is true. Likewise, (4.3.1) holds and hence the conclusion of Theorem 4.3.3 is true. Moreover, the following holds:
(i) There exists

$$\text{s}-C_\infty - \lim_{t\to\infty} \mathrm{e}^{\mathrm{i}tH}\frac{x}{t}\mathrm{e}^{-\mathrm{i}tH} =: P^+. \tag{4.4.2}$$

The vector of commuting self-adjoint operators P^+ is densely defined and commutes with the Hamiltonian H. It is called the asymptotic velocity.
(ii)

$$\text{s}-C_\infty - \lim_{t\to\infty} \mathrm{e}^{\mathrm{i}tH} D\mathrm{e}^{-\mathrm{i}tH}\mathbb{1}_{X\setminus\{0\}}(P^+) = P^+. \tag{4.4.3}$$

(iii)

$$H\mathbb{1}_{X\setminus\{0\}}(P^+) = \frac{1}{2}(P^+)^2.$$

(iv)

$$\mathbb{1}_{\{0\}}(P^+) = \mathbb{1}^{\text{pp}}(H).$$

(v)

$$\sigma(P^+, H) = \left\{\left(\xi, \tfrac{1}{2}\xi^2\right) \mid \xi\in X\right\}\cup\{0\}\times\sigma_{\text{pp}}(H). \tag{4.4.4}$$

Remark. The assumptions of this theorem are more general than the assumptions of Theorem 4.6.1 about the existence of short-range wave operators and than the assumptions of Theorem 4.7.1 about the existence of long-range wave operators.

Similarly as in the classical case, the space $\text{Ran}\mathbb{1}_{X\setminus\{0\}}(P^+)$ can be called the space of *scattering states.* By Theorem 4.4.1 *(iv)*, it coincides with the continuous spectral subspace. This fact should be compared with the case of classical Hamiltonians, where there may exist unbounded trajectories that are not scattering trajectories.

(4.4.4) gives a description of the joint spectrum of the asymptotic velocity and the Hamiltonian. It can be shown independently, as a consequence of Sects. 4.6 and 4.7, where we prove the existence and completeness of wave operators under additional assumptions on the potentials. Therefore, (4.4.4) can be considered as a very weak version of asymptotic completeness. Its proof will be given in the next section.

The proof of Theorem 4.4.1 will be divided into a series of lemmas and propositions, some of them of an independent interest. Throughout the section we assume (4.1.2) and (4.4.1).

We start with a rather technical lemma that deals with various properties of the potentials.

Lemma 4.4.2

$$(1-\Delta)^{-1}\nabla_x V(x)(1-\Delta)^{-1} \textit{ is compact,} \tag{4.4.5}$$

$$(1-\Delta)^{-1}x\nabla_x V(x)(1-\Delta)^{-1} \quad \textit{is compact.} \tag{4.4.6}$$

Moreover, if $J \in C^\infty(X)$ such that $J' \in C_0^\infty(X)$ and $J = 0$ on a neighborhood of zero, then

$$(1-\Delta)^{-1}\nabla_x V(x) J\left(\frac{x}{R}\right)(1-\Delta)^{-1} \in L^1(\mathrm{d}R). \tag{4.4.7}$$

Finally, if $\chi \in C_0^\infty(\mathbb{R})$, then

$$[D, \chi(H)]J(\tfrac{x}{R}) \in o(R^0) \cap L^1(\mathrm{d}R). \tag{4.4.8}$$

Proof. Writing

$$(1-\Delta)^{-1}\nabla_x V(x)(1-\Delta)^{-1} = (1-\Delta)^{-1}[D, iV(x)](1-\Delta)^{-1}$$

we see that (4.4.5) is true.

Let us prove now (4.4.7). We can assume that $\mathrm{supp} J \subset X \backslash B(1)$. Now (4.4.7) follows easily from the following identity:

$$\begin{aligned}
&(1-\Delta)^{-1}\nabla_x V(x) J\left(\tfrac{x}{R}\right)(1-\Delta)^{-1} \\
&= J\left(\tfrac{x}{R}\right)(1-\Delta)^{-1}\nabla_x V(x) \mathbb{1}_{[1,\infty[}(\tfrac{|x|}{R})(1-\Delta)^{-1} \\
&+(1-\Delta)^{-1}\left(\tfrac{1}{R} D\nabla J\left(\tfrac{x}{R}\right) + \tfrac{1}{R^2}\Delta J\left(\tfrac{x}{R}\right)\right)(1-\Delta)^{-1}\nabla_x V(x)\mathbb{1}_{[1,\infty[}(\tfrac{|x|}{R})(1-\Delta)^{-1}.
\end{aligned}$$

Both terms of the above expression are integrable in R on $[1, \infty[$. This completes the proof of (4.4.7).

Next let us prove (4.4.6). Choose $F \in C^\infty(\mathbb{R})$ such that $F = 0$ on $[0, 1]$ and $F = 1$ on $[2, \infty[$. Set

$$C_0 := \int_0^\infty F(1/s)\mathrm{d}s.$$

We first note that

$$\begin{aligned}&(1-\Delta)^{-1}\nabla_x V(x)\tfrac{x}{|x|}F\left(\tfrac{|x|}{R}\right)(1-\Delta)^{-1}\\ &=(1-\Delta)^{-1}\left[F\left(\tfrac{|x|}{R}\right)\tfrac{x}{|x|}D, V(x)\right](1-\Delta)^{-1}\end{aligned}$$

is compact. Moreover,

$$\begin{aligned}&\textstyle\int_1^\infty(1-\Delta)^{-1}\nabla_x V(x)\tfrac{x}{|x|}F\left(\tfrac{|x|}{R}\right)(1-\Delta)^{-1}\mathrm{d}R\\ &=(1-\Delta)^{-1}x\nabla_x V(x)f(|x|)(1-\Delta)^{-1},\end{aligned}\tag{4.4.9}$$

where

$$f(|x|)=\int_{1/|x|}^{\infty}F(1/s)\mathrm{d}s$$

and the integral on the left of (4.4.9) is norm convergent by (4.4.7). So (4.4.9) is compact. But $C_0-f(|x|)$ is a function of compact support. Hence

$$(1-\Delta)^{-1}x\nabla_x V(x)(C_0-f(|x|))(1-\Delta)^{-1}$$

is compact too. This immediately implies (4.4.6). (4.4.8) follows easily from (4.4.5) and (4.4.7). □

Our next proposition describes an important weak propagation estimate, which, in the more general N-body case, was first obtained by Graf [Gr].

Proposition 4.4.3
Let $\chi\in C_0^\infty(\mathbb{R})$ and $0<\theta_1<\theta_2$. Then

$$\int_1^\infty\left\|\mathbb{1}_{[\theta_1,\theta_2]}\left(\frac{|x|}{t}\right)\left(\frac{x}{t}-D\right)\chi(H)\phi_t\right\|^2\frac{\mathrm{d}t}{t}\le C\|\phi\|^2.\tag{4.4.10}$$

Proof. Let $\theta_2<\theta_3<\theta_4$ be such that $\mathrm{supp}\chi\subset]-\infty,\frac12\theta_3^2[$ and let $R\in C^\infty(X)$ be a function such that

$$\nabla^2R\ge0,\quad \nabla R=0\quad\text{on a neighborhood of zero,}\quad R(x)=\frac12x^2,\quad |x|>\theta_1.$$

Let $J\in C_0^\infty(\mathbb{R})$ such that $J=1$ on $[0,\theta_4]$.

Set

$$M(t):=\tfrac12\langle D-\tfrac{x}{t},\nabla R(\tfrac{x}{t})\rangle+\tfrac12\langle\nabla R(\tfrac{x}{t}),D-\tfrac{x}{t}\rangle+R(\tfrac{x}{t}).$$

We consider the following uniformly bounded propagation observable

$$\Phi(t):=\chi(H)J(\tfrac{x}{t})M(t)J(\tfrac{x}{t})\chi(H).$$

We compute:

$$\begin{aligned}\mathbf{D}\Phi(t)\ &=\chi(H)\left(\mathbf{D}J(\tfrac{x}{t})\right)M(t)J(\tfrac{x}{t})\chi(H)+\mathrm{hc}\\ &-\chi(H)J(\tfrac{x}{t})\nabla_xR(\tfrac{x}{t})\nabla_xV(x)J(\tfrac{x}{t})\chi(H)\\ &+\tfrac1t\chi(H)J(\tfrac{x}{t})\langle D-\tfrac{x}{t},\nabla^2R(\tfrac{x}{t})(D-\tfrac{x}{t})\rangle J(\tfrac{x}{t})\chi(H).\end{aligned}\tag{4.4.11}$$

The first term on the right-hand side of (4.4.11) can be written as

$$\frac{1}{t}\chi(H)\tilde{j}\left(\frac{x}{t}\right)B(t)\tilde{j}\left(\frac{x}{t}\right)\chi(H)+O(t^{-2})+\langle t\rangle^{-1}L^1(\mathrm{d}t)$$

for a certain uniformly bounded operator $B(t)$ and $\tilde{j}\in C_0^\infty(\mathbb{R}), \operatorname{supp}\tilde{j}\subset]\theta_3,\infty[$. Using Proposition 4.2.1, we see that this term gives an integrable contribution along the evolution.

The second term in the right-hand side of (4.4.11) is integrable in norm by Lemma 4.4.2.

We observe now that

$$\begin{aligned} J(\tfrac{x}{t})\tfrac{1}{t}\langle D-\tfrac{x}{t},\nabla^2R(\tfrac{x}{t})D-\tfrac{x}{t}\rangle J(\tfrac{x}{t}) &= \tfrac{1}{t}\langle D-\tfrac{x}{t},J^2(\tfrac{x}{t})\nabla^2R(\tfrac{x}{t})D-\tfrac{x}{t}\rangle+O(t^{-3})\\ &\geq \tfrac{1}{t}\langle(D-\tfrac{x}{t}),\mathbb{1}_{[\theta_1,\theta_2]}(\tfrac{|x|}{t})(D-\tfrac{x}{t})\rangle+O(t^{-3})\end{aligned}$$

and apply Lemma B.4.1 to complete the proof of the proposition. □

Lemma 4.4.4
Let $\chi\in C_0^\infty(\mathbb{R})$ and $0<\theta_1<\theta_2$. Then

$$\mathrm{s}-\lim_{t\to\infty}\mathbb{1}_{[\theta_1,\theta_2]}\left(\frac{|x|}{t}\right)\left(\frac{x}{t}-D\right)\chi(H)\mathrm{e}^{-itH}=0. \qquad (4.4.12)$$

Proof. Let $J\in C_0^\infty(X)$ such that $0\notin\operatorname{supp}J$ and $J(x)=1$ for $\theta_1\leq|x|\leq\theta_2$. Let $\tilde{\chi}\in C_0^\infty(\mathbb{R})$ such that $\chi\tilde{\chi}=\chi$. First note, using (4.4.8) that

$$\mathrm{s}-\lim_{t\to\infty}J\left(\tfrac{|x|}{t}\right)\left(\tfrac{x}{t}-D\right)\chi(H)\mathrm{e}^{-itH}=\mathrm{s}-\lim_{t\to\infty}\tilde{\chi}(H)J\left(\tfrac{|x|}{t}\right)\left(\tfrac{x}{t}-D\right)\chi(H)\mathrm{e}^{-itH}.$$

Set

$$\Phi(t):=\chi(H)\left(D-\frac{x}{t}\right)J\left(\frac{x}{t}\right)\tilde{\chi}^2(H)J\left(\frac{x}{t}\right)\left(D-\frac{x}{t}\right)\chi(H).$$

We have

$$\begin{aligned}&-\mathbf{D}\Phi(t)=2t^{-1}\Phi(t)\\ &+2t^{-1}\chi(H)\left(D-\tfrac{x}{t}\right)\langle D-\tfrac{x}{t},\nabla J\left(\tfrac{x}{t}\right)\rangle\tilde{\chi}^2(H)J\left(\tfrac{x}{t}\right)\left(D-\tfrac{x}{t}\right)\chi(H)+\mathrm{hc}\\ &+\chi(H)\nabla_xV(x)J(\tfrac{x}{t})\tilde{\chi}^2(H)J\left(\tfrac{x}{t}\right)\left(D-\tfrac{x}{t}\right)\chi(H)+\mathrm{hc}+O(t^{-2}).\end{aligned}$$

Clearly, the right-hand side of the above expression is integrable along the evolution by (4.4.7) and Proposition 4.4.3. Therefore, there exists the limit

$$\lim_{t\to\infty}(\phi_t|\Phi(t)\phi_t)\,. \qquad (4.4.13)$$

But, again by Proposition 4.4.3, we have

$$\int_1^\infty (\phi_t|\Phi(t)\phi_t)\frac{\mathrm{d}t}{t} < \infty. \tag{4.4.14}$$

Hence the limit (4.4.13) is zero. □

Proposition 4.4.5
Let $J \in C_\infty(\mathbb{R})$. Then there exists

$$\mathrm{s-}\lim_{t\to\infty} \mathrm{e}^{itH} J\left(\frac{x}{t}\right)\mathrm{e}^{-itH}. \tag{4.4.15}$$

Moreover, if $J(0) = 1$, then

$$\mathrm{s-}\lim_{R\to\infty}\left(\mathrm{s-}\lim_{t\to\infty} \mathrm{e}^{itH} J\left(\frac{x}{Rt}\right)\mathrm{e}^{-itH}\right) = 1. \tag{4.4.16}$$

If we define

$$\mathrm{s-}C_\infty - \lim_{t\to\infty} \mathrm{e}^{itH}\frac{x}{t}\mathrm{e}^{-itH} =: P^+, \tag{4.4.17}$$

then the vector of commuting self-adjoint operators P^+ is densely defined and commutes with the Hamiltonian H. Hence Theorem 4.4.1 (i) is true.

Proof. By density, we may assume that $J \in C_0^\infty(X)$ and that J is constant on a neighborhood of zero. It also suffices to prove the existence of

$$\begin{aligned}&\mathrm{s-}\lim_{t\to\infty} \mathrm{e}^{itH} J\left(\tfrac{x}{t}\right)\chi^2(H)\mathrm{e}^{-itH}\\ &= \mathrm{s-}\lim_{t\to\infty} \mathrm{e}^{itH}\chi(H) J\left(\tfrac{x}{t}\right)\chi(H)\mathrm{e}^{-itH}\end{aligned} \tag{4.4.18}$$

for any $\chi \in C_0^\infty(\mathbb{R})$, using Lemma 4.2.2 *(i)*.

As the first step, we will show that there exists

$$\begin{aligned}&\mathrm{s-}\lim_{t\to\infty} \mathrm{e}^{itH}\Phi(t)\mathrm{e}^{-itH},\\ &\text{where}\quad \Phi(t) := \chi(H)\left(J\left(\tfrac{x}{t}\right) + \left\langle D - \tfrac{x}{t}, \nabla J\left(\tfrac{x}{t}\right)\right\rangle\right)\chi(H).\end{aligned} \tag{4.4.19}$$

To see this, we note that

$$\begin{aligned}\mathbf{D}\Phi(t) &= -\chi(H)\nabla_x V(x)\nabla J(\tfrac{x}{t})\chi(H)\\ &+t^{-1}\chi(H)\langle D - \tfrac{x}{t}, \nabla^2 J\left(\tfrac{x}{t}\right)(D - \tfrac{x}{t})\rangle\chi(H) + O(t^{-2}).\end{aligned} \tag{4.4.20}$$

The first term on the right of (4.4.20) is integrable in norm by (4.4.7) and the second is integrable along the evolution by Lemma 4.4.3. Therefore the limit (4.4.19) exists.

It remains to show that (4.4.19) equals (4.4.18). But this follows from Lemma 4.4.4.

This completes the proof of the existence of (4.4.15). (4.4.16) follows immediately from Proposition 4.2.1 *(ii)*. The fact that P^+ exists as a densely defined commuting vector of operators follows then from Proposition B.2.1 and B.2.2. The fact that P^+ commutes with H follows easily from Lemma 4.2.2 *(i)*. □

As in Chap. 3, the asymptotic velocity of a scattering state is also the limit of its momentum.

Proposition 4.4.6
Let $g \in C_\infty(X)$. *Then*

$$\text{s}-\lim_{t\to\infty} \mathrm{e}^{itH} g(D) \mathrm{e}^{-itH} \mathbb{1}_{X\setminus\{0\}}(P^+) = g(P^+)\mathbb{1}_{X\setminus\{0\}}(P^+). \tag{4.4.21}$$

Hence Theorem 4.4.1 (ii) is true.

Proof. It is enough to assume that $g \in C_0^\infty(X)$ and to prove that

$$\text{s}-\lim_{t\to\infty} \mathrm{e}^{itH} \left(g(D) - g\left(\frac{x}{t}\right)\right) J\left(\frac{x}{t}\right) \chi(H)\mathrm{e}^{-itH} = 0 \tag{4.4.22}$$

for any $J \in C_0^\infty(X\setminus\{0\})$ and $\chi \in C_0^\infty(\mathbb{R})$.

We already know that the limit exists. Next we note that, by the Baker-Campbell-Hausdorff formula (3.2.28),

$$\begin{aligned} &\left(g(D) - g\left(\tfrac{x}{t}\right)\right) J\left(\tfrac{x}{t}\right) \chi(H)\mathrm{e}^{-itH} \\ &= B(t)\left(D - \tfrac{x}{t}\right) J\left(\tfrac{x}{t}\right) \chi(H)\mathrm{e}^{-itH} + O(t^{-1}), \end{aligned} \tag{4.4.23}$$

where $B(t)$ is uniformly bounded.

(4.4.23) converges to zero strongly by Lemma 4.4.4. Hence (4.4.22) is true. □

Next we will prove the so-called *low velocity estimate* in a version due to Graf [Gr].

Proposition 4.4.7
Let $\chi \in C_0^\infty(\mathbb{R})$, $\theta_0 > 0$ *and* $\mathrm{supp}\chi \subset]\frac{1}{2}\theta_0^2, \infty[\setminus \sigma_{\mathrm{pp}}(H)$. *Then*

$$\int_1^\infty \left\| \mathbb{1}_{[0,\theta_0]}\left(\frac{|x|}{t}\right) \chi(H)\phi_t \right\|^2 \frac{\mathrm{d}t}{t} \le C\|\phi\|^2. \tag{4.4.24}$$

Proof. Let $\theta_0 < \theta_3$ such that $\frac{1}{2}\theta_3^2 \notin \sigma_{\mathrm{pp}}(H)$. Let us also fix θ_1, θ_2, θ_4 such that $\theta_0 < \theta_1 < \theta_2 < \theta_3 < \theta_4$ and $\theta_2^2 > \theta_1\theta_4$.

By Proposition 4.3.1, we will find a function $\tilde{\chi} \in C_0^\infty(\mathbb{R})$ such that $\tilde{\chi} = 1$ on a neighborhood of $\frac{1}{2}\theta_3^2$, $\mathrm{supp}\tilde{\chi} \subset]\frac{1}{2}\theta_2^2, \frac{1}{2}\theta_4^2[$ and

$$\tilde{\chi}(H)[H, iA]\tilde{\chi}(H) \ge \theta_2^2 \tilde{\chi}^2(H). \tag{4.4.25}$$

Let $\chi \in C_0^\infty(\mathbb{R})$ such that $\tilde{\chi} = 1$ on $\mathrm{supp}\chi$.

We also choose $J, \tilde{J} \in C_0^\infty(X)$ such that $J = 1$ for $|x| \leq \theta_0$, $\tilde{J} = 1$ on supp$J$ and supp$\tilde{J}$, supp$J \subset \{x \mid |x| < \theta_1\}$.

Set

$$M(t) := J\left(\tfrac{x}{t}\right) + \langle D - \tfrac{x}{t}, \nabla J\left(\tfrac{x}{t}\right)\rangle,$$
$$\Phi(t) := \chi(H)M(t)\tilde{\chi}(H)\tfrac{A}{t}\tilde{\chi}(H)M^*(t)\chi(H).$$

Using the boundedness of $\langle x\rangle\tilde{\chi}(H)\langle x\rangle^{-1}$, we see that $\Phi(t)$ is uniformly bounded. We compute

$$\begin{aligned}\mathbf{D}\Phi(t) &= t^{-1}\chi(H)\langle D - \tfrac{x}{t}, \nabla^2 J\left(\tfrac{x}{t}\right)(D - \tfrac{x}{t})\rangle\tilde{\chi}(H)\tfrac{A}{t}\tilde{\chi}(H)M^*(t)\chi(H) + \mathrm{hc}\\ &-\chi(H)\nabla_x V(x)J(\tfrac{x}{t})\tilde{\chi}(H)\tfrac{A}{t}\tilde{\chi}(H)M^*(t)\chi(H) + \mathrm{hc}\\ &+t^{-1}\chi(H)M(t)\tilde{\chi}(H)\left([H, iA] - \tfrac{A}{t}\right)\tilde{\chi}(H)M^*(t)\chi(H)\\ &=: R_1(t) + R_2(t) + R_3(t).\end{aligned}$$

The term $R_1(t)$ can be written as

$$R_1(t) = t^{-1}\chi(H)(D - \tfrac{x}{t})\tilde{J}(\tfrac{x}{t})B(t)\tilde{J}(\tfrac{x}{t})(D - \tfrac{x}{t})\chi(H) + O(t^{-2}) + \langle t\rangle^{-1}L^1(\mathrm{d}t)$$

for a certain uniformly bounded operator $B(t)$. Using Proposition 4.4.3, we see that $R_1(t)$ is integrable along the evolution.

The term $R_2(t)$ is integrable in norm.

Let us now estimate the term $R_3(t)$. By (4.4.25), we have

$$\begin{aligned}&t^{-1}\chi(H)M(t)\tilde{\chi}(H)i[H, A]\tilde{\chi}(H)M^*(t)\chi(H)\\ &\geq \theta_2^2 t^{-1}\chi(H)M(t)\tilde{\chi}^2(H)M^*(t)\chi(H)\\ &= \theta_2^2 t^{-1}\chi(H)M(t)M^*(t)\chi(H) + O(t^{-2}) + \langle t\rangle^{-1}L^1(\mathrm{d}t).\end{aligned}$$

By Lemma 4.2.2 *(i)* and *(ii)*, we have

$$\begin{aligned}\|\tilde{J}(\tfrac{x}{t})\tilde{\chi}(H)\tfrac{A}{t}\tilde{\chi}(H)\tilde{J}(\tfrac{x}{t})\| &\leq \|\tfrac{|x|}{t}\tilde{J}(\tfrac{x}{t})\tilde{\chi}(H)\|\|D\tilde{J}(\tfrac{x}{t})\tilde{\chi}(H)\| + O(t^{-1})\\ &\leq \theta_1\theta_4 + o(t^0).\end{aligned}$$

So,

$$\begin{aligned}&-\tfrac{1}{t}\chi(H)M(t)\tilde{\chi}(H)\tfrac{A}{t}\tilde{\chi}(H)M^*(t)\chi(H)\\ &= -\tfrac{1}{t}\chi(H)M(t)\tilde{J}(\tfrac{x}{t})\tilde{\chi}(H)\tfrac{A}{t}\tilde{\chi}(H)\tilde{J}(\tfrac{x}{t})M^*(t)\chi(H) + O(t^{-2})\\ &\geq -\tfrac{1}{t}\theta_1\theta_4\chi(H)M(t)M^*(t)\chi(H) + O(t^{-2}).\end{aligned}$$

Therefore,

$$R_3(t) \geq C_0\chi(H)M(t)M^*(t)\chi(H) + O(t^{-2}) + \langle t\rangle^{-1}L^1(\mathrm{d}t),$$

where $C_0 := \theta_2^2 - \theta_1\theta_4 > 0$.

We write then

$$M(t) = J\left(\tfrac{x}{t}\right) + \langle D - \tfrac{x}{t}, \nabla J\left(\tfrac{x}{t}\right)\rangle =: M_1(t) + M_2(t).$$

We use now the inequality

$$(M_1 + M_2)(M_1^* + M_2^*) \geq (1-\epsilon)M_1M_1^* + (1-\epsilon^{-1})M_2M_2^*$$

to deduce that

$$\begin{aligned} R_3(t) \ &\geq (1-\epsilon)C_0t^{-1}\chi(H)J^2\left(\tfrac{x}{t}\right)\chi(H) \\ &+(1-\epsilon^{-1})C_0\chi(H)\langle D - \tfrac{x}{t}, \nabla J\left(\tfrac{x}{t}\right)\rangle\langle \nabla J\left(\tfrac{x}{t}\right), D - \tfrac{x}{t}\rangle\chi(H). \end{aligned} \tag{4.4.26}$$

The second term on the right-hand side of (4.4.26) is integrable along the evolution by Lemma 4.4.3. Hence, by Proposition 4.2.1, we obtain

$$\int_1^\infty \left\| J\left(\frac{x}{t}\right)\chi^2(H)\phi_t\right\|^2 \frac{dt}{t} \leq C\|\phi\|^2. \tag{4.4.27}$$

Thus we have shown our lemma for χ supported in a sufficiently small neighborhood of

$$\tfrac{1}{2}\theta_3^2 \in]\tfrac{1}{2}\theta_0^2, \infty[\backslash\sigma_{\rm pp}(H).$$

Now assume that χ is any C_0^∞ function supported in $]\frac{1}{2}\theta_0^2, \infty[\backslash\sigma_{\rm pp}(H)$. We can write

$$\chi = \sum_{j=1}^n \chi_j,$$

where χ_j are C_0^∞ functions with sufficiently small supports, such that

$$\int_1^\infty \left\| 1\!\!1_{[0,\theta_0]}\left(\frac{|x|}{t}\right)\chi_j(H)\phi_t\right\|^2 \frac{dt}{t} \leq C\|\phi\|^2,\ j = 1,\ldots,n.$$

By the Schwarz inequality,

$$\begin{aligned} &\textstyle\int_1^\infty \left(1\!\!1_{[0,\theta_0]}\left(\frac{|x|}{t}\right)\chi_i(H)\phi_t | 1\!\!1_{[0,\theta_0]}\left(\frac{|x|}{t}\right)\chi_j(H)\phi_t\right)\frac{dt}{t} \\ &\leq \left(\textstyle\int_1^\infty \left\|1\!\!1_{[0,\theta_0]}\left(\frac{|x|}{t}\right)\chi_i(H)\phi_t\right\|^2\frac{dt}{t}\right)^{\frac{1}{2}}\left(\textstyle\int_1^\infty \left\|1\!\!1_{[0,\theta_0]}\left(\frac{|x|}{t}\right)\chi_j(H)\phi_t\right\|^2\frac{dt}{t}\right)^{\frac{1}{2}} \\ &\leq C\|\phi\|^2. \end{aligned}$$

Hence

$$\int_1^\infty \left\| 1\!\!1_{[0,\theta_0]}\left(\frac{|x|}{t}\right)\chi(H)\phi_t\right\|^2 \frac{dt}{t} \leq n^2C\|\phi\|^2.$$

□

The following proposition shows that the states with zero asymptotic velocity coincide with the bound states, and hence Theorem 4.4.1 *(iv)* is true.

Proposition 4.4.8

$$\mathbb{1}_{\{0\}}(P^+) = \mathbb{1}^{\mathrm{pp}}(H).$$

Proof. Let $H\phi = \tau\phi$ and $J \in C_\infty(X)$. Then

$$\begin{aligned}&\lim_{t\to\infty} \mathrm{e}^{itH} J\left(\tfrac{x}{t}\right) \mathrm{e}^{-itH}\phi \\ &= J(0)\phi + \lim_{t\to\infty} \mathrm{e}^{it(H-\tau)} \left(J\left(\tfrac{x}{t}\right) - J(0)\right)\phi = J(0)\phi.\end{aligned} \tag{4.4.28}$$

This shows that $P^+\phi = 0$ and proves

$$\mathbb{1}_{\{0\}}(P^+) \geq \mathbb{1}^{\mathrm{pp}}(H).$$

Let us prove the opposite inequality. Let $\theta > 0$. Let $\chi \in C_0^\infty(]\frac{1}{2}\theta^2, \infty[\backslash\sigma_{\mathrm{pp}}(H))$ and $J \in C_0^\infty(\{x \mid |x| \leq \theta\})$. Then

$$\mathrm{s-}\lim_{t\to\infty} \mathrm{e}^{itH}\chi(H)J^2\left(\frac{x}{t}\right)\chi(H)\mathrm{e}^{-itH} = \chi^2(H)J^2(P^+). \tag{4.4.29}$$

By Proposition 4.4.7,

$$\int_1^\infty \left\| J\left(\frac{x}{t}\right)\chi(H)\phi_t \right\|^2 \frac{\mathrm{d}t}{t} < \infty. \tag{4.4.30}$$

Therefore, (4.4.29) is zero, which proves that

$$\mathbb{1}_{\{0\}}(P^+) \leq \mathbb{1}_{\overline{\{0\}\cup\sigma_{\mathrm{pp}}(H)}}(H). \tag{4.4.31}$$

But by (4.3.8) the right-hand side of (4.4.31) equals $\mathbb{1}^{\mathrm{pp}}(H)$. □

The following proposition proves that Theorem 4.4.1 *(iii)* is true.

Proposition 4.4.9

$$H\mathbb{1}_{X\backslash\{0\}}(P^+) = \frac{1}{2}(P^+)^2\mathbb{1}_{X\backslash\{0\}}(P^+).$$

Proof. Let $\chi \in C_0^\infty(\mathbb{R})$ and $J \in C_0^\infty(X\backslash\{0\})$. It is enough to show that

$$\chi(H)J(P^+) = \chi(\tfrac{1}{2}(P^+)^2)J(P^+). \tag{4.4.32}$$

We have

$$\begin{aligned}&\left(\chi(H) - \chi(\tfrac{1}{2}(P^+)^2)\right) J(P^+) \\ &= \mathrm{s-}\lim_{t\to\infty} \mathrm{e}^{itH}\left(\chi(H) - \chi(H_0)\right) J\left(\tfrac{x}{t}\right) \mathrm{e}^{-itH},\end{aligned} \tag{4.4.33}$$

by Lemma 4.4.6.

But $\chi(H) - \chi(H_0)$ is compact and $J\left(\frac{x}{t}\right)$ goes strongly to zero. Therefore, the right-hand side of (4.4.33) goes to zero in norm. □

4.5 Joint Spectrum of P^+ and H

In this section we prove part *(v)* of Theorem 4.4.1. It gives a complete description of the joint spectrum of the commuting operators P^+, H. This description can be shown independently as a consequence of the existence and completeness of (short-range or long-range) wave operators. But in order to show the completeness of wave operators, one has to strengthen the assumptions on the potentials, and the result about the joint spectrum, as we will see below, is true under the same assumptions as those we used in the theorem about the existence of the asymptotic velocity.

Lemma 4.5.1
Let $J \in C_0^\infty(X\backslash\{0\})$ and $g \in C_0^\infty(X)$. Then

$$\int_0^\infty \left\|(1-\Delta)^{-1}J\left(\frac{x}{t}\right)[V(x), g(D)](1-\Delta)^{-1}\right\| \mathrm{d}t < \infty.$$

Proof. Using the representation

$$g(D) = (2\pi)^{-n}\int \hat{g}(y)\mathrm{e}^{-i\langle y,D\rangle}\mathrm{d}y,$$

we obtain

$$\begin{aligned} I(t) &:= (1-\Delta)^{-1}J\left(\tfrac{x}{t}\right)[V(x), g(D)](1-\Delta)^{-1} \\ &= (2\pi)^{-n}\textstyle\int \hat{g}(y)y\mathrm{d}y \int_0^1 \mathrm{d}\tau R(t,\mathrm{d}\tau), \end{aligned}$$

where $\quad R(t,\tau) := (1-\Delta)^{-1}J\left(\dfrac{x}{t}\right)\mathrm{e}^{-i\tau\langle y,D\rangle}\nabla_x V(x)\mathrm{e}^{-i(1-\tau)\langle y,D\rangle}(1-\Delta)^{-1}.$

Let $J_1 \in C_0^\infty(X\backslash\{0\})$ such that $J_1 = 1$ on a neighborhood of $\mathrm{supp}J$. Commuting N times functions of $\frac{x}{t}$ we obtain

$$\begin{aligned} R(t,\tau) &= (1-\Delta)^{-1}J\left(\tfrac{x}{t}\right)\mathrm{e}^{-i\tau\langle y,D\rangle}J_1\left(\tfrac{x}{t}\right)\nabla_x V(x)\mathrm{e}^{-i(1-\tau)\langle y,D\rangle}(1-\Delta)^{-1} \\ &+J\left(\tfrac{x}{t}\right)B_N(t,y\tau)\nabla_x V(x)\mathrm{e}^{-i(1-\tau)\langle y,D\rangle}(1-\Delta)^{-1}, \end{aligned}$$

where

$$\|B_N(t,y\tau)(1-\Delta)\| \le C\left\langle\frac{\tau y}{t}\right\rangle^N.$$

Therefore, $I(t)$ can be estimated by

$$\begin{aligned} &C\textstyle\int|\hat{g}(y)y|\mathrm{d}y\int_0^1\mathrm{d}\tau\left\|(1-\Delta)^{-1}J_1\left(\frac{x}{t}\right)\nabla_x V(1-\Delta)^{-1}\right\| \\ &+C\textstyle\int|\hat{g}(y)y|\mathrm{d}y\int_0^1\mathrm{d}\tau\langle\frac{\tau y}{t}\rangle^N\left\|(1-\Delta)^{-1}\nabla_x V\mathrm{e}^{-i(1-\tau)\langle y,D\rangle}(1-\Delta)^{-1}\right\|. \end{aligned} \tag{4.5.1}$$

Now we use Lemma 4.4.2 (4.4.7) to see that the first term of (4.5.1) is integrable and (4.4.5) to see that the second term is $O(t^{-N})$. □

Proposition 4.5.2

$$\sigma(P^+) = X.$$

Proof. Let $\xi_0 \in X$ and $|\xi_0| = r_2 > r_1 > r_0 > 0$. Let $g \in C_0^\infty(B(\xi_0, r_0))$ such that $g(\xi_0) \neq 0$. Let $J \in C_0^\infty(]-\infty, r_2])$ such that $J = 1$ on $]-\infty, r_1]$ and $j := -J' \geq 0$. Clearly, we have

$$J(|\xi - \xi_0|)g(\xi) = g(\xi), \tag{4.5.2}$$

$$\begin{aligned} j(|\xi - \xi_0|)|\xi - \xi_0| &\geq r_1 j(|\xi - \xi_0|), \\ g(\xi)|\xi - \xi_0| &\leq r_0 g(\xi). \end{aligned} \tag{4.5.3}$$

Let $\chi \in C_0^\infty(\mathbb{R})$ such that $\chi(\frac{1}{2}\xi^2)g(\xi) = g(\xi)$. Consider the observable

$$\Phi(t) := \chi(H) J\left(\left|\frac{x}{t} - \xi_0\right|\right) g^2(D) J\left(\left|\frac{x}{t} - \xi_0\right|\right) \chi(H).$$

By Proposition 4.4.6, we have

$$\text{s-}\lim_{t\to\infty} \mathrm{e}^{itH}\Phi(t)\mathrm{e}^{-itH} = g^2(P^+).$$

Let us now compute:

$$\begin{aligned} \mathbf{D}\Phi(t) &= \chi(H)J\left(\left|\tfrac{x}{t} - \xi_0\right|\right)[V(x), g^2(D)]J\left(\left|\tfrac{x}{t} - \xi_0\right|\right)\chi(H) \\ &+\chi(H)\left(\mathbf{D}J\left(\left|\tfrac{x}{t} - \xi_0\right|\right)\right)g^2(D)J\left(\left|\tfrac{x}{t} - \xi_0\right|\right)\chi(H) + \text{hc}. \end{aligned} \tag{4.5.4}$$

We claim now that

$$\left(\mathbf{D}J\left(\left|\frac{x}{t} - \xi_0\right|\right)\right) g^2(D) J\left(\left|\frac{x}{t} - \xi_0\right|\right) + \text{hc} \geq -Ct^{-2}. \tag{4.5.5}$$

In fact, choose $\tilde{g} \in C_0^\infty(X)$ such that $\tilde{g}g = g$ and $\text{supp}\tilde{g} \subset B(\xi_0, r_0)$. Then, commuting functions of D and of $\frac{x}{t}$, we see that the left-hand side of (4.5.5) equals

$$\begin{aligned} &\tfrac{1}{t}g(D)(Jj)^{\frac{1}{2}}\left(\left|\tfrac{x}{t} - \xi_0\right|\right)\left\langle \tfrac{\frac{x}{t}-\xi_0}{|\frac{x}{t}-\xi_0|}, \xi_0 - D\right\rangle \tilde{g}^2(D)(Jj)^{\frac{1}{2}}\left(\left|\tfrac{x}{t} - \xi_0\right|\right) + \text{hc} \\ &+\tfrac{1}{t}g(D)\left|\tfrac{x}{t} - \xi_0\right|(Jj)\left(\left|\tfrac{x}{t} - \xi_0\right|\right)g(D) + O(t^{-2}) \\ &\geq (r_2 - r_1)\tfrac{1}{t}g(D)\left|\tfrac{x}{t} - \xi_0\right|(Jj)\left(\left|\tfrac{x}{t} - \xi_0\right|\right)g(D) + O(t^{-2}), \end{aligned}$$

which implies (4.5.5). Note that if we look for cancellations in the left-hand side of (4.5.5), we actually obtain

$$\left(\mathbf{D}J\left(\left|\frac{x}{t} - \xi_0\right|\right)\right) g^2(D) J\left(\left|\frac{x}{t} - \xi_0\right|\right) + \text{hc} \geq -Ct^{-3},$$

although it will not be needed in what follows.

By Lemma 4.5.1, the first term on the right-hand side of (4.5.4) is integrable in norm. Thus

$$\mathbf{D}\Phi(t) \geq R(t) \in L^1(\mathrm{d}t).$$

Let us now complete our proof. Using (4.5.5), for any $t_0 \in \mathbb{R}^+$, we have

$$\begin{aligned} g^2(P^+) &= \mathrm{e}^{it_0H}\Phi(t_0)\mathrm{e}^{-it_0H} + \textstyle\int_{t_0}^{\infty} \mathrm{e}^{itH}(\mathbf{D}\Phi(t))\mathrm{e}^{-itH}\mathrm{d}t \\ &\geq \mathrm{e}^{it_0H}\Phi(t_0)\mathrm{e}^{-it_0H} - \textstyle\int_{t_0}^{\infty} \|R(t)\|\mathrm{d}t. \end{aligned} \tag{4.5.6}$$

By choosing t_0 big enough, we can make the integral on the right-hand side of (4.5.6) as small as we wish. We claim that

$$\lim_{t_0\to\infty} \left\|\mathrm{e}^{it_0H}\Phi(t_0)\mathrm{e}^{-it_0H}\right\| \tag{4.5.7}$$

exists and is non-zero. To this end, note that

$$\begin{aligned} &\mathrm{e}^{it_0\xi_0D}\Phi(t_0)\mathrm{e}^{-it_0\xi_0D} \\ &= \chi\left(\tfrac{1}{2}D^2 + V(x+t_0\xi_0)\right) J\left(\tfrac{|x|}{t_0}\right) g^2(D) J\left(\tfrac{|x|}{t_0}\right) \chi\left(\tfrac{1}{2}D^2 + V(x+t_0\xi_0)\right). \end{aligned}$$

This goes strongly to $g^2(D)$, which is a non-zero operator. But

$$\|\mathrm{e}^{it_0H}\Phi(t_0)\mathrm{e}^{it_0H}\| = \|\mathrm{e}^{it_0\xi_0D}\Phi(t_0)\mathrm{e}^{-it_0\xi_0D}\|.$$

This shows that (4.5.7) is nonzero.

Therefore, $g(P^+) \neq 0$. Hence $\xi_0 \in \sigma(P^+)$. This completes the proof of the proposition. □

Corollary 4.5.3

$$\sigma(P^+, H) = \left\{\left(\xi, \frac{1}{2}\xi^2\right) \mid \xi \in X\right\} \cup \{0\} \times \sigma_{\mathrm{pp}}(H).$$

Proof. Proposition 4.4.9 shows that

$$\sigma(P^+, H) \cap (X\backslash\{0\}) \times \mathbb{R} \subset \left\{\left(\xi, \frac{1}{2}\xi^2\right) \mid \xi \in X\right\}. \tag{4.5.8}$$

It follows from Proposition 4.5.2 that we have the equality in (4.5.8).

Finally, Proposition 4.4.8 and (4.3.8) show that

$$\sigma(P^+, H) \cap (\{0\} \times \mathbb{R}) = \{0\} \times (\sigma_{\mathrm{pp}}(H) \cup \{0\}).$$

□

4.6 Short-Range Case

The asymptotic velocity constructed in Theorem 4.4.1 gives a classification of the states in $L^2(X)$ according to their asymptotic behavior under the evolution e^{-itH}. The states with zero asymptotic velocity coincide with the bound states of H. However, we would like to have a better understanding of the space of scattering states $\mathrm{Ran}\mathbb{1}_{X\setminus\{0\}}(P^+)$. It is natural to ask whether $\mathbb{1}_{X\setminus\{0\}}(P^+)P^+$ is unitarily equivalent to the momentum D. In this section we will assume that the potential $V(x)$ satisfies a *short-range* condition. In this case, one can give a positive answer to this question by constructing the *wave operators*. The main result of this section is the following theorem.

Theorem 4.6.1
Assume that (4.1.2) holds,

$$\int_0^\infty \left\| (1-\Delta)^{-1}\nabla V(x)\mathbb{1}_{[1,\infty[}\left(\frac{|x|}{R}\right)(1-\Delta)^{-1}\right\| \mathrm{d}R < \infty, \tag{4.6.1}$$

and, for a certain N,

$$\int_0^\infty \left\| (1-\Delta)^{-1}V(x)\mathbb{1}_{[1,\infty[}\left(\frac{|x|}{R}\right)(1-\Delta)^{-N}\right\| \mathrm{d}R < \infty. \tag{4.6.2}$$

Then there exist

$$\mathrm{s-}\lim_{t\to\infty} \mathrm{e}^{itH}\mathrm{e}^{-itH_0}, \tag{4.6.3}$$

$$\mathrm{s-}\lim_{t\to\infty} \mathrm{e}^{itH_0}\mathrm{e}^{-itH}\mathbb{1}^{\mathrm{c}}(H). \tag{4.6.4}$$

If we denote (4.6.3) by Ω_{sr}^+, then (4.6.4) equals $\Omega_{\mathrm{sr}}^{+*}$. One has

$$\Omega_{\mathrm{sr}}^{+*}\Omega_{\mathrm{sr}}^+ = \mathbb{1}, \qquad \Omega_{\mathrm{sr}}^+\Omega_{\mathrm{sr}}^{+*} = \mathbb{1}^{\mathrm{c}}(H).$$

Moreover, the hypothesis (4.4.1) holds, and hence the operator P^+ exists and one has

$$P^+ = \Omega_{\mathrm{sr}}^+ D \Omega_{\mathrm{sr}}^{+*}, \tag{4.6.5}$$

$$\mathbb{1}^{\mathrm{c}}(H)H = \Omega_{\mathrm{sr}}^+ H_0 \Omega_{\mathrm{sr}}^{+*}. \tag{4.6.6}$$

Remark. It is easy to see that just for the existence of wave operators the hypotheses (4.1.2) and (4.6.2) are sufficient.

Remark. Sometimes it will be convenient to strengthen the hypothesis (4.6.2) as follows:

$$\int_0^\infty \left\| (1-\Delta)^{-1}V(x)\mathbb{1}_{[1,\infty[}\left(\frac{|x|}{R}\right)(1-\Delta)^{-\frac{1}{2}}\right\| \mathrm{d}R < \infty. \tag{4.6.7}$$

Writing $\nabla_x V(x)$ as $[iD, V(x)]$ we see that then the hypothesis (4.6.1) follows from (4.6.7) and (4.1.2).

Proof. Let us first prove the existence of (4.6.3). Let $J \in C_0^\infty(X\backslash\{0\})$ and $\chi \in C_0^\infty(\mathbb{R})$. Denote

$$M(t) := J\left(\tfrac{x}{t}\right) + \langle D - \tfrac{x}{t}, \nabla J\left(\tfrac{x}{t}\right)\rangle,$$
$$\Phi(t) := \chi(H)M(t)\chi(H_0).$$

We will use the following easy identity consequence of (4.4.3):

$$\text{s-}\lim_{t\to\infty} \mathrm{e}^{itH_0}M(t)\chi(H_0)\mathrm{e}^{-itH_0} = J(D)\chi(H_0). \tag{4.6.8}$$

By a density argument, it suffices prove the existence of

$$\begin{aligned} \text{s-}\lim_{t\to\infty} \mathrm{e}^{itH}\mathrm{e}^{-itH_0}J(D)\chi^2(H_0) &= \text{s-}\lim_{t\to\infty} \mathrm{e}^{itH}\chi(H_0)M(t)\chi(H_0)\mathrm{e}^{-itH_0} \\ &= \text{s-}\lim_{t\to\infty} \mathrm{e}^{itH}\Phi(t)\mathrm{e}^{-itH_0}, \end{aligned} \tag{4.6.9}$$

using (4.1.2) and (4.6.8).

We compute:

$$\begin{aligned} &\tfrac{\mathrm{d}}{\mathrm{d}t}\Phi(t) + iH\Phi(t) - i\Phi(t)H_0 \\ &= \chi(H)\left\langle D - \tfrac{x}{t}, \nabla^2 J\left(\tfrac{x}{t}\right)\left(D - \tfrac{x}{t}\right)\right\rangle \chi(H_0) \\ &+\chi(H)V(x)M(t)\chi(H_0). \end{aligned} \tag{4.6.10}$$

The second term on the right of (4.6.10) is integrable in norm by hypothesis (4.6.2) and the first is integrable along the evolution using Lemma 4.4.3. This implies the existence of the limit (4.6.9).

To prove the existence of (4.6.4), we note first that by Theorem 4.4.1 and Lemma 4.4.4

$$\text{s-}\lim_{t\to\infty} \mathrm{e}^{itH}M(t)\mathrm{e}^{-itH} = J(P^+). \tag{4.6.11}$$

Using the fact that $\mathbb{1}^{\mathrm{c}}(H) = \mathbb{1}_{X\backslash\{0\}}(P^+)$ and a density argument, we see that it is enough to show the existence of

$$\text{s-}\lim_{t\to\infty} \mathrm{e}^{itH_0}\mathrm{e}^{-itH}J(P^+)\chi^2(H). \tag{4.6.12}$$

But by (4.6.11) and (4.1.2), the limit (4.6.12) equals

$$\text{s-}\lim_{t\to\infty} \mathrm{e}^{itH_0}\Phi^*(t)\mathrm{e}^{-itH}. \tag{4.6.13}$$

But the existence of the limit (4.6.13) follows by the same arguments as the existence of (4.6.9).

To show (4.6.5), let us consider $g \in C_\infty(X)$. Then by Theorem 4.4.1 *(ii)*

$$\begin{aligned} g(P^+)\mathbb{1}_{X\setminus\{0\}}(P^+) &= \text{s}-\lim_{t\to\infty} \mathrm{e}^{itH} g(D)\mathrm{e}^{-itH}\mathbb{1}_{X\setminus\{0\}}(P^+) \\ &= \text{s}-\lim_{t\to\infty} \mathrm{e}^{itH}\mathrm{e}^{-itH_0} g(D)\text{s}-\lim_{t\to\infty} \mathrm{e}^{itH_0}\mathrm{e}^{-itH}\mathbb{1}_{X\setminus\{0\}}(P^+) \\ &= \Omega^+_{\text{sr}} g(D)\Omega^{+*}_{\text{sr}}. \end{aligned}$$

Finally, to prove (4.6.6), we use (4.6.5) to obtain

$$\frac{1}{2}(P^+)^2 = \frac{1}{2}\Omega^+_{\text{sr}} D^2 \Omega^{+*}_{\text{sr}} = \Omega^+_{\text{sr}} H_0 \Omega^{+*}_{\text{sr}}.$$

Then we use Theorem 4.4.1 *(iii)* and *(iv)*. □

One can generalize the above theorem and define a wave operator that intertwines two Hamiltonians that differ by a short-range term.

Proposition 4.6.2
Let

$$H_i = \frac{1}{2}D^2 + V_i(x), \quad i = 1, 2,$$

where V_i are two potentials satisfying the hypotheses (4.1.2) and (4.4.1) of Theorem 4.4.1 and let P_i^+ be the asymptotic velocities associated with H_i. Assume that

$$\int_0^\infty \left\| (1-\Delta)^{-1}(V_1(x) - V_2(x))\mathbb{1}_{[1,\infty[}\left(\frac{|x|}{R}\right)(1-\Delta)^{-\frac{1}{2}} \right\| \mathrm{d}R < \infty. \qquad (4.6.14)$$

Then if $(k, j) = (1, 2)$ or $(k, j) = (2, 1)$, the limits

$$\text{s}-\lim_{t\to\infty} \mathrm{e}^{itH_k}\mathrm{e}^{-itH_j}\mathbb{1}^{\text{c}}(H_j) =: \Omega^+(H_k, H_j)$$

exist and

$$\begin{aligned} &\Omega^+(H_k, H_j)^* = \Omega^+(H_j, H_k), \\ &\Omega^+(H_k, H_j)\Omega^+(H_j, H_k) = \mathbb{1}^{\text{c}}(H_k), \\ &\Omega^+(H_j, H_k)P_k^+ = P_j^+\Omega^+(H_j, H_k), \\ &\Omega^+(H_j, H_k)H_k = H_j\Omega^+(H_j, H_k). \end{aligned}$$

Proof. Let J, χ and $M(t)$ be as in the proof of Theorem 4.6.1. Arguing as in the proof of this theorem we see that it is sufficient to show the existence of

$$\begin{aligned} &\text{s}-\lim_{t\to\infty} \mathrm{e}^{itH_k}\mathrm{e}^{-itH_j} J(P_j^+)\chi^2(H_j), \\ &= \text{s}-\lim_{t\to\infty} \mathrm{e}^{itH_k}\chi(H_k)M(t)\chi(H_j)\mathrm{e}^{-itH_j}. \end{aligned}$$

Then we mimic the arguments of the proof of Theorem 4.6.1. □

4.7 Long-Range Case

In this section we begin our study of scattering theory in the long-range case. In this case, the asymptotic velocity is well defined, although the usual wave operators used for short-range potentials typically do not exist. Nevertheless, one can construct modified wave operators, which intertwine the momentum and the asymptotic velocity on the space of scattering states.

In this section we will construct modified wave operators and we will show that they are complete. We will use the results of Sect. 4.3 to reduce ourselves to the case of long-range time-dependent potentials treated in Sect. 3.4. The main result of this section is the following theorem.

Theorem 4.7.1
Assume that $V(x)$ satisfies the assumptions of Theorem 4.4.1 and can be written as

$$V(x) = V_{\mathrm{s}}(x) + V_{\mathrm{l}}(x) \tag{4.7.1}$$

such that

$$\int_0^\infty \left\| (1-\Delta)^{-1} V_{\mathrm{s}}(x) \mathbb{1}_{[1,\infty[}\left(\frac{|x|}{R}\right)(1-\Delta)^{-\frac{1}{2}} \right\| \mathrm{d}R < \infty,$$

$$\begin{array}{l} \lim\limits_{|x|\to\infty} V_{\mathrm{l}}(x) = 0, \\ \int_0^\infty \sup\limits_{|x|>R} |\partial_x^\alpha V_{\mathrm{l}}(x)| \langle R\rangle^{|\alpha|-1} \mathrm{d}R < \infty, \quad |\alpha| = 1, 2. \end{array} \tag{4.7.2}$$

Then there exists a function $\tilde{S}(t,\xi)$ such that the limits

$$\mathrm{s-}\lim_{t\to\infty} \mathrm{e}^{itH} \mathrm{e}^{-i\tilde{S}(t,D)}, \tag{4.7.3}$$

$$\mathrm{s-}\lim_{t\to\infty} \mathrm{e}^{i\tilde{S}(t,D)} \mathrm{e}^{-itH} \mathbb{1}^{\mathrm{c}}(H) \tag{4.7.4}$$

exist. If we denote (4.7.3) by Ω_{lr}^+, then (4.7.4) equals $\Omega_{\mathrm{lr}}^{+*}$. Moreover, one has

$$\begin{array}{ll} \Omega_{\mathrm{lr}}^+ \Omega_{\mathrm{lr}}^{+*} = \mathbb{1}^{\mathrm{c}}(H), & \Omega_{\mathrm{lr}}^{+*} \Omega_{\mathrm{lr}}^+ = \mathbb{1}, \\ \Omega_{\mathrm{lr}}^+ P^+ = \Omega_{\mathrm{lr}}^+ D, & H\Omega_{\mathrm{lr}}^+ = \Omega_{\mathrm{lr}}^+ H_0. \end{array} \tag{4.7.5}$$

Before proving Theorem 4.7.1, let us explain how one can get rid of the short-range part $V_{\mathrm{s}}(x)$ of the potential. To this end, let us introduce the auxiliary Hamiltonian

$$H_{\mathrm{l}} := \tfrac{1}{2} D^2 + V_{\mathrm{l}}(x).$$

We also introduce the asymptotic velocity associated with H_{l}

$$P_{\mathrm{l}}^+ := \mathrm{s-}C_\infty - \lim_{t\to\infty} \mathrm{e}^{itH_{\mathrm{l}}} \tfrac{x}{t} \mathrm{e}^{-itH_{\mathrm{l}}}.$$

It follows immediately from Proposition 4.6.2 that the following lemma is true.

Lemma 4.7.2
There exists

$$\mathrm{s-}\lim_{t\to\infty} \mathrm{e}^{itH}\mathrm{e}^{-itH_1}\mathbb{1}^{\mathrm{c}}(H_1) =: \Omega_1^+, \tag{4.7.6}$$

$$\mathrm{s-}\lim_{t\to\infty} \mathrm{e}^{itH_1}\mathrm{e}^{-itH}\mathbb{1}^{\mathrm{c}}(H). \tag{4.7.7}$$

Moreover, (4.7.7) equals Ω_1^{+*} *and*

$$\Omega_1^{+*}\Omega_1^+ = \mathbb{1}^{\mathrm{c}}(H_1), \quad \Omega_1^+\Omega_1^{+*} = \mathbb{1}^{\mathrm{c}}(H),$$

$$\Omega_1^+ H_1 = H\Omega_1^+, \quad \Omega_1^+ P_1^+ = P^+\Omega_1^+.$$

By the above lemma, it is sufficient to show Theorem 4.7.1 assuming that $V_s(x) = 0$, which we will do in the remaining part of this section. In other words, in what follows, $V(x) = V_1(x)$.

Let us now explain how one can construct a function $\tilde{S}(t,\xi)$ that can be used in the definition of modified wave operators. This is somewhat complicated, because we want to show the theorem for a very large class of potentials (due essentially to Hörmander).

The most obvious candidate for this purpose is the function $S(t,\xi)$ constructed in Proposition 2.7.4 that satisfies asymptotically the Hamilton–Jacobi equation with the potential $V(x)$, more precisely, for every $\epsilon > 0$ there exists T_ϵ such that

$$\begin{aligned} &\partial_t S(t,\xi) = \tfrac{1}{2}\xi^2 + V(\nabla_\xi S(t,\xi)), \quad t > T_\epsilon, \quad |\xi| > \epsilon, \\ &\partial_\xi^\beta \left(S(t,\xi) - \tfrac{1}{2}t\xi^2\right) \in o(t), \quad \text{in} \quad |\xi| > \epsilon, \quad |\beta| \le 2. \end{aligned} \tag{4.7.8}$$

Unfortunately, this function can be used to define modified wave operators only under some additional assumptions on the potentials. In the general case, one needs a function whose construction is described in the following proposition.

Proposition 4.7.3
Let $V(x)$ *(4.7.2), that is,*

$$\int_0^\infty \langle R\rangle^{|\alpha|-1} \sup_{|x|\ge R} |\partial_x^\alpha V(x)| \mathrm{d}R < \infty, \quad |\alpha| = 1, 2.$$

Let $j \in C_0^\infty(X)$ *be a cutoff function such that*

$$\int j(y)\mathrm{d}y = 1, \quad \int yj(y)\mathrm{d}y = 0,$$

and let

$$\tilde{V}(t,x) := \int V(x + t^{\frac{1}{2}}y)j(y)\mathrm{d}y. \tag{4.7.9}$$

Then there exist a function $\tilde{S}(t,\xi)$ that satisfies the following properties:
(i) for every $\epsilon > 0$, there exists T_ϵ such that

$$\partial_t \tilde{S}(t,\xi) = \frac{1}{2}\xi^2 + \tilde{V}(t, \nabla_\xi \tilde{S}(t,\xi)), \quad t > T_\epsilon, \quad |\xi| > \epsilon.$$

(ii) For every $\epsilon > 0$,

$$\partial_\xi^\beta \left(\tilde{S}(t,\xi) - \tfrac{1}{2}t\xi^2\right) \in o(t), \qquad in \quad |\xi| > \epsilon, \quad |\beta| \le 2,$$
$$\partial_\xi^\beta \left(\tilde{S}(t,\xi) - \tfrac{1}{2}t\xi^2\right) \in o\left(t^{\frac{1}{2}|\beta|}\right), \qquad in \quad |\xi| > \epsilon, \quad |\beta| \ge 2.$$

Proof. Arguing as in the proof of Lemma 3.4.5, we see that, for any $\epsilon > 0$, the time-dependent potential $\tilde{V}(t,x)$ satisfies

$$\int_0^\infty \sup_{|x|\ge \epsilon t} |\partial_x^\alpha \tilde{V}(t,x)| \langle t\rangle^{|\alpha|-1}\mathrm{d}t < \infty, \quad |\alpha| = 1,2,$$
$$\int_0^\infty \sup_{|x|\ge \epsilon t} |\partial_x^\alpha \tilde{V}(t,x)| \langle t\rangle^{\frac{1}{2}|\alpha|}\mathrm{d}t < \infty, \quad |\alpha| \ge 2.$$

One can then follow the proof of Theorem 2.7.5 to prove the existence of $\tilde{S}(t,\xi)$. □

Now we can describe the modifiers that can be used to construct the wave operators.

Proposition 4.7.4
(i) The function $\tilde{S}(t,\xi)$ constructed in Proposition 4.7.3 using the potential $V_l(x)$ can always be used in Theorem 4.7.1 to construct modified wave operators.
(ii) Suppose that instead of (4.7.2) $V_l(x)$ satisfies one of the following hypotheses:

$$\int_0^\infty \langle R\rangle^{|\alpha|-1} \sup_{|x|\ge R} |\partial_x^\alpha V_l(x)|\mathrm{d}R < \infty, \quad |\alpha| = 1,2,3; \tag{4.7.10}$$

or

$$\begin{aligned} &\int_0^\infty \langle R\rangle^{1/2} \sup_{|x|\ge R} |\partial_x^\alpha V_l(x)|\mathrm{d}R < \infty, \quad |\alpha| = 1, \\ &\int_0^\infty \langle R\rangle \sup_{|x|\ge R} |\partial_x^\alpha V_l(x)|\mathrm{d}R < \infty, \quad |\alpha| = 2. \end{aligned} \tag{4.7.11}$$

Then, in Theorem 4.7.1, we can replace $\tilde{S}(t,\xi)$ with the function $S(t,\xi)$ constructed in Theorem 2.7.5 for the potential $V_l(x)$.

To prove the existence and completeness of modified wave operators, we will first reduce the problem to an effective time-dependent potential, and then we will apply the results of Chap. 3.

Let $\Theta \subset X\backslash\{0\}$ be a compact set. If the dimension of X is one, then we assume additionally that Θ is either to the right or to the left of 0. Let $J \in C_0^\infty(X\backslash\{0\})$ such that $J(x) = 1$ on a neighborhood of Θ. Fix also $x_0 \in \Theta$. Set

$$V_J(t,x) := (V(x) - V(tx_0))J\left(\frac{x}{t}\right) + V(tx_0).$$

The potential $V_J(t,x)$ will be called an *effective time-dependent potential.*

Proposition 4.7.5
(i) We have

$$V(t\xi) - V_J(t,t\xi) = 0 \textit{ for } \xi \textit{ in a certain neighborhood of } \Theta.$$

(ii) If $V(x)$ satisfies

$$\int_0^\infty \langle R\rangle^{|\alpha|-1} \sup_{|x|\geq R} |\partial_x^\alpha V(x)| \mathrm{d}R < \infty, \quad |\alpha| = 1,\ldots,n,$$

then $V_J(t,x)$ satisfies

$$\int_0^\infty \langle t\rangle^{|\alpha|-1} \|\partial_x^\alpha V_J(t,\cdot)\|_\infty \mathrm{d}t < \infty, \quad |\alpha| = 1,\ldots,n.$$

(iii) If $V(x)$ satisfies

$$\begin{array}{ll}\int_0^\infty \langle R\rangle^{1/2} \sup_{|x|\geq R} |\partial_x^\alpha V(x)| \mathrm{d}R < \infty, & |\alpha| = 1,\\ \int_0^\infty \langle R\rangle \sup_{|x|\geq R} |\partial_x^\alpha V(x)| \mathrm{d}R < \infty, & |\alpha| = 2,\end{array}$$

then $V_J(t,x)$ satisfies

$$\begin{array}{ll}\int_0^\infty \langle t\rangle^{1/2} \|\partial_x^\alpha V_J(t,\cdot)\|_\infty \mathrm{d}t < \infty, & |\alpha| = 1,\\ \int_0^\infty \langle t\rangle \|\partial_x^\alpha V_J(t,\cdot)\|_\infty \mathrm{d}t < \infty, & |\alpha| = 2.\end{array}$$

Proof. The property *(i)* is immediate.
To show *(ii)*, we write, for $|\alpha| \geq 1$,

$$\begin{array}{rcl}\partial_x^\alpha V_J(t,x) & = & \sum_{\beta\leq\alpha,\ 1\leq|\beta|} C_\beta \partial_x^\beta V(x) t^{|\beta|-|\alpha|} \partial^{\alpha-\beta} J(\frac{x}{t})\\ & & +(V(t,x) - V(tx_0))t^{-|\alpha|}\partial^\alpha J(\frac{x}{t}).\end{array}$$

The terms that contain the derivatives of $V(x)$ give clearly the correct decay. To estimate the last term, let us denote

$$r := \inf_{x\in\mathrm{supp}J} |x|.$$

Consider $x \in t\mathrm{supp}J$. Clearly, $tx_0 \in t\mathrm{supp}J$ too. Hence x and tx_0 can be joined by a curve of the length less than $\pi|x - tx_0|$ that lies entirely in $\{x \mid |x| > rt\}$. Therefore,

$$\begin{aligned}|(V(x) - V(tx_0))\partial^\alpha J(\tfrac{x}{t})| &\leq \pi|x - tx_0| \sup_{|x|>rt} |\nabla_x V(x)\partial^\alpha J(\tfrac{x}{t})| \\ &\leq Ct \sup_{|x|>rt} |\nabla_x V(x)|,\end{aligned}$$

where in the last step we used the compactness of $\mathrm{supp}J$. □

Let $S_J(t,\xi)$ be the solution of the Hamilton-Jacobi equation

$$\begin{cases} \partial_t S_J(t,\xi) = \frac{1}{2}\xi^2 + V_J(t, \nabla_\xi S_J(t,\xi)), \\ S_J(T_0, \xi) = 0, \end{cases}$$

constructed in Sect. 1.8 for some T_0 large enough.

Let

$$\tilde{V}_J(t,x) = \int V_J(t, x + t^{1/2}y)j(y)\mathrm{d}y,$$

where $j \in C_0^\infty(X)$ is the same function as in (4.7.9). For a T_0 sufficiently big, let $\tilde{S}_J(t,\xi)$ be the solution of the Hamilton-Jacobi equation

$$\begin{cases} \partial_t \tilde{S}_J(t,\xi) = \frac{1}{2}\xi^2 + \tilde{V}_J(t, \nabla_\xi \tilde{S}_J(t,\xi)), \\ \tilde{S}_J(T_0, \xi) = 0. \end{cases}$$

Lemma 4.7.6

Uniformly on Θ, there exists the following limits:

$$\lim_{t\to\infty} (S_J(t,\xi) - S(t,\xi)) =: \sigma^+(\xi), \tag{4.7.12}$$

$$\lim_{t\to\infty} \left(\tilde{S}_J(t,\xi) - \tilde{S}(t,\xi)\right) =: \tilde{\sigma}^+(\xi). \tag{4.7.13}$$

Proof. Let us show (4.7.12). By Proposition 4.7.5 *(i)*, for $t > T_1$ and for ξ in a certain neighborhood of Θ, both functions $S_J(t,\xi)$ and $S(t,\xi)$ satisfy the Hamilton-Jacobi equation with the same potential $V_J(t,x)$:

$$\begin{aligned} \partial_t S_J(t,\xi) &= \tfrac{1}{2}\xi^2 + V_J(t, \nabla_\xi S_J(t,\xi)), \\ \partial_t S(t,\xi) &= \tfrac{1}{2}\xi^2 + V_J(t, \nabla_\xi S(t,\xi)). \end{aligned}$$

They also satisfy, for ξ in a neighborhood of Θ, the estimates

$$\begin{aligned} &\partial_\xi^\beta (S_J(t,\xi) - \tfrac{1}{2}t\xi^2) \in o(t), \quad |\beta| \leq 2, \\ &\partial_\xi^\beta (S(t,\xi) - \tfrac{1}{2}t\xi^2) \in o(t), \quad |\beta| \leq 2, \end{aligned}$$

It follows then from Sect. 1.9 that

$$\lim_{t\to\infty}(S_J(t,\xi)-S(t,\xi))$$

exists for $\xi\in\Theta$. This ends the proof of (4.7.12).

To show (4.7.13), observe that, for ξ a certain neighborhood of Θ, and $t>T_2$

$$\tilde{V}(t,t\xi)=\tilde{V}_J(t,t\xi).$$

Then we argue as above. □

The following lemma follows immediately from Lemma 4.7.6.

Lemma 4.7.7
The following limits exist:

$$\lim_{t\to\infty}\mathrm{e}^{iS(t,D)}\mathrm{e}^{-iS_J(t,D)}\mathbb{1}_\Theta(D)=\mathrm{e}^{-i\sigma^+(D)},$$
$$\lim_{t\to\infty}\mathrm{e}^{i\tilde{S}(t,D)}\mathrm{e}^{-i\tilde{S}_J(t,D)}\mathbb{1}_\Theta(D)=\mathrm{e}^{-i\tilde{\sigma}^+(D)}.$$

We will denote by $U_J(t,s)$ the unitary evolution generated by

$$H_J(t):=\tfrac{1}{2}D^2+V_J(t,x).$$

We also introduce the asymptotic velocity (asymptotic momentum) associated with $H_J(t)$

$$\begin{aligned}D_J^+\ &:=\mathrm{s}-C_\infty-\lim_{t\to\infty}U_J(0,t)DU_J(t,0)\\ &=\mathrm{s}-C_\infty-\lim_{t\to\infty}U_J(0,t)\tfrac{x}{t}U_J(t,0).\end{aligned}$$

The Hamiltonian $H_J(t)$ belongs to the class considered in Chap. 3. Therefore, the following lemma is true.

Lemma 4.7.8
(i) There exist the limits

$$\mathrm{s}-\lim_{t\to\infty}U_J(0,t)\mathrm{e}^{-i\tilde{S}_J(t,D)},\tag{4.7.14}$$

$$\mathrm{s}-\lim_{t\to\infty}\mathrm{e}^{i\tilde{S}_J(t,D)}U_J(t,0).\tag{4.7.15}$$

If we denote (4.7.14) by $\tilde{\Omega}_{J,\mathrm{lr}}^+$, then (4.7.15) equals $\tilde{\Omega}_{J,\mathrm{lr}}^{+*}$. $\tilde{\Omega}_{J,\mathrm{lr}}^+$ is unitary and

$$\tilde{\Omega}_{J,\mathrm{lr}}^+D=D_J^+\tilde{\Omega}_{J,\mathrm{lr}}^+.\tag{4.7.16}$$

(ii) Under the additional assumptions (4.7.10) or (4.7.11), there exist the limits

$$\mathrm{s}-\lim_{t\to\infty}U_J(0,t)\mathrm{e}^{-iS_J(t,D)},\tag{4.7.17}$$

$$\text{s-}\lim_{t\to\infty} \mathrm{e}^{iS_J(t,D)} U_J(t,0). \tag{4.7.18}$$

If we denote (4.7.17) by $\Omega^+_{J,\mathrm{lr}}$, then (4.7.18) equals $\Omega^{+*}_{J,\mathrm{lr}}$. $\Omega^+_{J,\mathrm{lr}}$ is unitary and

$$\Omega^+_{J,\mathrm{lr}} D = D^+_J \Omega^+_{J,\mathrm{lr}}. \tag{4.7.19}$$

Lemma 4.7.9
There exist the limits

$$\text{s-}\lim_{t\to\infty} \mathrm{e}^{itH} U_J(0,t) \mathbb{1}_\Theta(D^+_J), \tag{4.7.20}$$

$$\text{s-}\lim_{t\to\infty} U_J(0,t) \mathrm{e}^{-itH} \mathbb{1}_\Theta(P^+). \tag{4.7.21}$$

If we denote (4.7.20) by $\omega^+_{J,\Theta}$, then (4.7.21) equals $\omega^{+*}_{J,\Theta}$. Moreover,

$$\omega^+_{J,\Theta}\omega^{+*}_{J,\Theta} = \mathbb{1}_\Theta(P^+), \quad \omega^{+*}_{J,\Theta}\omega^+_{J,\Theta} = \mathbb{1}_\Theta(D^+_J),$$
$$\omega^+_{J,\Theta} D^+_J = P^+ \omega^+_{J,\Theta}.$$

Proof. It is enough to show the existence of

$$\text{s-}\lim_{t\to\infty} \mathrm{e}^{itH} U_J(t,0) J_0(D^+_J)\chi^2\left(\tfrac{1}{2}(D^+_J)^2\right) \tag{4.7.22}$$

for any $J_0 \in C_0^\infty(\Theta)$ and $\chi \in C_0^\infty(\mathbb{R})$. Set

$$M(t) = J_0(\tfrac{x}{t}) + \langle D - \tfrac{x}{t}, \nabla J_0(\tfrac{x}{t})\rangle.$$

Then, by Theorem 3.2.2,

$$J_0(D^+_J)\chi^2\left(\tfrac{1}{2}(D^+_J)^2\right) = \text{s-}\lim_{t\to\infty} U_J(0,t)\chi(H_0)M(t)\chi(H_0)U_J(t,0).$$

Therefore, (4.7.22) equals

$$\text{s-}\lim_{t\to\infty} \mathrm{e}^{itH}\chi(H)M(t)\chi(H_0)U_J(t,0). \tag{4.7.23}$$

Define

$$\Phi(t) := \chi(H)M(t)\chi(H_0).$$

We have

$$\begin{aligned} \tfrac{\mathrm{d}}{\mathrm{d}t}\Phi(t) + iH\Phi(t) - i\Phi(t)H_J(t) &= \tfrac{1}{t}\chi(H)\langle D - \tfrac{x}{t}, \nabla^2 J_0(\tfrac{x}{t})\left(D - \tfrac{x}{t}\right)\rangle\chi(H_0) \\ &\quad + \chi(H)\nabla J_0(\tfrac{x}{t})\nabla_x V_J(t,x)\chi(H_0) \\ &\quad + \chi(H)M(t)[\chi(H_0), V_J(t,x)]. \end{aligned} \tag{4.7.24}$$

The first term on the right-hand side of (4.7.24) is integrable along the evolution by Lemma 4.4.3 and Proposition 3.2.4. The second term is clearly integrable in norm. The third term is seen to be integrable in norm using Proposition 4.7.5 *(ii)* and Lemma C.1.2. Therefore the limit (4.7.23) exists. □

Proof of Theorem 4.7.1. First consider the case with the additional assumptions (4.7.10) or (4.7.11). We apply Lemmas 4.7.8, 4.7.9, 4.7.7 and the chain rule:

$$\begin{aligned} \omega_{J,\Theta}^{+}\Omega_{J,\mathrm{lr}}^{+}\mathrm{e}^{-i\sigma^{+}(D)}1\!\!1_{\Theta}(D) &= \mathrm{s-}\lim_{t\to\infty} \mathrm{e}^{itH}U_J(t,0)1\!\!1_{\Theta}(D_J^{+}) \\ &\times \mathrm{s-}\lim_{t\to\infty} U_J(0,t)\mathrm{e}^{-iS_J(t,D)} \\ &\times \lim_{t\to\infty} \mathrm{e}^{iS_J(t,D)}\mathrm{e}^{-iS(t,D)}1\!\!1_{\Theta}(D) \\ &= \mathrm{s-}\lim_{t\to\infty} \mathrm{e}^{itH}\mathrm{e}^{-iS(t,D)}1\!\!1_{\Theta}(D). \end{aligned} \tag{4.7.25}$$

Note that we used (4.7.16) to pass $1\!\!1_{\Theta}(D_J^{+})$ through $\Omega_{J,\mathrm{lr}}^{+}$ and to change it into $1\!\!1_{\Theta}(D)$. This ends the proof of the existence of (4.7.3).

Using Lemmas 4.7.8, 4.7.9, 4.7.7 and the chain rule, we obtain

$$\begin{aligned} \mathrm{e}^{i\sigma^{+}(D)}1\!\!1_{\Theta}(D)\Omega_{J,\mathrm{lr}}^{+*}\omega_{J,\Theta}^{+*} &= \lim_{t\to\infty} \mathrm{e}^{iS(t,D)}\mathrm{e}^{-iS_J(t,D)}1\!\!1_{\Theta}(D) \\ &\times \mathrm{s-}\lim_{t\to\infty} \mathrm{e}^{iS_J(t,D)}U_J(t,0) \\ &\times \mathrm{s-}\lim_{t\to\infty} U_J(0,t)\mathrm{e}^{-itH}1\!\!1_{\Theta}(P^{+}) \\ &= \mathrm{s-}\lim_{t\to\infty} \mathrm{e}^{iS(t,D)}\mathrm{e}^{-itH}1\!\!1_{\Theta}(P^{+}). \end{aligned} \tag{4.7.26}$$

Since Θ was an arbitrary compact subset of X disjoint from 0, this gives us the existence of

$$\mathrm{s-}\lim_{t\to\infty} \mathrm{e}^{iS(t,D)}\mathrm{e}^{-itH}1\!\!1_{X\backslash\{0\}}(P^{+}).$$

But, by Theorem 4.4.1, we have

$$1\!\!1^{\mathrm{c}}(H) = 1\!\!1_{X\backslash\{0\}}(P^{+}). \tag{4.7.27}$$

This ends the proof of the existence of (4.7.4).

The proof of the general case, that is, without the additional assumptions (4.7.10) or (4.7.11) is similar. The only difference is that we use $\tilde{S}_J(t,\xi)$, $\tilde{\Omega}_{J,\mathrm{lr}}^{+}$, instead of $S_J(t,\xi)$, $\Omega_{J,\mathrm{lr}}^{+}$. □

4.8 Dollard Wave Operators

As in Chap. 3, it is often convenient to consider other modified free dynamics, like the Dollard and Buslaev-Matveev dynamics. Among them, the Dollard dynamics is of particular interest. In fact, using the Dollard dynamics, one can give a rather elementary proof of asymptotic completeness for a class of long-range potentials that are only in $C^{0,1}(X)$. Besides, in the case of the Dollard dynamics, we can easily take into account the presence of additional degrees of freedom.

We assume now that our Hilbert space is $L^2(X) \otimes \mathcal{H}_1$, where $\mathcal{H}_1$ is a certain auxiliary Hilbert space describing the additional degrees of freedom. Suppose that

$$H = \tfrac{1}{2}D^2 \otimes \mathbb{1}_{\mathcal{H}_1} + V(x),$$

$$\text{where} \quad V^*(x) = V(x) \in B(\mathcal{H}_1) \text{ for almost all } x \in X.$$

We assume that

$$V(x)(1-\Delta)^{-1} \text{ is compact on } L^2(X) \otimes \mathcal{H}_1,$$

$$\int_0^\infty \left\| (1-\Delta)^{-1} \nabla_x V(x) \mathbb{1}_{[1,\infty[}\left(\frac{|x|}{R}\right)(1-\Delta)^{-1} \right\| \mathrm{d}R < \infty,$$

and $V(x)$ can be written as

$$V(x) = V_{\mathrm{s}}(x) + V_{\mathrm{l}}(x),$$

such that

$$\int_0^\infty \left\| (1-\Delta)^{-1} V_{\mathrm{s}}(x) \mathbb{1}_{[1,\infty[}\left(\frac{|x|}{R}\right)(1-\Delta)^{-\frac{1}{2}} \right\| \mathrm{d}R < \infty,$$

$$\int_0^\infty \langle R\rangle^{\frac{1}{2}} \sup_{|x|>R} \|\partial_x^\alpha V_{\mathrm{l}}(x)\| \mathrm{d}R < \infty, \quad |\alpha| = 1.$$

Theorem 4.8.1
Define

$$U_{\mathrm{D}}(t) := \mathrm{e}^{-itH_0} T\left(\mathrm{e}^{-i\int_0^t V_{\mathrm{l}}(sD)\mathrm{d}s}\right),$$

where T denotes the time ordering. Then the limits

$$\mathrm{s-}\lim_{t\to\infty} \mathrm{e}^{itH} U_{\mathrm{D}}(t), \tag{4.8.1}$$

$$\mathrm{s-}\lim_{t\to\infty} U_{\mathrm{D}}^*(t) \mathrm{e}^{-itH} \mathbb{1}^{\mathrm{c}}(H) \tag{4.8.2}$$

exist. If we denote (4.8.1) by Ω_{D}^+, then (4.8.2) equals Ω_{D}^{+*}. One has

$$\begin{aligned} \Omega_{\mathrm{D}}^{+*}\Omega_{\mathrm{D}}^+ &= \mathbb{1}, \qquad \Omega_{\mathrm{D}}^+\Omega_{\mathrm{D}}^{+*} = \mathbb{1}^{\mathrm{c}}(H), \\ P^+\Omega_{\mathrm{D}}^+ &= \Omega_{\mathrm{D}}^+ D, \qquad H\Omega_{\mathrm{D}}^+ = \Omega_{\mathrm{D}}^+ H_0. \end{aligned} \tag{4.8.3}$$

Proof. First we get rid of the $V_{\mathrm{s}}(x)$ part of the potential in e^{-itH}. Having done this, we can assume that $V_{\mathrm{s}}(x) = 0$.

Then we introduce $\Theta \subset X$, $V_J(t,x)$, $H_J(t)$ and $U_J(s,t)$ as we did in the in the proof of Theorem 4.7.1. We set

$$U_{\mathrm{D},J}(t) := \mathrm{e}^{-itH_0} T\left(\mathrm{e}^{-i\int_0^t V_J(sD)\mathrm{d}s}\right).$$

Then we know from Chap. 3 that

$$\mathrm{s-}\lim_{t\to\infty} U_J(0,t)U_{\mathrm{D},J}(t), \quad \mathrm{s-}\lim_{t\to\infty} U^*_{\mathrm{D},J}(t)U_J(t,0)$$

exist. Besides, obviously

$$U_{\mathrm{D},J}(t)\mathbb{1}_\Theta(D) = U_\mathrm{D}(t)\mathbb{1}_\Theta(D).$$

Hence the theorem follows by the chain rule. □

4.9 Isozaki-Kitada Construction

In this section we introduce another construction of long-range wave operators due to Isozaki-Kitada [IK1]. This construction uses a time-independent modifier associated with a solution of the eikonal equation. We will prove that, for a correct choice of the solution of the eikonal equation, the two notions of wave operators coincide.

We will assume that

$$\begin{array}{l}\lim_{|x|\to\infty} V(x) = 0,\\ \int_0^\infty \langle R\rangle^{|\alpha|-1} \sup_{|x|\geq R} |\partial_x^\alpha V(x)|\mathrm{d}R < \infty, \quad |\alpha| \geq 1.\end{array} \tag{4.9.1}$$

Note that these conditions imply the hypotheses of Theorem 4.4.1 about the existence of the asymptotic velocity and the hypotheses of Theorem 4.7.1 about the existence and completeness of modified wave operators.

In Proposition 2.7.3, for any $\epsilon_0 > 0$, $\sigma_0 > -1$ and R_0 large enough, we constructed a function $S(t,x,\xi)$ on the outgoing region $\mathbb{R}^+ \times \Gamma^+_{R_0,\epsilon_0,\sigma_0}$ that solves the Hamilton-Jacobi equation

$$\begin{cases} \partial_t S(t,x,\xi) = \frac{1}{2}\xi^2 + V(\nabla_\xi S(t,x,\xi)), \\ S(0,x,\xi) = \langle x,\xi\rangle. \end{cases}$$

Moreover, in Theorem 2.7.5, we constructed a function $S(t,\xi)$ such that, for any $\epsilon > 0$, one has

$$\partial_t S(t,\xi) = \frac{1}{2}\xi^2 + V(\nabla_\xi S(t,\xi)), \quad |\xi| \geq \epsilon, \quad t \geq T_\epsilon.$$

Finally, in Proposition 2.8.2, we proved that the limit

$$\Phi^+_{\mathrm{lr}}(x,\xi) := \lim_{t\to\infty}(S(t,x,\xi) - S(t,\xi))$$

exists on $\Gamma^+_{R_0,\epsilon_0,\sigma_0}$ and solves there the eikonal equation

$$\frac{1}{2}\xi^2 = \frac{1}{2}(\nabla_x \Phi^+_{\mathrm{lr}}(x,\xi))^2 + V(x).$$

The function $\Phi^+_{\mathrm{lr}}(x,\xi)$ satisfies the estimates

$$\partial^\alpha_x \partial^\beta_\xi (\Phi^+_{\mathrm{lr}}(x,\xi) - \langle x,\xi\rangle) \in o(\langle x\rangle^{1-|\alpha|}), \quad |\alpha| + |\beta| \geq 0, \tag{4.9.2}$$

uniformly for $(x,\xi) \in \Gamma^+_{R_0,\epsilon_0,\sigma_0}$. We know from Sect. 4.7 that

$$\mathrm{s-}\lim_{t\to\infty} \mathrm{e}^{itH}\mathrm{e}^{-iS(t,D)} = \Omega^+_{\mathrm{lr}}$$

exists and has the properties described in Theorem 4.7.1. Below we will show how to construct Ω^+_{lr} in a different way.

Let us fix some constants $\epsilon_0 > 0$, $\sigma > -1$ and R_0 such that $S(t,x,\xi)$ is well defined on $\mathbb{R}^+ \times \Gamma^+_{R_0,\epsilon_0,\sigma_0}$. Choose also $\epsilon > \epsilon_0$, $\sigma > \sigma_0$ and $R > R_0$. Choose functions $F_0, F \in C^\infty(\mathbb{R})$ and $\chi_0 \in C_0^\infty(\mathbb{R})$ such that

$$\begin{aligned} &\chi_0(s) = 0, \quad s < \tfrac{1}{2}\epsilon_0^2, \quad &&\chi_0(s) = 1, \quad s > \tfrac{1}{2}\epsilon^2, \\ &F(s) = 0, \quad s < R_0, &&F(s) = 1, \quad s > R, \\ &F_0(s) = 0, \quad s < \sigma_0, &&F_0(s) = 1, \quad s > \sigma. \end{aligned} \tag{4.9.3}$$

We set

$$q^+(x,\xi) := F(|x|)\chi_0(\tfrac{1}{2}\xi^2)F_0\left(\tfrac{\langle x,\xi\rangle}{|x||\xi|}\right).$$

Let us note that $q^+(x,\xi) \in S(1, g_0)$. We define next the following operator:

$$J^+_{\mathrm{lr}}\phi(x) = (2\pi)^{-n}\int q^+(x,\xi)\mathrm{e}^{i\Phi^+_{\mathrm{lr}}(x,\xi) - i\langle y,\xi\rangle}\phi(y)\mathrm{d}\xi\mathrm{d}y. \tag{4.9.4}$$

Note that, by the remark after Theorem D.15.1, J^+_{lr} is a bounded operator on $L^2(X)$.

The following theorem shows how we can use J^+_{lr} to construct the modified wave operator.

Theorem 4.9.1
We have

$$\mathrm{s-}\lim_{t\to\infty} \mathrm{e}^{itH}J^+_{\mathrm{lr}}\mathrm{e}^{-itH_0} = \Omega^+_{\mathrm{lr}}\chi_0(H_0), \tag{4.9.5}$$

$$\mathrm{s-}\lim_{t\to\infty} \mathrm{e}^{itH_0}J^{+*}_{\mathrm{lr}}\mathrm{e}^{-itH}\mathbb{1}^{\mathrm{c}}(H) = \Omega^{+*}_{\mathrm{lr}}\chi_0(H). \tag{4.9.6}$$

As in Sect. 4.7, the proof will be based on a certain auxiliary time-dependent Hamiltonian. The notation will be completely analogous to that of Sect. 4.7. We define $V_J(t,x)$, $S_J(t,\xi)$, $H_J(t)$, $U_J(t,s)$, D^+_J and $\Omega^+_{J,\mathrm{lr}}$ as in Sect. 4.7.

Note that
$$\int_0^\infty \langle t\rangle^{|\alpha|-1}\|\partial_x^\alpha V_J(t,\cdot)\|_\infty \mathrm{d}t<\infty,\quad |\alpha|\geq 1,$$
and hence $V_J(t,x)$ satisfies the smooth slow-decaying assumption of Chaps. 1 and 3.

Recall from Theorem 1.8.1 that, for T_0 big enough and $T_0\leq s\leq t$, there exists a unique function $S_J(s,t,x,\xi)$ that solves the two-sided Hamilton-Jacobi equation for the potential $V_J(t,x)$. As in Chaps. 1 and 3, we set

$$\begin{aligned}S_J(t,\xi) &:= S_J(T_0,t,0,\xi),\\ \Phi_{J,\mathrm{lr}}^+(s,x,\xi) &:= \lim_{t\to\infty}(S_J(s,t,x,\xi)-S_J(t,\xi)).\end{aligned}$$

Recall that

$$\begin{aligned}&\partial_s\Phi_{J,\mathrm{lr}}^+(s,x,\xi)=\tfrac{1}{2}(\nabla_x\Phi_{J,\mathrm{lr}}^+(s,x,\xi))^2+V_J(s,x),\quad s\geq T_0,\\ &\partial_x^\alpha\partial_\xi^\beta(\Phi_{J,\mathrm{lr}}^+(s,x,\xi)-\langle x,\xi\rangle-\tfrac{1}{2}\xi^2)\in o(s^{1-|\alpha|}),\quad |\alpha|+|\beta|\geq 1.\end{aligned}$$

So $\Phi_{J,\mathrm{lr}}^+(s,x,\xi)$ satisfies the estimates (3.7.2). So we can define, as in Sect. 3.7,

$$J_{J,\mathrm{lr}}^+(s)\phi(x):=(2\pi)^{-n}\int\int \mathrm{e}^{i\Phi_{J,\mathrm{lr}}^+(s,x,\xi)-i\langle y,\xi\rangle}\phi(y)\mathrm{d}\xi\mathrm{d}y,\tag{4.9.7}$$

which is a bounded operator on $L^2(X)$ with norm $O(1)$.

Recall from Chap. 3 that the following alternative definition of the wave operator $\Omega_{J,\mathrm{lr}}^+$ is possible:

$$\Omega_{J,\mathrm{lr}}^+=\lim_{t\to\infty}U_J(0,t)J_J^+(t).\tag{4.9.8}$$

Next we show that the functions $\Phi_{\mathrm{lr}}^+(x,\xi)$ and $\Phi_{J,\mathrm{lr}}^+(s,x,\xi)$ are closely related. We recall that the function σ^+ was defined in Lemma 4.7.6.

Lemma 4.9.2
Let Θ_1 be a convex subset of Θ. Then there exists T_0 such that, for $T_0<t<t'$ and $\frac{x}{t},\xi\in\Theta_1$, we have

$$\Phi_{J,\mathrm{lr}}^+(t,x,\xi)=\Phi_{\mathrm{lr}}^+(x,\xi)-\frac{1}{2}t\xi^2+\sigma^+(\xi).$$

Proof. Let $\tilde{y}_J(s,t,t',x,\xi)$ denote the (unique) trajectory for the potential $V_J(t,x)$ that satisfies the boundary conditions as in Chap. 1. Clearly, for T_0 big enough, $T_0\leq t\leq s\leq t'$ and $\frac{x}{t},\xi\in\Theta_1$, $\tilde{y}(s,t,t',x,\xi)$ is also a trajectory for the potential $V(x)$. Consequently, for $T_0\leq t\leq t'$ and $\frac{x}{t},\xi\in\Theta_1$,

$$S_J(t,t',x,\xi)=S(t'-t,x,\xi).$$

We also observe that

$$\begin{aligned} S(t'-t,\xi)-S(t',\xi) &= -\int_{t'-t}^{t'}(\tfrac{1}{2}\xi^2+V(\nabla_\xi S(s,\xi)))\mathrm{d}s \\ &= -\tfrac{1}{2}t\xi^2+O(t)\sup_{s\geq t'-t}|V(\nabla_\xi S(s,\xi))|. \end{aligned}$$

So we have

$$\lim_{t'\to\infty}(S(t'-t,\xi)-S(t',\xi))=-\frac{1}{2}t\xi^2.$$

Now we compute:

$$\begin{aligned} \Phi_{J,\mathrm{lr}}^{+}(t,x,\xi)=\lim_{t'\to\infty}S_J(t,t',x,\xi)-S_J(t',\xi) &= \lim_{t'\to\infty}(S_J(t'-t,x,\xi)-S(t'-t,\xi)) \\ &+\lim_{t'\to\infty}(S(t'-t,\xi)-S(t',\xi)) \\ &+\lim_{t'\to\infty}(S(t',\xi)-S_J(t',\xi)) \\ &= \Phi_{\mathrm{lr}}^{+}(x,\xi)-\tfrac{1}{2}t\xi^2+\sigma^{+}(\xi). \end{aligned}$$

□

The following lemma says that, in some sense, we can localize J_{lr}^{+} in momentum and in position.

Lemma 4.9.3
(i) Let $f,\tilde{f}\in C_0^\infty(X)$ and $\tilde{f}=1$ on a neighborhood of supp$f$. *Then*

$$\left(\tilde{f}\left(\frac{x}{t}\right)-1\right)J_{\mathrm{lr}}^{+}f\left(\frac{x}{t}\right)\in O(t^{-\infty}). \tag{4.9.9}$$

(ii) If $g,\tilde{g}\in C_0^\infty(X)$ such that $\tilde{g}=1$ on a neighborhood of supp$g$, *then*

$$g(D)J_{\mathrm{lr}}^{+}(1-\tilde{g}(D))\in\Psi(\langle x\rangle^{-\infty}). \tag{4.9.10}$$

Proof. *(i)* follows by the non-stationary phase method described in Proposition D.12.1. *(ii)* follows by Proposition D.15.3 *(ii)*. □

The following lemma allows us to reduce the Isozaki-Kitada construction to the case of effective time-dependent potentials.

Lemma 4.9.4

$$\mathrm{s-}\lim_{t\to\infty}U_J(0,t)J_{\mathrm{lr}}^{+}\mathrm{e}^{-\mathrm{i}tH_0}1\!\!1_\Theta(D)=\Omega_{J,\mathrm{lr}}^{+}\mathrm{e}^{-\mathrm{i}\sigma^{+}(D)}1\!\!1_\Theta(D)\chi_0(H_0), \tag{4.9.11}$$

$$\mathrm{s-}\lim_{t\to\infty}\mathrm{e}^{\mathrm{i}tH_0}J_{\mathrm{lr}}^{+*}U_J(t,0)1\!\!1_\Theta(D_J^{+})=\mathrm{e}^{\mathrm{i}\sigma^{+}(D)}1\!\!1_\Theta(D)\chi_0(H_0)\Omega_{J,\mathrm{lr}}^{+*}. \tag{4.9.12}$$

Proof. Let Θ_1 be a convex set contained in Θ. Let $f,\tilde{f},g,\tilde{g}\in C_0^\infty(\Theta_1)$ such that $f=1$, $\tilde{g}=1$ on a neighborhood of supp$g$ and $\tilde{f}=1$ on a neighborhood of supp$f$. The proof follows from the following identity, which is true for $t>T_0$:

$$\tilde{f}(\tfrac{x}{t})J_{\rm lr}^{+}\mathrm{e}^{-itH_0}g(D) = \tilde{f}(\tfrac{x}{t})J_{J,\rm lr}^{+}(t)g(D)\mathrm{e}^{i\sigma^{+}(D)}\chi_0(H_0). \tag{4.9.13}$$

This identity follows immediately from Lemma 4.9.2 and

$$\tilde{f}\left(\tfrac{x}{t}\right)q^{+}(x,\xi)g(\xi) = \tilde{f}\left(\tfrac{x}{t}\right)g(\xi)\chi_0\left(\tfrac{1}{2}\xi^2\right).$$

Let us show (4.9.11). Clearly,

$$g(D) = \mathrm{s-}\lim_{t\to\infty}\mathrm{e}^{itH_0}f(\tfrac{x}{t})g(D)\mathrm{e}^{-itH_0}.$$

Now,

$$\begin{aligned}\mathrm{s-}\lim_{t\to\infty}U_J(0,t)J_{\rm lr}^{+}\mathrm{e}^{-itH_0}g(D) &= \mathrm{s-}\lim_{t\to\infty}U_J(0,t)J_{\rm lr}^{+}f(\tfrac{x}{t})g(D)\mathrm{e}^{-itH_0}\\ &= \mathrm{s-}\lim_{t\to\infty}U_J(0,t)\tilde{f}(\tfrac{x}{t})J_{\rm lr}^{+}f(\tfrac{x}{t})g(D)\mathrm{e}^{-itH_0}\\ &= \mathrm{s-}\lim_{t\to\infty}U_J(0,t)\tilde{f}(\tfrac{x}{t})J_{\rm lr}^{+}g(D)\mathrm{e}^{-itH_0}\\ &= \mathrm{s-}\lim_{t\to\infty}U_J(0,t)\tilde{f}(\tfrac{x}{t})J_{J,\rm lr}^{+}(t)\mathrm{e}^{-i\sigma^{+}(D)}\chi_0(H_0)g(D)\\ &= \tilde{f}(D_J^{+})\mathrm{s-}\lim_{t\to\infty}U_J(0,t)J_{J,\rm lr}^{+}(t)\mathrm{e}^{-i\sigma^{+}(D)}\chi_0(H_0)g(D)\\ &= \tilde{f}(D_J^{+})\Omega_{J,\rm lr}^{+}\mathrm{e}^{-i\sigma^{+}(D)}\chi_0(H_0)g(D)\\ &= \Omega_{J,\rm lr}^{+}\mathrm{e}^{-i\sigma^{+}(D)}\chi_0(H_0)g(D).\end{aligned}$$

We used: (4.9.13), Lemma 4.9.3 *(i)*, at the last step, we used

$$\tilde{f}(D_J^{+})\Omega_{J,\rm lr}^{+} = \Omega_{J,\rm lr}^{+}\tilde{f}(D).$$

Likewise,

$$g(D_J^{+}) = \mathrm{s-}\lim_{t\to\infty}U_J(0,t)g(D)f(\tfrac{x}{t})U_J(t,0),$$

and, by similar arguments, we obtain

$$\begin{aligned}\mathrm{s-}\lim_{t\to\infty}\mathrm{e}^{itH_0}J_{\rm lr}^{+*}U_J(t,0)g(D_J^{+}) &= \mathrm{s-}\lim_{t\to\infty}\mathrm{e}^{itH_0}J_{\rm lr}^{+*}g(D)f(\tfrac{x}{t})U_J(t,0)\\ &= \mathrm{s-}\lim_{t\to\infty}\mathrm{e}^{itH_0}\tilde{g}(D)J_{\rm lr}^{+*}g(D)f(\tfrac{x}{t})U_J(t,0)\\ &= \mathrm{s-}\lim_{t\to\infty}\mathrm{e}^{itH_0}\tilde{g}(D)J_{\rm lr}^{+*}f(\tfrac{x}{t})g(D)U_J(t,0)\\ &= \mathrm{e}^{i\sigma^{+}(D)}\tilde{g}(D)\chi_0(H_0)\mathrm{s-}\lim_{t\to\infty}J_{J,\rm lr}^{+*}(t)U_J(t,0)g(D_J^{+})\\ &= \mathrm{e}^{i\sigma^{+}(D)}\tilde{g}(D)\chi_0(H_0)\Omega_{J,\rm lr}^{+*}g(D_J^{+})\\ &= \mathrm{e}^{i\sigma^{+}(D)}\mathbb{1}_{\Theta}(D)\chi_0(H_0)\Omega_{J,\rm lr}^{+*}g(D_J^{+}).\end{aligned}$$

□

Proof of Theorem 4.9.1. We apply (4.9.8), Lemma 4.9.4 and the chain rule, and we get

$$\mathrm{s-}\lim_{t\to\infty}\mathrm{e}^{itH}J_{\rm lr}^{+}\mathrm{e}^{-itH_0}\mathbb{1}_{\Theta}(D) = \omega_{J,\Theta}^{+}\Omega_{J,\rm lr}^{+}\mathrm{e}^{-i\sigma^{+}(D)}\mathbb{1}_{\Theta}(D)\chi_0(H_0), \tag{4.9.14}$$

$$\mathrm{s-}\lim_{t\to\infty}\mathrm{e}^{itH_0}J_{\rm lr}^{+*}\mathrm{e}^{-itH}\mathbb{1}_{\Theta}(P^{+}) = \mathrm{e}^{i\sigma^{+}(D)}\mathbb{1}_{\Theta}(D)\chi_0(H_0)\Omega_{J,\rm lr}^{+*}\omega_{J,\Theta}^{+*}. \tag{4.9.15}$$

But, by (4.7.25), the limit (4.9.14) equals $\Omega_{\rm lr}^{+}\chi_0(H_0)$ and, by (4.7.26), the limit (4.9.15) equals $\chi_0(H_0)\Omega_{\rm lr}^{+*}$. □

4.10 Counterexamples to Asymptotic Completeness

In this section we construct a class of time-independent potentials for which the asymptotic velocity P^+ and the short-range wave operator Ω_{sr}^+ are well defined but the asymptotic completeness fails, i.e. $\mathrm{Ran}\Omega_{\mathrm{sr}}^+ \neq \mathcal{H}_{\mathrm{c}}(H)$. We will also see that P^+ restricted to $\mathcal{H}_{\mathrm{c}}(H)$ will not be unitarily equivalent to D, therefore the asymptotic completeness breaks down for any definition of modified wave operators. Such examples were first constructed by Yafaev [Yaf2], and the construction that we will give is based on Yafaev's. They are related to the time-dependent counterexamples based on the adiabatic approximation given in Subsect. 3.8.2.

We start this section with an abstract version of the Born-Oppenheimer approximation in scattering theory, which will be the key idea used to construct the counterexamples.

4.10.1 The Born-Oppenheimer Approximation – an Abstract Setting

Let $\mathcal{H}$ be a Hilbert space and H a self-adjoint operator on $\mathcal{H}$. Let P be an orthogonal projection. Suppose that we want to approximate the Hamiltonian H with another Hamiltonian that commutes with P. There exists a natural choice of such an approximation, which is described below.

Definition 4.10.1
We define the Born-Oppenheimer Hamiltonian *as*

$$\begin{aligned} H_{\mathrm{BO}} &:= PHP + (1-P)H(1-P) \\ &= H + [P,[H,P]] \\ &= H - PH(1-P) - (1-P)HP. \end{aligned}$$

It is clear from the first formula defining H_{BO} that $[H_{\mathrm{BO}}, P] = 0$.

Remark. The adiabatic evolution $U_{\mathrm{ad}}(t,s)$ introduced in Sect. 3.8 was a unitary evolution that preserved a family of spectral projections $P(t)$. Similarly, the Born-Oppenheimer evolution $\mathrm{e}^{-itH_{\mathrm{BO}}}$ commutes with P.

We would like to know under what conditions the Born-Oppenheimer evolution approximates the exact evolution on $\mathrm{Ran}P$ for large times. To express this property, we will investigate the existence of the limit

$$\lim_{t\to\infty} \mathrm{e}^{itH}\mathrm{e}^{-itH_{\mathrm{BO}}}\psi. \tag{4.10.1}$$

for $\psi \in \mathrm{Ran}P$.

Let $\mathcal{H}_{\text{eff}}$ be an auxiliary Hilbert space, and let $I : \mathcal{H}_{\text{eff}} \to \mathcal{H}$ be an isometry such that

$$II^* = P.$$

Let us define

$$h_{\text{eff}} := I^* H I = I^* H_{\text{BO}} I.$$

Clearly, $H_{\text{BO}} I = I h_{\text{eff}}$. Hence, instead of (4.10.1), we can look at the existence of

$$\lim_{t\to\infty} \mathrm{e}^{itH} \mathrm{e}^{-itH_{\text{BO}}} I\phi = \lim_{t\to\infty} \mathrm{e}^{itH} I \mathrm{e}^{-ith_{\text{eff}}}\phi \tag{4.10.2}$$

for $\phi \in \mathcal{H}_{\text{eff}}$.

In order to prove the existence of (4.10.2), one has to start with a study of the dynamics generated by h_{eff}. To this end, let us introduce the notation

$$\mathbf{D}_{h_{\text{eff}}} := \frac{\mathrm{d}}{\mathrm{d}t} + i[h_{\text{eff}}, \cdot],$$

In what follows, we suppose that $\mathcal{D}$ is a dense subset in $\mathcal{H}_{\text{eff}}$ and $B(t) \in B(\mathcal{H}_{\text{eff}})$ satisfy

$$\int_0^\infty \|(\mathbf{D}_{h_{\text{eff}}} B(t))\mathrm{e}^{-ith_{\text{eff}}}\phi\|\mathrm{d}t < \infty, \quad \phi \in \mathcal{D}. \tag{4.10.3}$$

Then there exists the limit

$$B^+ := \text{s}-\lim_{t\to\infty} \mathrm{e}^{ith_{\text{eff}}} B(t) \mathrm{e}^{-ith_{\text{eff}}}.$$

We will try to give a criterion that guarantees the existence of

$$\text{s}-\lim_{t\to\infty} \mathrm{e}^{itH} I \mathrm{e}^{-ith_{\text{eff}}} B^+. \tag{4.10.4}$$

Proposition 4.10.2
Assume that L is a self-adjoint operator such that

$$LP = PL = 0, \quad \text{Ran} L \subset \text{Ran}(1-P).$$

Let

$$K := -L^{-1}(1-P)[H, P]P$$

be bounded. Let $t \to B(t) \in B(\mathcal{H}_{\text{eff}})$ satisfy and

$$\lim_{t\to\infty} KIB(t) = 0, \tag{4.10.5}$$

$$\int_0^\infty \|[H-L, K]IB(t)\|\mathrm{d}t < \infty. \tag{4.10.6}$$

Then the limit (4.10.4) exists.

Proof. We first note that

$$\begin{aligned} \text{s-}\lim_{t\to\infty} \mathrm{e}^{itH} I \mathrm{e}^{-ith_{\mathrm{eff}}} B^+ &= \text{s-}\lim_{t\to\infty} \mathrm{e}^{itH} I B(t) \mathrm{e}^{-ith_{\mathrm{eff}}} \\ &= \text{s-}\lim_{t\to\infty} \mathrm{e}^{itH} (1+K) I B(t) \mathrm{e}^{-ith_{\mathrm{eff}}}. \end{aligned} \tag{4.10.7}$$

Now,

$$\begin{aligned} \tfrac{\mathrm{d}}{\mathrm{d}t} \mathrm{e}^{itH}(1+K)IB(t)\mathrm{e}^{-ith_{\mathrm{eff}}} &= \mathrm{e}^{itH}(H(1+K)-(1+K)H_{\mathrm{BO}})IB(t)\mathrm{e}^{-ith_{\mathrm{eff}}} \\ &+ \mathrm{e}^{itH}(1+K)I(\mathbf{D}_{h_{\mathrm{eff}}}B(t))\mathrm{e}^{-ith_{\mathrm{eff}}} =: R_1(t)+R_2(t). \end{aligned}$$

But

$$\begin{aligned} &R_1(t) = \mathrm{e}^{itH}[H-L,K]IB(t)\mathrm{e}^{-ith_{\mathrm{eff}}} \in L^1(\mathrm{d}t), \\ &R_2(t)\phi \in L^1(\mathrm{d}t), \quad \phi \in \mathcal{D}, \end{aligned}$$

by (4.10.6) and (4.10.2) respectively. Hence (4.10.7) exists. □

4.10.2 The Born-Oppenheimer Approximation for Schrödinger Operators

Let us now be more specific about our Hamiltonians. Let $\mathcal{G}$ be an auxiliary Hilbert space and $X := \mathbb{R}^n$. Let $\mathcal{H}_{\mathrm{eff}} = L^2(X)$ and $\mathcal{H} = L^2(X,\mathcal{G}) = L^2(X)\otimes\mathcal{G}$. Let $X \ni x \mapsto G(x)$ be a family of self-adjoint operators with a fixed domain $\mathcal{D} \subset \mathcal{G}$ that is C^∞ in the norm-resolvent sense. Let $x \mapsto E(x) \in B(\mathcal{G})$ be another family of self-adjoint operators. Let $\phi(x)$ be a normalized eigenfunction of $G(x)$ for the eigenvalue $\lambda(x)$. Let $P(x)$ denote the projection onto $\phi(x)$. All the dependence on x is assumed to be C^∞. We set as operators on $\mathcal{H}$

$$\begin{aligned} H &:= \tfrac{1}{2}D_x^2 + \textstyle\int_X^\oplus (G(x)+E(x))\mathrm{d}x, \\ P &:= \textstyle\int_X^\oplus P(x)\mathrm{d}x, \\ L &:= \textstyle\int_X^\oplus (G(x)-\lambda(x))\mathrm{d}x. \end{aligned}$$

We also define an operator from $\mathcal{H}_{\mathrm{eff}}$ to $\mathcal{H}$:

$$I\psi = \int_X^\oplus \psi(x)\phi(x)\mathrm{d}x.$$

(Note that, in the above formula, $X \ni x \mapsto \psi(x)$ is a L^2 function with values in $\mathbb{C}$ and $X \ni x \mapsto \phi(x)$ is a smooth function with values in unit vectors in $\mathcal{G}$.)

In this case, the effective Hamiltonian is equal to

$$h_{\mathrm{eff}} = \tfrac{1}{2}D_x^2 + \tfrac{1}{2}(D_x a(x) + a(x)D_x) + e(x) + \lambda(x), \tag{4.10.8}$$

where

$$\begin{aligned} a(x) &:= -i(\nabla_x\phi(x)|\phi(x)), \\ e(x) &:= (\phi(x)|E(x)\phi(x)) + \tfrac{1}{2}(\nabla_x\phi(x)|\nabla_x\phi(x)). \end{aligned}$$

We need to know some properties of the evolution generated by $h_{\rm eff}$. Unfortunately, because of the first order term, this operator does not fall into the class that we considered in this chapter. Nevertheless, we easily see that methods of this chapter after minor modifications apply to operators of this form.

The following lemma follows easily by the methods of this chapter.

Lemma 4.10.3
Suppose that

$$h = \frac{1}{2}D_x^2 + \frac{1}{2}(D_x a(x) + a(x)D_x) + v(x),$$

where $a(x) \in C^\infty(X, X)$, $v(x) \in C^\infty(X, \mathbb{R})$ *are real valued functions such that*

$$\lim_{|x|\to\infty} a(x),\ x\nabla_x a(x),\ v(x),\ x\nabla_x v(x) = 0,$$

$$|\partial_x^\alpha a(x)| \leq C_\alpha \langle x\rangle^{-|\alpha|},$$
$$|\partial_x^\alpha v(x)| \leq C_\alpha \langle x\rangle^{-|\alpha|}.$$

Then the Hamiltonian h *has the following properties:*
(i) $\sigma_{\rm ess}(h) = [0, \infty[$ *and, for any* $\lambda_1 \leq \lambda_2$ *such that* $0 \notin [\lambda_1, \lambda_2]$,

$$\dim \mathbb{1}^{\rm pp}(h)\mathbb{1}_{[\lambda_1,\lambda_2]}(h) < \infty.$$

(ii) If $\chi \in C_0^\infty(\mathbb{R})$, $\mathrm{supp}\chi \cap (\sigma_{\rm pp}(h) \cup \{0\}) = \emptyset$, $J \in C_0^\infty(X)$ *and* $J = 1$ *on a neighborhood of* $\{x \mid \frac{1}{2}x^2 \in \mathrm{supp}\chi\}$, *then for any* N,

$$\left(1 - J\left(\frac{x}{t}\right)\right) \mathrm{e}^{-ith}\chi(h)\langle x\rangle^{-N} \in O(t^{-N}).$$

Now let us formulate a criterion for the existence of the effective wave operator for Schrödinger operators.

Proposition 4.10.4
Suppose that the function $g \in C^\infty(X, \mathbb{R})$ *satisfies*

$$\begin{aligned} &|\partial_x^\alpha g(x)| \leq C_\alpha \langle x\rangle^{-|\alpha|} g(x),\\ &\lim_{|x|\to\infty} \tfrac{g^2(x)}{|x|} = 0,\\ &\textstyle\int_1^\infty \sup_{|x|>R} g^2(x)\frac{\mathrm{d}R}{R^2} < \infty. \end{aligned} \tag{4.10.9}$$

Moreover, suppose that

$$\begin{aligned} &\|\partial_x^\alpha P(x)\| \leq C_\alpha \langle x\rangle^{-|\alpha|},\\ &\|[\partial_x^\alpha P(x), \partial_x^\beta E(x)]\| \leq C_\alpha \langle x\rangle^{-1-|\alpha|-|\beta|},\\ &\|\partial_x^\alpha (1 - P(x))L^{-1}(x)\| \leq C_\alpha g^2(x)\langle x\rangle^{-|\alpha|},\\ &\|[E(x), (1 - P(x))L^{-1}(x)]\| \leq C g^2(x)\langle x\rangle^{-1}. \end{aligned} \tag{4.10.10}$$

Then there exists the limit

$$\lim_{t\to\infty} \mathrm{e}^{itH} I \mathrm{e}^{-ith_{\text{eff}}} 1\!\!1^{\mathrm{c}}(h_{\text{eff}}). \tag{4.10.11}$$

Proof. Let $\chi \in C_0^\infty(\mathbb{R})$, $\mathrm{supp}\chi \cap (\sigma_{\mathrm{pp}}(h_{\text{eff}}) \cup \{0\}) = \emptyset$. By density and Lemma 4.10.3 *(i)*, it is enough to show the existence of

$$\lim_{t\to\infty} \mathrm{e}^{itH} I \mathrm{e}^{-ith_{\text{eff}}} \chi(h_{\text{eff}}).$$

Let $J \in C_0^\infty(X)$ and $J = 1$ on a neighborhood of $\{x \mid \frac{1}{2}x^2 \in \mathrm{supp}\chi\}$. Then, by Lemma 4.10.3 *(ii)*,

$$\left(\mathbf{D}_{h_{\text{eff}}} J\left(\frac{x}{t}\right) \chi(h_{\text{eff}})\right) \mathrm{e}^{-ith_{\text{eff}}} \langle x\rangle^{-N} \in O(t^{-N}).$$

Therefore, it follows easily by the methods of this chapter that

$$\mathrm{s-}\lim_{t\to\infty} \mathrm{e}^{ith_{\text{eff}}} J\left(\frac{x}{t}\right) \chi(h_{\text{eff}}) \mathrm{e}^{-ith_{\text{eff}}} = \chi(h_{\text{eff}}).$$

We will take

$$B(t) := J\left(\frac{x}{t}\right) \chi(h_{\text{eff}})$$

and we will apply Proposition 4.10.2.

Note also that in our case

$$\begin{aligned} H - L &= \tfrac{1}{2}D_x^2 + E(x), \\ K &= L^{-1}(1-P)[\tfrac{1}{2}D_x^2 + E(x), P]P. \end{aligned}$$

Therefore, we need to show that

$$KIJ(\tfrac{x}{t})\chi(h_{\text{eff}}) \in o(t^0), \tag{4.10.12}$$

$$\left[\tfrac{1}{2}D_x^2 + E(x), K\right] IJ(\tfrac{x}{t})\chi(h_{\text{eff}}) \in L^1(\mathrm{d}t). \tag{4.10.13}$$

But

$$IJ\left(\frac{x}{t}\right) \chi\left(h_{\text{eff}}\right) = J\left(\frac{x}{t}\right) \chi\left(H_{\mathrm{BO}}\right) PI.$$

Therefore, we can replace $IJ\left(\frac{x}{t}\right)\chi\left(h_{\text{eff}}\right)$ in (4.10.12) by $J\left(\frac{x}{t}\right)\langle D_x\rangle^{-N}P$. In estimating (4.10.12), we move D_x to the right until it hits $J\left(\frac{x}{t}\right)\langle D_x\rangle^{-N}P$. In this way, the first expression of (4.10.12) can be estimated by

$$\sup_{|x|>\epsilon t} \|L^{-1}(x)(1-P(x))\|(\|P'(x)\| + \|P''\| + \|[E(x), P(x)]\|).$$

This is less than $C\sup_{|x|>\epsilon t} g^2(x)\langle x\rangle^{-1}$ and thus it converges to zero.

The estimation of (4.10.13) is based on the same principle, except that it involves a much larger number of terms. In order to handle this, denote by ad either ∇_x or $\mathrm{ad}_{E(x)}$. Then

$$\left[\frac{1}{2}D_x^2 + E(x), L^{-1}(x)(1-P(x))\left[\frac{1}{2}D_x^2 + E(x), P(x)\right]P(x)\right]$$

is a sum of terms

$$\mathrm{ad}^{\alpha}\left(L^{-1}(x)(1-P(x))\right)\mathrm{ad}^{\beta_1}P(x)\mathrm{ad}^{\beta_2}P(x)D_x^{\gamma},$$

where $|\alpha|+|\beta_1|+|\beta_2|\geq 2$. This (modulo D_x^{γ}, which is controllable by $\langle D_x\rangle^{-N}$,) can be estimated by $g^2(x)\langle x\rangle^{-2}$. This shows that (4.10.13) and (4.10.12) hold, and completes the proof of the proposition. □

Remark. The propagation estimate of Lemma 4.10.3 *(ii)* is an example of a strong propagation estimate. A similar estimate will be shown later in Sect. 4.13. Of course, in order to show Proposition 4.10.4, it is enough to use Lemma 4.10.3 *(ii)* just for $N>1$.

We think that it is not very elegant to use strong propagation estimates to show existence of wave operators, as we did that above. In fact, in the counterexample that we will give in the following subsection, it would be possible to avoid strong propagation estimates and to use just weak propagation estimates. This, in fact, would be more in the spirit of our general approach. However, referring to strong propagation estimates makes it possible to avoid giving conditions on the dynamics e^{-itH} in Proposition 4.10.4.

4.10.3 Counterexample to Asymptotic Completeness

We will now use Proposition 4.10.4 to construct examples of potentials for which the asymptotic completeness of wave operators is violated.

We will assume that $\mathcal{G} = L^2(Y)$, where $Y=\mathbb{R}^m$ and $\mu>0$. We take $W\in S(\langle y\rangle^{-\mu}, \langle y\rangle^{-2}\mathrm{d}y^2)$. We set

$$g(x) = \langle x\rangle^{\frac{1}{2}}(\log\langle x\rangle)^{-\epsilon}$$

for $\epsilon>1/2$. Let $q\in C_0^{\infty}(\mathbb{R})$ be a cutoff function equal to 1 in a neighborhood of 0. Our basic Hamiltonian will be

$$H = \tfrac{1}{2}D_x^2 + \tfrac{1}{2}D_y^2 + V(x,y) \quad \text{on} \quad L^2(X\times Y),$$

$$\text{where} \quad V(x,y) := g^{-2}(x)q\left(\tfrac{|y|}{\langle x\rangle}\right)W\left(\tfrac{y}{g(x)}\right).$$

We also introduce the auxiliary Hamiltonian

$$G(x) = \frac{1}{2}D_y^2 + g^{-2}(x)W\left(\frac{y}{g(x)}\right).$$

Note that if

$$G_0 = \tfrac{1}{2}D_y^2 + W(y),$$
$$A = \tfrac{1}{2}(\langle y, D_y\rangle + \langle D_y, y\rangle),$$

then

$$G(x) := g^{-2}(x)g^{-iA}(x)G_0 g^{iA}(x).$$

We choose an isolated eigenvalue λ_0 of the operator G_0. Let ϕ_0 be a corresponding normalized eigenvector and P_0 the projection onto ϕ_0. Let $L_0 := G_0 - \lambda_0$. We have

$$G(x) = g^{-2}(x)g^{-iA}(x)G_0 g^{iA}(x),$$
$$\phi(x) = g^{-iA}(x)\phi_0,$$
$$P_0(x) = g^{-iA}(x)P_0 g^{iA}(x),$$
$$\lambda(x) = g^{-2}(x)\lambda_0,$$
$$E(x,y) = g^{-2}(x)\left(1 - q(\tfrac{|y|}{\langle x\rangle})\right) W(\tfrac{y}{g(x)}),$$
$$L(x) = g^{-2}(x)g^{-iA}(x)L_0 g^{iA}(x).$$

We now have

$$a(x) = \nabla_x g(x) g^{-1}(x)(\phi_0|A\phi_0),$$
$$e(x) = \int E(x,y)\left|\phi_0\left(\tfrac{y}{g(x)}\right)\right|^2 \tfrac{\mathrm{d}y}{g^n(x)} + \tfrac{1}{2}\tfrac{|\nabla_x g(x)|^2}{g^2(x)}\|A\phi_0\|^2.$$

Recall that the effective Hamiltonian h_{eff} was defined in (4.10.8).

Let us state a theorem that describes the properties of the Hamiltonian H.

Theorem 4.10.5
(i) The asymptotic velocity P^+ for H exists and

$$\mathbb{1}^{\text{pp}}(H) = \mathbb{1}_{\{0\}\times\{0\}}(P^+).$$

(ii) There exists the short-range wave operator

$$\Omega_{\text{sr}}^+ := \text{s-}\lim_{t\to\infty} \mathrm{e}^{itH}\mathrm{e}^{-itH_0};$$

it satisfies $P^+\Omega_{\text{sr}}^+ = \Omega_{\text{sr}}^+ D$ and

$$\text{Ran}\,\Omega_{\text{sr}}^+ = \text{Ran}\,\mathbb{1}_{X\times(Y\setminus\{0\})}(P^+).$$

(iii) There exists the Born-Oppenheimer effective wave operator

$$\mathrm{e}^{itH} I \mathrm{e}^{-ith_{\text{eff}}}\mathbb{1}^{\text{c}}(h_{\text{eff}}) =: \Omega_{\text{eff}}^+. \tag{4.10.14}$$

Moreover,

$$\text{Ran}\,\Omega_{\text{eff}}^+ \subset \text{Ran}\,\mathbb{1}_{X\times\{0\}}(P^+).$$

Before we show Theorem 4.10.5 let us give the estimates satisfied by the potential $V(x,y)$.

Lemma 4.10.6
(i) $V(x,y)$ satisfies the estimates

$$|\partial_x^\gamma \partial_y^\delta V(x,y)| \le C_{\gamma,\delta}(\langle x\rangle + \langle y\rangle)^{-1-\frac{|\gamma|+|\delta|}{2}} (\log(\langle x\rangle + \langle y\rangle))^{(2+|\gamma|+|\delta|)\epsilon}.$$

(ii) For any $C_0 > 0$, on $\{(x,y) \mid |x| < C_0|y|\}$, the potential $V(x,y)$ satisfies the estimates

$$|V(x,y)| \le C(\langle x\rangle + \langle y\rangle)^{-1-\frac{\mu}{2}} (\log(\langle x\rangle + \langle y\rangle))^{\epsilon(2-\mu)}.$$

(iii)

$$|\partial_y^\beta E(x,y)| \le C_\beta g^{-2}(x)\langle x\rangle^{-|\beta|}, \ |\beta| \ge 0.$$

Proof. Let us show *(i)*. It is enough to consider

$$V_1(x,y) := g^{-2}(x) W(\frac{y}{g(x)}).$$

The derivative $\partial_x^\gamma \partial_y^\delta V_1(x,y)$ is a linear combination of terms

$$(\partial_x^{\gamma_1} g^{-1}(x)) \cdots (\partial_x^{\gamma_n} g^{-1}(x))\, y^k \partial^\delta \nabla^k W(\tfrac{y}{g(x)})$$

where $\gamma = \gamma_1 + \cdots + \gamma_n$ and $n = k + |\delta| + 2$. This can be estimated by

$$\langle x\rangle^{-|\gamma|} g^{-k-|\delta|-2}(x)|y|^k \left|\partial^\delta \nabla^k W\left(\frac{y}{g(x)}\right)\right|. \tag{4.10.15}$$

Using

$$\left||y|^{k+|\delta|} g^{-k-|\delta|}(x) \partial^\delta \nabla^k W\left(\frac{y}{g(x)}\right)\right| \le C,$$

we obtain that, on $\{g(x) \le C_0|y|\}$, (4.10.15) is less than

$$C\langle x\rangle^{-|\gamma|} g^{-|\delta|-2}(x) = C\langle x\rangle^{-|\gamma|-|\delta|/2-1} (\log\langle x\rangle)^{\epsilon|\delta|+2\epsilon}. \tag{4.10.16}$$

Using

$$\left|\partial^\delta \nabla^k W\left(\frac{y}{g(x)}\right)\right| \le C,$$

we obtain that, on $\{g(x) \ge C_0|y|\}$, (4.10.15) is also less than (4.10.16). This ends the proof of *(i)*.

To show *(ii)*, we note that on $\{(x,y) \mid |x| < C_0|y|\}$ one has

$$\begin{aligned}|V(x,y)| &\le Cg^{-2}(x)\left|\frac{y}{g(x)}\right|^{-\mu} \\ &\le Cg^{-2+\mu}(x)\langle x\rangle^{-\mu} = C\langle x\rangle^{-1-\mu/2}\left(\log\langle x\rangle\right)^{\epsilon(2-\mu)}.\end{aligned}$$

Finally *(iii)* is easy and left to the reader. □

Proof of Theorem 4.10.5. $V(x,y)$ satisfies the assumptions of Theorem 4.4.1. Hence *(i)* is true.

By Lemma 4.10.6 *(ii)*, $V(x,y)$ satisfies the short-range condition outside any conical neighborhood of $X \times \{0\}$; hence *(ii)* follows by an obvious modification of Theorem 4.6.1.

Let us show *(iii)*. We will show that the hypotheses of Proposition 4.10.4 are satisfied.

First note that the function $g(x)$ satisfies the conditions of Proposition 4.10.4, in particular,

$$|\partial_x^\alpha g(x)| \le C_\alpha g(x)\langle x\rangle^{-|\alpha|}. \tag{4.10.17}$$

Next note that our operator G_0 is "dilation smooth", that is,

$$\|\mathrm{ad}_A^n (G_0 - z)^{-1}\| < \infty, \quad z \notin \sigma(G_0).$$

Hence

$$\|\mathrm{ad}_A^n P_0\| < \infty, \quad \|\mathrm{ad}_A^n (1 - P_0)G_0^{-1}\| < \infty.$$

Therefore, if we take into account (4.10.17), then we get

$$\begin{aligned}&\|\partial_x^\alpha P(x)\| \le C_\alpha \langle x\rangle^{-|\alpha|}, \\ &\|\partial_x^\alpha (1 - P(x))L^{-1}(x)\| \le C_\alpha g^2(x)\langle x\rangle^{-|\alpha|}.\end{aligned}$$

If we take into account the exponential decay of ϕ_0 (cf Thm. 4.3.4), then we see that

$$\|\partial_x^\alpha P(x)\partial_x^\beta E(x)\| \le C_N \langle x\rangle^{-N}, \ N > 0.$$

Finally, let us write

$$\begin{aligned}&[E(x), (1 - P(x))L^{-1}(x)] \\ &= -[E(x), P(x)](1 - P(x))L^{-1}(x) + \mathrm{hc} \\ &+(1 - P(x))L^{-1}(x)[\tfrac{1}{2}D_y^2, E(x)]L^{-1}(x)(1 - P(x)).\end{aligned}$$

The first term of the right-hand side is $O(\langle x\rangle^{-\infty})$. To deal with the second term, we use

$$\begin{aligned}&|\nabla_y E(x,y)| \le Cg^{-2}(x)\langle x\rangle^{-1}, \\ &\|(L(x) + i)^{-1}D_y\| \le C, \\ &\|(1 - P(x))L^{-1}(x)(L(x) + i)\| \le Cg^2(x), \\ &\|(1 - P(x))L^{-1}(x)\| \le Cg^2(x).\end{aligned}$$

Thus we see that

$$\|[E(x),(1-P(x))L^{-1}(x)]\| \leq Cg^2(x)\langle x\rangle^{-1}.$$

Therefore, the assumptions of Proposition 4.10.4 are satisfied. Hence the limit (4.10.14) exists.

We also see, using the exponential decay of ϕ_0, that, for any $C_0 > 0$,

$$\text{s-}\lim_{t\to\infty} \mathbb{1}_{[C_0,\infty[}(\tfrac{|y|}{t}) I \mathrm{e}^{-ith_{\mathrm{eff}}} = 0.$$

This shows that $\mathrm{Ran}\Omega_{\mathrm{eff}}^+ \subset \mathrm{Ran}\mathbb{1}_{X\times\{0\}}(P^+)$. □

4.11 Strong Large Velocity Estimates

In our proof of the existence of the asymptotic velocity and of the existence and completeness of wave operators, the main tool that we used were the so-called weak propagation estimates. In the remaining part of this chapter, a major role will be played by strong propagation estimates.

In spite of their name, strong propagation estimates do not imply the corresponding weak propagation estimates. They say that certain observables decay quite rapidly, but only on vectors in a certain dense subset of the Hilbert space.

The first strong propagation estimates that we prove are the strong large velocity estimates. Their intuitive content is similar to that of the weak large velocity estimates of Sect. 4.2. Note that strong large velocity estimates hold under rather weak assumptions: it is enough if the potential is form bounded. They were discovered by Sigal and Soffer [SS3].

Proposition 4.11.1
Assume (4.1.3). Suppose that $\chi \in C_0^\infty(\mathbb{R})$, *and* $\mathrm{supp}\chi \subset]-\infty, \frac{1}{2}\theta_0^2[$. *Then, for any* $s \geq 1$,

$$\int_1^\infty \left\| |\,|x| - t\theta_0|_+^{(s-1)/2} \chi(H)\phi_t \right\|^2 \mathrm{d}t \leq C\|\langle x\rangle^{s/2}\phi\|^2, \tag{4.11.1}$$

and, for any $s \geq 0$,

$$\left\| |\,|x| - t\theta_0|_+^{s/2} \chi(H)\phi_t \right\|^2 \leq C\|\langle x\rangle^{s/2}\phi\|^2. \tag{4.11.2}$$

Proof of Proposition 4.11.1. It is enough to prove the proposition for $s = n \in \mathbb{N}$, and then to use interpolation. The proof uses induction with respect to n. We assume that our proposition is proven with n replaced with $n-1$ (unless $n = 1$ when we do not assume anything at all). Let us fix some constants $\theta_1 > \theta_2 \cdots > \theta_4$ with $\theta_0 > \theta_1$ and $\mathrm{supp}\chi \subset]-\infty, \frac{1}{2}\theta_4^2]$. Using Lemma A.4.1, we can construct a function $J \in C^\infty(\mathbb{R})$ such that $J' \geq 0$,

$$J(s) = \begin{cases} 0, & s \le \theta_2, \\ 1, & s \ge \theta_1, \end{cases}$$

and $\sqrt{J} \in C^\infty(\mathbb{R})$, $\sqrt{J'} \in C_0^\infty(\mathbb{R})$. We set

$$F(s) := J(s)|s - \theta_3|^n.$$

Note that

$$F(s) = \begin{cases} 0, & s \le \theta_2, \\ |s-\theta_3|^n, & s \ge \theta_1, \end{cases} \tag{4.11.3}$$

and

$$-nF(s) + F'(s)(s - \theta_3) = J'(s)|s - \theta_3|^{n+1} \ge 0.$$

We also have

$$F'(s) = J'(s)|s - \theta_3|^n + n|s - \theta_3|^{n-1}J(s) =: f_1^2(s) + f_2^2(s)$$

for $f_1 \in C_0^\infty(\mathbb{R})$, $f_2 \in C^\infty(\mathbb{R})$. Note that

$$f_2(s) = \begin{cases} 0, & s \le \theta_2, \\ n^{1/2}|s-\theta_3|^{(n-1)/2}, & s \ge \theta_1. \end{cases} \tag{4.11.4}$$

Let us note that the only reason for splitting F' into f_1^2 and f_2^2 is the need to guarantee the smoothness of f_1, f_2, which in the case of such a splitting follows immediately from Lemma A.4.1. With a little more care we could avoid it.

We consider the following (unbounded) propagation observable

$$\Phi(t) = \chi(H)F\left(\frac{|x|}{t}\right)\chi(H)t^n.$$

We compute:

$$\begin{aligned} -\mathbf{D}\Phi(t) &= \chi(H)\left(-nF\left(\tfrac{|x|}{t}\right) + F'\left(\tfrac{|x|}{t}\right)\tfrac{|x|}{t}\right)\chi(H)t^{n-1} \\ &\quad -\chi(H)\left(D\tfrac{x}{|x|}F'\left(\tfrac{|x|}{t}\right) + \mathrm{hc}\right)\chi(H)t^{n-1} \\ &\ge \theta_3\chi(H)F'\left(\tfrac{|x|}{t}\right)\chi(H)t^{n-1} \\ &\quad - \textstyle\sum_{i=1}^2 \chi(H)\left(D\tfrac{x}{|x|}f_i^2\left(\tfrac{|x|}{t}\right) + \mathrm{hc}\right)\chi(H)t^{n-1}. \end{aligned} \tag{4.11.5}$$

Let us now choose two cutoff functions $\tilde{\chi} \in C_0^\infty(\mathbb{R})$ and $\tilde{f} \in C^\infty(\mathbb{R})$ such that $\tilde{\chi} = 1$ on a neighborhood of $\mathrm{supp}\chi$, $\mathrm{supp}\tilde{\chi} \subset]-\infty, \frac{1}{2}\theta_4^2[$ and $\tilde{f} = 1$ on a neighborhood of $\mathrm{supp} f_i$, $\mathrm{supp}\tilde{f} \subset [\theta_3, \infty[$. We have

$$\begin{aligned} &-\chi(H)\left(D\tfrac{x}{|x|}f_i^2\left(\tfrac{|x|}{t}\right) + \mathrm{hc}\right)\chi(H)t^{n-1} \\ &= -\chi(H)f_i\left(\tfrac{|x|}{t}\right)\left(D\tfrac{x}{|x|}\tilde{f}\left(\tfrac{|x|}{t}\right) + \mathrm{hc}\right)f_i\left(\tfrac{|x|}{t}\right)\chi(H)t^{n-1}. \end{aligned} \tag{4.11.6}$$

Applying the commutator expansion lemma C.3.1 for $A = x, B = \tilde{\chi}(H)$, we obtain

$$\chi(H)f_i\left(\frac{|x|}{t}\right) = \sum_{|\alpha|<N} t^{-|\alpha|}\chi(H)\partial^\alpha f_i(\tfrac{|x|}{t})\tilde{\chi}(H) + O(t^{-N}).$$

Therefore, (4.11.6) equals

$$\begin{aligned}
&-\chi(H)f_i\left(\tfrac{|x|}{t}\right)\tilde{\chi}(H)\left(D\tfrac{x}{|x|}\tilde{f}\left(\tfrac{|x|}{t}\right) + \mathrm{hc}\right)\tilde{\chi}(H)f_i\left(\tfrac{|x|}{t}\right)\chi(H)t^{n-1}\\
&+\sum_{1\leq|\alpha_1|+|\alpha_2|<N}\chi(H)\partial^{\alpha_1}f_i(\tfrac{|x|}{t})B_{\alpha_1,\alpha_2}(t)\partial^{\alpha_2}f_i(\tfrac{|x|}{t})\chi(H)t^{n-1-|\alpha_1|-|\alpha_2|}\\
&+O(t^{n-1-N}),
\end{aligned} \tag{4.11.7}$$

where $B_{\alpha_1,\alpha_2}(t)$ are uniformly bounded. Using then Lemma 4.2.2 to deal with the first term of (4.11.7) and the inequality

$$B_1^*B_2 + B_2^*B_1 \geq -\delta B_1^*B_1 - \delta^{-1}B_2^*B_2$$

to deal with terms with $|\alpha_1| = 0$ or $|\alpha_2| = 0$, we see that, for any $\delta > 0$, (4.11.7) is greater than or equal to

$$\begin{aligned}
&(-\theta_5 - o(t^0) - C\delta)\chi(H)f_i^2(\tfrac{|x|}{t})\chi(H)\\
&-C(1+\delta^{-1})\,||x| - \theta_3 t|_+^{n-2}\,t^{n-2} + O(t^{-2}).
\end{aligned} \tag{4.11.8}$$

Collecting (4.11.5) and (4.11.8), we obtain, for $t \geq t_0$,

$$\begin{aligned}
-\mathbf{D}\Phi(t) \;&\geq C_0\chi(H)f_2^2\left(\tfrac{|x|}{t}\right)\chi(H)t^{n-1}\\
&-Ct^{n-2}\chi(H)\,||x| - \theta_3 t|_+^{n-2}\,\chi(H)t^{n-2} + O(t^{-2}),
\end{aligned} \tag{4.11.9}$$

where $C_0 := \theta_3 - \theta_4 > 0$. Integrating (4.11.9) from t_0 to t_1, we obtain

$$\begin{aligned}
(\phi_{t_0}, \Phi(t_0)\phi_{t_0}) \;&\geq (\phi_{t_1}, \Phi(t_1)\phi_{t_1})\\
&+C_0\int_{t_0}^{t_1}\left\|f_2\left(\tfrac{|x|}{t}\right)\chi(H)\phi_t\right\|^2 t^{n-1}\mathrm{d}t\\
&-C\int_{t_0}^{t_1}\left\|\,||x| - \theta_3 t|_+^{\frac{n-2}{2}}\chi(H)\phi_t\right\|^2 t^{n-2}\mathrm{d}t - C\|\phi\|^2.
\end{aligned} \tag{4.11.10}$$

We know, by the induction assumption, that

$$\int_{t_0}^{t_1}\left\|\,||x| - \theta_3 t|_+^{\frac{n-2}{2}}\chi(H)\phi_t\right\|^2 t^{n-2}\mathrm{d}t \leq C\left\|\langle x\rangle^{\frac{n-1}{2}}\phi\right\|^2.$$

Moreover,

$$(\phi_{t_0}, \Phi(t_0)\phi_{t_0}) \leq C\|\langle x\rangle^{\frac{n}{2}}\phi\|^2,$$

Therefore, (4.11.10) implies

$$(\phi_{t_1}, \Phi(t_1)\phi_{t_1}) + C_0\int_{t_0}^{t_1}\left\|f_2\left(\frac{|x|}{t}\right)\chi(H)\phi_t\right\|^2 t^{n-1}\mathrm{d}t \leq C\|\langle x\rangle^{\frac{n}{2}}\phi\|^2,$$

which, using (4.11.3), (4.11.4), completes the proof of the proposition. □

4.12 Strong Propagation Estimates for the Generator of Dilations

Our next goal is the proof of strong minimal velocity estimates. They are much more difficult to show than strong maximal velocity estimates and require stronger assumptions on the potentials.

As a first step towards their proof, we will show a family of strong propagation estimates about the generator of dilations A due to Sigal-Soffer [SS3].

Note that various estimates for the generator of dilations are interesting and useful for their own sake. They are closely related to microlocal propagation estimates of Sect. 4.16.

Theorem 4.12.1
Suppose that the hypotheses of Theorem 4.3.1 hold and

$$\|\mathrm{ad}_A^n (H+i)^{-1}\| < \infty,\ n \in \mathbb{N}. \tag{4.12.1}$$

Suppose that $\lambda_0 > 0$, $\chi \in C_0^\infty(\mathbb{R})$, *and* $\mathrm{supp}\chi \subset [\frac{1}{2}\lambda_0, \infty[\backslash \sigma_{\mathrm{pp}}(H)$. *Then, for any* $s \geq 1$,

$$\int_1^\infty \left\| |A - t\lambda_0|_-^{(s-1)/2} \chi(H)\phi_t \right\|^2 \mathrm{d}t \leq C \|\langle A\rangle_-^{s/2}\phi\|^2, \tag{4.12.2}$$

and, for any $s \geq 0$,

$$\left\| |A - t\lambda_0|_-^{s/2} \chi(H)\phi_t \right\|^2 \leq C \|\langle A\rangle_-^{s/2}\phi\|^2. \tag{4.12.3}$$

Proof. It is sufficient to show the proposition for $s = n \in \mathbb{N}$ and then to use interpolation. We will use induction on n. We assume that (4.12.2) and (4.12.3) hold for all $m < n$, unless $n = 1$ where we do not assume anything. Let us choose constants $\lambda_0 < \lambda_1 \leq \lambda_2 \cdots \leq \lambda_5$. We assume that $\tilde{\chi} \in C_0^\infty(\mathbb{R})$, $\tilde{\chi} = 1$ on a neighborhood of λ_5 and

$$\tilde{\chi}(H) i[H, A] \tilde{\chi}(H) \geq \lambda_4 \tilde{\chi}^2(H). \tag{4.12.4}$$

Using Lemma A.4.1, we can construct a function $J \in C^\infty(\mathbb{R})$ such that $J' \leq 0$ and

$$J(s) = \begin{cases} 1, & s \leq \lambda_1, \\ 0, & s \geq \lambda_2, \end{cases}$$

and $\sqrt{J} \in C^\infty(\mathbb{R})$, $\sqrt{-J'} \in C_0^\infty(\mathbb{R})$. We set

$$F(s) := J(s)|s - \lambda_3|^n.$$

Note that $F \in S(\langle x\rangle^n, \langle x\rangle^{-2}\mathrm{d}x^2)$,

$$F(s) = \begin{cases} |s-\lambda_3|^n, & s \le \lambda_1, \\ 0, & s \ge \lambda_2, \end{cases} \tag{4.12.5}$$

and

$$nF(s) - F'(s)(s-\lambda_3) = J'(s)|s-\lambda_3|^{n+1} \le 0.$$

We also have

$$F'(s) = J'(s)|s-\lambda_3|^n - n|s-\lambda_3|^{n-1}J(s) =: -f_1^2(s) - f_2^2(s)$$

for $f_1 \in C_0^\infty(\mathbb{R})$, $f_2 \in S(\langle x\rangle^{(n-1)/2}, \langle x\rangle^{-2}\mathrm{d}x^2)$. Note that

$$f_2(s) = \begin{cases} n^{1/2}|s-\lambda_3|^{(n-1)/2}, & s \le \lambda_1, \\ 0, & s \ge \lambda_2. \end{cases} \tag{4.12.6}$$

We consider the following positive unbounded propagation observable

$$\Phi(t) = \chi(H)F\left(\frac{A}{t}\right)\chi(H)t^n,$$

and compute

$$\begin{aligned} -\mathbf{D}\Phi(t) &= \chi(H)\left(F'\left(\tfrac{A}{t}\right)\tfrac{A}{t} - nF\left(\tfrac{A}{t}\right)\right)\chi(H)t^{n-1} \\ &\quad -\chi(H)\left[H, iF\left(\tfrac{A}{t}\right)\right]\chi(H)t^n \\ &\ge \lambda_3\chi(H)F'\left(\tfrac{A}{t}\right)\chi(H)t^{n-1} - \chi(H)i\left[H, F\left(\tfrac{A}{t}\right)\right]\chi(H)t^n. \end{aligned} \tag{4.12.7}$$

Let us now estimate the second term in the last line of (4.12.7).

From (4.12.1) and the commutator expansion lemma C.3.1, if N is large enough and $m \in \mathbb{R}$ such that $H + m > 0$, we obtain

$$\begin{aligned} &[(H+m)^{-1}, iF(\tfrac{A}{t})] \\ &= \textstyle\sum_1^N \frac{1}{j!}\frac{1}{t^j}F^{(j)}(\tfrac{A}{t})\mathrm{ad}_A^j(H+m)^{-1} + O(t^{-N-1}). \end{aligned}$$

We write $F^{(1)} = -f_1^2 - f_2^2$; we also split $F^{(k)} = f_{k,1}f_{k,2}$ with $f_{k,1}, f_{k,2} \in S(\langle x\rangle^{(n-k)/2}, \langle x\rangle^{-2}\mathrm{d}x^2)$. After applying the commutator expansion lemma several times we arrive at the following expression:

$$\begin{aligned} &[(H+m)^{-1}, iF(\tfrac{A}{t})] \\ &= -\textstyle\sum_{i=1}^2 \frac{1}{t}f_i(\tfrac{A}{t})(H+m)^{-1}[H, iA](H+m)^{-1}f_i(\tfrac{A}{t}) \\ &+\tfrac{1}{t^2}|\tfrac{A}{t} - \lambda_3|_-^{(n-2)/2}B(t)|\tfrac{A}{t} - \lambda_3|_-^{(n-2)/2} + O(t^{-N-1}), \end{aligned} \tag{4.12.8}$$

where $B(t)$ is uniformly bounded.

We observe now that, using hypothesis (4.12.1) and formula C.2.2, we have

$$\|\mathrm{ad}_A^n \chi(H)\| < \infty, \quad n \in \mathbb{N}, \quad \chi \in C_0^\infty(\mathbb{R}). \tag{4.12.9}$$

Using (4.12.9) and Lemma C.3.1, we obtain

$$\begin{aligned} &\chi(H)(H+m)f_i(\tfrac{A}{t})(H+m)^{-1} \\ &= \chi(H)f_i(\tfrac{A}{t})\tilde{\chi}(H) \\ &+ \textstyle\sum_{j=1}^N t^{-j} f_i^{(j)}(\tfrac{A}{t})B_j(t) + O(t^{-N-1}), \end{aligned} \tag{4.12.10}$$

where $B_j(t)$ are uniformly bounded. Using (4.12.10) and (4.12.8), we obtain, for some uniformly bounded $\tilde{B}(t)$,

$$\begin{aligned} &-\chi(H)[H, iF(\tfrac{A}{t})]\chi(H) \\ &= -\chi(H)(H+m)[(H+m)^{-1}, iF(\tfrac{A}{t})](H+m)\chi(H) \\ &= \textstyle\sum_{i=1}^2 \frac{1}{t}\chi(H)f_i(\tfrac{A}{t})\tilde{\chi}(H)[H, iA]\tilde{\chi}(H)f_i(\tfrac{A}{t})\chi(H) \\ &+\tfrac{1}{t^2}\chi(H)(H+m)|\tfrac{A}{t}-\lambda_3|_-^{(n-2)/2}\tilde{B}(t)|\tfrac{A}{t}-\lambda_3|_-^{(n-2)/2}(H+m)\chi(H) + O(t^{-N-1}) \\ &\geq \lambda_4 \textstyle\sum_{i=1}^2 \frac{1}{t}\chi(H)f_i(\tfrac{A}{t})\tilde{\chi}^2(H)f_i(\tfrac{A}{t})\chi(H) \\ &-C\tfrac{1}{t^2}\chi(H)(H+m)|\tfrac{A}{t}-\lambda_3|_-^{n-2}(H+m)\chi(H) + O(t^{-N-1}) \\ &\geq \lambda_4 \textstyle\sum_{i=1}^2 \frac{1}{t}\chi(H)f_i^2(\tfrac{A}{t})\chi(H) \\ &-C\tfrac{1}{t^2}\chi(H)(H+m)|\tfrac{A}{t}-\lambda_3|_-^{n-2}(H+m)\chi(H) + O(t^{-N-1}). \end{aligned}$$

Plugging this into (4.12.7), we finally get

$$\begin{aligned} -\mathbf{D}\Phi(t) &\geq t^{n-1}C_0\chi(H)f_2^2(\tfrac{A}{t})\chi(H) \\ &-C\chi(H)(H+m)|A-\lambda_3 t|_-^{n-2}(H+m)\chi(H) \\ &+O(t^{n-N-1}), \end{aligned} \tag{4.12.11}$$

where $C_0 := \lambda_4 - \lambda_3$. Integrating from t_0 to t_1, we deduce from (4.12.11) and (4.12.6) that

$$\begin{aligned} (\phi_{t_0}|\Phi(t_0)\phi_{t_0}) &\geq (\phi_{t_1}|\Phi(t_1)\phi_{t_1}) + \textstyle\int_{t_0}^{t_1} t^{n-1}\|f_2(\tfrac{A}{t})\chi(H)\phi_t\|^2 \mathrm{d}t \\ &-C\textstyle\int_{t_0}^{t_1} \||A-\lambda_3 t|_-^{\frac{n-2}{2}}(H+m)\chi(H)\phi_t\|^2 \mathrm{d}t - C\|\phi\|^2. \end{aligned}$$

Now we note that, by the induction assumption,

$$\textstyle\int_1^\infty \||A-\lambda_3 t|_-^{(n-2)/2}(H+m)\chi(H)\phi_t\|^2 \mathrm{d}t \leq C\|\langle A\rangle_-^{(n-1)/2}\phi\|^2.$$

By an application of the commutator expansion lemma, we have

$$|(\phi_{t_0}|\Phi(t_0)\phi_{t_0})| \leq C\|\langle A\rangle_-^{\frac{n}{2}}\phi\|^2.$$

Therefore,

$$(\phi_{t_1}|\Phi(t_1)\phi_{t_1}) + \textstyle\int_{t_0}^{t_1} t^{n-1}\|f_2(\tfrac{A}{t})\chi(H)\phi_t\|^2 \mathrm{d}t \leq C\|\langle A\rangle_-^{n/2}\phi\|^2.$$

Therefore, (4.12.2) and (4.12.3) are true for n, which completes the proof of the induction step. □

4.13 Strong Low Velocity Estimates

In this section, following [SS3], we will show how the strong propagation estimates for the generator of dilations imply strong low velocity estimates.

Theorem 4.13.1
Suppose that the hypotheses of Theorem 4.12.1 hold. Suppose that $\lambda_0 > 0$, $\chi \in C_0^\infty(\mathbb{R})$, *and* $\mathrm{supp}\chi \subset [\frac{1}{2}\lambda_0, \infty[$. *Then, for any* $s \geq 1$,

$$\int_1^\infty \left\| 1\!\!1_{[0,\lambda_0]}\left(\frac{x^2}{t^2}\right)\chi(H)\mathrm{e}^{-itH}\phi\right\|^2 t^{s-1}\mathrm{d}t \leq C\|\langle A\rangle_-^{s/2}\phi\|^2, \tag{4.13.1}$$

and, for any $s \geq 0$,

$$\left\| 1\!\!1_{[0,\lambda_0]}\left(\frac{x^2}{t^2}\right)\chi(H)\mathrm{e}^{-itH}\phi\right\|^2 \leq Ct^{-s}\|\langle A\rangle_-^{s/2}\phi\|^2. \tag{4.13.2}$$

Proof. We will use the constants $\lambda_1, \dots, \lambda_5$ and the functions h, F and f_i introduced in the proof of Proposition 4.12.1.

Consider the observable

$$\Phi(t) = \chi(H)F\left(\frac{x^2}{t^2}\right)\chi(H)t^n.$$

Then

$$\begin{aligned} -\mathbf{D}\Phi(t) &= 2\chi(H)\left(F'\left(\tfrac{x^2}{t^2}\right)\tfrac{x^2}{t^2} - \tfrac{n}{2}F\left(\tfrac{x^2}{t^2}\right)\right)\chi(H)t^{n-1} \\ &+2\textstyle\sum_{i=1}^2 \chi(H)f_i\left(\tfrac{x^2}{t^2}\right)\tfrac{A}{t}f_i\left(\tfrac{x^2}{t^2}\right)\chi(H)t^{n-1} \\ &\geq 2\textstyle\sum_{i=1}^2 \chi(H)f_i\left(\tfrac{x^2}{t^2}\right)\left(\tfrac{A}{t} - \lambda_3\right)f_i\left(\tfrac{x^2}{t^2}\right)\chi(H)t^{n-1}. \end{aligned} \tag{4.13.3}$$

Choose a function $g \in C^\infty(\mathbb{R})$ such that $1 \geq g \geq 0$ and

$$g(s) = \begin{cases} 1, & s \leq \lambda_4, \\ 0, & s \geq \lambda_5. \end{cases}$$

Then the right-hand side of (4.13.3) is greater than

$$\begin{aligned} &2\textstyle\sum_{i=1}^2 C_0\chi(H)f_i^2\left(\tfrac{x^2}{t^2}\right)\chi(H)t^{n-1} \\ &+2\textstyle\sum_{i=1}^2 \chi(H)f_i\left(\tfrac{x^2}{t^2}\right)\left(\tfrac{A}{t} - \lambda_4\right)g^2\left(\tfrac{A}{t}\right)f_i\left(\tfrac{x^2}{t^2}\right)\chi(H)t^{n-1}, \end{aligned} \tag{4.13.4}$$

where $C_0 := \lambda_4 - \lambda_3 > 0$. We note that

$$\mathrm{ad}_A^k f_i\left(\frac{x^2}{t^2}\right) \in O(t^0),\ k \in \mathbb{N},$$

(because, effectively, f_i can be considered as a function in C_0^∞).

Using the commutator expansion lemma C.3.1 to move $g\left(\frac{A}{t}\right)$ to the right and to the left, we transform (4.13.4) into a sum of terms of the form

$$\chi(H)g_1\left(\frac{A}{t}\right)B(t)g_2\left(\frac{A}{t}\right)\chi(H)t^{n-1-j}+O(t^{-2}), \tag{4.13.5}$$

where $B(t)$ is bounded uniformly in t and the functions g_i are bounded and supported in $]-\infty,\lambda_5]$ and $j\geq 0$.

Thus we have shown that

$$\begin{aligned}-\mathbf{D}\Phi(t) &\geq t^{n-1}C_0\chi(H)f_2^2(\tfrac{x^2}{t^2})\chi(H)\\ &-C\chi(H)\mathbb{1}_{[-\infty,\lambda_4]}(\tfrac{A}{t})\chi(H)t^{n-1}+O(t^{-2}).\end{aligned}$$

Hence,

$$\begin{aligned}(\phi_{t_0}|\Phi(t_0)\phi_{t_0}) &\geq (\phi_{t_1}|\Phi(t_1)\phi_{t_1}+C_0\textstyle\int_{t_0}^{t_1}\|f_2(\frac{x^2}{t^2})\chi(H)\phi_t\|^2t^{n-1}\mathrm{d}t\\ &-C\textstyle\int_{t_0}^{t_1}\|\mathbb{1}_{[-\infty,\lambda_4]}(\frac{A}{t})\chi(H)\phi_t\|^2t^{n-1}\mathrm{d}t-C\|\phi\|^2.\end{aligned}$$

But by Proposition 4.12.1, we have

$$\int_1^\infty\left\|\mathbb{1}_{[-\infty,\lambda_4]}\left(\frac{A}{t}\right)\chi(H)\phi(t)\right\|^2t^{n-1}\mathrm{d}t\leq C\|\langle A\rangle_-^{\frac{n}{2}}\phi\|^2.$$

Obviously,

$$(\phi_{t_0}|\Phi(t_0)\phi t_0)\leq C\|\phi\|^2.$$

Thus we obtain

$$(\phi_{t_1}|\Phi(t_1)\phi_{t_1})+C_0\int_{t_0}^{t_1}\|f_2(\frac{x^2}{t^2})\chi(H)\phi_t\|^2t^{n-1}\mathrm{d}t\leq C\|\langle A\rangle_-^{n/2}\phi\|^2.$$

This implies (4.13.1) and (4.13.2) and completes the proof of the theorem. □

Corollary 4.13.2
Under the assumptions of Theorem 4.13.1, for any $s\geq 1$, we have

$$\int_1^\infty\left\|\mathbb{1}_{[0,\lambda_0]}\left(\frac{x^2}{t^2}\right)\chi(H)\phi_t\right\|^2t^{s-1}\mathrm{d}t\leq C\|\langle x\rangle^{s/2}\phi\|^2, \tag{4.13.6}$$

and, for $s\geq 0$,

$$\left\|\mathbb{1}_{[0,\lambda_0]}\left(\frac{x^2}{t^2}\right)\chi(H)\phi_t\right\|^2\leq C\langle t\rangle^{-s}\|\langle x\rangle^{s/2}\phi\|^2. \tag{4.13.7}$$

Proof. Clearly, in the estimates of Theorem 4.13.1 we can replace $\|\langle A\rangle_-^{s/2}\phi\|^2$ with $\|\langle A\rangle_-^{s/2}\chi_1(H)\phi\|^2$ for any $\chi_1\in C_0^\infty(\mathbb{R})$ such that $\chi\chi_1=\chi$. But it is easy to see that

$$\|\langle A\rangle_-^{s/2}\chi_1(H)\phi\|^2\leq\|\langle x\rangle^{s/2}\phi\|^2 \tag{4.13.8}$$

for $s\in\mathbb{N}$. By interpolation, (4.13.8) can be extended to $s\geq 0$. □

4.14 Schrödinger Operators as Pseudo-differential Operators

In this section we will collect some properties of H and its functions related to the pseudo-differential calculus. We use the notations in Sect. D.8. The first proposition follows immediately from Proposition D.11.2.

Proposition 4.14.1
Assume that $V(x) \in S(1, \langle x\rangle^{-2}\mathrm{d}x^2)$. Then for any $z \notin \sigma(H)$ and $\chi \in C_0^\infty(\mathbb{R})$, we have

$$(z - H)^{-1}, \chi(H) \in \Psi(1, g_1).$$

Next we describe a simple consequence of the commutator expansion lemma.

Lemma 4.14.2
Assume that $V(x) \in S(1, \langle x\rangle^{-2}\mathrm{d}x^2)$. Let $\chi \in C_0^\infty(\mathbb{R})$ and $n \in \mathbb{N}$. Then

$$\mathrm{e}^{-itH}\chi(H)\langle x\rangle^n = \sum_{j=0}^{n} \langle x\rangle^{n-j}\langle t\rangle^j B_j(t), \tag{4.14.1}$$

with $B_j(t)$ uniformly bounded. Besides, for any m,

$$\langle x\rangle^{-n}\mathrm{e}^{-itH}\chi(H)\langle x\rangle^n\langle D\rangle^m \in O(\langle t\rangle^n). \tag{4.14.2}$$

Proof. Let us show (4.14.1). For $n \in \mathbb{N}$, set

$$A_n(t) := \mathrm{e}^{-itH}\chi(H)\langle x\rangle^n \mathrm{e}^{itH}.$$

By a direct computation, we check that

$$\frac{\mathrm{d}^k}{\mathrm{d}t^k}A_n(0)\langle x\rangle^{k-n} = (-i)^k\chi(H)\left(\mathrm{ad}_H^k\langle x\rangle^n\right)\langle x\rangle^{k-n}$$

is bounded for $0 \le k \le n$. Using Taylor's formula, we obtain

$$A_n(t) = \sum_{k=0}^{n-1} \frac{t^k}{k!}\frac{\mathrm{d}^k}{\mathrm{d}t^k}A_n(0) + O(t^n),$$

which implies (4.14.1).

(4.14.2) follows easily from (4.14.1). □

Lemma 4.14.3
Assume that $V(x) \in S(1, \langle x\rangle^{-2}\mathrm{d}x^2)$ and

$$\lim_{|x|\to\infty} V(x) = 0.$$

Then for any $\chi, \chi_0 \in C^\infty(\mathbb{R})$ *such that* $\chi', \chi_0' \in C_0^\infty(\mathbb{R})$ *and* $\mathrm{supp}\chi \cap \mathrm{supp}\chi_0 = \emptyset$,

$$\chi_0(H_0)\chi(H) \in \Psi(\langle x\rangle^{-\infty}).$$

Proof. We can split the potential

$$V(x) = V_1(x) + V_2(x)$$

such that $V_1(x) \in \Psi(\langle x\rangle^{-\infty})$ and $|V_2(x)| < \mathrm{dist}(\mathrm{supp}\chi_0, \mathrm{supp}\chi)$. We set

$$H_1 = \frac{1}{2}D^2 + V_1(x).$$

By Proposition D.11.3, we have

$$\chi_0(H_0) - \chi_0(H_1) \in \Psi(\langle x\rangle^{-\infty}).$$

By Proposition D.11.4, we obtain

$$\chi_0(H_1)\chi(H) \in \Psi(\langle x\rangle^{-\infty}).$$

□

4.15 Improved Isozaki-Kitada Modifiers

The goal of this section is to construct certain Fourier integral operators similar to the Isozaki-Kitada modifiers. They will turn out to be approximations to the wave operators in the outgoing region (although we will have to wait until Sect. 4.17 to see a proof of this fact).

We would like to treat the short-range case in a way parallel to the long-range case. Therefore, let us recall from Proposition 2.8.1 that if we assume that the potential $V(x)$ satisfies the *smooth short-range condition*

$$\int_0^\infty \sup_{|x|>R} |\partial_x^\alpha V(x)| R^{|\alpha|} \mathrm{d}R < \infty, \ |\alpha| \geq 0, \tag{4.15.1}$$

then, for $\epsilon_0 > 0$, $\sigma_0 > -1$ and $R_0 > 0$ large enough, and for $(x, \xi) \in \Gamma^+_{\epsilon_0,\sigma_0,R_0}$, there exists the limit

$$\lim_{t\to\infty} \left(S(t,x,\xi) - \tfrac{1}{2}t\xi^2\right) =: \Phi^+_{\mathrm{sr}}(x,\xi). \tag{4.15.2}$$

This function satisfies on $\Gamma^+_{R_0,\epsilon_0,\sigma_0}$

$$\begin{aligned} &\tfrac{1}{2}\xi^2 = \tfrac{1}{2}(\nabla_x \Phi^+_{\mathrm{sr}}(x,\xi))^2 + V(x), \\ &\partial_x^\alpha \partial_\xi^\beta (\Phi^+_{\mathrm{sr}}(x,\xi) - \langle x, \xi\rangle) \in o(\langle x\rangle^{-|\alpha|}), \quad : \alpha, \beta \in \mathbb{N}^n. \end{aligned} \tag{4.15.3}$$

In the following lemma we state the basic properties of the Isozaki-Kitada modifiers for the short-range and long-range case. *(ii)* is just a reminder of Theorem 4.9.1. So, it is enough to prove *(i)*.

Lemma 4.15.1
Let $q^+(x,\xi) \in S(1, g_1)$ and $\chi_0 \in C_0^\infty(\mathbb{R})$ be the functions introduced in Sect. 4.9.
(i) Assume the smooth short-range condition (4.15.1) and set

$$J_{\rm sr}^+ = J(\Phi_{\rm sr}^+, q^+), \qquad \Omega_{\rm sr}^+ = \mathrm{s-}\lim_{t\to\infty} \mathrm{e}^{itH}\mathrm{e}^{-itH_0}.$$

Then

$$\Omega_{\rm sr}^+ \chi_0(H_0) = \mathrm{s-}\lim_{t\to\infty} \mathrm{e}^{itH} J_{\rm sr}^+ \mathrm{e}^{-itH_0}. \tag{4.15.4}$$

(ii) Assume the smooth long-range condition (4.9.1) and set

$$J_{\rm lr}^+ = J(\Phi_{\rm lr}^+, q^+), \qquad \Omega_{\rm lr}^+ = \mathrm{s-}\lim_{t\to\infty} \mathrm{e}^{itH}\mathrm{e}^{-iS(t,D)}.$$

Then

$$\Omega_{\rm lr}^+ \chi_0(H_0) = \mathrm{s-}\lim_{t\to\infty} \mathrm{e}^{itH} J_{\rm lr}^+ \mathrm{e}^{-itH_0}.$$

Proof. It is enough to show that, for $g \in C_0^\infty(X)$ with a sufficiently small support,

$$\mathrm{s-}\lim_{t\to\infty} (\chi_0(H_0) - J_{\rm sr}^+)\mathrm{e}^{-itH_0} g(D) = 0.$$

Let $J, J_1 \in C_0^\infty(X\backslash\{0\})$ such that $J = 1$ on a neighborhood of supp$g$ and $J_1 = 1$ on a neighborhood of supp$J$. We choose the supports of g, J, J_1 small enough such that

$$J_1\left(\tfrac{x}{t}\right) q^+(x,\xi) g(\xi) = J_1\left(\tfrac{x}{t}\right) \chi_0(\tfrac{1}{2}\xi^2) g(\xi).$$

Then

$$\begin{aligned}
\mathrm{s-}\lim_{t\to\infty} (\chi_0(H_0) - J_{\rm sr}^+)\mathrm{e}^{-itH_0} g(D) &= \mathrm{s-}\lim_{t\to\infty} (\chi_0(H_0) - J_{\rm sr}^+) J(\tfrac{x}{t}) \mathrm{e}^{-itH_0} g(D) \\
&= \mathrm{s-}\lim_{t\to\infty} J_1(\tfrac{x}{t})(\chi_0(H_0) - J_{\rm sr}^+) J(\tfrac{x}{t}) \mathrm{e}^{-itH_0} g(D) \\
&= \mathrm{s-}\lim_{t\to\infty} J_1(\tfrac{x}{t})(\chi_0(H_0) - J_{\rm sr}^+)\mathrm{e}^{-itH_0} g(D).
\end{aligned}$$

But

$$J_1(\frac{x}{t})(\chi_0(H_0) - J_{\rm sr}^+) g(D) \tag{4.15.5}$$

is a pseudo-differential operator with symbol

$$\begin{aligned}
&J_1(\tfrac{x}{t}) \left(\chi_0(\tfrac{1}{2}\xi^2) - q^+(x,\xi)\mathrm{e}^{i\Phi_{\rm sr}^+(x,\xi) - i\langle x,\xi\rangle}\right) g(\xi) \\
&= J_1(\tfrac{x}{t}) \left(1 - \mathrm{e}^{i\Phi_{\rm sr}^+(x,\xi) - i\langle x,\xi\rangle}\right) \chi_0(\tfrac{1}{2}\xi^2) g(\xi)
\end{aligned}$$

and by (4.15.3), all the semi-norms in $S(1, g_0)$ of this symbol go to zero as $t \to \infty$. Therefore (4.15.5) converges in norm to zero. □

Let us note that FIO's with phase $\Phi^+_{\rm sr}(x,\xi)$ are pseudo-differential operators. More exactly, the following fact is true.

Lemma 4.15.2
If $a \in S(1, g_1)$ with supp$a \in \Gamma^+_{R_0,\epsilon_0,\sigma_0}$, *then an operator $A = J(\Phi^+_{\rm sr}, a)$ belongs to $\Psi(1, g_1)$. Moreover, if $\Gamma \subset X \times X'$, then $A \in \Psi(\langle x\rangle^{-\infty})$ on Γ iff $A \in \Psi(\langle x\rangle^{-\infty})$ on Γ.*

Proof. We have $J(\Phi^+_{\rm sr}, a) = b(x, D)$ for

$$b(x,\xi) = a(x,\xi)\mathrm{e}^{i\langle x,\xi\rangle - i\Phi^+_{\rm sr}(x,\xi)}.$$

By (4.15.3),

$$\mathrm{e}^{i\langle x,\xi\rangle - i\Phi^+_{\rm sr}(x,\xi)} \in S(1, g_1).$$

So $b \in S(1, g_1)$. Moreover, $b \in S(\langle x\rangle^{-\infty})$ on Γ iff $a \in S(\langle x\rangle^{-\infty})$ on Γ. □

In what follows, we will write $\Phi^+(x,\xi)$ instead of $\Phi^+_{\rm lr}(x,\xi)$ or $\Phi^+_{\rm sr}(x,\xi)$, and J^+ for $J^+_{\rm lr}$ or $J^+_{\rm sr}$. We denote by Ω^+ the short-range or long-range wave operators $\Omega^+_{\rm lr}$ and $\Omega^+_{\rm sr}$.

Let us list the properties of these functions that we will use:

$$\Omega^+\chi_0(H_0) = \mathrm{s-}\lim_{t\to\infty} \mathrm{e}^{itH} J^+ \mathrm{e}^{-itH_0},$$
$$\tfrac{1}{2}\xi^2 = \tfrac{1}{2}(\nabla_x \Phi^+(x,\xi))^2 + V(x),$$

and, uniformly for $(x,\xi) \in \Gamma^+_{R_0,\epsilon_0,\sigma_0}$, we have

$$\partial_x^\alpha \partial_\xi^\beta(\Phi^+(x,\xi) - \langle x,\xi\rangle) \in o(\langle x\rangle^{1-|\alpha|}),\ |\alpha| + |\beta| \geq 0. \tag{4.15.6}$$

Proposition 4.15.3
Let $\epsilon_0 > 0$, $-1 < \sigma_0$ and $R_0 > 0$ be such that $\Phi^+(x,\xi)$ is defined on $\Gamma^+_{R_0,\epsilon_0,\sigma_0}$. Let $\epsilon > \epsilon_0$, $\sigma > \sigma_0$. Then there exist functions $i^+(x,\xi) \in S(1, g_1)$, $r^+(x,\xi) \in S(\langle x\rangle^{-1}, g_1)$ such that suppi^+, supp$r^+ \subset \Gamma^+_{R_0,\epsilon_0,\sigma_0}$, *$r^+ \in S(\langle x\rangle^{-\infty})$ on $\Gamma^+_{\epsilon,\sigma}$, and the operators*

$$I^+ := J(\Phi^+, i^+),\quad R^+ := J(\Phi^+, r^+)$$

satisfy

$$HI^+ - I^+H_0 = R^+,$$
$$\Omega^+\chi_0(H_0) = \mathrm{s-}\lim_{t\to\infty} \mathrm{e}^{itH} I^+ \mathrm{e}^{-itH_0}.$$

Moreover, I^+I^{+*} is elliptic on a conical neighborhood of $\Gamma^+_{\epsilon,\sigma}$.

Proof. Note that the following identity is true:

$$\left(\nabla_x\Phi^+(x,\xi)\nabla_x+\tfrac{1}{2}\Delta_x\Phi^+(x,\xi)\right)\left(\det\nabla_x\nabla_\xi\Phi^+(x,\xi)\right)^{1/2}=0. \tag{4.15.7}$$

Indeed, by differentiating with respect to x and ξ the eikonal equation

$$\tfrac{1}{2}(\nabla_x\Phi^+(x,\xi))^2+V(x)=\tfrac{1}{2}\xi^2,$$

we get

$$\nabla_x\Phi^+\nabla_x\nabla_x\nabla_\xi\Phi^+=-\nabla_x^2\Phi^+\nabla_x\nabla_\xi\Phi^+.$$

Using then the identity

$$(\det A(s))^{-1}\frac{\mathrm{d}}{\mathrm{d}s}\det A(s)=\mathrm{Tr}\left(A^{-1}(s)\frac{\mathrm{d}A(s)}{\mathrm{d}s}\right),$$

we obtain

$$\begin{aligned}&\nabla_x\Phi^+\nabla_x\det(\nabla_x\nabla_\xi\Phi^+)^{\frac{1}{2}}\\&=\tfrac{1}{2}(\det\nabla_x\nabla_\xi\Phi^+)^{\frac{1}{2}}\mathrm{Tr}\left((\nabla_x\nabla_\xi\Phi^+)^{-1}\nabla_x\Phi^+\nabla_x(\nabla_x\nabla_\xi\Phi^+)\right)\\&=-\tfrac{1}{2}\det(\nabla_x\nabla_\xi\Phi^+)^{\frac{1}{2}}\Delta_x\Phi^+,\end{aligned}$$

which proves (4.15.7).

The amplitudes of I^+ and R^+ are related by the identity

$$\left(\nabla_x\Phi^+(x,\xi)\nabla_x+\tfrac{1}{2}\Delta_x\Phi^+(x,\xi)-\tfrac{i}{2}\Delta_x\right)i^+(x,\xi)=r^+(x,\xi). \tag{4.15.8}$$

For any $c(x,\xi)\in S(\langle x\rangle^m,g_1)$, we set

$$\begin{aligned}\mathcal{L}c(x,\xi)\ &:=\nabla_x\Phi^+(x,\xi)\nabla_x c(x,\xi)\\&\quad-\tfrac{i}{2}\left(\det\nabla_x\nabla_\xi\Phi^+(x,\xi)\right)^{-1/2}\Delta_x\left(\det\nabla_x\nabla_\xi\Phi^+(x,\xi)\right)^{1/2}c(x,\xi).\end{aligned}$$

Note that $\mathcal{L}$ maps $S(\langle x\rangle^m,g_1)$ into $S(\langle x\rangle^{m-1},g_1)$. Putting

$$\begin{aligned}b^+(x,\xi)&=\left(\det\nabla_x\nabla_\xi\Phi^+(x,\xi)\right)^{-1/2}i^+(x,\xi),\\p^+(x,\xi)&=\left(\det\nabla_x\nabla_\xi\Phi^+(x,\xi)\right)^{-1/2}r^+(x,\xi),\end{aligned}$$

we can rewrite equation (4.15.8) as

$$\mathcal{L}b^+(x,\xi)=p^+(x,\xi). \tag{4.15.9}$$

We will find a sequence of $(\epsilon_1,\sigma_1,R_1)$, $(\epsilon_2,\sigma_2,R_2),\dots$ such that $\epsilon_0<\epsilon_1<\dots<\epsilon_\infty<\epsilon$, $\sigma_0<\sigma_1<\dots<\sigma_\infty<\sigma$ and $R_0<R_1<\dots<R_\infty<R$, and if $(x,\xi)\in\Gamma^+_{\epsilon_j,\sigma_j,R_j}$ and $0<s$, then we have

$$(\tilde{y}(s,\infty,x,\xi),\xi)\in\Gamma^+_{\epsilon_{j+1},\sigma_{j+1},R_{j+1}},$$

where $\tilde{y}(s,\infty,x,\xi)$ are the trajectories defined in Chap. 2. Then we define inductively the following functions:

$$\begin{aligned}
\tilde b_0^+(x,\xi) &= 1, && (x,\xi)\in\Gamma^+_{\epsilon_0,\sigma_0,R_0},\\
\tilde b_m^+(x,\xi) &= \textstyle\int_0^\infty \tilde p_{m-1}(\tilde y(u,\infty,x,\xi),\xi)\mathrm{d}u, && (x,\xi)\in\Gamma^+_{\epsilon_m,\sigma_m,R_m},\\
\tilde i_m^+(x,\xi) &= (\det\nabla_x\nabla_\xi\Phi(x,\xi))^{1/2}\,\tilde b_m^+(x,\xi), && (x,\xi)\in\Gamma^+_{\epsilon_m,\sigma_m,R_m},\\
\tilde p_m^+(x,\xi) &= \tfrac{i}{2}\,(\det\nabla_x\nabla_\xi\Phi^+(x,\xi))^{-1/2}\,\Delta_x\tilde i_m^+(x,\xi), && (x,\xi)\in\Gamma^+_{\epsilon_m,\sigma_m,R_m}.
\end{aligned}$$

Recall that

$$|\tilde y(u,\infty,x,\xi)|\geq C(|x|+u|\xi|).$$

Using this, we easily see that $\tilde b_m^+(x,\xi),\tilde i_m^+(x,\xi)\in S(\langle x\rangle^{-m},g_1)$ and $\tilde p_m^+(x,\xi)\in S(\langle x\rangle^{-m-2},g_1)$ on $\Gamma^+_{\epsilon_m,\sigma_m,R_m}$. Moreover, for $(x,\xi)\in\Gamma^+_{\epsilon_m,\sigma_m,R_m}$,

$$\mathcal{L}\sum_{j=0}^{m}\tilde b_j^+(x,\xi)=\tilde p_m(x,\xi). \tag{4.15.10}$$

We can assume that the function $q^+(x,\xi)\in S(1,g_1)$ constructed in Sect. 4.9 is supported in $\Gamma^+_{\epsilon_\infty,\sigma_\infty,R_\infty}$. By the Borel Lemma (see Lemma D.9.3), we can find $b^+(x,\xi)\in S(1,g_1)$ such that, for any $m\in\mathbb{N}$,

$$b^+(x,\xi)-\sum_{j=0}^{m}q^+(x,\xi)\tilde b_j^+(x,\xi)=:c_m^+(x,\xi)\in S(\langle x\rangle^{-m-1},g_1). \tag{4.15.11}$$

It follows from (4.15.10) and (4.15.11) that

$$p^+(x,\xi):=\mathcal{L}b^+(x,\xi)=q^+(x,\xi)\tilde p_m^+(x,\xi)+w_m(x,\xi)+\mathcal{L}c_m^+(x,\xi),$$

where $w_m(x,\xi)\in S(\langle x\rangle^{-1},g_1)$ and $\mathrm{supp}w_m\subset\mathrm{supp}\nabla q^+\subset\Gamma^+_{\epsilon_\infty,\sigma_\infty,R_\infty}\backslash\Gamma^+_{\epsilon,\sigma,R}$. Since m was arbitrary, this shows that $\mathcal{L}b^+\in S(\langle x\rangle^{-1},g_1)$, $\mathrm{supp}\mathcal{L}b^+\subset\Gamma^+_{\epsilon_\infty,\sigma_\infty,R_\infty}$ and $\mathcal{L}b^+\in S(\langle x\rangle^{-\infty})$ on $\Gamma^+_{\epsilon,\sigma}$.

Finally, we set

$$\begin{aligned}
i^+(x,\xi) &:= (\det\nabla_x\nabla_\xi\Phi^+(x,\xi))^{-1/2}\,b^+(x,\xi),\\
r^+(x,\xi) &:= (\det\nabla_x\nabla_\xi\Phi^+(x,\xi))^{-1/2}\,p^+(x,\xi).
\end{aligned}$$

We have $r^+(x,\xi)\in S(\langle x\rangle^{-1},g_1)$, $\mathrm{supp}r^+(x,\xi)\subset\Gamma^+_{\epsilon_\infty,\sigma_\infty,R_\infty}$ and $r^+\in S(\langle x\rangle^{-\infty})$ on $\Gamma^+_{\epsilon,\sigma}$. □

4.16 Microlocal Propagation Estimates

Throughout this section we will assume the smooth long-range condition (4.9.1). Let us note that the smooth short-range condition (4.15.1) implies (4.9.1). Moreover, by Lemma A.1.3, it follows from (4.9.1) that

$$\partial_x^\alpha V(x)\in o(\langle x\rangle^{-|\alpha|}),\quad \alpha\in\mathbb{N}^n. \tag{4.16.1}$$

Therefore, the results of Sect. 4.14 on the pseudo-differential properties of H and of Sect. 4.13 on strong propagation estimates hold. Moreover, by Theorem 4.3.4, the operator H has no positive eigenvalues.

The main results of this section are certain microlocal propagation estimates for the propagator e^{-itH}. These estimates intuitively mean that the evolution of a state localized in an outgoing region is very close to the free evolution and are related to the estimates of Theorem 2.3.3 *(iii)* for classical 2-body Hamiltonians. They can be shown in a number of different ways. They were originally obtained by Isozaki-Kitada [IK4] using estimates on the resolvent $(H-\lambda)^{-1}$ and the Fourier transform. Our proof is based on the time-dependent approach. We first show these estimates for the free evolution, where they follow from the non-stationary phase method. Then we use the improved Isozaki-Kitada modifiers to extend these estimates to the case of the full evolution.

Theorem 4.16.1
Assume the smooth long-range condition (4.9.1). Let $\epsilon > \epsilon_0 > 0$, $\sigma > \sigma_0 > -1$. *Let* $\chi \in C_0^\infty(\mathbb{R}\backslash\{0\})$, $p_+(x,\xi) \in S(1,g_1)$ *such that* $p_+ \in S(\langle x\rangle^{-\infty}, g_1)$ *outside* $\Gamma^+_{\epsilon,\sigma}$. *Let* $p_- \in S(1,g_1)$ *such that* $p_- \in S(\langle x\rangle^{-\infty})$ *on* $\Gamma^+_{\epsilon_0,\sigma_0}$. *Then the following results hold:*
(i) There exists $\delta_0 > 0$ *such that*

$$\mathbb{1}_{[0,\delta_0]}(\tfrac{|x|}{t})\chi(H)e^{-itH}\langle x\rangle^{-N} \in O(\langle t\rangle^{-N}),\ t \geq 0,\ N \in \mathbb{N}.$$

In the case when $V(x) = 0$, *we can improve this result: there exist* $\delta_0, \delta_1 > 0$ *such that*

$$\mathbb{1}_{[0,\delta_0]}(\tfrac{|x|}{t})\chi(H)e^{-itH}\mathbb{1}_{[0,\delta_1]}(\tfrac{|x|}{t}) \in O(t^{-\infty}),\ t \geq 0,\ N \in \mathbb{N}.$$

(ii)

$$\langle x\rangle^{-N}\chi(H)e^{-itH}\langle x\rangle^{-N} \in O(\langle t\rangle^{-N}),\ t \geq 0,\ N \in \mathbb{N}.$$

(iii) There exists $\delta_0 > 0$ *such that*

$$\mathbb{1}_{[0,\delta_0]}(\tfrac{|x|}{t})\chi(H)e^{-itH}p_+(x,D)\langle x\rangle^{N} \in O(t^{-\infty}),\ t \geq 0,\ N \in \mathbb{N}.$$

(iv)

$$\langle x\rangle^{-M}\chi(H)e^{-itH}p_+(x,D)\langle x\rangle^{N} \in O(t^{N-M}),\ t \geq 0,\ N < M.$$

(v)

$$\langle x\rangle^{N}p_-(x,D)\chi(H)e^{-itH}p_+(x,D)\langle x\rangle^{N} \in O(t^{-\infty}),\ t \geq 0,\ N \in \mathbb{N}.$$

Let us show first a special case of this theorem, namely the propagation estimates for the free Hamiltonian.

Proof of Theorem 4.16.1 for $V(x) = 0$. *(i)*, *(iii)* and *(v)* follow easily by the non-stationary phase method from Proposition D.12.1.

To see *(ii)*, we write

$$\begin{aligned}&\langle x\rangle^{-N}\chi(H_0)\mathrm{e}^{-itH_0}\langle x\rangle^{-N}\\&=1_{[0,\delta_0]}(\tfrac{|x|}{t})\langle x\rangle^{-N}\chi(H_0)\mathrm{e}^{-itH_0}\langle x\rangle^{-N}1_{[0,\delta_1]}(\tfrac{|x|}{t})+O(\langle t\rangle^{-N}).\end{aligned} \tag{4.16.2}$$

The first term on the right of (4.16.2) is $O(t^{-\infty})$ for a sufficiently small $\delta_0 > 0$ by *(i)*.

To show *(iv)*, we choose $\delta_0 > 0$ as in *(iii)* and write

$$\begin{aligned}\langle x\rangle^{-M}\chi(H_0)\mathrm{e}^{-itH_0}p_+(x,D)\langle x\rangle^{N} &= 1_{[0,\delta_0]}(\tfrac{|x|}{t})\langle x\rangle^{-M}\chi(H_0)\mathrm{e}^{-itH_0}p_+(x,D)\langle x\rangle^{N}\\&\quad+1_{[\delta_0,\infty]}(\tfrac{|x|}{t})\langle x\rangle^{-M}\chi(H_0)\mathrm{e}^{-itH_0}p_+(x,D)\langle x\rangle^{N}.\end{aligned}$$

The first term is $O(t^{-\infty})$ by *(iii)*. The second term we write using (4.14.1) as

$$1_{[\delta_0,\infty]}(\tfrac{|x|}{t})\sum_{j=0}^{N}\langle x\rangle^{-M}\langle x\rangle^{N-j}\langle t\rangle^{j}B_j(t),$$

where $B_j(t)$ are uniformly bounded. This is clearly $O(\langle t\rangle^{N-M})$. □

Next we would like to show the propagation estimates for the full Hamiltonian. We will use the same conventions and constructions as in the previous section. In particular, we will write $\Phi^+(x,\xi)$ and we will use the functions $i^+(x,\xi)$, $r^+(x,\xi)$ that we constructed in Proposition 4.15.3.

Lemma 4.16.2
Let $\epsilon > 0$ and $\sigma > -1$. Let $p_+(x,\xi) \in S(1,g_1)$ such that $p_+ \in S(\langle x\rangle^{-\infty})$ on a conical neighborhood of $X \times X'\backslash\Gamma^+_{\epsilon,\sigma}$. Let I^+ be as in Proposition 4.15.3. Let $\chi, \chi_1 \in C_0^\infty(\mathbb{R})$ such that $\chi\chi_1 = \chi$. Then there exists $\tilde{p}_+(x,\xi) \in S(1,g_1)$ and $r_{-\infty}(x,\xi) \in S(\langle x\rangle^{-\infty})$ such that $\tilde{p}_+(x,\xi) \in S(\langle x\rangle^{-\infty})$ on a conical neighborhood of $X \times X'\backslash\Gamma^+_{\epsilon,\sigma}$ and

$$\chi(H_0)p_+(x,D)\langle x\rangle^{N} = I^+\chi_1(H_0)\tilde{p}_+(x,D)\langle x\rangle^{N}I^{+*} + r_{-\infty}(x,D).$$

Proof. The proof is based on the fact that I^+I^{+*} is elliptic on a conical neighborhood of $\Gamma^+_{\epsilon,\sigma}$. Therefore, we can write

$$\chi_1(H_0)p_+(x,D)\langle x\rangle^{N} = I^+I^{+*}\tilde{p}_{+,0}(x,D)\langle x\rangle^{N}I^+I^{+*} + r_{-\infty,0}(x,D)$$

with $\tilde{p}_{+,0}(x,\xi) \in S(1,g_1)$ and $\tilde{p}_{+,0} \in S(\langle x\rangle^{-\infty})$ on a conical neighborhood of $X \times X'\backslash\Gamma^+_{\epsilon,\sigma}$ and $r_{-\infty,0}(x,\xi) \in S(\langle x\rangle^{-\infty})$. Define

$$\tilde{p}_{+,1}(x,D) = I^{+*}\tilde{p}_{+,0}(x,D)\langle x\rangle^{N}I^+\langle x\rangle^{-N}.$$

Clearly, $\tilde{p}_{+,1}(x,\xi) \in S(1,g_1)$ and $\tilde{p}_{+,1}(x,D) \in \Psi(\langle x\rangle^{-\infty})$ on a conical neighborhood of $X \times X'\backslash\Gamma^+_{\epsilon,\sigma}$. Therefore we can find $\tilde{p}_+(x,\xi) \in S(1,g_1)$ such that $\tilde{p}_+(x,D) \in \Psi(\langle x\rangle^{-\infty})$ on a conical neighborhood of $X \times X'\backslash\Gamma^+_{\epsilon,\sigma}$. $r_{-\infty,1} \in S(\langle x\rangle^{-\infty})$ and

$$\tilde{p}_{+,1}(x,D) = \chi_1(H_0)\tilde{p}_+(x,D) + r_{-\infty,1}(x,D).$$

□

Proof of Theorem 4.16.1. *(i)* is essentially a reformulation (under stronger assumptions on the potentials) of Theorem 4.13.1.

(ii) follows from *(i)* by the argument similar to the one used in the $V(x) = 0$ case.

Let us show *(iii)*. Our basic tool will be the identity

$$\mathrm{e}^{-itH}I^+ = I^+\mathrm{e}^{-itH_0} + i\int_0^t \mathrm{e}^{-i(t-s)H}R^+\mathrm{e}^{-isH_0}\mathrm{d}s. \tag{4.16.3}$$

We can assume that $p^+ \in S(\langle x\rangle^{-\infty})$ on a conical neighborhood of $X\times X'\backslash\Gamma^+_{\epsilon,\sigma}$. Let $\chi_1 \in C_0^\infty(\mathbb{R}\backslash\{0\})$ such that $\chi\chi_1 = \chi$. First we note that by Lemma 4.16.2 and Lemma 4.14.3

$$\begin{aligned} &\chi(H)p_+(x,D)\langle x\rangle^N \\ &= \chi(H)I^+\chi_1(H_0)\tilde{p}_+(x,D)\langle x\rangle^N I^{+*} + \chi(H)R_{-\infty,1}, \end{aligned} \tag{4.16.4}$$

where $R_{-\infty,1} \in \Psi(\langle x\rangle^{-\infty})$, $\tilde{p}_+(x,\xi) \in S(1,g_1)$ and $\tilde{p}_+ \in S(\langle x\rangle^{-\infty}, g_1)$ on a conical neighborhood of $X \times X'\backslash\Gamma^+_{\epsilon,\sigma}$.

Now, by (4.16.3) and (4.16.4),

$$\begin{aligned} &\mathbb{1}_{[0,\delta_0]}(\tfrac{|x|}{t})\chi(H)\mathrm{e}^{-itH}\tilde{p}_+(x,D)\langle x\rangle^N \\ &= \mathbb{1}_{[0,\delta_0]}(\tfrac{|x|}{t})\chi(H)\mathrm{e}^{-itH}R_{-\infty,1} \\ &+\mathbb{1}_{[0,\delta_0]}(\tfrac{|x|}{t})\chi(H)I^+\mathrm{e}^{-itH_0}\chi_1(H_0)\tilde{p}_+(x,D)\langle x\rangle^N I^{+*} \\ &+ \textstyle\int_0^t \mathbb{1}_{[0,\delta_0]}(\tfrac{|x|}{t})\chi(H)\mathrm{e}^{-i(t-s)H}R^+\mathrm{e}^{-isH_0}\chi_1(H_0)\tilde{p}_+(x,D)\langle x\rangle^N I^{+*}\mathrm{d}s \\ &=: I_0(t) + I_1(t) + I_2(t). \end{aligned}$$

By *(i)*, $I_0(t) \in O(t^{-\infty})$.

To estimate $I_1(t)$, we note that, for $\delta_1 > \delta_0$,

$$\begin{aligned} &\mathbb{1}_{[0,\delta_0]}(\tfrac{|x|}{t})\chi(H)I^+\langle x\rangle^N \\ &= \mathbb{1}_{[0,\delta_0]}(\tfrac{|x|}{t})\chi(H)I^+\mathbb{1}_{[0,\delta_1]}(\tfrac{|x|}{t})\langle x\rangle^N + O(t^{-\infty}). \end{aligned}$$

Using then *(iii)* for the free Hamiltonian, we get $I_1(t) \in O(t^{-\infty})$.

To estimate $I_2(t)$, we first note that we can assume that $r^+ \in S(\langle x\rangle^{-\infty})$ on a conical neighborhood of $\Gamma^+_{\epsilon,\sigma}$. Therefore, by Proposition D.15.2, for $M \in \mathbb{N}$,

$$R^+ = \langle x\rangle^{-M} B_M \langle x\rangle^M \tilde{p}_-(x,D) + r_{-\infty,2}(x,D), \tag{4.16.5}$$

where B_M is bounded, $\tilde{p}_- \in S(1,g_1)$, $\tilde{p}_- \in S(\langle x\rangle^{-\infty}, g_1)$ on a conical neighborhood of $\Gamma^+_{\epsilon,\sigma}$ and $r_{-\infty,2} \in S(\langle x\rangle^{-\infty})$. We cut the integral into pieces $0 \le s \le t/2$ and $t/2 \le s \le t$. To the first piece we apply *(v)* for the free Hamiltonian, and to the second piece we apply *(i)* for the full Hamiltonian. We get $I_2(t) \in O(t^{-\infty})$. This ends the proof of *(iii)*.

(iv) follows from *(iii)* and (4.14.1) similarly as in the case $V(x) = 0$.

To show *(v)*, we also use (4.16.4) and (4.16.3) and we write

$$\begin{aligned}
&\langle x\rangle^N p_-(x,D)\chi(H)\mathrm{e}^{-itH} p_+(x,D)\langle x\rangle^N \\
&= \langle x\rangle^N p_-(x,D)\chi(H)\mathrm{e}^{-itH} R_{-\infty,1} \\
&+\langle x\rangle^N p_-(x,D)\chi(H) I^+ \mathrm{e}^{-itH_0}\chi_1(H_0)\tilde{p}_+(x,D)\langle x\rangle^N I^{+*} \\
&+ \int_0^t \langle x\rangle^N p_-(x,D)\chi(H)\mathrm{e}^{-i(t-s)H} R^+ \mathrm{e}^{-itH_0}\chi_1(H_0) q_+(x,D)\langle x\rangle^N I^{+*} \\
&=: I_0(t) + I_1(t) + I_2(t).
\end{aligned}$$

Using the estimate corresponding to *(iv)* for $t \le 0$, we get that $I_0(t) \in O(t^{-\infty})$.

Using Propositions D.15.2 and D.15.3, we have

$$\langle x\rangle^N p_-(x,D)\chi(H) I^+ = B\langle x\rangle^N \tilde{p}_{-,1}(x,D) + r_{-\infty,2}(x,D)$$

where B is bounded, $\tilde{p}_{-,1}(x,\xi) \in S(1,g_1)$, $\tilde{p}_{-,1} \in S(\langle x\rangle^{-\infty})$ on a conical neighborhood of $\Gamma^+_{\epsilon,\sigma}$ and $r_{-\infty,2} \in S(\langle x\rangle^{-\infty})$. Using *(iv)* and *(v)* for the free Hamiltonian, we get $I_1(t) \in O(t^{-\infty})$.

To estimate $I_2(t)$, we use again (4.16.5). We cut the integral into pieces $0 \le s \le t/2$ and $t/2 \le s \le t$. We apply *(v)* for the free Hamiltonian to the first piece and to the second piece the analog of *(iv)* for $t \le 0$. Thus we obtain $I_2(t) \in O(t^{-\infty})$, which proves *(v)*. □

4.17 Wave Operators with Outgoing Cutoffs

Wave operators for potentials satisfying smooth long- or short-range assumptions have good regularity properties. In this section we will show one of them. We will prove that if we multiply the wave operator with a pseudo-differential cutoff supported in an outgoing region and with the energy bounded away from zero, then we obtain a pseudo-differential operator in the short-range case and a Fourier integral operator in the long-range case. It will turn out that the wave operator with such a cutoff is essentially equal to the operator I^+ constructed in Proposition 4.15.3.

Theorem 4.17.1
Let $\epsilon > \epsilon_0 > 0$ and $\sigma > \sigma_0 > -1$. Let $p_+(x,\xi) \in S(1,g_1)$ such that $p_+ \in S(\langle x\rangle^{-\infty})$ outside $\Gamma^+_{\epsilon,\sigma}$ and $\chi \in C_0^\infty(\mathbb{R}\backslash\{0\})$.
(i) Assume the smooth short-range condition (4.15.1). Then there exist $a_1, a_2 \in S(1,g_1)$ with $a_1, a_2 \in S(\langle x\rangle^{-\infty})$ outside $\Gamma^+_{\epsilon_0,\sigma_0}$ such that

$$\Omega^+_{\rm sr}\chi(H_0)p_+(x,D) = a_1(x,D),$$
$$\Omega^{+*}_{\rm sr}\chi(H)p_+(x,D) = a_2(x,D).$$

(ii) Assume the smooth long-range condition (4.9.1). Then there exist $c_1, c_2 \in S(1,g_1)$ such that $c_1, c_2 \in S(\langle x\rangle^{-\infty})$ outside $\Gamma^+_{\epsilon_0,\sigma_0}$, and $r_{-\infty,1}, r_{-\infty,2} \in S(\langle x\rangle^{-\infty})$ such that

$$\Omega^+_{\rm lr}\chi(H_0)p_+(x,D) = J(\Phi^+_{\rm lr}, c_1) + r_{-\infty,1}(x,D),$$
$$\Omega^{+*}_{\rm lr}\chi(H)p_+(x,D) = J(\Phi^+_{\rm lr}, c_2)^* + r_{-\infty,2}(x,D).$$

We will use the same conventions and constructions as in the previous section. In particular, we will write $\Phi^+(x,\xi)$, and we will use the functions $i^+(x,\xi)$, $r^+(x,\xi)$ and the operators I^+, R^+ that we constructed in Proposition 4.15.3.

Now Theorem 4.17.1 follows from the following theorem.

Theorem 4.17.2
Let $\epsilon, \sigma, p^+, \chi$ be as in Theorem 4.17.1 and let i^+ be as in Proposition 4.15.3. Let $\chi_1 \in C_0^\infty(\mathbb{R})$ such that $\chi\chi_1 = \chi$. Then under either the smooth long-range or smooth short-range assumptions there exists $r_{-\infty,i} \in \Psi(\langle x\rangle^{-\infty})$ such that

$$\begin{aligned}\Omega^+\chi(H_0)p_+(x,D) &= \chi_1(H)I^+\chi(H_0)p_+(x,D) + r_{-\infty,1}(x,D),\\ \Omega^{+*}\chi(H)p_+(x,D) &= \chi_1(H_0)I^{+*}\chi(H)p_+(x,D) + r_{-\infty,2}(x,D).\end{aligned} \tag{4.17.1}$$

Proof. Let us show the first identity of (4.17.1). The proof will be based on

$$\begin{aligned}\Omega^+\chi_0(H_0) &= \text{s}-\lim_{t\to\infty} \mathrm{e}^{itH}I^+\mathrm{e}^{-itH_0}\\ &= I^+ + i\int_0^\infty \mathrm{e}^{itH}R^+\mathrm{e}^{-itH_0}\mathrm{d}t.\end{aligned}$$

We have

$$\begin{aligned}\Omega^+\chi(H_0)p_+(x,D) &= \chi_1(H)\Omega^+\chi(H_0)p_+(x,D)\\ &= \chi_1(H)I^+\chi(H_0)p_+(x,D) + \int_0^\infty R_{-\infty,1}(t)\mathrm{d}t,\end{aligned}$$

where

$$R_{-\infty,1}(t) = i\chi_1(H)\mathrm{e}^{itH}R^+\mathrm{e}^{-itH_0}\chi(H_0)p_+(x,D).$$

Let us prove that, for any N, M,

$$\langle D\rangle^M\langle x\rangle^N R_{-\infty,1}(t)\langle x\rangle^N\langle D\rangle^M \in O(t^{-\infty}). \tag{4.17.2}$$

This will imply that

$$\|\langle D\rangle^M\langle x\rangle^N \textstyle\int_0^\infty R_{-\infty,1}\mathrm{d}t\langle x\rangle^N\langle D\rangle^M\| < \infty, \quad N, M \in \mathbb{N}$$

and hence $\int_0^\infty R_{-\infty,1}(t)\mathrm{d}t \in \Psi(\langle x\rangle^{-\infty})$.

First note that, for any M, N, by (4.14.2),

$$\langle D\rangle^M\langle x\rangle^N\chi_1(H)\mathrm{e}^{itH}\langle x\rangle^{-N} \in O(t^N).$$

Next, by Proposition D.15.2,

$$\langle x\rangle^N R^+ = B\langle x\rangle^N\tilde{p}_-(x,D) + r_{-\infty,0}(x,D),$$

where $\tilde{p}_-(x,\xi) \in S(\langle x\rangle^{-1}, g_1)$ such that $p_- \in S(\langle x\rangle^{-\infty})$ on a conical neighborhood of $\Gamma^+_{\epsilon,\sigma}$, $r_{-\infty,0} \in S(\langle x\rangle^{-\infty})$ and B is bounded. Now,

$$(\langle x\rangle^N\tilde{p}_-(x,D) + r_{-\infty,0}(x,D))\mathrm{e}^{-itH_0}\chi(H_0)p_+(x,D)\langle x\rangle^N\langle D\rangle^M \in O(t^{-\infty}),$$

by Theorem 4.16.1 *(iv)* and *(v)* applied to the free Hamiltonian H_0. Therefore (4.17.2) is true. This ends the proof of the first equality of (4.17.1).

To show the second equality of (4.17.1), we use the same arguments, switching the roles of H and H_0. □

4.18 Wave Operators on Weighted Spaces

The goal of this section is to study wave operators multiplied by an energy cutoff with support away from zero as maps on weighted L^2 spaces.

The main result of this section is the following theorem.

Theorem 4.18.1
Suppose that $\chi \in C_0^\infty(\mathbb{R})$ and $0 \notin \mathrm{supp}\chi$ and $s' < s$ or $0 < s' = s$.
(i) Assume the smooth short-range assumption (4.15.1). Then

$$\langle x\rangle^{-s}\Omega^+_{\mathrm{sr}}\chi(H_0)\langle x\rangle^{s'} \in B(L^2(X)). \tag{4.18.1}$$

(ii) Assume the smooth long-range assumption (4.9.1). Then

$$\langle x\rangle^{-s}\Omega^+_{\mathrm{lr}}\chi(H_0)\langle x\rangle^{s'} \in B(L^2(X)). \tag{4.18.2}$$

Theorem 4.18.1 for $0 < s' = s$ is due to Isozaki [I3]. For general $s' < s$, it is due to Jensen-Nakamura [JN] and Herbst-Skibsted [HeSk1].

Proof of Theorem 4.18.1 in the case $s' < s \leq 0$. We will consider the short- and the long-range cases at the same time, using the unified notation introduced in Sect. 4.15. We will use the improved Isozaki-Kitada modifiers.

It is enough to assume that $-s =: n \in \mathbb{N}$, and then to use interpolation. We will write $m := -s' \in \mathbb{R}$, where $0 \leq n < m$.

Let $\chi_1 \in C_0^\infty(\mathbb{R})$ with $\chi_1 \chi = \chi$. We have

$$\begin{aligned} \langle x\rangle^n \Omega^+ \chi(H_0) \langle x\rangle^{-m} &= \langle x\rangle^n \chi_1(H) \Omega^+ \chi(H_0) \langle x\rangle^{-m} \\ &= \langle x\rangle^n \chi_1(H) I^+ \chi(H_0) \langle x\rangle^{-m} \\ &+ i \int_0^\infty \langle x\rangle^n \chi_1(H) \mathrm{e}^{itH} R^+ \mathrm{e}^{-itH_0} \chi(H_0) \langle x\rangle^{-m} \mathrm{d}t. \end{aligned} \tag{4.18.3}$$

The first term on the right-hand side of (4.18.3) is obviously bounded. Let us consider the second term. By Lemma 4.14.2,

$$\langle x\rangle^n \mathrm{e}^{-itH} \chi_1(H) = \sum_{j=0}^{n} B_j(t) \langle t\rangle^j \langle x\rangle^{n-j},$$

where $B_j(t)$ are uniformly bounded. On the other hand,

$$\langle x\rangle^{n-j} R^+ = \tilde{B}_{n-j} \langle x\rangle^{n-j-1} \tilde{p}_-(x, D) + r_{-\infty}(x, D),$$

where $\tilde{B}_{n-j}$ is bounded, $\tilde{p}_- \in S(1, g_1)$ with $\tilde{p}_- \in S(\langle x\rangle^{-\infty})$ on a conical neighborhood of $\Gamma_{\epsilon,\sigma}^+$ and $r_{-\infty} \in S(\langle x\rangle^{-\infty})$. Using the analog of Theorem 4.16.1 *(ii)*, *(iv)* for $t < 0$ and the free Hamiltonian, we get

$$\langle t\rangle^j \langle x\rangle^{n-j} R^+ \mathrm{e}^{itH_0} \chi(H_0) \langle x\rangle^{-m} \in O(\langle t\rangle^{n-1-m}),$$

which is integrable and yields the boundedness of the second term (4.18.3). □

The proof of the $s = s'$ case will be more direct, without the use of improved Isozaki-Kitada modifiers. We will deal separately with the short- and long-range case. The proof of the short-range case with $s = s' \geq 0$ is based on the following estimate:

Proposition 4.18.2
Assume (4.15.1). Let $\chi \in C_0^\infty(\mathbb{R})$, $0 \notin \mathrm{supp}\chi$, $J_0 \in C_0^\infty(X \backslash \{0\})$, $J_0 = 1$ *on* $\left\{x : \frac{1}{2}x^2 \in \mathrm{supp}\chi\right\}$ *and* $s \geq 0$. *Then*

$$\left\| |x - tD|^s J_0\left(\frac{x}{t}\right) \chi(H) \phi_t \right\|^2 \leq C \|\langle x\rangle^s \phi\|^2. \tag{4.18.4}$$

Proof. We will show the estimate for $s =: n \in \mathbb{N}$, and then extend it by interpolation. We need to show that

$$f_n(t) := \left\| |x - tD|^n J_0\left(\tfrac{x}{t}\right) \chi(H) \phi_t \right\| \leq C_n \|\langle x\rangle^n \phi\|, \ n \geq 0. \tag{4.18.5}$$

Clearly, (4.18.5) is true for $n = 0$.

Now suppose that we know that

$$f_m(t) \leq C_m \|\langle x\rangle^m \phi\|, \quad 0 \leq m \leq n-1.$$

Choose $J \in C_0^\infty(X)$ such that $J_0 J = J_0$ and $0 \notin \mathrm{supp} J$. We set

$$V_J(t,x) := J\left(\tfrac{x}{t}\right) V(x),$$

and note that

$$\langle t\rangle^{|\alpha|} \|\partial_x^\alpha V_J(t,.)\|_\infty \in L^1(\mathrm{d}t), \quad |\alpha| \geq 0. \qquad (4.18.6)$$

Let us compute the derivative of $f_n^2(t)$. We have

$$\begin{aligned} \tfrac{\mathrm{d}}{\mathrm{d}t} f_n^2(t) &= \textstyle\sum_{j=0}^{2n-1} \left(\phi_t | \chi(H) J_0\left(\tfrac{x}{t}\right)(x-tD)^j\right. \\ &\times t\nabla_x V_J(t,x)(x-tD)^{2n-j-1} J_0\left(\tfrac{x}{t}\right)\chi(H)\phi_t\Big) \qquad (4.18.7) \\ &+ \left(\phi_t|\chi(H)\left(\mathbf{D} J_0\left(\tfrac{x}{t}\right)\right)(x-tD)^{2n} J_0\left(\tfrac{x}{t}\right)\chi(H)\phi_t\right) + \mathrm{cc}. \end{aligned}$$

Let $j \in C_0^\infty(X)$ such that $j\nabla J_0 = \nabla J_0$ and the support of j is disjoint of $\{0\} \cup \left\{x : \tfrac{1}{2}x^2 \in \mathrm{supp}\chi\right\}$. Then the second term of the right-hand side of (4.18.7) can be written as

$$g_0(t) := (\phi_t|\chi(H) j\left(\tfrac{x}{t}\right) B(t) j\left(\tfrac{x}{t}\right)\chi(H)\phi_t) t^{2n-1} + O(t^{-2})$$

for some uniformly bounded operator $B(t)$. By Theorem 4.13.1, this is integrable.

The first term of the right-hand side of (4.18.7) can be estimated by

$$C \textstyle\sum_{j=1}^n f_{n-j}(t) t^j \|\nabla_x^j V_J(t,\cdot)\|_\infty f_n(t) = g_1(t) f_n(t).$$

Using the induction assumption and (4.18.6), we see that $g_1(t) \in L^1(\mathrm{d}t)$. Thus we have

$$\left|\tfrac{\mathrm{d}}{\mathrm{d}t} f_n^2(t)\right| \leq g_0(t) + \tfrac{1}{2} g_1(t) + \tfrac{1}{2} g_1(t) f_n^2(t).$$

Applying Gronwall's inequality (see Proposition A.1.1), we obtain that $f_n^2(t) \leq C_n f_n^2(0)$, which completes the proof of the proposition. □

Proof of Theorem 4.18.1 *(i)* **in the case** $s' = s \geq 0$. Let J_0 be as in Proposition 4.18.2. We know that

$$\mathrm{s-}\lim_{t\to\infty} \mathrm{e}^{itH}\mathrm{e}^{-itH_0}\chi(H_0) = \mathrm{s-}\lim_{t\to\infty} \mathrm{e}^{itH}\chi(H) J_0\left(\tfrac{x}{t}\right)\mathrm{e}^{-itH_0}.$$

We have

$$\begin{aligned} &\left(\phi_t|\chi(H) J_0\left(\tfrac{x}{t}\right)\mathrm{e}^{-itH_0}|x|^{2s}\mathrm{e}^{itH_0} J_0\left(\tfrac{x}{t}\right)\chi(H)\phi_t\right) \\ &= \left(\phi_t|\chi(H) J_0\left(\tfrac{x}{t}\right)|x-tD|^{2s} J_0\left(\tfrac{x}{t}\right)\chi(H)\phi_t\right). \end{aligned} \qquad (4.18.8)$$

Using Proposition 4.18.2, we see that (4.18.8) is bounded by $C\|\langle x\rangle^s\phi\|^2$. Therefore,

$$\Omega_{\mathrm{sr}}^+ \chi(H_0)|x|^{2s}\chi(H_0)\Omega_{\mathrm{sr}}^{+*} \leq C\langle x\rangle^{2s}. \qquad (4.18.9)$$

This shows the boundedness of (4.18.2). □

Next we would like to treat the long-range case of Theorem 4.18.1 with $s' = s \geq 0$.

Let us fix $\chi \in C_0^\infty(\mathbb{R})$ with $0 \notin \mathrm{supp}\chi$, $J, J_0 \in C_0^\infty(X\backslash\{0\})$ such that $J_0 = 1$ on $\left\{x \mid \frac{1}{2}x^2 \in \mathrm{supp}\chi\right\}$ $J = 1$ on a neighborhood of supp$J$. We will use $V_J(t,x)$, $H_J(t)$, $U_J(t,s)$ and $S_J(t,\xi)$ introduced in Sect. 4.7. Recall that

$$\begin{aligned}&\left(\tfrac{\mathrm{d}}{\mathrm{d}t} + [iH_J,\cdot]\right)(x - \nabla_\xi S_J(t,D))\\ &= \left[(x - \nabla_\xi S_J(t,D)), (V_J(t,x) - V_J(t, \nabla_\xi S(t,D)))\right].\end{aligned} \tag{4.18.10}$$

Besides, Lemma 3.5.2 implies that

$$V_J(t,x) - V_J(t,\nabla_\xi S(t,D)) = (x - \nabla_\xi S_J(t,D))G_J(t) + R_J(t),$$

where the operators $G_J(t)$ and $R_J(t)$ have the property that

$$\mathrm{ad}^n_{(x-\nabla_\xi S_J(t,D))}G_J(t),\ \ \mathrm{ad}^n_{(x-\nabla_\xi S_J(t,D))}R_J(t)\ \in L^1(\mathrm{d}t),\ n \geq 0. \tag{4.18.11}$$

We have the following estimate.

Proposition 4.18.3
Let $S_J(t,\xi)$, $\chi \in C_0^\infty(\mathbb{R})$ and $J_0 \in C_0^\infty(X)$ be as above. Let $s \geq 0$. Then

$$\left\| |x - \nabla_\xi S_J(t,D)|^s J_0\left(\tfrac{x}{t}\right)\chi(H)\phi_t \right\|^2 \leq C\|\langle x\rangle^s\phi\|^2. \tag{4.18.12}$$

Proof. We will prove the proposition by induction on $s = n \in \mathbb{N}$, and then apply an interpolation argument. Let

$$f_n(t) := \left\| |x - \nabla_\xi S_J(t,D)|^n J_0\left(\frac{x}{t}\right)\chi(H)\phi_t \right\|,$$

and let us compute the derivative of $f_n^2(t)$. Using (4.18.10), we obtain

$$\begin{aligned}\tfrac{\mathrm{d}}{\mathrm{d}t}f_n^2(t) &= \textstyle\sum_{j=0}^{2n-1}\left(\phi_t|\chi(H)J_0\left(\tfrac{x}{t}\right)(x - \nabla_\xi S_J(t,D))^j\right.\\ &\quad\times \left[(x - \nabla_\xi S_J(t,D)), i(V_J(t,x) - V_J(t,\nabla_\xi S_J(t,D))\right]\\ &\quad\times \left.(x - \nabla_\xi S_J(t,D))^{2n-j-1}J_0\left(\tfrac{x}{t}\right)\chi(H)\phi_t\right)\\ &\quad+ \left(\phi_t|\chi(H)\left(\mathbf{D}J_0\left(\tfrac{x}{t}\right)\right)(x - \nabla_\xi S_J(t,D))^{2n}J_0\left(\tfrac{x}{t}\right)\chi(H)\phi_t\right) + \mathrm{cc.}\end{aligned} \tag{4.18.13}$$

The term on the last line of (4.18.13), which we call $g_0(t)$, is integrable by Theorem 4.13.1 (see the proof of Proposition 4.18.2).

Commuting factors of $x - \nabla_\xi S_J(t,D)$ through $G_J(t)$ and $R_J(t)$, we see that the sum in (4.18.13) equals

$$\sum_{j=0}^{n} C_j \left(\phi_t | \chi(H) J_0\left(\tfrac{x}{t}\right)(x-\nabla_\xi S_J(t,D))^{n-j} \mathrm{ad}^j_{(x-\nabla_\xi S_J(t,D))} G_J(t)\right.$$
$$\times \left.(x-\nabla_\xi S_J(t,D))^n J_0\left(\tfrac{x}{t}\right)\chi(H)\phi_t\right) + \mathrm{cc}$$
$$+\sum_{j=0}^{n-1} C_j \left(\phi_t | \chi(H) J_0\left(\tfrac{x}{t}\right)(x-\nabla_\xi S_J(t,D))^{n-1-j} \mathrm{ad}^j_{(x-\nabla_\xi S_J(t,D))} R_J(t)\right.$$
$$\times \left.(x-\nabla_\xi S_J(t,D))^n J_0\left(\tfrac{x}{t}\right)\chi(H)\phi_t\right) + \mathrm{cc}.$$

This can be estimated by

$$\sum_{j=0}^{n} C_j f_n(t) f_{n-j}(t) \|\mathrm{ad}^j_{(x-\nabla_\xi S_J(t,D))} G_J(t)\|$$
$$+\sum_{j=1}^{n-1} C_j f_n(t) f_{n-1-j}(t) \left\|\mathrm{ad}^j_{(x-\nabla_\xi S_J(t,D))} R_J(t)\right\| = g_1(t) f_n(t) + g_2(t) f_n^2(t),$$

where, by the induction hypothesis and by (4.18.11), $g_1, g_2 \in L^1$. Therefore,

$$\left|\tfrac{\mathrm{d}}{\mathrm{d}t} f_n^2(t)\right| \leq g_0(t) + \tfrac{1}{2} g_1(t) + \left(\tfrac{1}{2} g_1(t) + g_2(t)\right) f_n^2(t).$$

Finally, we apply Gronwall's inequality, and we see that $f_n(t)$ is bounded by $C_n \|\langle x\rangle^n \phi\|$, which completes the proof of the proposition. □

Proof of Theorem 4.18.1 *(i)* **in the case** $s' = s \geq 0$. We know that

$$\begin{aligned}
\Omega_{\mathrm{lr}}^{+} \chi(H_0) &:= \mathrm{s-}\lim_{t\to\infty} \mathrm{e}^{itH} \mathrm{e}^{-iS(t,D)} \chi(H_0) \\
&= \mathrm{s-}\lim_{t\to\infty} \mathrm{e}^{itH} J_0\left(\tfrac{x}{t}\right) \mathrm{e}^{-iS(t,D)} \chi(H_0) \\
&= \mathrm{s-}\lim_{t\to\infty} \mathrm{e}^{itH} \chi(H) J_0\left(\tfrac{x}{t}\right) \mathrm{e}^{-iS_J(t,D)} \mathrm{e}^{i\sigma^+(D)} \tilde{\chi}(H_0),
\end{aligned}$$

for $\tilde{\chi} \in C_0^\infty(\mathbb{R}\backslash\{0\})$, $\tilde{\chi}\chi = \chi$ and a smooth function σ^+. But $\langle x\rangle^s \mathrm{e}^{i\sigma^+(D)} \tilde{\chi}(H_0) \langle x\rangle^{-s}$ is bounded for $s \in \mathbb{R}$, so it suffices to show the boundedness of

$$\langle x\rangle^{-s} \mathrm{s-}\lim_{t\to\infty} \mathrm{e}^{itH} \chi(H) J_0\left(\frac{x}{t}\right) \mathrm{e}^{-iS_J(t,D)} \langle x\rangle^s. \tag{4.18.14}$$

We have

$$\begin{aligned}
&\left(\phi_t | \chi(H) J_0\left(\tfrac{x}{t}\right) \mathrm{e}^{-iS_J(t,D)} |x|^{2s} \mathrm{e}^{iS_J(t,D)} J_0\left(\tfrac{x}{t}\right) \chi(H)\phi_t\right) \\
&= \left(\phi_t | \chi(H) J_0\left(\tfrac{x}{t}\right) |x-\nabla_\xi S_J(t,D)|^{2s} J_0\left(\tfrac{x}{t}\right) \chi(H)\phi_t\right).
\end{aligned} \tag{4.18.15}$$

Using the bound of Proposition 4.18.3, we see that (4.18.15) is bounded by $C\|\langle x\rangle^s \phi\|^2$. Therefore, (4.18.14) is bounded. □

5. Classical N-Body Hamiltonians

5.0 Introduction

A system of N non-relativistic particles moving in Euclidean space $\mathbb{R}^\nu$ is described with phase space $\mathbb{R}^{N\nu} \times \mathbb{R}^{N\nu}$, with the coordinates

$$(x_1, \cdots, x_N, \xi_1, \cdots, \xi_N),$$

where (x_i, ξ_i) are the position and momentum of the i-th particle. Its motion is described by the Hamiltonian

$$H(x, \xi) = \sum_{i=1}^{N} \frac{\xi_i^2}{2m_i} + \sum_{i<j} V_{ij}(x_i - x_j), \tag{5.0.1}$$

where m_i is the mass of the i-th particle and $V_{ij}(x)$ is the interaction potential between particles i and j. The most important case of such a Hamiltonian is the one encountered in celestial mechanics where

$$V_{ij}(x) = \frac{-m_i m_j}{|x|}. \tag{5.0.2}$$

Typical assumptions that we will keep in mind in this chapter are

$$|\partial_x^\alpha V_{ij}(x)| \le C\langle x\rangle^{-\mu-|\alpha|}, \ \mu > 0, \quad |\alpha| = 0, 1. \tag{5.0.3}$$

This and the next chapter will be devoted to the scattering theory of such systems. In this chapter we will discuss the classical case and in the next chapter the quantum case. In this chapter we will also introduce basic concepts and constructions that are used in the description of N-body configuration space. They are common to classical and quantum N-body systems. They will be used both in this and the next chapter.

As in Chap. 2, where 2-body Hamiltonians were considered, the aim of scattering theory for N-body Hamiltonians is to give a classification of the asymptotic behavior for large times of all trajectories of $H(x, \xi)$.

A number of articles contained in the literature deals with problems connected with the singularity at the origin of the potential (5.0.2) (see e.g. [Sa1] and references therein). This class of questions will not concern us. In fact, throughout this chapter we will assume that the potentials are bounded.

From our point of view, the literature on the classical N-body theory is rather limited (see works of Saari [Sa2], Marchal-Saari [MarSa] and Hunziker [Hu2]). On the other hand, the quantum N-body problem was studied much more extensively. There are some reasons for this. The results that can be obtained in the quantum N-body case are usually more complete and satisfactory than those that describe the classical N-body case. In fact, the article [De7], which is the main source for this chapter, was directly inspired by the results of [De8], which concerned the quantum case. Nevertheless, from the pedagogical and logical point of view it seems that it is better to look at classical N-body systems first. In particular, some of the important results about classical N-body scattering have simpler and more transparent proofs than their quantum counterparts.

Let us briefly describe the contents of this chapter. In Sect. 5.1 we introduce basic concepts used to describe N-body systems. We will use the formalism of *generalized N-body systems*. To our knowledge, this formalism was first used by Hörmander in unpublished lecture notes and it was then described by Agmon in [Ag2]. Generalized N-body Hamiltonians are functions on the phase space $X \times X'$ of the form

$$H(x,\xi) := \frac{1}{2}\xi^2 + \sum_{b\in\mathcal{B}} v^b(x^b),$$

where, for every $b \in \mathcal{B}$, x^b is the projection of the vector x on a certain subspace X^b. They are sometimes called *Agmon Hamiltonians*. We will usually call them simply N-body Hamiltonians. They provide us with a mathematical framework that makes it possible to describe N-body Hamiltonians of the form (5.0.1) and also some other similar Hamiltonians in a particularly convenient way. At the end of Sect. 5.1 we explain how this formalism is related to usual N-body Hamiltonians of the form (5.0.1).

In Sect. 5.2 we introduce various functions on the configuration space that are very important in some of the proofs of both classical and quantum N-body scattering theory. With these functions, one constructs certain observables on phase space whose Poisson bracket with the Hamiltonian is approximately positive. Functions with an approximately positive Poisson bracket were an important ingredient of a number of papers on scattering theory, notably [Mo1, SS1]. But only in [Gr] it was discovered how to distort these functions to make them more adapted to the N-body problem (see also [De6, Yaf5]). Our presentation and details of the construction come from [De6, De8]. Sects. 5.1 and 5.2 are prerequisites for the next chapter.

In Sect. 5.3 we prove among other things that the union of all trapping energies for all subsystems is a closed set. The proof of this fact is based on the so-called classical Mourre estimate, which says that a certain observable has a positive Poisson bracket with the Hamiltonian in a certain subset of phase space. The Mourre estimate was first introduced in the quantum case in [Mo1] and [PSS]. Its classical counterpart comes from [Ge1] in the 3-body case and from [Wa2] in the N-body case. The construction of the observable that is used in the Mourre estimate is quite technical and can be skipped on the first reading.

Section 5.4 presents a proof of the existence of the asymptotic velocity, that is, the limit

$$\lim_{t\to\infty} \frac{x(t,y,\eta)}{t} =: \xi^+(y,\eta).$$

This is a very important construction, because it provides us with a nontrivial parametrization of all possible trajectories. Unfortunately, this parametrization is far from being one to one. It has a quantum analog, which actually was discovered first and can be used as an intermediate step in the proof of asymptotic completeness.

Historically, it was the proof of the asymptotic completeness of N-body short-range systems that was first (see [SS1] and a later proof in [Gr]). The proof contained in [Gr] inspired the construction of the asymptotic velocity in the quantum case, first in [De6], and then in a somewhat different formalism in [De8]. Only afterwards, in [De7], it was realized that the existence of the asymptotic velocity in the classical case follows by essentially the same arguments.

In physical terms, one expects that every N-body system for large time will break into a certain number of almost independent clusters. If we denote the corresponding cluster decomposition by a, then we say that such a trajectory is a-clustered. The existence of the asymptotic velocity implies that every trajectory is a-clustered for a certain cluster decomposition a. Note that in the case of a-clustered trajectories it is natural too look separately at its internal (intra-cluster) motion, described by the component $x^a(t)$ and the external (inter-cluster) motion, denoted by $x_a(t)$.

At this point we know two functions that describe the asymptotics of trajectories: the full energy $H(y,\eta)$ and the asymptotic velocity $\xi^+(y,\eta)$. One can ask if it is possible to describe the joint image of these two functions inside of $\mathbb{R} \times X$. The main result of Sect. 5.5 says that the closure of this image is contained in a certain set described in terms of trapping energies of subsystems. In Sect. 5.6 we will show that the closure of this image contains another similar set described in terms of the so-called *regular trapping energies.* This pair of inclusions has a quantum analog, which in fact is more satisfactory: the inclusions are replaced with an identity, the joint image with the joint spectrum and the trapping energies with the thresholds – see [De6, De8] and the next chapter.

The a-clustered trajectories whose internal coordinate is bounded for $t \to \infty$ can have various behaviors. Some of them are very unstable. The category that has particularly good properties is the set of trajectories that end up in a "well" of the potential $V^a(x^a)$. For these so-called regular trajectories, it is possible to show certain regularity properties of asymptotic quantities $\xi^+(y,\eta)$, $H^{a,+}(y,\eta)$. These results are also presented in Sect. 5.6.

In the two-body case, we showed that zero-energy trajectories satisfy the following a priori bound

$$|x(t)| \leq C\langle t\rangle^{2/(2+\mu)}.$$

It turns out that this bound has an N-body analog, which we show in Sect. 5.7. Namely, we prove that every a-clustered trajectory satisfies

$$|x^a(t)| \leq C\langle t\rangle^{2/(2+\mu)}.$$

This result comes from [De7] and its quantum analog is an important step in the proof of asymptotic completeness in the long-range quantum case contained in [De8].

Classical trajectories that end up in the free region can be described in a much more satisfactory way than those that move in other directions. In particular, it is possible to define free region wave transformations that fully classify all the asymptotically free trajectories. This can be done both in the short-range and in the long-range cases with $\mu > 0$. It is described in Sect. 5.8.

Outside of the free region a full description of trajectories seems to be impossible without very severe restrictions on the potentials. In the case of the so-called a-clustered trajectories, it is usually more difficult to describe the asymptotics of the internal coordinate $x^a(t)$. On the other hand, the external coordinate $x_a(t)$ is much better behaved. If $\mu > 1$ in (5.0.3), then there exists the limit

$$\lim_{t\to\infty}(x_a(t) - t\xi_a(t)) := x^+_{\mathrm{sr},a}.$$

In the long-range case, if $S_a(t,\xi_a)$ is a solution of the appropriate Hamilton-Jacobi equation, the limit

$$\lim_{t\to\infty}(x_a(t) - \nabla_{\xi_a}S_a(t,\xi_a(t))) := x^+_{\mathrm{lr},a}$$

exists only under some additional assumptions. In particular, it always exists if $\mu > \sqrt{3}-1$. These facts are shown in Sect. 5.9 and they were first proven in [De7]. Note that this is the same borderline as for the proof of asymptotic completeness in quantum N-body scattering (see [De8] and the next chapter). In fact, one can argue that the existence of $x^+_{\mathrm{lr},a}$ is the correct classical analog of the quantum asymptotic completeness. Nevertheless, the result in the quantum case is much more satisfactory, because it gives a complete classification of states in the Hilbert space, whereas the classification given by $(H(y,\eta),\xi^+(y,\eta),x^+_{\mathrm{sr},a}(y,\eta))$ or $(H(y,\eta),\xi^+(y,\eta),x^+_{\mathrm{lr},a}(y,\eta))$ is only partial.

In the case of regular trajectories, one can show certain regularity properties of the asymptotic quantities $x^+_{\mathrm{sr},a}(y,\eta), x^+_{\mathrm{lr},a}(y,\eta)$, which we indicate in Subsect. 5.9.3.

Hunziker showed in [Hu2] that if the potentials are of compact support, then not only the external motion of a-clustered trajectories has a good asymptotics, but also the internal motion is asymptotic to a bounded trajectory of the internal Hamiltonian. He proposed to call this property the asymptotic completeness of classical N-body scattering. In Sect. 5.10 we prove a closely related result. We assume that the potentials decay faster than any exponential. Then the internal motion is asymptotic to a trajectory of the internal Hamiltonian with a zero asymptotic velocity. This can be thought of as another property that can be called the asymptotic completeness of N-body classical systems.

Let us mention that in the case of 3-body systems with radial potentials it is possible to obtain quite detailed understanding of classical scattering theory (see [Ge3]).

5.1 N-Body Systems

In this section we introduce a generalization of N-particle Hamiltonians that is originally due to Hörmander and Agmon [Ag2]. We will define various geometric concepts and auxiliary Hamiltonians that are useful in N-body scattering theory.

We will see at the end of this section how standard N-particle Hamiltonians fit into the more general class of generalized N-body Hamiltonians. The reader not familiar with N-particle Hamiltonians should first go to the end of the section to get a intuitive feeling for the various definitions that we introduce.

Although the class of generalized N-body Hamiltonians was first introduced in the quantum case (see [Ag2]), it is natural to consider it also in the classical case. Actually, the material in this section will also be used to describe quantum Hamiltonians in Chap. 6.

We denote by X a finite dimensional Euclidean space. The Euclidean norm of a vector $x \in X$ will be denoted $|x|$. Moreover, we will put

$$\langle x \rangle := \sqrt{x^2 + 1}.$$

Suppose that

$$\{X_b \mid b \in \mathcal{B}\} \tag{5.1.1}$$

is a finite family of subspaces of X. Let

$$\{X_a \mid a \in \mathcal{A}\} \tag{5.1.2}$$

be the smallest family of subspaces of X satisfying the following conditions:
(1) X belongs to (5.1.2);
(2) the family (5.1.2) is closed with respect to intersection;
(3) the family (5.1.1) is contained in (5.1.2).

Subspaces X_a will be sometimes called *collision planes.*

We endow the finite set $\mathcal{A}$ with a *semi-lattice structure* by

$$a \leq b \ \text{ if } \ X_a \supset X_b. \tag{5.1.3}$$

Clearly, there exist unique minimal and maximal elements in $\mathcal{A}$ denoted by $a_{\min}$ and $a_{\max}$. In fact,

$$X_{a_{\min}} = X, \qquad X_{a_{\max}} = \bigcap_{a \in \mathcal{A}} X_a.$$

One often assumes that

$$X_{a_{\max}} = \{0\},$$

but this additional condition is not necessary. Clearly, for $a_1, a_2 \in \mathcal{A}$, there exist a unique element bigger than a_1 and a_2. This element will be denoted by $a_1 \vee a_2$. One has

$$X_{a_1 \vee a_2} = X_{a_1} \cap X_{a_2}.$$

A *chain* is an ordered sequence $\{a_1, \cdots, a_k\}$ of elements of $\mathcal{A}$ with

$$a_1 < \cdots < a_k.$$

The chain is said to *connect* $a = a_1$ to $b = a_k$. The *length* of the chain is k. A chain is *maximal* if one cannot insert a new element into it.

For $a \in \mathcal{A}$, we denote by $\#a$ the maximal length of a maximal chain connecting a to $a_{\max}$. The number $\#a$ is called the *height* of a.

For $a \in \mathcal{A}$, we denote by X^a the space $X_a^{\perp}$. We denote by π^a and π_a the orthogonal projections of X onto X^a and X_a respectively. We will often write x^a and x_a instead of $\pi^a x$ and $\pi_a x$. If $a \leq b$ then we define

$$\pi_a \pi^b x = \pi^b \pi_a x =: x_a^b.$$

The following sets will be important later:

$$Z_a := X_a \setminus \bigcup_{b \not\leq a} X_b,$$
$$Y_a := X \setminus \bigcup_{b \not\leq a} X_b.$$

The following elementary proposition describes the "stratified structure" of the family $\{Z_a\}_{a\in\mathcal{A}}$.

Proposition 5.1.1
The family $\{Z_a\}_{a\in\mathcal{A}}$ is a partition of X. It means that

$$Z_a \cap Z_b = \emptyset, \quad a \neq b, \quad \textit{and} \quad X = \bigcup_{a\in\mathcal{A}} Z_a.$$

Moreover, the family $\{Z_b\}_{b\leq a}$ is a partition of Y_a. The set Y_a is open.

If $\epsilon, \delta > 0$ then it is useful to define the following sets:

$$\begin{aligned} X_a^{\epsilon} &:= \{x \in X \mid \mathrm{dist}(x, X_a) < \epsilon\}, \\ Z_a^{\delta} &= X_a \setminus \bigcup_{b \not\leq a} X_b^{\delta}, \\ Z_a^{\epsilon,\delta} &:= X_a^{\epsilon} \setminus \bigcup_{b \not\leq a} X_b^{\delta}, \\ Y_a^{\delta} &:= X \setminus \bigcup_{b \not\leq a} X_b^{\delta}. \end{aligned} \tag{5.1.4}$$

The following class of functions will be especially useful in our study of N-body systems:

Definition 5.1.2
Let f be a function $X \ni x \mapsto f(x) \in \mathbb{C}$. We say that $f \in \mathcal{F}$ if, for any $a \in \mathcal{A}$, there exists a neighborhood $\mathcal{U}_a$ of X_a such that f depends within $\mathcal{U}_a$ only on x_a.

Let us now introduce the definition of a (generalized) many-body Hamiltonian.

Definition 5.1.3
Let X be a Euclidean space and let $\{X_b \mid b \in \mathcal{B}\}$ be a finite family of linear subspaces. For each $b \in \mathcal{B}$, let $v^b \in C^{1,1}(X^b)$ be a real function such that

$$\lim_{|x^b|\to\infty} v^b(x^b) = 0. \tag{5.1.5}$$

Then the classical Hamiltonian on $X \times X'$

$$H(x,\xi) := \frac{1}{2}\xi^2 + \sum_{b\in\mathcal{B}} v^b(x^b)$$

is called a (generalized) many-body Hamiltonian.

Let $\mathcal{A}$ be the corresponding lattice of subspaces and

$$N := \#a_{\min}.$$

*Then $H(x,\xi)$ is also called an N-*body Hamiltonian.

For any $a \in \mathcal{A}$, we set

$$V(x) := \sum_{b\in\mathcal{B}} v^b(x^b), \quad \text{and} \quad V^a(x^a) := \sum_{b\le a} v^b(x^b).$$

We define

$$H_a(x,\xi) := \frac{1}{2}\xi^2 + V^a(x^a).$$

Clearly, $H = H_{a_{\max}}$. Note that

$$H_a(x,\xi) = \frac{1}{2}\xi_a^2 + H^a(x^a,\xi^a), \quad \text{where} \quad H^a(x^a,\xi^a) := \frac{1}{2}(\xi^a)^2 + V^a(x^a).$$

Clearly, H^a is a many-body Hamiltonian on the space X^a. The corresponding lattice of collision planes is indexed by $\mathcal{A}^a := \{b \le a\}$.

If we set

$$I_a(x) := V(x) - V^a(x^a),$$

then we have

$$H(x,\xi) = H_a(x,\xi) + I_a(x).$$

Since $X^{a_{\min}} = \{0\}$, it is convenient to assume that

$$v^{a_{\min}}(x^{a_{\min}}) = 0.$$

We then have

$$V^{a_{\min}}(x^{a_{\min}}) = 0 \quad \text{and} \quad H^{a_{\min}}(x^{a_{\min}}, \xi^{a_{\min}}) = 0$$

on the phase space $\{0\} \times \{0\}$.

We will also define the Liouville derivative

$$\mathbf{D} := \frac{\mathrm{d}}{\mathrm{d}t} + \{H(x,\xi), \cdot\}.$$

To end this section, we will explain how standard N-body Hamiltonians fit into the class of generalized N-particle ones.

The configuration space of a system of N ν-dimensional particles is

$$X_1 \times \cdots \times X_N,$$

where $X_1 = \cdots = X_N = \mathbb{R}^\nu$. An N-particle Hamiltonian is a function on $X_1 \times \cdots \times X_N \times X_1' \times \cdots \times X_N'$ given by

$$H(x,\xi) = \sum_{i=1}^{N} \frac{\xi_i^2}{2m_i} + \sum_{i<j} V_{ij}(x_i - x_j). \tag{5.1.6}$$

It describes the motion of N particles of masses m_i interacting through pair potentials $V_{ij}(x_i - x_j)$, which are usually assumed to go to 0 at infinity.

Let us now identify various concepts from the formalism of generalized N-body Hamiltonians in the case of (5.1.6). We have

$$X = X_1 \times \cdots \times X_N,$$

where the scalar product is defined by the quadratic form

$$\sum_{i=1}^{N} m_i x_i^2. \tag{5.1.7}$$

The set $\mathcal{B}$ is the set of pairs in $\{1, \ldots, N\}$. The set $\mathcal{A}$ is the set of partitions of $\{1, \ldots, N\}$, whose elements a will also be written as

$$a = \{C_1, \ldots, C_k\}.$$

The sets C_i are traditionally called *clusters* and $a \in \mathcal{A}$ are called cluster decompositions. The set $\mathcal{A}$ is endowed with its natural lattice structure by saying that $a \le b$ if a is *finer* than b, that is, all clusters of a are included in clusters of b. Note that the height $\#a$ is equal to the number of clusters in a.

For a pair i, j of indices, we will denote by (i, j) the smallest partition having $\{i, j\}$ as one of its clusters. Then we define

$$X_a = \{x \in X \mid x_i = x_j, \quad (i,j) \le a\}. \tag{5.1.8}$$

In other words,

$$X_a = \{x \in X \mid \text{ for every } \ m = 1, \ldots, k, \ \text{ if } \ i, j \in C_m, \ \text{ then } \ x_i = x_j\}.$$

It is immediate to verify that the family $\{X_a\}_{a \in \mathcal{A}}$ defined in (5.1.8) satisfies the conditions specified at the beginning of the section. Moreover, the lattice structure defined by (5.1.3) coincides with the one introduced above.

A special role is played by the maximal cluster decomposition $a_{\max} = \{1, \ldots, N\}$. Clearly,

$$X_{a_{\max}} = \{x \mid x_1 = \cdots = x_N\},$$
$$X^{a_{\max}} = \{x \mid m_1 x_1 + \cdots + m_N x_N = 0\}.$$

Note that $X_{a_{\max}}$ is the subspace of the center-of-mass motion and $H^{a_{\max}}$ is the full Hamiltonian without the center-of-mass motion. The space $X^{a_{\max}}$ is sometimes called the reduced configuration space of the system and $H^{a_{\max}}$ is called the reduced Hamiltonian. We have

$$H(x, \xi) = \frac{1}{2}\xi_{a_{\max}}^2 + H^{a_{\max}}(x^{a_{\max}}, \xi^{a_{\max}}),$$

therefore the reduced Hamiltonian contains the whole nontrivial information about the system. For an arbitrary a, the Hamiltonian $H^a(x^a, \xi^a)$ describes the internal motion of the clusters $C_1, \ldots C_k$ and $\frac{1}{2}\xi_a^2$ describes the kinetic energy of the relative motion of the clusters $C_1 \ldots, C_k$.

Below we give some pictures that show typical low-dimensional configuration spaces of many-body systems. In the case of a 2-body system, the full configuration space is 2-dimensional, and there are just two collision planes corresponding to $a_{\min} = \{\{1\}, \{2\}\}$ and $a_{\max} = \{1, 2\}$.

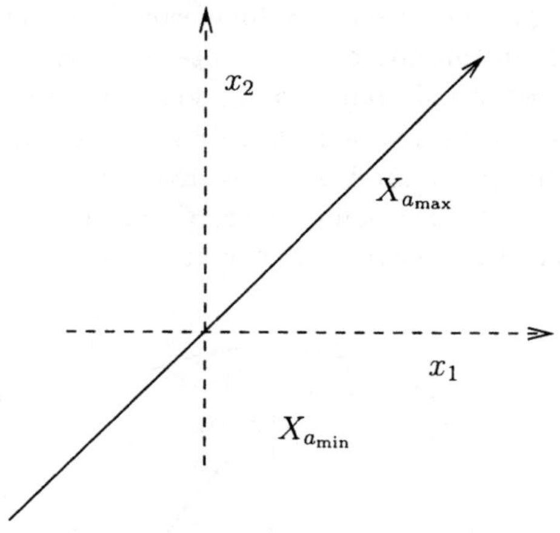

Fig. 5.1. Configuration space of a 2-body system.

The configuration space of a system of 3 one-dimensional particles is 3-dimensional. On the following picture, we show the reduced configuration space of such a system, which is 2-dimensional. The picture shows the collision planes corresponding to all five cluster decompositions:

$$\begin{aligned} &a_{\max} = \{\{1,2,3\}\}, && \# = 1,\\ &(1,2),\ \ (2,3),\ \ (3,1), && \# = 2,\\ &a_{\min} = \{\{1\},\{2\},\{3\}\}, && \# = 3. \end{aligned}$$

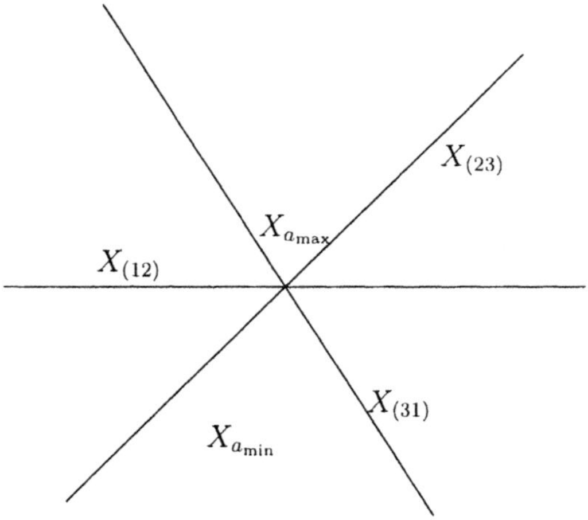

Fig. 5.2. Reduced configuration space of a 3-body system

Probably the most complicated N-body configuration space whose picture it is still possible to draw is the reduced space of a system of 4 one-dimensional particles. This configuration space is 3-dimensional. In the following picture, we show the unit sphere in this space and its intersections with collision planes.

In the case of a reduced 4-body system, 3-cluster decompositions are labeled by pairs of particles. They correspond to 2-dimensional collision planes and their intersections with the sphere are big circles. Every two 3-cluster collision planes intersect along a line that corresponds to a certain 2-cluster decomposition. Every such a line intersects the sphere at two points, one of which is shown in the picture.

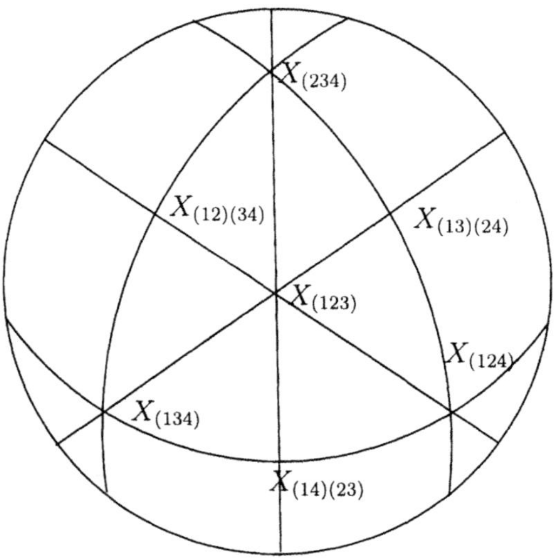

Fig. 5.3. A sphere in the reduced configuration space of a 4-body system

To finish the section, let us mention some other Hamiltonians slightly different from the standard ones that can be considered as generalized N-particle Hamiltonians. The first example is obtained by allowing particles not only with pair interactions but also with triple, quadruple, etc. interactions. In the quantum case, such models are quite common in nuclear physics. For example, a potential describing a triple interaction between particles i, j, k would be of the form

$$V_{ijk}(x_i - x_j, x_j - x_k),$$

where $V_{ijk}(x, y)$ goes to 0 when $|x|$ or $|y|$ go to ∞. Such a potential is of the form $v^a(x^a)$ for

$$a = \{\{i, j, k\}, \{i_1\}, \dots, \{i_{N-3}\}\}.$$

Another example is obtained by adding some particles of infinite masses. One then obtains a Hamiltonian of the form

$$H(x, \xi) = \sum_{i=1}^{N} \frac{\xi_i^2}{2m_i} + \sum_{i<j} V_{ij}(x_i - x_j) + \sum_i V_i(x_i). \tag{5.1.9}$$

Note that in this case $H(x, \xi)$ is no more translation invariant and we have $X_{a_{\max}} = \{0\}$.

The configuration space of such a system is $X = X_1 \times \cdots \times X_n$, where X_j are configuration spaces of particles having a finite mass, so it is the same as in the system of N particles, but the family of collision planes is different. Namely, to the usual family of subspaces corresponding to cluster decompositions we have to add the subspaces $\{(x_1, \dots, x_N) \mid x_j = 0\}$ and their intersections.

Below we give a picture that shows the configuration space of a system of 2 particles interacting with a particle of infinite mass.

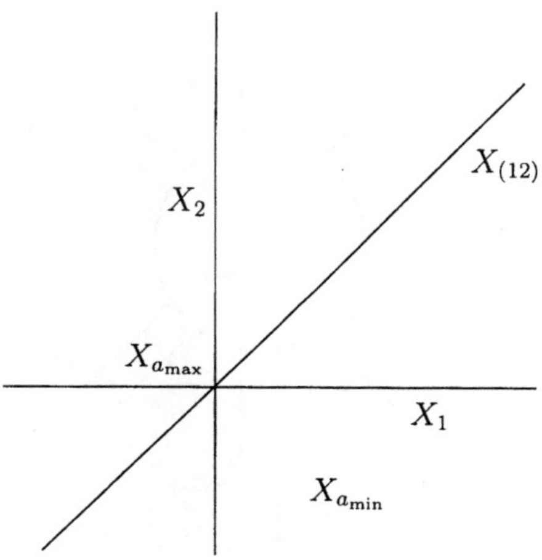

Fig. 5.4. A 3-body system with one particle of infinite mass

5.2 Some Special Observables

One of the most important techniques of scattering theory is the use of observables having an approximately positive Poisson bracket with the Hamiltonian. An especially successful example of such an observable is the so-called Graf's vector field. This section will be devoted mostly to a construction of this vector field, or actually, of its "potential" – a scalar function on the configuration space. Graf's vector field appeared originally in [Gr]. The function $R(x)$, whose gradient is essentially Graf's vector field, first appeared in [De6]. Its construction that we are going to present is taken from [De6, De8].

The function $R(x)$ is a distortion of $x^2/2$. At the end of this section we will study another useful observable, $r(x)$, which is a distortion of $|x|$. This observable first appeared in [Yaf5], and was then used in [De8] to prove the asymptotic completeness of long-range N-body systems.

Note also that similar functions having an approximately positive Poisson bracket with the Hamiltonian were used before [Gr], especially in [SS1]. The main discovery of [Gr] was the idea how to distort these functions in a vicinity of collision planes, which considerably simplified the study of N-body systems.

Let $\rho = \{\rho_a, a \in \mathcal{A}\}$ be a sequence of numbers indexed with elements of $\mathcal{A}$ such that $\rho_{a_{\min}} = 0$. We define

$$\Xi_a^\rho := \{x \in X \mid x_a^2 + \rho_a \geq x_b^2 + \rho_b, \ \forall b \neq a\}. \tag{5.2.1}$$

We will say that a sequence ρ is *admissible* iff there exists $\delta > 0$ such that

$$\Xi_a^\rho \cap X_b^\delta = \emptyset, \quad b \not\leq a.$$

In applications to N-body systems, we will need a family of sets Ξ_a^ρ with an admissible ρ. Such a family of sets we will call a *Graf partition*. On the following picture we show a typical Graf partition for a 3-body system.

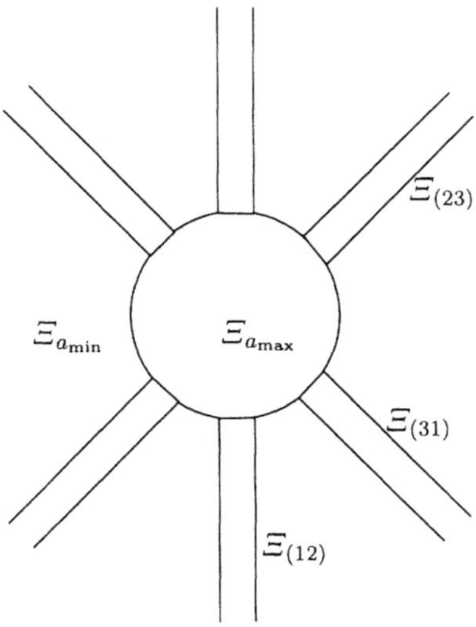

Fig. 5.5. Graf partition on the configuration space of a 3-body system

The following picture of the configuration space of a simplified 4-body system will be used to illustrate the properties of admissible partitions of unity.

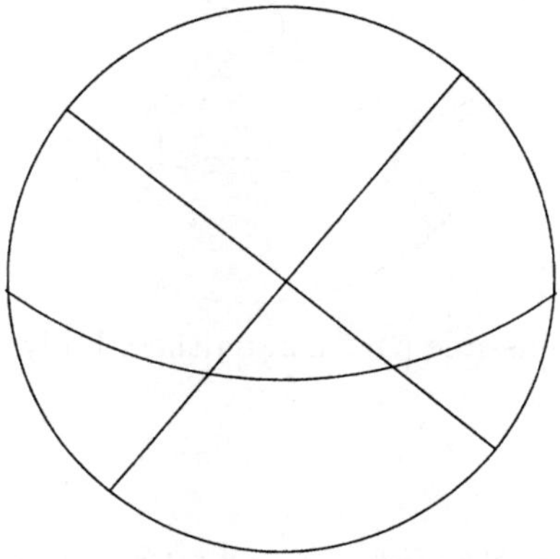

Fig. 5.6. A sphere in the configuration space of a simplified 4-body system

If we choose a non-admissible ρ, we still get a partition of the configuration space with a positive Poisson bracket, but it does not have the properties that we need. An example is shown on the picture below.

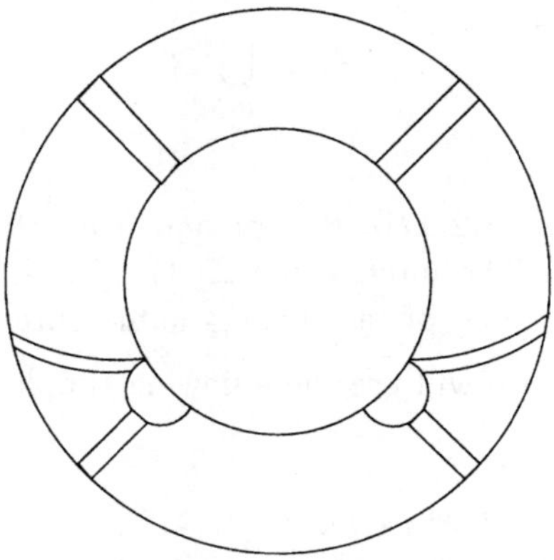

Fig. 5.7. A non-admissible partition on the 4-body configuration space

The next picture shows an admissible partition of unity for the simplified 4-body system.

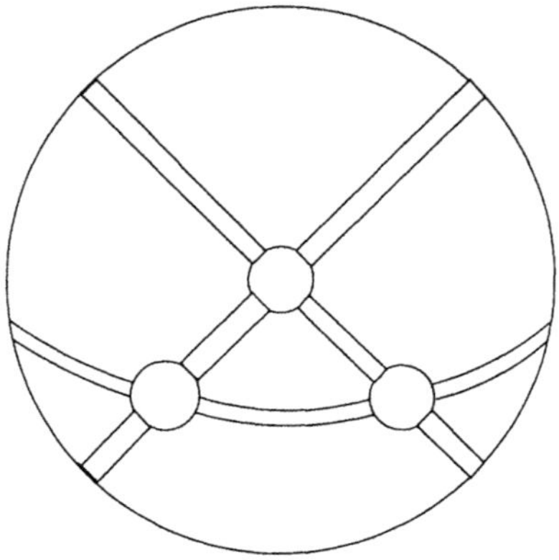

Fig. 5.8. An admissible partition on the 4-body configuration space

Proposition 5.2.1
(i) For $a \neq b$, the intersection $\Xi_a^\rho \cap \Xi_b^\rho$ is a set of measure zero. The sets $\{\Xi_a^\rho\}_{a\in\mathcal{A}}$ are closed and

$$\bigcup_{a\in\mathcal{A}} \Xi_a^\rho = X. \tag{5.2.2}$$

(ii) There exist admissible sequences ρ. More precisely, for any $\epsilon > 0$, there exists $\delta > 0$ and a sequence ρ such that

$$\Xi_a^\rho \subset Z_a^{\epsilon,\delta} = X_a^\epsilon \setminus \bigcup_{b\not\leq a} X_b^\delta. \tag{5.2.3}$$

Moreover, such ρ satisfies

$$X_a^\delta \subset \bigcup_{a\leq b} \Xi_b^\rho. \tag{5.2.4}$$

Proof of Proposition 5.2.1*(i)*. It is obvious that the sets Ξ_a^ρ are closed and that their union is X. The intersection $\Xi_a^\rho \cap \Xi_b^\rho$ is a zero set of a non-trivial quadratic polynomial, hence of zero Lebesgue measure. □

In the proof of *(ii)* we will use the following simple geometric observation.

Lemma 5.2.2
There exists $M > 0$ such that we have

$$M|x^{a\vee b}|^2 \leq |x^a|^2 + |x^b|^2,\ a, b \in \mathcal{A}. \tag{5.2.5}$$

Proof. For $a, b \in \mathcal{A}$, we have

$$X^{a\vee b} = X^a \oplus X^b.$$

Therefore $\sqrt{|x^a|^2 + |x^b|^2}$ is a norm on $X^{a\vee b}$, which is clearly equivalent to the usual norm $|x^{a\vee b}|$, because the space is finite dimensional. Therefore, there exists a constant $M > 0$ such that (5.2.5) is true. □

Now suppose that ρ is an arbitrary sequence with $\rho_{a_{\min}} = 0$. We split our study of the set Ξ_a^ρ in a series of lemmas.

Lemma 5.2.3
If $a \neq a_{\min}$, then

$$\Xi_a^\rho \subset X_a^{\sqrt{\rho_a}}.$$

Proof. If $x \in \Xi_a^\rho$, then

$$x_a^2 + \rho_a \geq x^2.$$

Therefore,

$$|x^a|^2 \leq \rho_a.$$

□

Lemma 5.2.4
For any $a, b \in \mathcal{A}$,

$$\Xi_a^\rho \cap X_b^{\sqrt{\rho_b - \rho_a}} = \emptyset.$$

Proof. We have

$$x^2 + \rho_a \geq x_a^2 + \rho_a \geq x_b^2 + \rho_b.$$

Therefore

$$|x^b|^2 \geq \rho_b - \rho_a.$$

□

Lemma 5.2.5
For any $a, b \in \mathcal{A}$,

$$\Xi_a^\rho \cap X_b^{\sqrt{M(\rho_{a\vee b} - \rho_a) - \rho_a}} = \emptyset.$$

Proof. We have

$$x^2 + \rho_a \geq x_a^2 + \rho_a \geq x_{a\vee b}^2 + \rho_{a\vee b}.$$

Therefore

$$|x^{a\vee b}|^2 \geq \rho_{a\vee b} - \rho_a.$$

By Lemma 5.2.2, this yields

$$|x^a|^2 + |x^b|^2 \geq M(\rho_{a\vee b} - \rho_a).$$

By Lemma 5.2.3, this implies

$$|x^b|^2 \geq M(\rho_{a\vee b} - \rho_a) - \rho_a.$$

□

Proof of Proposition 5.2.1. Let $\gamma > 0$. Choose ρ such that

$$\begin{aligned} &\rho_a \leq \epsilon, \ a \in \mathcal{A}, \\ &\rho_b > \tfrac{M+1+\gamma}{M}\rho_a, \ \#b < \#a, \ a, b \in \mathcal{A}. \end{aligned}$$

If $\#b < \#a$, then

$$\begin{aligned} \rho_b - \rho_a &> \tfrac{M+1+\gamma}{M}\rho_a - \rho_a \\ &= (1+\gamma)M^{-1}\rho_a =: \delta_{a,1}^2 > 0. \end{aligned}$$

If $\#b \geq \#a$ and $b \not\leq a$, then, clearly, $\#(a \vee b) < \#a$. Hence

$$\begin{aligned} M(\rho_{a\vee b} - \rho_a) - \rho_a &> M\left(\tfrac{M+1+\gamma}{M} - 1\right)\rho_a - \rho_a \\ &= \gamma\rho_a =: \delta_{a,2}^2 > 0. \end{aligned}$$

Therefore, by Lemmas 5.2.3, 5.2.4 and 5.2.5, we have

$$\Xi_a^\rho \subset X_a^{\rho_a} \setminus \left(\bigcup_{\#b<\#a} X_b^{\delta_{a,1}} \cup \bigcup_{\#b\geq\#a, \ b\not\leq a} X_b^{\delta_{a,2}} \right).$$

This implies (5.2.3).

Let us now prove that (5.2.3) implies (5.2.4). Observe that if $a \not\leq b$, then

$$\Xi_b^\rho \subset X \setminus X_a^\delta.$$

Hence

$$\bigcup_{a\not\leq b} \Xi_b^\rho \subset X \setminus X_a^\delta.$$

By (5.2.2), this implies (5.2.4). □

Let us now put

$$q_a^\rho(x) := \mathbb{1}_{\Xi_a^\rho}(x), \tag{5.2.6}$$

$$R^\rho(x) := \tfrac{1}{2}\max_{a\in\mathcal{A}}\{x_a^2 + \rho_a\}. \tag{5.2.7}$$

The next proposition describes some properties of R^ρ and $\{q_a^\rho\}_{a\in\mathcal{A}}$. Note that the first identity of this proposition is only valid almost everywhere. Besides, all the derivatives in this proposition are in the distributional sense.

On the following picture, we show the vector field $\nabla_x R(x)$ for a system of 3 particles. $\nabla_x R(x)$ is often called a *Graf vector field*.

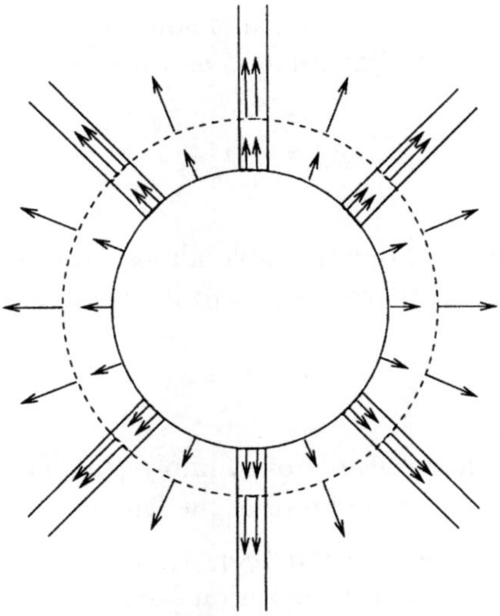

Fig. 5.9. Level sets and gradient of R^ρ on the 3-body configuration space

Proposition 5.2.6
R^ρ is a continuous convex function. Moreover, one has

$$(i)\quad R^\rho(x) = \sum_{a\in\mathcal{A}} \tfrac{1}{2} q_a^\rho(x)(x_a^2 + \rho_a),$$

$$(ii)\quad \nabla_x R^\rho(x) = \sum_{a\in\mathcal{A}} x_a q_a^\rho(x),$$

$$(iii)\quad \nabla_x^2 R^\rho(x) \geq \sum_{a\in\mathcal{A}} \pi_a q_a^\rho(x),$$

$$(iv)\quad \bar{\xi}\nabla^2 R^\rho(x)\xi - \bar{\xi}\nabla R^\rho(x) - \nabla R^\rho(x)\xi + 2R^\rho(x)$$

$$\geq \sum_{a\in\mathcal{A}} q_a^\rho(x)|\xi_a - x_a|^2, \quad \xi \in X + iX,$$

$$(v)\quad \max\{x^2, C_1\} \leq 2R^\rho(x) \leq x^2 + C_2, \quad \textit{for some} \quad C_1, C_2 > 0,$$

(vi) *if ρ is admissible, then $R^\rho, q_a^\rho \in \mathcal{F}$.*

Proof. The function $R^\rho(x)$ is, clearly, continuous, as the supremum of a finite family of continuous functions. It is convex, as the supremum of a family of convex functions. In other words, its second distributional derivative is a measure with values in positive matrices. Inside $\mathrm{Int}\,\Xi_a^\rho$ we have

$$\nabla_x^2 R^\rho = \pi_a.$$

This proves *(iii)*.

The set $X\backslash\bigcup_{a\in\mathcal{A}}\mathrm{Int}\Xi_a^\rho$ is the union of subsets of zero sets of quadratic polynomials. Therefore it has codimension 1, and hence it is of measure zero. Inside $\mathrm{Int}\Xi_a^\rho$ we have

$$R_a^\rho = \frac{1}{2}(x_a^2+\rho_a).$$

This implies *(i)*.

Any convex function is differentiable almost everywhere in the usual sense and its distributional derivative is equal to its usual derivative. Inside $\mathrm{Int}\Xi_a^\rho$ we have

$$\nabla_x R^\rho = x_a.$$

Therefore *(ii)* is true.

(i), *(ii)*, *(iii)* and the positivity of ρ_a imply *(iv)*. The properties *(v)* and *(vi)* are obvious. This completes the proof of the lemma. □

We will now smooth out R^ρ and $\{q_a^\rho\}_{a\in\mathcal{A}}$.

It follows from the proof of Proposition 5.2.1 that we can find two sequences ρ^- and ρ^+ such that $\rho_{a_{\min}}^- = \rho_{a_{\min}}^+ = 0$, $\rho_a^- < \rho_a^+$ for $a\neq a_{\min}$ and if $\rho_a^-\leq\rho_a\leq\rho_a^+$, $a\in\mathcal{A}$, then ρ is admissible. We fix two such sequences. We also fix a function

$$f\in C_0^\infty\left(\times_{a\neq a_{\min}}[\rho_a^-,\rho_a^+]\right)\qquad\text{such that}\quad f\geq 0,\ \int f(\rho)\mathrm{d}\rho = 1,$$

where $\mathrm{d}\rho = \otimes_{a\neq a_{\min}}\mathrm{d}\rho_a$. We define

$$\begin{aligned} R(x) &:= \textstyle\int f(\rho)R^\rho(x)\mathrm{d}\rho,\\ q_a(x) &:= \textstyle\int f(\rho)q_a^\rho(x)\mathrm{d}\rho, \end{aligned}$$

Lemma 5.2.7
$R(x)$ is a smooth convex function. Moreover one has:

(i) $\max\{x^2,C_1\}\leq 2R(x)\leq x^2+C_2$, *for some* $C_1,C_2>0$,

(ii) $\nabla_x R(x) = \sum_{a\in\mathcal{A}} x_a q_a(x)$,

(iii) $\nabla_x^2 R(x) = \sum_{a\in\mathcal{A}}\pi_a q_a(x) - x^a\nabla_x q_a(x)\geq\sum_{a\in\mathcal{A}}\pi_a q_a(x)$,

(iv) $\bar\xi\nabla^2 R(x)\xi - \bar\xi\nabla R(x) - \nabla R(x)\xi + 2R(x)\geq\sum_{a\in\mathcal{A}} q_a(x)|\xi_a - x_a|^2$,

$\xi\in X+iX$,

(v) $R, q_a\in\mathcal{F}$, $a\in\mathcal{A}$,

(vi) *the following functions are bounded:*

$\partial_x^\alpha(2R(x)-x^2)$, $\alpha\in\mathbb{N}^n$,

$\partial_x^\alpha(x\nabla_x R(x)-x^2)$, $\alpha\in\mathbb{N}^n$,

$\partial_x^\alpha(x\nabla_x^2 R(x)x-x^2)$, $\alpha\in\mathbb{N}^n$.

Proof. The properties *(i)–(v)* of Lemma 5.2.7 are immediate consequences of Lemmas 5.2.6.

Let us now prove property *(vi)* of Lemma 5.2.7. We will actually prove the following more general estimate:

$$|\partial_x^\alpha (x\nabla_x)^k (2R(x) - x^2)| \le C_{\alpha,k},\ k \in \mathbb{N},\ \alpha \in \mathbb{N}^n. \tag{5.2.8}$$

Note that

$$x\nabla_x R(x) - x^2 = x\nabla_x (R(x) - \tfrac{1}{2}x^2),$$
$$x\nabla_x^2 R(x)x - x^2 = ((x\nabla_x)^2 - x\nabla_x)\left(\tfrac{1}{2}R(x) - \tfrac{1}{4}x^2\right).$$

Therefore (5.2.8) will imply *(vi)*.

Note that

$$2R(x) - x^2 = \sum_{a\in\mathcal{A}} \int (|x^a|^2 + \rho_a) q_a^\rho(x) f(\rho) d\rho. \tag{5.2.9}$$

Let us fix $a \ne a_{\min}$ and a subset $\{a_1, a_2, \ldots, a_n\} \subset \mathcal{A}\backslash\{a, a_{\min}\}$. Let us also label the elements of $\mathcal{A}\backslash\{a, a_{\min}, a_1, \ldots, a_n\}$ as $\{b_1, \ldots, b_m\}$.

Our aim will be to estimate the derivatives of the ath term in the sum (5.2.9) for $x \in X$ such that

$$x_a^2 + \rho_a^- \le x_{a_j}^2 + \rho_{a_j}^+,\ j = 1, \ldots, n,$$

and

$$x_a^2 + \rho_a^- > x_{b_k}^2 + \rho_{b_k}^+,\ k = 1, \ldots, m.$$

Obviously, any $x \in X$ satisfies such conditions for some subset $\{a_1, a_2, \ldots, a_n\} \subset \mathcal{A}\backslash\{a, a_{\min}\}$.

For $j = 1, \ldots, n$, we set

$$\tilde{\rho}_{a_j} := \rho_{a_j} - \rho_a + x_{a_j}^2 - x_a^2$$

and

$$\tilde{f}(\rho_a, \rho_{a_1}, \ldots, \rho_{a_n}) = \int f(\rho_a, \rho_{a_1}, \ldots, \rho_{a_n}, \rho_{b_1}, \ldots, \rho_{b_m}) \mathrm{d}\rho_{b_1} \ldots \mathrm{d}\rho_{b_m}.$$

We have the following identity

$$\begin{aligned}
&\textstyle\int (|x^a|^2 + \rho_a) q_a^\rho(x) f(\rho) \mathrm{d}\rho \\
&= \textstyle\int (|x^a|^2 + \rho_a) q_a^\rho(x) \tilde{f}(\rho_a, \rho_{a_1}, \ldots, \rho_{a_n}) \mathrm{d}\rho_a \mathrm{d}\rho_{a_1} \cdots \mathrm{d}\rho_{a_n} \\
&= \textstyle\int_{-\infty}^{\infty} \mathrm{d}\rho_a \int_{-\infty}^0 \mathrm{d}\tilde{\rho}_{a_1} \cdots \int_{-\infty}^0 \mathrm{d}\tilde{\rho}_{a_n} \\
&(|x^a|^2 + \rho_a) \tilde{f}(\rho_a, \tilde{\rho}_{a_1} + \rho_a - x_{a_1}^2 + x_a^2, \ldots, \tilde{\rho}_{a_n} + \rho_a - x_{a_n}^2 + x_a^2).
\end{aligned} \tag{5.2.10}$$

Now note that on the support of the above integral we have

$$|x^a|^2 \leq \rho_a^+.$$

Therefore, $|x - x_a| \leq C_1$.

Moreover,

$$x_a^2 - x_{a_j}^2 \leq \rho_{a_j}^+ - \rho_a^-. \tag{5.2.11}$$

Therefore,

$$x^2 - x_{a_j}^2 = x^2 - x_a^2 + x_a^2 - x_{a_j}^2 \leq \rho_a^+ + \rho_{a_j}^+ - \rho_a^-.$$

Hence $|x - x_{a_j}| \leq C_2$.

Thus, on the support of the integrand we have

$$|x_a - x_{a_j}| \leq |x_a - x| + |x - x_{a_j}| \leq C_1 + C_2. \tag{5.2.12}$$

Now we remark that

$$\partial_x^\alpha (x\nabla_x)^k \tilde{f}(\rho_a, \tilde{\rho}_{a_1} + \rho_a + x_{a_1}^2 - x_a^2, \dots, \tilde{\rho}_{a_{n-1}} + \rho_a + x_{a_n}^2 - x_a^2)$$

is a linear combination of terms of the form

$$(x_{a_1}^2 - x_a^2)^{k_1} \dots (x_{a_{n-1}}^2 - x_a^2)^{k_n} (x_{a_1} - x_a)^{\alpha_1} \dots (x_{a_{n-1}} - x_a)^{\alpha_n}$$
$$\times \partial_{\tilde{\rho}_{a_1}}^{\beta_1} \dots \partial_{\tilde{\rho}_{a_n}}^{\beta_n} \tilde{f}(\rho_a, \tilde{\rho}_{a_1} + \rho_a + x_{a_1}^2 - x_a^2, \dots, \tilde{\rho}_{a_n} + \rho_a + x_{a_n}^2 - x_a^2).$$

By (5.2.11) and (5.2.12), this is bounded. □

Proposition 5.2.8
The set $\mathcal{F} \cap C_0^\infty(X)$ is dense in $C_\infty(X)$ in the uniform topology.

Proof. It is enough to show that one can approximate functions of $C_0^\infty(X)$ with elements of $\mathcal{F} \cap C_0^\infty(X)$.

Let $f \in C_0^\infty(X)$. Clearly, f is uniformly continuous. Therefore given $\delta > 0$ there exists $\epsilon > 0$ such that

$$|x_1 - x_2| < \epsilon \quad \text{implies} \quad |f(x_1) - f(x_2)| < \delta.$$

We construct a partition of unity q_a as in Lemma 5.2.7 with the requirement that $\text{supp}q_a \subset X_a^\epsilon$. We set

$$\tilde{f}(x) := \sum_{a \in \mathcal{A}} q_a(x) f(x_a).$$

Clearly, $\tilde{f} \in \mathcal{F} \cap C_0^\infty(X)$. Moreover,

$$f(x) - \tilde{f}(x) = \sum_{a \in \mathcal{A}} (f(x_a) - f(x)) q_a(x).$$

Therefore,

$$|f(x) - \tilde{f}(x)| \le \sum_{a\in\mathcal{A}} |f(x_a) - f(x)| q_a(x) \le \delta \sum_{a\in\mathcal{A}} q_a(x) \le \delta.$$

□

Another function on the configuration space that will play an important role in our considerations is

$$r(x) := \sqrt{2R(x)}.$$

The following lemma describes the properties of $r(x)$.

Lemma 5.2.9
$r(x)$ is a smooth convex function. It belongs to $\mathcal{F}$. It satisfies

$$\max\{|x|, C_1\} \le r(x) \le |x| + C_2,$$

for some $C_1, C_2 > 0$. Moreover,

$$\begin{aligned} &|\partial_x^\alpha (r(x) - \langle x\rangle)| \le C_\alpha \langle x\rangle^{-1}, \ \alpha \in \mathbb{N}^n, \\ &|\partial_x^\alpha (x\nabla_x r(x) - \langle x\rangle)| \le C_\alpha \langle x\rangle^{-1}, \ \alpha \in \mathbb{N}^n, \\ &|\partial_x^\alpha (x\nabla_x^2 r(x) x)| \le C_\alpha \langle x\rangle^{-1}, \ \alpha \in \mathbb{N}^n. \end{aligned} \tag{5.2.13}$$

Proof. Note that, by (5.2.7), the polynomial

$$\mathbb{R} \ni t \mapsto t^2 \xi\nabla^2 R(x)\xi - t2\nabla R(x)\xi + 2R(x)$$

is nonnegative. Hence its discriminant is negative, which means that

$$4(\xi\nabla R(x))^2 - 8\xi\nabla^2 R(x)\xi R(x) \le 0.$$

Consequently,

$$\xi\nabla^2 r(x)\xi = \xi\frac{1}{\sqrt{2R(x)}}\left(\nabla^2 R(x) - \frac{\nabla R(x)\nabla R(x)}{2R(x)}\right)\xi \ge 0.$$

This proves the convexity of $r(x)$.

Now we would like to show (5.2.13). Actually, we will prove a more general estimate

$$|\partial_x^\alpha (x\nabla_x)^k (r(x) - \langle x\rangle)| \le C_{\alpha,k} \langle x\rangle^{-1}, \ k \in \mathbb{N}, \ \alpha \in \mathbb{N}^n. \tag{5.2.14}$$

If we note that

$$\begin{aligned} &x\nabla_x r(x) - |x| = x\nabla_x (r(x) - |x|), \\ &x\nabla_x^2 r(x) x = ((x\nabla_x)^2 - x\nabla_x)(r(x) - |x|), \end{aligned}$$

then we see that (5.2.14) implies (5.2.13).

By (5.2.8), we clearly have

$$|\partial_x^\alpha (x\nabla_x)^k R(x)| \le C\langle x\rangle^{\max(2-|\alpha|,0)},\ k\in\mathbb{N},\ \alpha\in\mathbb{N}^n.$$

Therefore,

$$|\partial_x^\alpha (x\nabla_x)^k r(x)| \le C\langle x\rangle^{\max(1-|\alpha|,-1)},\ k\in\mathbb{N},\ \alpha\in\mathbb{N}^n.$$

Obviously,

$$|\partial_x^\alpha (x\nabla_x)^k \langle x\rangle| \le C\langle x\rangle^{1-|\alpha|},\ k\in\mathbb{N},\ \alpha\in\mathbb{N}^n.$$

The above two estimates imply easily

$$|\partial_x^\alpha (x\nabla_x)^k (r(x)+\langle x\rangle)^{-1}| \le C\langle x\rangle^{\max(-1-|\alpha|,-3)},\ k\in\mathbb{N},\ \alpha\in\mathbb{N}^n.$$

This is actually a stronger inequality than what we need, for our purposes it would suffice to know that

$$|\partial_x^\alpha (x\nabla_x)^k (r(x)+\langle x\rangle)^{-1}| \le C\langle x\rangle^{-1},\ k\in\mathbb{N},\ \alpha\in\mathbb{N}^n. \tag{5.2.15}$$

Now (5.2.14) follows from the identity

$$r(x)-\langle x\rangle = \frac{2R(x)-\langle x\rangle^2}{r(x)+\langle x\rangle},$$

from the estimate (5.2.15) and from (5.2.8). □

5.3 Bounded Trajectories and the Classical Mourre Estimate

In this section we begin our study of N-body Hamiltonians. We assume in this section that the potential satisfy

$$\lim_{|x^b|\to\infty} \langle x^b\rangle |\nabla_{x^b} v^b(x^b)| = 0,\ b\in\mathcal{B}. \tag{5.3.1}$$

We will study bounded and unbounded trajectories of N-body systems. One of the concepts that we will use will be that of *trapping energies*, that is, the energies at which bounded trajectories exist. We will show a certain relationship between the set of trapping energies of the full system and the sets of trapping energies of subsystems.

The main tool used in this section will be the so-called classical Mourre estimate. This estimate says that a certain observable, which is a modification of $\langle x,\xi\rangle$, has a positive Poisson bracket with the Hamiltonian in a certain region of phase space. This construction is based on an analogous construction of the quantum case [Mo1, PSS] and in the classical case was first given in [Ge1] in the 3-body case and in [Wa2] in the N-body case.

Let us first introduce some definitions.

Definition 5.3.1
For $a \in \mathcal{A}$, we define $\mathcal{B}^{a,+}$ to be the set of $(y^a, \eta^a) \in X^a \times X^a$ such that $x^a(t, y^a, \eta^a)$ is bounded for $t > 0$, where (exceptionally) we denote by $x^a(t, y^a, \eta^a)$ the trajectory generated by $H^a(x^a, \xi^a)$ with the initial conditions (y^a, η^a).

The set

$$\sigma^a := H^a(\mathcal{B}^{a+})$$

is called the set of trapping energy levels *of $H^a(x^a, \xi^a)$. Note that $\sigma^{a_{\min}} = \{0\}$. We also define*

$$\tau^a := \bigcup_{b<a} \sigma^b.$$

We will denote simply by σ, τ the sets $\sigma^{a_{\max}}, \tau^{a_{\max}}$.

Note that τ is an analog of the set of thresholds of the quantum Hamiltonian H.

Let us first notice an immediate property of the sets τ and σ.

Proposition 5.3.2
(i)

$$[\min_{a \neq a_{\max}} \inf V^a(x^a), 0] \subset \tau,$$

(ii)

$$[\inf V(x), \min_{a \neq a_{\max}} \inf V^a(x^a)[\subset \sigma.$$

Proof. Let $\inf V(x) < \lambda < \min_{a \neq a_{\max}} \inf V^a(x^a)$. Then $V^{-1}(\{\lambda\})$ is a non-empty set. The trajectory with initial conditions $(y, 0)$ for some $y \in V^{-1}(\{\lambda\})$ is confined to $V^{-1}(]-\infty, \lambda])$, which is a compact set. Hence $\lambda \in \sigma$. This proves *(ii)*.

To prove *(i)*, we note that

$$\bigcup_{a \neq a_{\max},\ a \neq a_{\min}} [\inf V^a, \min_{b<a} V^b[\cup \{0\} = [\min_{a \neq a_{\max}} \inf V^a(x^a), 0].$$

□

In particular, the set τ of "classical thresholds" usually contains some intervals. On the contrary, in the quantum case, we will see that the set of thresholds is much smaller.

The main results of this section are the following two theorems.

Theorem 5.3.3
Let $[\lambda_1, \lambda_2] \subset \mathbb{R} \backslash \tau$. Then there exist constants C_1 and $C_0 > 0$ such that, for any $(y, \eta) \in H^{-1}([\lambda_1, \lambda_2])$, either

$$|x(t, y, \eta)| \leq C_1, \tag{5.3.2}$$

or, for some C,

$$|x(t, y, \eta)| \geq C_0 t - C. \tag{5.3.3}$$

Theorem 5.3.4
The sets τ *and* $\sigma \cup \tau$ *are closed.*

As we mentioned, in order to prove the above theorems, we will construct a certain observable having a positive Poisson bracket with the Hamiltonian $H(x, \xi)$ in some region of phase space. This observable extends the one constructed in the 2-body case in Propositions 2.3.3 and 2.4.2.

Proposition 5.3.5
(i) Let $[\lambda_1, \lambda_2] \subset \mathbb{R} \setminus \tau$. *Then there exists a function* $G(x, \xi) \in C^{0,1}(X \times X')$, R_0 *and* $C_0 > 0$ *such that*

$$|\partial_x^\alpha \partial_\xi^\beta (G(x, \xi) - \langle x, \xi \rangle)| \leq C, \quad |\alpha| + |\beta| \leq 1, \ (x, \xi) \in H^{-1}([\lambda_1, \lambda_2]),$$
$$\{H, G\}(x, \xi) \geq C_0, \quad (x, \xi) \in H^{-1}([\lambda_1, \lambda_2]), \ |x| \geq R_0.$$

(ii) If moreover $[\lambda_1, \lambda_2] \subset \mathbb{R} \setminus (\tau \cup \sigma)$, *then we can choose* $G(x, \xi)$ *in such a way that it will additionally satisfy*

$$\{H, G\}(x, \xi) \geq C_0, \quad (x, \xi) \in H^{-1}([\lambda_1, \lambda_2]).$$

Proof of Theorem 5.3.3 given Proposition 5.3.5. First we construct a function $G(x, \xi)$ that satisfies *(i)* of Proposition 5.3.5.

Note that we have the following estimates:

$$G(x, \xi) \leq C_1, \ (x, \xi) \in H^{-1}([\lambda_1, \lambda_2]), \ |x| < R_0,$$
$$C_2 \langle x \rangle \geq G(x, \xi), \ (x, \xi) \in H^{-1}([\lambda_1, \lambda_2]), \ \text{ for some } \ C_2,$$
$$\{H, G\}(x, \xi) \geq C_0, \ (x, \xi) \in H^{-1}([\lambda_1, \lambda_2]), \ |x| \geq R_0, \ \text{ for some } \ C_0 > 0.$$

Now suppose that $(x(t), \xi(t))$ is a trajectory in $H^{-1}([\lambda_1, \lambda_2])$. We consider separately 3 cases.
Case (1) If

$$\liminf_{t \to \infty} |x(t)| \geq R_0,$$

then, for $t > T$,

$$\frac{\mathrm{d}}{\mathrm{d}t} G(x(t), \xi(t)) \geq C_0 > 0. \tag{5.3.4}$$

Therefore,

$$G(x(t), \xi(t)) \geq C_3 + C_0 t.$$

Hence

$$\langle x(t)\rangle \geq C_2^{-1}(C_3 + C_0 t),$$

which shows that (5.3.3) is satisfied.

Case (2) Let

$$\liminf_{t\to\infty} |x(t)| \leq R_0, \qquad \limsup_{t\to\infty} |x(t)| \geq R_0.$$

Suppose that, for $t \in [T_1, T_2]$,

$$|x(t)| \geq R_0 \quad \text{and} \quad |x(T_1)| = |x(T_2)| = R_0.$$

Then for $t \in [T_1, T_2]$, we have

$$\frac{\mathrm{d}}{\mathrm{d}t} G(x(t), \xi(t)) \geq C_0,$$

and hence

$$G(x(T_2), \xi(T_2)) - G(x(T_1), \xi(T_1)) \geq C_0(T_2 - T_1). \tag{5.3.5}$$

Therefore,

$$2C_2\langle R_0\rangle \geq C_0(T_2 - T_1).$$

This gives a bound on the time that can be spent without interruption outside a ball of radius R_0. By the finiteness of the velocity, this means that the trajectory $x(t)$ is bounded.

Case (3)

$$\limsup_{t\to\infty} |x(t)| \leq R_0.$$

In this case, the trajectory is obviously bounded by R_0. This ends the proof of the theorem. □

Proof of Theorem 5.3.4 given Theorem 5.3.3. The proof uses the induction with respect to $a \in \mathcal{A}$. We will explain the last step of the induction.

Assume that we know that, for $a < a_{\max}$, the set $\sigma^a \cup \tau^a$ is closed. This implies that τ is closed, because

$$\tau = \bigcup_{a \neq a_{\max}} (\sigma^a \cup \tau^a).$$

Let now $\lambda \notin \sigma \cup \tau$. By the closedness of τ, there exist $\lambda_1 < \lambda < \lambda_2$ such that $[\lambda_1, \lambda_2] \subset \mathbb{R}\backslash\tau$. By Theorem 5.3.3, we will find R_0 such that all the bounded trajectories of $H^{-1}(]\lambda_1, \lambda_2[)$ are confined to the ball of radius R_0.

Suppose now that there exist $\lambda_n \in \sigma$ such that $\lambda_n \to \lambda$. Let the trajectories $(x_n(t), \xi_n(t))$ be bounded and $H(x_n(t), \xi_n(t)) = \lambda_n$. Then they will stay inside a compact set and, by passing to a subsequence, we can assume that there exists

$$\lim_{n\to\infty} (x_n(0), \xi_n(0)) =: (y, \eta).$$

By the continuity of the flow, the trajectory $(x(t, y, \eta), \xi(t, y, \eta))$ is bounded. Clearly, $H(y, \eta) = \lambda$. Therefore, $\lambda \in \sigma$. □

Now we are going to construct an observable with the properties described in Proposition 5.3.5. The construction of this observable is rather tedious. It consists in cutting and pasting functions in various regions of phase space. Moreover, one has to use an induction argument on the number of particles N. We start with a lemma that describes such a construction close to Z_a.

Lemma 5.3.6
Let $\lambda \in \mathbb{R}$, $a \in \mathcal{A}$, $[\tilde{\lambda}_1, \tilde{\lambda}_2] \subset \mathbb{R}\backslash(\sigma^a \cup \tau^a)$ and $\epsilon, \gamma > 0$. Assume that Proposition 5.3.5 holds for the Hamiltonian $H^a(x^a, \xi^a)$ and the energy interval $[\tilde{\lambda}_1, \tilde{\lambda}_2]$.

Then for any large enough $\delta > 0$, there exists a function $g_a \in C^{0,1}(X \times X')$ such that

$$|\partial_x^\alpha \partial_\xi^\beta g_a(x,\xi)| \leq C, \quad |\alpha| + |\beta| \leq 1, \quad (x,\xi) \in H^{-1}(]-\infty, \lambda]),$$

$$\{H, g_a\}(x,\xi) \geq \mathbb{1}_{Z_a^{\epsilon,\delta}}(x) \mathbb{1}_{[\tilde{\lambda}_1, \tilde{\lambda}_2]}(H^a(x^a, \xi^a)) - \gamma, \quad (x,\xi) \in H^{-1}(]-\infty, \lambda]).$$

Proof. Let $\chi \in C_0^\infty(\mathbb{R})$ such that $0 \leq \chi \leq 1$ and $\chi = 1$ on a neighborhood of zero. Set

$$f_{a,R}(x) := \chi\left(\tfrac{|x^a|}{R}\right) \prod_{b \not\leq a} \left(1 - \chi\left(\tfrac{|x^b|}{R}\right)\right).$$

Let $m \in C_0^\infty(X^a)$ such that $m \geq 0$, and $m = 1$ on $\{x^a \mid |x^a| < \epsilon\}$. We choose $h \in C_0^\infty(\mathbb{R})$ such that $h = 1$ on $[\tilde{\lambda}_1, \tilde{\lambda}_2]$ and supp$h \subset \mathbb{R}\backslash\sigma^a$. We define

$$M(y^a, \eta^a) := -h(H^a(y^a, \eta^a)) \int_0^\infty m(x^a(t, y^a, \eta^a)) \mathrm{d}t, \tag{5.3.6}$$

where $x^a(t, y^a, \eta^a)$ is the trajectory generated by $H^a(x^a, \xi^a)$ with the initial conditions (y^a, η^a). We observe then that since by hypothesis Proposition 5.3.5 holds for the Hamiltonian H^a for the energies in supp$h$, we can apply Theorem 5.3.3 to H^a. Consequently, every such a trajectory with the energy H^a from supp$h$ spends a finite amount of time in supp$m$. Therefore the integral in (5.3.6) is well defined. By the compactness argument, we see that $M(x^a, \xi^a)$ is bounded together with its derivative for $|x^a| \leq R$. We set

$$g_{a,R}(x,\xi) := f_{a,R}(x) M(x^a, \xi^a).$$

We compute:

$$\begin{aligned} \{H, g_{a,R}\}(x,\xi) &= \xi \nabla_x f_{a,R}(x) M(x^a, \xi^a) \\ &+ f_{a,R}(x) \nabla_{x^a} I_a(x) \nabla_{\xi^a} M(x^a, \xi^a) \\ &+ f_{a,R}(x) h(H^a(x^a, \xi^a)) m(x^a). \end{aligned} \tag{5.3.7}$$

The first term on the right of (5.3.7) is bounded by $O(R^{-1})$ for $(x,\xi) \in H^{-1}(]-\infty, \lambda])$. The second term is bounded by $C \sup_{x \in \mathrm{supp} f_{a,R}} |\nabla I_a(x)|$. By choosing R big enough, we can make the sum of these terms less than γ. Enlarging R and choosing δ big enough, we can make sure that the third term is greater than or equal to

$$\mathbb{1}_{Y_a^{\epsilon,\delta}}(x)\mathbb{1}_{[\tilde{\lambda}_1,\tilde{\lambda}_2]}(H^a(x^a,\xi^a)).$$

We set

$$g_a(x,\xi) := g_{a,R}(x,\xi)$$

for such R. □

Lemma 5.3.7
Let $\lambda \in \mathbb{R}$. (i) Let $[\tilde{\lambda}_1,\tilde{\lambda}_2] \subset \mathbb{R}\backslash\tau$. Then for any $\gamma_1,\gamma_2 > 0$, there exists a sequence $\rho = \{\rho_a \mid a \in \mathcal{A}\}$ and, for any $a \neq a_{\max}$, a function $g_a(x,\xi)$ such that if $(x,\xi) \in H^{-1}(]-\infty,\lambda])$, then

$$\begin{aligned} |\partial_x^\alpha \partial_\xi^\beta g_a(x,\xi)| \leq C, \quad |\alpha|+|\beta| \leq 1, \\ \{H,g_a\}(x,\xi) \geq q_a^\rho(x)\mathbb{1}_{[\tilde{\lambda}_1,\tilde{\lambda}_2]}(H^a(x^a,\xi^a)) - \gamma_1, \\ |q_a^\rho(x)I_a(x)| < \gamma_2. \end{aligned} \tag{5.3.8}$$

(ii) Let $[\tilde{\lambda}_1,\tilde{\lambda}_2] \subset \mathbb{R}\backslash(\tau\cup\sigma)$. Then for any $\gamma_1,\gamma_2 > 0$, there exists a sequence $\rho = \{\rho_a \mid a \in \mathcal{A}\}$ and, for any $a \in \mathcal{A}$, a function $g_a(x,\xi)$ such that (5.3.8) is satisfied.

Proof. The proof uses a decreasing induction with respect to $n = \#a$. At the nth step of the induction we construct a sequence ρ^n and functions $g_a(x,\xi)$ with $\#a = n$.

We start the induction by setting $\rho_a^{N+1} = 0$, $a \in \mathcal{A}$. (There are no a's with $\#a = N+1$, hence no functions $g_a(x,\xi)$ are available yet).

Let $n = N, N-1, \ldots$. Suppose that we have already constructed $g_b(x,\xi)$ for $\#b > n$ and we have chosen a sequence ρ^{n+1} such that, for $\#b > n$ and $(x,\xi) \in H^{-1}(]-\infty,\lambda])$, we have

$$\begin{aligned} |\partial_x^\alpha \partial_\xi^\beta g_b(x,\xi)| \leq C, \quad |\alpha|+|\beta| \leq 1, \\ \{H,g_b\}(x,\xi) \geq q_b^{\rho^{n+1}}(x)\mathbb{1}_{[\tilde{\lambda}_1,\tilde{\lambda}_2]}(H^b(x^b,\xi^b)) - \gamma_1, \\ |q_b^{\rho^{n+1}}(x)I_b(x)| < \gamma_2. \end{aligned} \tag{5.3.9}$$

Let $\epsilon > 0$ have the property

$$X_a^\epsilon \supset \Xi_a^{\rho^{n+1}}, \quad \#a = n. \tag{5.3.10}$$

Then there exists δ_1 such that

$$|I_a(x)\mathbb{1}_{Y_a^{\epsilon,\delta_1}}(x)| < \gamma_2, \quad \#a = n.$$

Using Lemma 5.3.6 for this ϵ determined in (5.3.10), we find $\delta > \delta_1$ and, for any $a \in \mathcal{A}$ such that $\#a = n$, we construct $g_a(x,\xi) \in C^{0,1}(X,X')$ with the following property:

$$|\partial_x^\alpha \partial_\xi^\beta g_a(x,\xi)| \leq C, \quad |\alpha| + |\beta| \leq 1, \quad (x,\xi) \in H^{-1}(]-\infty,\lambda]),$$
$$\{H, g_a\}(x,\xi) \geq \mathbb{1}_{Y_a^{\epsilon,\delta}}(x)\mathbb{1}_{[\tilde\lambda_1,\tilde\lambda_2]}(H^a(x^a,\xi^a)) - \gamma_1, \quad (x,\xi) \in H^{-1}(]-\infty,\lambda]).$$

We choose a new sequence ρ^n such that

$$\rho_b^{n+1} = \rho_b^n, \ \#b \geq n, \qquad \rho_b^{n+1} \leq \rho_b^n, \ \#b < n,$$

so as to guarantee that, for $\#a = n$,

$$\mathbb{1}_{Y_a^{\epsilon,\delta}} \geq q_a^{\rho^n}.$$

This implies that, for $\#a = n$ and $(x,\xi) \in H^{-1}(]-\infty,\lambda])$, we have

$$\begin{aligned} \{H, g_a\}(x,\xi) \geq q_a^{\rho^n}(x)\mathbb{1}_{[\tilde\lambda_1,\tilde\lambda_2]}(H^a(x^a,\xi^a)) - \gamma_1, \\ |I_a(x) q_a^{\rho^n}(x)| < \gamma_2. \end{aligned} \tag{5.3.11}$$

Note also that

$$q_b^{\rho^{n+1}} \geq q_b^{\rho^n}, \ \#b \geq n.$$

Hence the induction assumption (5.3.9) implies that (5.3.11) is true also for $\#a > n$.

In the case of the proof of *(i)*, we repeat this construction until we arrive at $n = 2$. We put $\rho := \rho^2$. Note that in this case we have constructed g_a for all $\#a \geq 2$, which means for all $a \neq a_{\max}$.

In the case of *(ii)*, we stop at $n = 1$ and we put $\rho := \rho^1$. In this case, we have constructed g_a for all $\#a \geq 1$, which means for all $a \in \mathcal{A}$. □

Proof of Proposition 5.3.5. We choose $\gamma_1, \gamma_2 > 0$, $\tilde\lambda_1$, $\tilde\lambda_2$ and $1/2 \geq C_1 > 0$ such that

$$\begin{aligned} &[\tilde\lambda_1, \tilde\lambda_2] \cap \sigma \cup \tau = \emptyset, \\ &\lambda_1 = \tilde\lambda_1 + C_1 + \gamma_1, \quad \lambda_2 = \tilde\lambda_2 - \gamma_1, \\ &C_1 - M\gamma_2 \geq C_2 > 0, \end{aligned}$$

where M is the number of elements of $\mathcal{A}$. We construct the functions g_a and a sequence ρ as in Lemma 5.3.7. We set

$$G(x,\xi) := \langle \nabla_x R^\rho(x), \xi \rangle + \sum_{a\in\mathcal{A}} g_a(x,\xi).$$

For $(x,\xi) \in H^{-1}(]-\infty,\lambda])$, we have

$$\begin{aligned} \{H, G\}(x,\xi) &= \langle \nabla_x^2 R^\rho(x)\xi, \xi\rangle + \textstyle\sum_{a\in\mathcal{A}} \{H, g_a\}(x,\xi) \\ &\geq \sum_{a\in\mathcal{A}} q_a^\rho(x)\left(\xi_a^2 + \mathbb{1}_{[\tilde\lambda_1,\tilde\lambda_2]}(H^a(x^a,\xi^a))\right) - M\gamma_1. \end{aligned} \tag{5.3.12}$$

It is easy to see using $1/2 \geq C_1$ that

$$\xi_a^2 + \mathbb{1}_{[\tilde\lambda_1,\tilde\lambda_2]}(H^a(x^a,\xi^a)) \geq 2C_1 \mathbb{1}_{[\tilde\lambda_1+C_1,\tilde\lambda_2]}(H_a(x,\xi)).$$

Besides, on the support of $q_a^\rho(x)$ we have

$$|H_a(x,\xi) - H(x,\xi)| < \gamma_1.$$

Therefore,

$$q_a^\rho(x)\mathbb{1}_{[\tilde\lambda_1+C_1,\tilde\lambda_2]}(H_a(x,\xi)) \geq q_a^\rho(x)\mathbb{1}_{[\lambda_1,\lambda_2]}(H(x,\xi)).$$

Hence

$$q_a^\rho(x)(\xi_a^2 + \mathbb{1}_{[\tilde\lambda_1,\tilde\lambda_2]}(H^a(x^a,\xi^a)) \geq 2C_1 q_a^\rho(x)\mathbb{1}_{[\lambda_1,\lambda_2]}(H(x,\xi)).$$

Therefore,

$$\{H,G\}(x,\xi) \geq C_1 - M\gamma_1, \quad (x,\xi) \in H^{-1}([\lambda_1,\lambda_2]).$$

By choosing $\gamma_1 > 0$ small enough, we can guarantee that $2C_1 - M\gamma_1 = C_0 > 0$. This proves *(ii)*.

The proof of *(i)* is similar except we sum just over $a \neq a_{\max}$. □

5.4 Asymptotic Velocity

In this section we introduce the basic asymptotic quantity for classical N-particle Hamiltonians, the *asymptotic velocity.* This result is similar to the one already obtained for 2-body Hamiltonians. It is, however, much deeper.

The proof of the existence of the asymptotic velocity is taken from [De7]. It was inspired by arguments used in [Gr] to prove the asymptotic completeness of N-body quantum short-range systems.

Throughout this section we will assume that

$$\int_0^\infty \sup_{|x^b|\geq R} |\nabla_{x^b} v^b(x^b)| \mathrm{d}R < \infty, \; b \in \mathcal{B}. \tag{5.4.1}$$

The following theorem describes the main result of this section.

Theorem 5.4.1
Assume the hypotheses (5.4.1). Then the following properties hold:
(i) For any $(y,\eta) \in X \times X'$, *the following limit exists*

$$\lim_{t\to\infty} \frac{x(t,y,\eta)}{t} =: \xi^+(y,\eta). \tag{5.4.2}$$

This limit is called the asymptotic velocity.
(ii) If $\xi^+(y,\eta) \in Y_a$, *then*

$$\lim_{t\to\infty} \xi_a(t,y,\eta) = \xi_a^+(y,\eta).$$

(iii) If $\xi^+(y,\eta) \in Y_a$, *then there exists*

$$\lim_{t\to\infty} H^a(x(t,y,\eta),\xi(t,y,\eta)) =: H^{a,+}(y,\eta). \tag{5.4.3}$$

(iv)

$$H(y,\eta) = \frac{1}{2}(\xi_a^+)^2(y,\eta) + H^{a,+}(y,\eta). \tag{5.4.4}$$

Theorem 5.4.1 gives a rough classification of all trajectories of an N-body system. In this classification, two functions are available: the full energy

$$X \times X' \ni (y,\eta) \mapsto H(y,\eta) \in \mathbb{R}$$

and the asymptotic velocity

$$X \times X' \ni (y,\eta) \mapsto \xi^+(y,\eta) \in X.$$

It is sometimes convenient to use another asymptotic quantity, the asymptotic internal energy of the subsystem a defined in (5.4.3):

$$(\xi^+)^{-1}(Y_a) \ni (y,\eta) \mapsto H^{a,+}(y,\eta) \in \mathbb{R}.$$

However, as follows from (5.4.4), this quantity is a function of $H(y,\eta)$ and $\xi^+(y,\eta)$.

The set

$$(\xi^+)^{-1}(Z_a)$$

will be called the set of $a-$*clustered trajectories.*

In physical terms, Theorem 5.4.1 means that, for large times, any system of classical particles separates into subsystems. The centers of masses of the subsystems have a separation of order $O(t)$. The size of the subsystems is of order $o(t)$. We will see in Sect. 5.7 that one can get a more precise result about the size of the subsystems.

We start with a simple lemma about the boundedness of the velocity.

Lemma 5.4.2
Assume the hypotheses (5.4.1). Then for any $(y,\eta) \in X \times X'$, there exists C such that

$$|x(t,y,\eta)| \le C\langle t\rangle, \quad |\xi(t,y,\eta)| \le C.$$

Proof. Since $V(x)$ is bounded and $H(x,\xi)$ is constant on a trajectory, we get that

$$\frac{1}{2}\xi^2(t) = H(x(t),\xi(t)) - V(x(t)) \le C.$$

Using then the identity $\dot{x}(t) = \xi(t)$, we obtain the desired result. □

The main tools of this section are the so-called propagation estimates. They were first used in quantum scattering theory, notably in [SS1]. The next two propositions are classical analogs of quantum propagation estimates due to Graf [Gr]. The abstract argument that is used in the proof of classical propagation estimates is explained in Lemma A.5.1.

Proposition 5.4.3
Assume the hypotheses (5.4.1). Then for any $\epsilon > 0$, $a \in \mathcal{A}$ and any trajectory $(x(t), \xi(t))$, one has

$$\int_1^\infty \mathbb{1}_{Y_a^\epsilon}(\tfrac{x(t)}{t})(\tfrac{x_a(t)}{t} - \xi_a(t))^2 \tfrac{dt}{t} < \infty. \tag{5.4.5}$$

Proof. We consider the observable

$$\begin{aligned} \Phi(t,x,\xi) &:= \mathbf{D}tR(\tfrac{x}{t}) \\ &= R(\tfrac{x}{t}) + \langle \nabla R(\tfrac{x}{t}), (\xi - \tfrac{x}{t})\rangle. \end{aligned}$$

We compute

$$\begin{aligned} \mathbf{D}\Phi(t,x,\xi) &= t^{-1}\langle(\xi - \tfrac{x}{t}), \nabla^2 R(\tfrac{x}{t})(\xi - \tfrac{x}{t})\rangle - \langle \nabla R(\tfrac{x}{t}), \nabla_x V(x)\rangle \\ &\geq \sum_{a\in\mathcal{A}} \tfrac{1}{t} q_a(\tfrac{x}{t})(\tfrac{x_a}{t} - \xi_a)^2 - \langle \nabla R(\tfrac{x}{t}), \nabla_x V(x)\rangle. \end{aligned} \tag{5.4.6}$$

We observe that, by Lemma 5.4.2, the function $\Phi(t, x(t), \xi(t))$ is uniformly bounded in t. Moreover, the second term on the right-hand side of (5.4.6) is integrable along the trajectory, using Lemma 5.2.7 *(v)* and hypothesis (5.4.1). Therefore, by Lemma A.5.1, for any $a \in \mathcal{A}$,

$$\int_1^\infty q_a(\tfrac{x(t)}{t})(\tfrac{x_a(t)}{t} - \xi_a(t))^2 \tfrac{dt}{t} < \infty. \tag{5.4.7}$$

Let us also note that

$$\sum_{b\leq a} \tfrac{1}{t} q_b(\tfrac{x}{t})(\tfrac{x_b}{t} - \xi_b)^2 \geq \tfrac{1}{t} \sum_{b\leq a} q_b(\tfrac{x}{t})(\tfrac{x_a}{t} - \xi_a)^2 \tag{5.4.8}$$

Given $\epsilon > 0$ we can choose the functions $R(x)$ and $q_a(x)$ such that

$$\sum_{b\leq a} q_b(x) \geq \mathbb{1}_{Y_a^\epsilon}(x) \tag{5.4.9}$$

Now (5.4.7), (5.4.8) and (5.4.9) imply the proposition. □

Proposition 5.4.4
Assume the hypotheses (5.4.1). Then for any $\epsilon > 0$, $a \in \mathcal{A}$ and any trajectory $(x(t), \xi(t))$, one has

$$\int_1^\infty \mathbb{1}_{Y_a^\epsilon}(\tfrac{x(t)}{t}) \left|\tfrac{x_a(t)}{t} - \xi_a(t)\right| \tfrac{dt}{t} < \infty. \tag{5.4.10}$$

Proof. Let $J \in \mathcal{F} \cap C_0^\infty(X)$ such that supp$J \subset Y_a$. We consider the observable

$$\Phi(t,x,\xi) := J\left(\frac{x}{t}\right)\left|\frac{x_a}{t} - \xi_a\right|$$

and we compute

$$\begin{aligned} \mathbf{D}\Phi(t,x,\xi) &= -\tfrac{1}{t}J(\tfrac{x}{t})|\tfrac{x_a}{t} - \xi_a| \\ &+\tfrac{1}{t}\langle \nabla J(\tfrac{x}{t}), \xi - \tfrac{x}{t}\rangle |\tfrac{x_a}{t} - \xi_a| \\ &+J(\tfrac{x}{t})\langle \nabla_x I_a(x), \tfrac{x_a}{t} - \xi_a\rangle |\tfrac{x_a}{t} - \xi_a|^{-1}. \end{aligned} \tag{5.4.11}$$

We observe, using Lemma 5.4.2, that the function $\Phi(t, x(t), \xi(t))$ is bounded uniformly in t. The second term on the right of (5.4.11), for an appropriate admissible ρ, equals

$$\sum_{b\le a} \tfrac{1}{t} q_b^\rho(x)\langle \nabla J(\tfrac{x}{t}), \xi_b - \tfrac{x_b}{t}\rangle |\tfrac{x_a}{t} - \xi_a| \le C \sum_{b\le a} \tfrac{1}{t} q_b^\rho(x)|\xi_b - \tfrac{x_b}{t}|^2,$$

which is integrable along the flow by Proposition 5.4.3. Moreover, the third term on the right of (5.4.11) is also integrable along the flow. Therefore, by Lemma A.5.1,

$$\int_1^\infty J(\tfrac{x(t)}{t})|\tfrac{x_a(t)}{t} - \xi_a(t)|\tfrac{\mathrm{d}t}{t} < \infty. \tag{5.4.12}$$

□

Proof of Theorem 5.4.1. Consider first a function $J \in C_0^\infty(X) \cap \mathcal{F}$. We have

$$\tfrac{\mathrm{d}}{\mathrm{d}t} J(\tfrac{x(t)}{t}) = -\tfrac{1}{t}\langle \nabla_x J(\tfrac{x(t)}{t}), (\tfrac{x(t)}{t} - \xi(t))\rangle \tag{5.4.13}$$

For an appropriate admissible ρ, (5.4.13) equals

$$-\tfrac{1}{t}\sum_{a\in\mathcal{A}} q_a^\rho(\tfrac{x(t)}{t})\langle \nabla_x J(\tfrac{x(t)}{t}), \tfrac{x_a(t)}{t} - \xi_a(t)\rangle,$$

which is integrable by Proposition 5.4.4. So the limit

$$\lim_{t\to\infty} J\left(\frac{x(t)}{t}\right) \tag{5.4.14}$$

exists for any $J \in C_0^\infty(X) \cap \mathcal{F}$. Since $C_0^\infty(X) \cap \mathcal{F}$ is dense in $C_0(X)$, we see that the limit (5.4.14) also exists for $J \in C_0(X)$. For a given (y,η), we can choose a function $J \in C_0(X)$ such that

$$J\left(\frac{x(t)}{t}\right) = \frac{x(t)}{t}, \quad t \ge T_0.$$

Hence there exists (5.4.2), which proves *(i)*.

Let us now prove *(ii)*. Let $(y,\eta) \in X \times X'$ such that $x = \xi^+(y,\eta) \in Y_a$. Note first that

$$\nabla_{x_a} I_a(x(t)) \in L^1(\mathrm{d}t),$$

by (5.4.1). Next we compute

$$\tfrac{\mathrm{d}}{\mathrm{d}t}(x_a(t) - t\xi_a(t)) = t\nabla_{x_a} I_a(x(t)).$$

So

$$x_a(t) - t\xi_a(t) = y - \textstyle\int_0^t s\nabla_{x_a} I_a(x(s))\mathrm{d}s \in o(t),$$

and hence

$$\frac{x_a(t)}{t} = \xi_a(t) + o(t^0),$$

which completes the proof of *(ii)*.

To prove *(iii)*, we just note that if $\xi^+(y,\eta) \in Y_a$, then

$$\frac{\mathrm{d}}{\mathrm{d}t} H^a(x(t), \xi(t)) = \langle \nabla_x I_a(x(t)), \xi(t) \rangle,$$

which is integrable.

To prove *(iv)*, note that, by the conservation of energy,

$$\begin{aligned} H(y,\eta) &= H(x(t,y,\eta), \xi(t,y,\eta)) \\ &= \tfrac{1}{2}\xi_a^2(t,y,\eta) + I_a(x(t,y,\eta)) + H^a(x^a(t,y,\eta), \xi^a(t,y,\eta)). \end{aligned}$$

Now we use

$$\lim_{t\to\infty} \frac{1}{2}\xi_a^2(t,y,\eta) = (\xi_a^+(y,\eta))^2, \quad \lim_{t\to\infty} I_a(x(t,y,\eta)) = 0.$$

□

5.5 Joint Localization of the Energy and the Asymptotic Velocity

In this section we will make the same assumptions on the potentials as in the previous section, that is, (5.4.1). Note that, by Proposition 2.5.1, (5.4.1) implies the assumptions (5.3.1). Therefore all the results of Sect. 5.3 on the bounded trajectories and trapping energies are true under the hypothesis (5.4.1). In this section we will present their refinements that take into account the asymptotic velocity.

First let us present a refinement of Theorem 5.3.3.

Theorem 5.5.1
Let $[\lambda_1, \lambda_2] \subset \mathbb{R}\backslash\tau^a$, and let $\Theta \subset Y_a$ be compact. Then there exist constants C_1 and $C_0 > 0$ such that, for any $(y,\eta) \in (\xi^+, H^{a,+})^{-1}(\Theta \times [\lambda_1, \lambda_2])$ and some T, C, either

$$|x^a(t,y,\eta)| \le C_1, \ t > T, \tag{5.5.1}$$

or

$$|x^a(t,y,\eta)| \geq C_0(t-T) - C, \ t > T, \tag{5.5.2}$$

Definition 5.5.2
Let us denote by $\mathcal{E}^{a,+}$ the set of $(y,\eta) \in (\xi^+)^{-1}(Z_a)$ such that $x^a(t,y,\eta)$ is bounded for $t > 0$, where, as usual, $x(t,y,\eta)$ denotes the trajectory generated by $H(x,\xi)$ with the initial conditions (y,η).

Let us note that $\mathcal{E}^{a_{\max},+} = \mathcal{B}^{a_{\max},+}$.
By definition, we have

$$\mathcal{E}^{a,+} \subset (\xi^+)^{-1}(Z_a). \tag{5.5.3}$$

In general, however, we cannot replace the inclusion in (5.5.3) by the equality. Using Theorem 5.5.1, we can prove the following inclusions.

Theorem 5.5.3
The following inclusions are true:

$$(\xi^+, H^{a,+})^{-1}(Z_a \times (\sigma^a \backslash \tau^a)) \subset \mathcal{E}^{a,+}, \tag{5.5.4}$$

$$(\xi^+, H^{a,+})^{-1}(Z_a \times (\sigma^a \cup \tau^a)) = (\xi^+)^{-1}(Z_a), \tag{5.5.5}$$

$$\overline{(\xi^+, H)(X \times X')} \subset \bigcup_{a \in \mathcal{A}} \{(\xi_a, \lambda + \frac{1}{2}\xi_a^2) \mid \xi_a \in X_a, \ \lambda \in \sigma^a \cup \tau^a\}. \tag{5.5.6}$$

Proof of Theorem 5.5.1. First we construct a function $G^a(x^a,\xi^a)$ that satisfies *(i)* of Proposition 5.3.5 with $H(x,\xi)$ replaced with $H^a(x^a,\xi^a)$.

Let $(x(t),\xi(t))$ be a trajectory in $(\xi^+, H)^{-1}(\Theta \times [\lambda_1,\lambda_2])$. Then

$$\begin{aligned} \tfrac{\mathrm{d}}{\mathrm{d}t} G^a(x^a(t),\xi^a(t)) &= \{H^a, G^a\}(x^a(t),\xi^a(t)) \\ &\quad - \nabla_{\xi^a} G^a(x^a(t),\xi^a(t)) \nabla_{x^a} I_a(x(t)). \end{aligned} \tag{5.5.7}$$

The second term on the right-hand side of (5.5.7) is $o(t^0)$. Moreover, for large enough time, $(x^a(t),\xi^a(t)) \in (H^a)^{-1}([\tilde\lambda_1,\tilde\lambda_2])$ for some $[\tilde\lambda_1,\tilde\lambda_2] \subset \mathbb{R}\backslash\tau^a$ such that $[\lambda_1,\lambda_2] \subset]\tilde\lambda_1,\tilde\lambda_2[$. Therefore, for $t > T$, $|x^a| \geq R_0$ and T large enough, we have

$$\frac{\mathrm{d}}{\mathrm{d}t} G^a(x^a(t),\xi^a(t)) \geq \tilde C_0 > 0. \tag{5.5.8}$$

Next we argue as in the proof of Theorem 5.3.3. □

Proof of Theorem 5.5.3. The first and second inclusions follow immediately from Theorem 5.5.1.

To prove (5.5.6), note that (5.5.5) implies

$$(\xi^+, H^{a,+}) \left((\xi^+)^{-1}(Z_a)\right) \subset Z_a \times (\sigma^a \cup \tau^a).$$

Hence

$$(\xi^+, H) \left((\xi^+)^{-1}(Z_a)\right) \subset \{(\xi_a, \frac{1}{2}\xi_a^2 + \lambda) \mid \xi_a \in Z_a,\ \lambda \in \sigma^a \cup \tau^a\}.$$

Finally,

$$\overline{(\xi^+, H) \left((\xi^+)^{-1}(Z_a)\right)} \subset \{(\xi_a, \frac{1}{2}\xi_a^2 + \lambda) \mid \xi_a \in X_a,\ \lambda \in \sigma^a \cup \tau^a\},$$

since the right-hand side is easily seen to be closed, using Theorem 5.3.4. □

5.6 Regular a-Trajectories

Various asymptotic quantities that can be defined in N-body scattering tend to be ill-behaved. They are often discontinuous, it is difficult to predict their range. Nevertheless, there are some regions in phase space where they are well-behaved. This is for example the case of the free region $(\xi^+)^{-1}(Z_{a_{\min}})$, where scattering is essentially as regular as possible, as we will see in Section 5.8. In this section we will describe some regions where the asymptotic internal energy and the asymptotic velocity are continuous and the closure of their joint range fills up the set that one would expect by heuristic arguments. These regions correspond to the "wells" of the potentials $V^a(x^a)$, which can trap a-clustered trajectories for large enough time when the effects of the interaction become negligible.

Definition 5.6.1
For any $a \in \mathcal{A}$, we define

$$\begin{aligned}
\mathcal{W}^a(\lambda) &:= \bigcup \textit{bounded connected components of } (V^a)^{-1}(]-\infty, \lambda[),\\
\sigma^a_{\mathrm{reg}} &:= \bigcup_{\lambda \in \mathbb{R}} V^a(\mathcal{W}^a(\lambda))\\
&= [\inf V^a(x^a), \sup\{\lambda \mid \mathcal{W}^a(\lambda) \neq \emptyset\}[,\\
\mathcal{E}^{a,+}_{\mathrm{reg}} &:= \bigcup_{\lambda \in \mathbb{R}} \{(y, \eta) \in (\xi^+)^{-1}(Z_a) \mid\ x^a(t, y, \eta) \in \mathcal{W}^a(\lambda),\ \ t\ \ \textit{big enough}\}.
\end{aligned}$$

The set $\mathcal{E}^{a,+}_{\mathrm{reg}}$ is the union of a–clustered trajectories that end up in a well of the potential V^a. It turns out that scattering inside $\mathcal{E}^{a,+}_{\mathrm{reg}}$ is quite regular in comparison with the general case. Indeed, we have the following theorem:

Theorem 5.6.2
The following inclusions are true:

$$\sigma^a_{\mathrm{reg}} \subset \sigma^a, \qquad \mathcal{E}^{a,+}_{\mathrm{reg}} \subset \mathcal{E}^{a,+}. \tag{5.6.1}$$

The sets $\mathcal{E}_{\rm reg}^{a,+}$ *are open. Moreover, the function*

$$\mathcal{E}_{\rm reg}^{a,+} \ni (y,\eta) \mapsto (\xi^+(y,\eta), H^{a,+}(y,\eta)) \tag{5.6.2}$$

is continuous. Finally,

$$\overline{(\xi^+, H^{a,+})(\mathcal{E}_{\rm reg}^{a,+})} \supset X_a \times \sigma_{\rm reg}^a. \tag{5.6.3}$$

The proof of Theorem 5.6.2 is based on the following lemma.

Lemma 5.6.3
Let $\xi_a^+ \in Z_a$, $\delta > 0$, $\lambda_1 \in \sigma_{\rm reg}^a$. *Let* U_1 *be a certain compact connected component of* $(V^a)^{-1}(]-\infty,\lambda_1[)$. *Then there exist* $\epsilon > 0$ *and* T_0 *such that if* (y,η) *satisfies*

$$\begin{aligned}
&|y_a - T_0\xi_a^+| \leq \epsilon T_0,\\
&|\eta_a - \xi_a^+| \leq \epsilon,\\
&H^a(y^a,\eta^a) < \lambda_1,\\
&y^a \in U_1,
\end{aligned}$$

then one has, for $t \geq T_0$,

$$\begin{aligned}
&|x_a(t,y,\eta) - y_a - t\eta_a| \leq \delta|t|,\\
&|\xi_a(t,y,\eta) - \eta_a| \leq \delta,\\
&|H^a(x^a(t,y,\eta),\xi^a(t,y,\eta)) - H^a(y^a,\eta^a)| \leq \delta,\\
&x^a(t,y,\eta) \in U_1.
\end{aligned} \tag{5.6.4}$$

Proof. Let $J \in C_0^\infty(X)$ a cutoff function equal to 1 near ξ_a^+ and supported in Y_a. We will denote by

$$(x(t,s,y,\eta), \xi(t,s,y,\eta))$$

the trajectories satisfying the initial conditions $(x(s,s,y,\eta),\xi(s,s,y,\eta)) = (y,\eta)$ associated with the time-dependent force

$$-\nabla_{x^a} V^a(x^a) + F_J(t,x), \tag{5.6.5}$$

where

$$F_J(t,x) := -J\left(\frac{x}{t}\right)\nabla_x I_a(x).$$

This force satisfies the assumptions of Chap. 1. By Proposition 1.3.2, these trajectories, uniformly for $s \leq t$, satisfy

$$|x_a(t,s,y,\eta) - y_a - (t-s)\eta_a| \in o(s^0)|t-s|, \tag{5.6.6}$$

$$|\xi_a(t,s,y,\eta) - \eta_a| \in o(s^0), \tag{5.6.7}$$

$$|H^a(x^a(t,s,y,\eta),\xi^a(t,s,y,\eta)) - H^a(y^a,\eta^a)| \in o(s^0). \tag{5.6.8}$$

Suppose now that (y,η) satisfy the assumptions of the lemma. We see that, for s large enough,

$$H^a(x^a(t,s,y,\eta),\xi^a(t,s,y,\eta)) < \lambda_1.$$

By continuity of the flow, we also have

$$x^a(t,s,y,\eta) \in U_1. \tag{5.6.9}$$

By (5.6.6) and (5.6.7), we can choose s large enough such that, for $s \le t$,

$$|x_a(t,s,y,\eta) - y_a - (t-s)\eta_a| \le \delta|t-s|, \tag{5.6.10}$$

$$|\xi_a(t,s,y,\eta) - \eta_a| \le \delta. \tag{5.6.11}$$

We have

$$\begin{aligned} \left|\frac{x(t,s,y,\eta)}{t} - \xi_a^+\right| &\le \frac{|x^a(t,s,y,\eta)|}{t} + \frac{|x_a(t,s,y,\eta)-y_a-(t-s)\eta_a|}{t} \\ &+ \frac{|y_a - s\xi_a^+|}{t} + \frac{(t-s)|\eta_a - \xi_a^+|}{t}. \end{aligned} \tag{5.6.12}$$

The first term on the right of (5.6.12) is less that C/t because of (5.6.9). The second is $o(s^0)$ by (5.6.6). The last two terms are less than 2ϵ. Therefore, for any given δ_1, by choosing s big enough and $\epsilon > 0$ small enough, we can make sure that

$$\left|\frac{x(t,s,y,\eta)}{t} - \xi_a^+\right| < \delta_1.$$

In particular, we can demand that this is true for δ_1 satisfying $J = 1$ on the ball $B(\xi_a^+,\delta_1)$. Therefore, for $t \ge s$,

$$F_J(t,x(t,s,y,\eta)) = -\nabla_x I_a(x(t,s,y,\eta)).$$

Hence, for T_0 large enough and $t > 0$, we have

$$(x(t+T_0,T_0,y,\eta),\xi(t+T_0,T_0,x,\eta)) = (x(t,y,\eta),\xi(t,y,\eta)). \tag{5.6.13}$$

Now the estimates (5.6.4) follow from (5.6.6), (5.6.7) and (5.6.8) □

Proof of Theorem 5.6.2. The inclusions (5.6.1) are obvious.

Let us prove the openness of $\mathcal{E}_{\text{reg}}^{a,+}$. Let $(y^0,\eta^0) \in \mathcal{E}_{\text{reg}}^{a,+}$ with such that

$$\begin{aligned} &\xi^+(y^0,\eta^0) = \xi_a^+, \\ &H^{a,+}(y^0,\eta^0) = \lambda_0 < \lambda_1, \\ &x^a(t,y^0,\eta^0) \in U_1,\ t > T_1, \end{aligned}$$

where U_1 is a compact connected component of $(V^a)^{-1}(]-\infty,\lambda_1[)$.

Let $\epsilon > 0$ and $T_0 > T_1$ be chosen as in Lemma 5.6.3. By enlarging T_0 if needed, we can make sure that

$$\begin{aligned} &|x(T_0, y^0, \eta^0) - T_0\xi_a^+| < \epsilon T_0, \\ &|\xi_a(T_0, y^0, \eta^0) - \xi_a^+| < \epsilon, \\ &H^a(x^a(T_0, y^0, \eta^0), \xi^a(t, y^0, \eta^0)) < \lambda_1. \end{aligned} \tag{5.6.14}$$

By the continuity of the flow, we can find a neighborhood $\mathcal{U}$ of (y^0, η^0) such that if $(y^1, \eta^1) \in \mathcal{U}$, then

$$\begin{aligned} &\left|\tfrac{x(T_0, y^1, \eta^1)}{T_0} - \xi_a^+\right| < \epsilon, \\ &|\xi_a(T_0, y^1, \eta^1) - \xi_a^+| < \epsilon, \\ &x^a(T_0, y^1, \eta^1) \in U_1, \\ &H^a(x^a(T_0, y^1, \eta^1), \xi(T_0, y^1, \eta^1)) < \lambda_1. \end{aligned} \tag{5.6.15}$$

Thus the conditions of Lemma 5.6.3 are satisfied for all

$$(y, \eta) = (x(T_0, y^1, \eta^1), \xi(T_0, y^1, \eta^1)), \ (y^1, \eta^1) \in \mathcal{U}.$$

This implies that $\mathcal{U} \subset \mathcal{E}_{\text{reg}}^{a,+}$ and proves that $\mathcal{E}_{\text{reg}}^{a,+}$ is open.

Let us now prove the continuity of ξ^+ inside $\mathcal{E}_{\text{reg}}^{a,+}$. With the above notation, we have, for $(y_1, \eta_1) \in \mathcal{U}$,

$$\begin{aligned} &|\xi^+(y^1, \eta^1) - \xi^+(y^0, \eta^0)| = \lim_{t\to\infty} |\xi_a(t, y^1, \eta^1) - \xi_a^+| \\ &\le \limsup_{t\to\infty} |\xi_a(t, y^1, \eta^1) - \ \xi_a(T_0, y^1, \eta^1)| + |\xi_a(T_0, y^1, \eta^1) - \xi_a^+| \le \delta + \epsilon, \end{aligned}$$

Since ϵ, δ can be arbitrary small, this implies the continuity of ξ^+ at (y^0, η^0).

A similar argument proves (5.6.3). □

The following corollary of Theorem 5.6.2 should be compared with the inclusion (5.5.6) of Theorem 5.5.3.

Corollary 5.6.4

$$\overline{(\xi^+, H)(X \times X')} \supset \bigcup_{a\in\mathcal{A}} \left\{\left(\xi_a, \lambda + \frac{1}{2}\xi_a^2\right) \mid \xi_a \in X_a, \ \lambda \in \underset{b\le a}{\cup} \sigma_{\text{reg}}^b\right\}.$$

5.7 Upper Bound on the Size of Clusters

The existence of the asymptotic velocity implies that if $(y, \eta) \in (\xi^+)^{-1}(Z_a)$, then the system separates into clusters of size $o(t)$. From Theorem 5.7.2 below, we shall see that the size of the clusters can be estimated in a more accurate way, for instance if $|\nabla_x v^b(x^b)| \leq C\langle x^b\rangle^{-1-\mu}$, then the size of clusters can be bounded by $C\langle t\rangle^{2(2+\mu)^{-1}}$. This theorem can be viewed as a generalization of Proposition 2.2.1 to N-particle systems. A quantum analog of this estimate plays a big role in the proof of the asymptotic completeness of quantum long-range systems [De8]. The results of this section are taken from [De7].

In our estimate we will use certain auxiliary functions that measure the rate of decay of the potentials.

Definition 5.7.1
For $a \in \mathcal{A}$, let

$$g^a(s) := \sup_{b\leq a, |x^b|\geq s} |\nabla_x v^b(x^b)|,$$

$$g_a(s) := \sup_{b\not\leq a, |x^b|\geq s} |\nabla_x v^b(x^b)|.$$

We denote by $w^a(t)$ the unique solution of

$$\begin{cases} \ddot{w}^a(t) = -g^a(w^a(t)), \\ w^a(0) = 0, \ \lim_{t\to\infty} \frac{w^a(t)}{t} = 0, \end{cases}$$

and by $w_a(t)$ the unique solution of

$$\begin{cases} \ddot{w}_a(t) = -g_a(t), \\ w_a(0) = 0, \ \lim_{t\to\infty} \frac{w_a(t)}{t} = 0. \end{cases}$$

Note that if we put

$$G^a(s) := \int_s^\infty g^a(s_1)ds_1,$$

then $w^a(t)$ is also the unique solution of

$$\begin{cases} \frac{1}{2}(\dot{w}^a(t))^2 = G^a(w^a(t)), \\ w^a(0) = 0, \ \dot{w}^a(0) \geq 0. \end{cases}$$

The function $w^a(t)$ can be computed exactly as in Sect. 2.2.

The function $w_a(t)$ can be computed from the formula

$$w_a(t) = \int_0^t \mathrm{d}s \int_s^\infty g_a(s_1)\mathrm{d}s_1.$$

In the following theorem we state the main result of this section.

Theorem 5.7.2
Assume (5.4.1). Suppose that $x(t, y, \eta)$ is a trajectory such that $(y, \eta) \in (\xi^+)^{-1}(Z_a)$. Then there exists $\epsilon > 0$ and C such that

$$|x^a(t,y,\eta)| \le Cw^a(t) + Cw_a(\epsilon t). \tag{5.7.1}$$

Proof. Let us consider a function $r \in C^\infty(X^a)$ constructed as in Sect. 5.2 with the space X replaced by the space X^a. We put

$$r_1(t,x) := w^a(t) r\left(\frac{x^a}{w^a(t)}\right).$$

We compute

$$\mathbf{D} r_1(t,x,\xi) = \dot{w}^a(t) r(\tfrac{x^a}{w^a(t)}) + \langle \nabla r(\tfrac{x^a}{w^a(t)}), (\xi^a - \tfrac{\dot{w}^a(t)x^a}{w^a(t)})\rangle,$$

$$\begin{aligned}\mathbf{D}^2 r_1(t,x,\xi) &= \tfrac{1}{w^a(t)}\langle(\xi^a - \tfrac{x^a\dot{w}^a(t)}{w^a(t)}), \nabla^2 r(\tfrac{x^a}{w^a(t)})(\xi^a - \tfrac{x^a\dot{w}^a(t)}{w^a(t)})\rangle \\ &\quad -\nabla r(\tfrac{x^a}{w^a(t)})\nabla_x V(x) + \ddot{w}^a(t)\left(r(\tfrac{x^a}{w^a(t)}) - \tfrac{x^a}{w^a(t)}\nabla r(\tfrac{x^a}{w^a(t)})\right) \\ &\ge -\nabla r(\tfrac{x^a}{w^a(t)})\nabla_x V(x) + \ddot{w}^a(t)\left(r(\tfrac{x^a}{w^a(t)}) - \tfrac{x^a}{w^a(t)}\nabla r(\tfrac{x^a}{w^a(t)})\right).\end{aligned}$$

Using the properties of $r(x^a)$ described in Lemma 5.2.9 and the fact that $(y,\eta) \in (\xi^+)^{-1}(Z_a)$, we get that, for some $\epsilon > 0$,

$$\begin{aligned}-\nabla r(\tfrac{x^a(t)}{w^a(t)})\nabla_{x^a} V(x(t)) &= -\nabla r(\tfrac{x^a(t)}{w^a(t)})\nabla_{x^a} V^a(x^a(t)) - \nabla r(\tfrac{x^a(t)}{w^a(t)})\nabla_{x^a} I_a(x(t)) \\ &\le Cg^a(w^a(t)) + Cg_a(\epsilon t).\end{aligned}$$

On the other hand, using again Lemma 5.2.9, we get

$$\left|\left(r(\tfrac{x^a}{w^a(t)}) - \tfrac{x^a}{w^a(t)}\nabla_{x^a} r(\tfrac{x^a}{w^a(t)})\right)\right| \le C.$$

Therefore,

$$\ddot{r}_1(t,x(t)) \ge -C|\ddot{w}^a(t)| - Cg^a(w^a(t)) - Cg_a(\epsilon t). \tag{5.7.2}$$

Since $\ddot{w}^a \le 0$ and $\ddot{w}^a(t) = -g^a(w^a(t))$, we deduce from (5.7.2) that

$$\frac{\mathrm{d}^2}{\mathrm{d}t^2}(r_1(t,x(t)) - 2Cw^a(t) - Cw_a(\epsilon t)) \ge 0. \tag{5.7.3}$$

From Lemma 5.2.9 and the fact that $|x^a(t)| \in o(t)$, we deduce

$$r_1(t,x(t)) \in o(t).$$

So we have

$$r_1(t,x(t)) - 2Cw^a(t) - Cw_a(\epsilon t) \in o(t).$$

This together with (5.7.3) implies

$$\frac{\mathrm{d}}{\mathrm{d}t}(r_1(t,x(t)) - 2Cw^a(t) - Cw_a(\epsilon t)) \le 0,$$

and hence

$$r_1(t,x(t)) \le 2Cw^a(t) + Cw_a(\epsilon t) + C_1,$$

which by Lemma 5.2.9 implies (5.7.1). □

Let us give examples of the functions $w_a(t)$, $w^a(t)$ for various rates of decay of the pair potentials.

Corollary 5.7.3
(i) Assume that

$$\int_0^\infty \sup_{|x^b|\ge R} |\nabla_{x^b} v^b(x^b)| R^\mu \mathrm{d}R < \infty, \quad b \in \mathcal{B}, \quad \mu \ge 0.$$

Then

$$w^a(t) \in o(t^{2/(2+\mu)}), \quad w_a(t) \in \begin{cases} o(t^{1-\mu}), & \mu < 1 \\ O(1), & \mu \ge 1. \end{cases}$$

Therefore, in this case, if $(y,\eta) \in (\xi^+)^{-1}(Z_a)$, *then*

$$|x^a(t,y,\eta)| \le C\langle t\rangle^{2/(2+\mu)}.$$

(ii) Assume that

$$|\nabla_{x^b} v^b(x^b)| \le C\mathrm{e}^{-C|x^b|} \quad b \in \mathcal{B}.$$

Then

$$w^a(t) \in O(\log t), \; w_a(t) \in O(1).$$

Therefore, in this case, if $(y,\eta) \in (\xi^+)^{-1}(Z_a)$, *then*

$$|x^a(t,y,\eta)| \le C\langle \log t\rangle.$$

(iii) Assume that, for any $b \in \mathcal{B}$, *the support of* $v^b(x^b)$ *is compact in* X^b. *Then*

$$w^a(t) \in O(1), \; w_a(t) \in O(1).$$

Therefore, in this case, $(\xi^+)^{-1}(Z_a) \subset \mathcal{E}^{a,+}$, *or in other words*

$$|x^a(t,y,\eta)| \le C.$$

5.8 Free Region Scattering

Scattering in the free region corresponds in physical terms to all particles moving apart without forming clusters. It is very similar to scattering for 2-body Hamiltonians. In particular, one is able to give a complete classification of trajectories. We have

$$\mathcal{E}^{a_{\min},+} = \mathcal{E}^{a_{\min},+}_{\mathrm{reg}} = (\xi^+)^{-1}(Z_{a_{\min}}).$$

We will see in next sections that classical scattering in other regions of phase space is not so well understood.

We will study separately the short-range and the long-range case.

5.8.1 Short-Range Free Region Case

In this subsection we will study the trajectories in $(\xi^+)^{-1}(Z_{a_{\min}})$ for short-range interactions. This case is almost identical to the case of 2-body potentials considered in Chap. 2. It is based on comparing the full dynamics with the free dynamics:

$$\phi_0(t)(x,\xi) = (x + t\xi, \xi).$$

Our first theorem is analogous to Theorems 2.5.2 and 2.6.3.

Theorem 5.8.1
(i) Assume

$$\int_0^\infty \langle R\rangle \sup_{|x^b|\geq R} |\nabla_x v^b(x^b)| \mathrm{d}R < \infty, \quad b \in \mathcal{B}. \tag{5.8.1}$$

Then there exist the limits

$$\lim_{t\to\infty} (x(t,y,\eta) - t\xi^+(t,y,\eta)) =: x^+_{\mathrm{sr},a_{\min}}(y,\eta), \tag{5.8.2}$$

$$\lim_{t\to\infty} \phi_0(-t)\phi(t) \tag{5.8.3}$$

uniformly on compact sets of $(\xi^+)^{-1}(Z_{a_{\min}})$. *Moreover, we have*

$$\lim_{t\to\infty} \phi_0(-t)\phi(t)(y,\eta) \;=\; (x^+_{\mathrm{sr},a_{\min}}(y,\eta), \xi^+(y,\eta)).$$

and the limit (5.8.3) is continuous from $(\xi^+)^{-1}(Z_{a_{\min}})$ *into* $X \times Z_{a_{\min}}$.
(ii) Assume, in addition, that

$$\int_0^\infty \langle R\rangle \sup_{|x^b|\geq R} |\partial_x^\alpha \nabla_x v^a(x^b)| \mathrm{d}R < \infty, \quad |\alpha| = 1, \quad b \in \mathcal{B}.$$

Then there exists the limit

$$\lim_{t\to\infty} \phi(-t)\phi_0(t) \;=:\; \mathcal{F}^+_{\mathrm{sr},a_{\min}} \tag{5.8.4}$$

uniformly on compact sets of $X \times Z_{a_{\min}}$. *The map*

$$\mathcal{F}^+_{\mathrm{sr},a_{\min}} : X \times Z_{a_{\min}} \;\to (\xi^+)^{-1}(Z_{a_{\min}})$$

is bijective, continuous and called the free region wave transformation. *Moreover, (5.8.3) is equal to* $(\mathcal{F}^+_{\mathrm{sr},a_{\min}})^{-1}$.
(iii) If $(y,\eta) = \mathcal{F}^+_{\mathrm{sr},a_{\min}}(x,\xi)$, *one has*

$$|\phi(t)(y,\eta) - \phi_0(t)(x,\xi)| \to 0 \text{ when } t \to \infty. \tag{5.8.5}$$

(iv) The mapping $\mathcal{F}^+_{\mathrm{sr},a_{\min}}$ *is symplectic.*
(v) The free region wave transformation intertwines the full and the free dynamics

$$H \circ \mathcal{F}_{\mathrm{sr},a_{\min}} = H_0,$$

$$\phi(t) \circ \mathcal{F}_{\mathrm{sr},a_{\min}} = \mathcal{F}^{+}_{\mathrm{sr},a_{\min}} \circ \phi_0(t).$$

Proof. Let us define

$$|x|^{a_{\min}} := \min_{a \neq a_{\min}} |x^a|,$$

By the continuity of ξ^+ on $(\xi^+)^{-1}(Z_{a_{\min}})$ proven in Theorem 5.6.2, we obtain that if K is a compact set included in $(\xi^+)^{-1}(Z_{a_{\min}})$, then there exist c_0, T such that

$$|x|^{a_{\min}}(t, y, \eta) \geq c_0(t - T), \quad (y, \eta) \in K, \quad t \geq 0. \tag{5.8.6}$$

We may then use (5.8.6), as we used Theorem 2.3.3 *(iv)* and *(iii)*, to introduce an effective time-dependent force

$$F_J(t, x) := -J\left(\frac{x}{t}\right) \nabla_x V(x),$$

where $J \in C^\infty(X)$ such that

$$\mathrm{supp} J \subset \left\{x \in X \mid |x|^{a_{\min}} \geq \frac{1}{2}c_0\right\}, \; J = 1 \text{ on } \left\{x \in X \mid |x|^{a_{\min}} \geq c_0\right\},$$

and apply the results about time-decaying forces. □

5.8.2 Long-Range Free Region Case

The study of long-range scattering for trajectories in the free region $(\xi^+)^{-1}(Z_{a_{\min}})$ can also be reduced to the case of 2-body potentials.

First note that we can construct a solution of the Hamilton-Jacobi equation in a similar way as we did in Theorem 2.7.5. We have the following analog of Theorem 2.7.5.

Theorem 5.8.2
Under the hypothesis

$$\int_0^\infty \sup_{|x^b| \geq R} |\partial^\alpha_{x^b} v^b(x^b)| \langle R \rangle^{|\alpha|-1} \mathrm{d}R, \qquad b \in \mathcal{B}, \quad |\alpha| = 1, 2, \tag{5.8.7}$$

there exists a function

$$\mathbb{R} \times X \ni (t, \xi) \mapsto S_{a_{\min}}(t, \xi) \in \mathbb{R}$$

that has the following property: for any $\epsilon > 0$, there exists T_ϵ such that

$$\partial_t S_{a_{\min}}(t, \xi) = \frac{1}{2}\xi^2 + V(\nabla_\xi S_{a_{\min}}(t, \xi)), \qquad \xi \in Z^\epsilon_{a_{\min}}, \quad t \geq T_\epsilon.$$

For any $\epsilon > 0$, the function $S_{a_{\min}}(t,\xi)$ satisfies, uniformly for $\xi \in Z^{\epsilon}_{a_{\min}}$,

$$\partial_\xi^\beta \left(S_{a_{\min}}(t,\xi) - \frac{1}{2}t\xi^2 \right) \in o(t), \qquad |\beta| \le 2.$$

Proof. The proof is almost identical to the one of Theorem 2.7.5, except that we replace sets $\{|\xi\ | > 2^{-n}\}$ with $Z^{2^{-n}}_{a_{\min}}$. □

We define now the *modified free flow* by

$$\begin{aligned} X \times Z_{a_{\min}} \ni (x,\xi) \mapsto \quad & \phi_{\mathrm{lr},a_{\min}}(t)(x,\xi) \\ & := (x + \nabla_\xi S_{a_{\min}}(t,\xi), \xi) \in X \times X'. \end{aligned} \tag{5.8.8}$$

We have now the following analog of Theorem 2.7.11.

Theorem 5.8.3
(i) Assume (5.8.7). Then the following limit exists uniformly on compact sets of $X \times Z_{a_{\min}}$

$$\lim_{t\to\infty} \phi(-t)\phi_{\mathrm{lr},a_{\min}}(t) \ =: \ \mathcal{F}^+_{\mathrm{lr},a_{\min}}, \tag{5.8.9}$$

and the following limits exist uniformly on compact sets of $(\xi^+)^{-1}(Z_{a_{\min}})$

$$\lim_{t\to\infty} (x(t,y,\eta) - \nabla_\xi S_{a_{\min}}(t,\xi(t,y,\eta))) = x^+_{\mathrm{lr},a_{\min}}(y,\eta), \tag{5.8.10}$$

$$\lim_{t\to\infty} \phi_{\mathrm{lr},a_{\min}}(t)\phi(t). \tag{5.8.11}$$

The limit in (5.8.11) is equal to $(\mathcal{F}^+_{\mathrm{lr},a_{\min}})^{-1}$. Moreover,

$$(\mathcal{F}^+_{\mathrm{lr},a_{\min}})^{-1}(y,\eta) \ := \ \left(x^+_{\mathrm{lr},a_{\min}}(y,\eta), \xi^+(y,\eta) \right).$$

The map

$$\mathcal{F}^+_{\mathrm{lr},a_{\min}} : X \times Z_{a_{\min}} :\longrightarrow (\xi^+)^{-1}(Z_{a_{\min}})$$

is bijective, continuous and called the modified free region wave transformations.
(ii) The mapping $\mathcal{F}^+_{\mathrm{lr},a_{\min}}$ is symplectic.
(iii) The modified free region wave transformation intertwines the full and free dynamics

$$\begin{aligned} H \circ \mathcal{F}_{\mathrm{lr},a_{\min}} &= H_0, \\ \phi(t) \circ \mathcal{F}_{\mathrm{lr},a_{\min}} &= \mathcal{F}^+_{\mathrm{lr},a_{\min}} \circ \phi_0(t). \end{aligned}$$

5.9 Existence of the Asymptotic External Position

The results about the asymptotic behavior of trajectories in $(\xi^+)^{-1}(Z_a)$ with $a \neq a_{\min}$ are rather limited. The components of a-clustered trajectories that are usually better behaved are x_a components. The results of this section, taken from [De7], can be viewed as poor substitutes for asymptotic completeness. (Note that asymptotic completeness can be shown in the quantum case under similar conditions on potentials as those used in this section).

5.9.1 Asymptotic External Position in the Short-Range Case

Theorem 5.9.1
Assume (5.8.1). Then for $(y,\eta) \in (\xi^+)^{-1}(Z_a)$, *there exists the limit*

$$\lim_{t\to\infty}(x_a(t,y,\eta) - t\xi_a(t,y,\eta)) \;=:\; x^+_{\mathrm{sr},a}(y,\eta). \tag{5.9.1}$$

We also have

$$\lim_{t\to\infty}(x_a(t,y,\eta) - t\xi^+_a(y,\eta)) \;=\; x^+_{\mathrm{sr},a}(y,\eta).$$

The proof is completely analogous to the proof of Theorem 2.6.3.
The observable $x^+_{\mathrm{sr},a}$ is called the *asymptotic external position.*
Note that Theorem 5.9.1 can be interpreted as follows.

Corollary 5.9.2
Let $(y,\eta) \in (\xi^+)^{-1}(Z_a)$. *Then there exists a unique trajectory* $\tilde{x}_a(t)$ *of the Hamiltonian* $\frac{1}{2}\xi_a^2$ *such that*

$$x_a(t,y,\eta) - \tilde{x}_a(t) \in o(t^0).$$

Proof. In fact, we set $\tilde{x}_a(t) = x^+_{\mathrm{sr},a}(y,\eta) + t\xi^+(y,\eta)$. □

5.9.2 Asymptotic External Position in the Long-Range Case

The external position in the long-range case is not asymptotic to free motion. It is possible to describe this asymptotics in a number of ways.

Probably the most canonical way to describe the asymptotics of the x_a components of a-clustered trajectories is to compare them with the motion generated by the following Hamiltonian on $X_a \times X'_a$:

$$h_a(x_a,\xi_a) := \frac{1}{2}\xi_a^2 + I_a(x_a).$$

This is also a generalized N-body Hamiltonian, but much simpler than $H(x,\xi)$. In particular, the set Z_a is the "free region" for $h_a(x_a,\xi_a)$. Therefore, the trajectories for $h_a(x_a,\xi_a)$ with the asymptotic velocity in Z_a are "asymptotically free" and well understood by the results of Subsect. 5.8.2.

Below we will show that the motion of an a-clustered trajectory is asymptotic to an asymptotically free trajectory generated by $h_a(x_a,\xi_a)$.

Theorem 5.9.3
Assume that

$$\int_0^\infty \sup_{|x^b|\geq R} |\partial^\alpha_{x^b} v^b(x^b)|\langle R\rangle^{|\alpha|-1+\mu}\mathrm{d}R, \qquad b\in\mathcal{B}, \quad |\alpha| = 1,2. \tag{5.9.2}$$

Assume either one of the following conditions:
(i) $(y,\eta) \in \mathcal{E}_a^+$ *and* $\mu = 0$, *or*
(ii) $(y,\eta) \in (\xi^+)^{-1}(Z_a)$ *and* $\mu = \sqrt{3}-1$.
Then there exists a unique trajectory $\tilde{x}_a(t)$ *in* X_a *for the Hamiltonian* $h_a(x_a,\xi_a)$ *such that*

$$x_a(t,y,\eta) - \tilde{x}_a(t) \in o(t^0).$$

Proof. Let $(y,\eta) \in (\xi^+)^{-1}(Z_a)$ and denote $x(t) = x(t,y,\eta)$. Let $\tilde{x}_a^1(t)$ be an arbitrary trajectory in X_a for the Hamiltonian $h_a(x_a,\xi_a)$ such that

$$\lim_{t\to\infty} \frac{\tilde{x}_a^1(t)}{t} = \xi^+(y,\eta). \tag{5.9.3}$$

Such a trajectory exists by Theorem 5.8.3, since $\xi^+(y,\eta) \in Z_a$ and Z_a is the free region for $h_a(x_a,\xi_a)$. We compute

$$\begin{aligned} \tfrac{\mathrm{d}^2}{\mathrm{d}t^2}\left(x_a(t) - \tilde{x}_a^1(t)\right) &= -\nabla_{x_a} I_a(x(t)) + \nabla_{x_a} I_a(\tilde{x}_a^1(t)) \\ &= -\left(\nabla_{x_a} I_a(x_a(t)) - \nabla_{x_a} I_a(\tilde{x}_a^1(t))\right) \\ &\quad - \left(\nabla_{x_a} I_a(x(t)) - \nabla_{x_a} I_a(x_a(t))\right). \end{aligned} \tag{5.9.4}$$

We set

$$z_a(t) := x_a(t) - \tilde{x}_a^1(t),$$

From (5.9.3) and (5.9.4), we obtain

$$|\ddot{z}_a| \le f(t)|z_a(t)| + g(t),$$
$$\lim_{t\to\infty} \dot{z}_a(t) = 0.$$

where $g(t) = f(t)|x^a(t)|$. Using (5.8.7), we obtain $\langle t\rangle f(t) \in L^1(\mathrm{d}t)$.

If $(y,\eta) \in \mathcal{E}_a^+$, then $|x^a(t)|$ is uniformly bounded. Therefore, $\langle t\rangle g(t) \in L^1(\mathrm{d}t)$.

In general, if $(y,\eta) \in (\xi^+)^{-1}(Z_a)$, then we have by Theorem 5.7.2

$$|x^a(t)| \le C\langle t\rangle^{2/(2+\mu)}.$$

We also have

$$\langle t\rangle^{1+\mu} f(t) \in L^1(\mathrm{d}t),$$

so, for $\mu = \sqrt{3}-1$, we have

$$\langle t\rangle g(t) \in L^1(\mathrm{d}t).$$

Summarizing, in both cases we have $\langle t\rangle g(t) \in L^1(\mathrm{d}t)$. We deduce from Lemma A.1.2 that

$$\lim_{t\to\infty} z_a(t)$$

exists. Using now the fact that Z_a is the free region for $h_a(x_a, \xi_a)$ and

$$\lim_{t\to\infty} \frac{\tilde{x}_a^1(t)}{t} \in Z_a,$$

we can find, as in Subsect. 5.8.2, a unique trajectory $\tilde{x}_a(t)$ for $h_a(x_a, \xi_a)$ such that

$$\lim_{t\to\infty} (\tilde{x}_a(t) - \tilde{x}_a^1(t)) = \lim_{t\to\infty} z_a(t).$$

This completes the proof of the theorem. □

An alternative method that can be applied to describe the external components of a-clustered trajectories uses solutions of the Hamilton-Jacobi equation.

If we apply Theorem 5.8.2 replacing $H(x, \xi)$ with $h_a(x_a, \xi_a)$, then we will convince ourselves that there exists a function

$$\mathbb{R} \times X_a' \ni (t, \xi_a) \mapsto S_a(t, \xi_a) \in \mathbb{R}$$

with the following property: for any $\epsilon > 0$, there exists T_ϵ such that

$$\begin{aligned} &\partial_t S_a(t, \xi_a) = \tfrac{1}{2}\xi_a^2 + I_a(\nabla_{\xi_a} S_a(t, \xi_a)), \quad t \geq T_\epsilon, \quad \xi_a \in Z_a^\epsilon, \\ &\partial_{\xi_a}^\beta (S_a(t, \xi_a) - \tfrac{1}{2}t\xi_a^2) \in o(t), \quad \xi_a \in Z_a^\epsilon, \quad |\beta| \leq 2. \end{aligned} \tag{5.9.5}$$

Theorem 5.9.4
Assume the hypotheses of Theorem 5.9.3. Then, for any $(y, \eta) \in (\xi^+)^{-1}(Z_a)$, there exists the limit

$$\lim_{t\to\infty} (x_a(t, y, \eta) - \nabla_{\xi_a} S_a(t, \xi_a(t, y, \eta))) =: x_{\mathrm{lr},a}^+(y, \eta).$$

The observable $x_{\mathrm{lr},a}^+(y, \eta)$ is called the asymptotic modified external position.

Proof. Using the Hamilton-Jacobi equation (5.9.5), we compute

$$\begin{aligned} &\tfrac{\mathrm{d}}{\mathrm{d}t}(x_a(t) - \nabla_{\xi_a} S_a(t, \xi_a(t))) \\ &= \nabla_{\xi_a}^2 S_a(t, \xi_a(t))\, (\nabla_{x_a} I_a(\nabla_{\xi_a} S_a(t, \xi_a(t)))) - (\nabla_{x_a} I_a(x(t))) \\ &= -\nabla_{\xi_a}^2 S_a(t, \xi_a(t))\, (\nabla_{x_a} I_a(x_a(t)) - \nabla_{x_a} I_a(\nabla_{\xi_a} S_a(t, \xi_a(t)))) \\ &\quad -\nabla_{\xi_a}^2 S_a(t, \xi_a(t))\, (\nabla_{x_a} I_a(x(t)) - \nabla_{x_a} I_a(x_a(t))) . \end{aligned} \tag{5.9.6}$$

As in the proof of Theorem 5.9.3, we deduce from (5.9.6) that, for

$$z_a(t) := x_a(t) - \nabla_{\xi_a} S_a(t, \xi_a(t)),$$

one has

$$|\dot{z}_a(t)| \leq f(t)|z_a(t)| + g(t) \tag{5.9.7}$$

for $f(t), g(t) \in L^1(\mathrm{d}t)$. Applying then Gronwall's inequality (A.1.2), we obtain the theorem. □

Note that an alternative way of proving Theorem 5.9.4 is to apply Theorem 5.9.3, and then Theorem 5.8.3 with $H(x, \xi)$ replaced with $h_a(x_a, \xi_a)$.

5.9.3 External Position for Regular a-Trajectories

We already saw in Sect. 5.6 that regular a-trajectories are better behaved than general a–trajectories. Here we study the regularity property of the external position.

Theorem 5.9.5
(i) Assume (5.8.1). Then the function

$$\mathcal{E}_{\mathrm{reg}}^{a,+} \ni (y,\eta) \mapsto x_{\mathrm{sr},a}^{+}(y,\eta) \tag{5.9.8}$$

is continuous and

$$\overline{(x_{\mathrm{sr},a}^{+}, \xi^{+}, H^{a,+})(\mathcal{E}_{\mathrm{reg}}^{a,+})} \supset X_a \times X_a \times \sigma_{\mathrm{reg}}^{a} \tag{5.9.9}$$

(ii) Assume (5.8.7). Then the function

$$\mathcal{E}_{\mathrm{reg}}^{a,+} \ni (y,\eta) \mapsto x_{\mathrm{lr},a}^{+}(y,\eta) \tag{5.9.10}$$

is continuous and

$$\overline{(x_{\mathrm{lr},a}^{+}, \xi^{+}, H^{a,+})(\mathcal{E}_{\mathrm{reg}}^{a,+})} \supset X_a \times X_a \times \sigma_{\mathrm{reg}}^{a}. \tag{5.9.11}$$

The proof is similar to the one of Theorem 5.6.2 and left to the reader.

5.10 Potentials of Super-Exponential Decay

In this section we consider the case of potentials that decay faster than any exponential. Only in this case, one can get a rather complete classification of all trajectories. In fact, the result of Theorem 5.10.1 below is close to what we would like to call the existence and completeness of wave transformations. This result comes from [De8]. It was inspired by a similar result about compactly supported N-body potentials that was proven in [Hu2].

Let us first introduce some notation. For any $a \in \mathcal{A}$, we denote by $\phi_{H_a}(t)$ the flow on $X \times X'$ generated by $H_a(x,\xi)$. By $\xi_{H_a}^{+}$ we denote the corresponding asymptotic velocity.

We will assume in this section that, for any $\theta > 0$, there exists C_θ such that

$$\begin{aligned} &|\nabla_{x^a} v^a(x^a)| \leq C_\theta \mathrm{e}^{-\theta|x^a|}, \quad a \in \mathcal{A}, \\ &|\nabla_{x^a}^2 v^a(x^a)| \leq C. \end{aligned} \tag{5.10.1}$$

The following theorem gives a complete classification of trajectories for pair potentials satisfying (5.10.1):

Theorem 5.10.1
Assume that (5.10.1) holds. Then one has the following results:
(i) For $(y,\eta) \in (\xi^+)^{-1}(Z_a)$*, there exists the limit*

$$\lim_{t\to\infty} \phi_{H_a}(-t)\phi(t)(y,\eta). \tag{5.10.2}$$

(ii) For $(x,\xi) \in (\xi^+_{H_a})^{-1}(Z_a)$*, there exists the limit*

$$\lim_{t\to\infty} \phi(-t)\phi_{H_a}(t)(x,\xi) \;=:\; \mathcal{F}_a^+(x,\xi). \tag{5.10.3}$$

The mapping

$$\mathcal{F}_a^+ : (\xi^+_{H_a})^{-1}(Z_a) \to (\xi^+)^{-1}(Z_a)$$

is called the $a-$region wave transformations. *The mapping (5.10.2) is equal to* $(\mathcal{F}_a^+)^{-1}$.
(iii) If $(y,\eta) = \mathcal{F}_a^+(x,\xi)$*, then one has*

$$|\phi(t)(y,\eta) - \phi_{H_a}(t)(x,\xi)| \le C_\theta e^{-\theta t}, \quad \theta > 0.$$

(iv) $\mathcal{F}_a^+$ *intertwines the dynamics of* $H(x,\xi)$ *and* $H_a(x,\xi)$

$$H \circ \mathcal{F}_a^+ = H_a, \qquad \phi(t)\circ\mathcal{F}_a^+ = \mathcal{F}_a^+ \circ \phi_{H_a}(t).$$

Remark. Let $\phi_{H^a}(t)$ denote the flow on $X^a \times X^{a'}$ generated by $H^a(x^a,\xi^a)$. By $\xi^+_{H^a}$, we denote the corresponding asymptotic velocity. Clearly, we have

$$\begin{aligned}
(\phi_{H_a}(t)(y,\eta))^a &= \phi_{H^a}(t)(y^a,\eta^a),\\
(\phi_{H_a}(t)(y,\eta))_a &= (y_a + t\eta_a, \eta_a),\\
\xi^+_{H_a}(y,\eta) &= \eta_a + \xi^+_{H^a}(y^a,\eta^a).
\end{aligned}$$

Thus Theorem 5.10.1 gives a complete classification of the $a-$clustered trajectories. Namely, they are classified by an *almost bounded or bounded* trajectory of $H^a(x^a;\xi^a)$ and a free trajectory of $\frac{1}{2}\xi_a^2$.

If we assume instead of (5.10.1) that the pair potentials have compact support, then, by Sect. 5.7, there are no almost-bounded trajectories of $H^a(x^a,\xi^a)$. In this case, the internal part of a trajectory in $(\xi^+)^{-1}(Z_a)$ is asymptotic to a *bounded* trajectory of $H^a(x^a,\xi^a)$. Of course, this result can also be deduced directly from Theorem 5.4.1. This result is due to Hunziker [Hü2], who called it the asymptotic completeness of classical N-body systems, although, as we saw, there are a number of other properties that can be called by this name.

Proof of Theorem 5.10.1. Let $(y,\eta) \in (\xi^+)^{-1}(Z_a)$. From (5.10.1) we deduce that, for any $\theta > 0$,

$$\lim_{t\to\infty} \mathrm{e}^{\theta t} \int_t^\infty |\nabla_x I_a(x(t,y,\eta))| \mathrm{d}t = 0. \tag{5.10.4}$$

From (5.10.4) we see that we can apply Corollary A.6.2 with

$$\begin{aligned} &F_1(x) = -\nabla_x V(x),\ F_2(x) = -\nabla_x V^a(x^a), \\ &x_1(t) := x(t,y,\eta), \end{aligned}$$

since $F_1(x) - F_2(x) = -\nabla_x I_a(x)$. This proves *(i)*.

Let now $(x,\xi) \in (\xi^+_{H_a})^{-1}(Z_a)$, and let $x_2(t)$ the trajectory for $H_a(x,\xi)$ starting from (x,ξ). Again from (5.10.1) we obtain

$$\lim_{t\to\infty} \mathrm{e}^{\theta t} \int_t^\infty |\nabla_x I_a(x_2(t))| \mathrm{d}t = 0.$$

We obtain now *(ii)* by exactly the same argument, exchanging the roles of F_1 and F_2.

Finally, *(iii)* follows from Corollary A.6.2 and *(iv)* is an immediate consequence of *(ii)*. □

6. Quantum N-Body Hamiltonians

6.0 Introduction

A system of N non-relativistic distinguishable particles moving in the Euclidean space $\mathbb{R}^\nu$ is described by the Hilbert space $L^2(\mathbb{R}^{N\nu})$. Its evolution is usually described by a Hamiltonian of the form

$$H := \sum_{j=1}^{N} \frac{1}{2m_j} D_j^2 + \sum_{1 \le i < j \le N} V_{ij}(x_i - x_j). \tag{6.0.1}$$

If some of the particles are identical, then only a certain subspace of $L^2(\mathbb{R}^{N\nu})$ carrying an appropriate (bosonic or fermionic) statistics describes physical states. We will not consider this question; let us only mention that the results described in this chapter can be easily modified to take into account particle statistics [De9].

If the particles are point charges, then the interaction is described by Coulomb potentials

$$V_{ij}(x) = \frac{Z_i Z_j}{|x|},$$

where Z_i are charges of the i-th particle. This class of potentials is the most important from the physical point of view; nevertheless other, usually short-range interactions are also of interest (e.g. Yukawa, van der Waals, dipole interactions).

One of the main goals of mathematical scattering theory is to give an asymptotic description of the evolution e^{-itH} for $t \to \infty$. As we will see, for a very large class of potentials, including the physically important Coulomb case, we possess a very deep and satisfactory understanding of this asymptotics.

This understanding was first reached on a heuristic level by physicists, who conjectured that an N-body system evolving according to (6.0.1) will eventually break up into independent clusters and each cluster will evolve as a bound state of the corresponding cluster Hamiltonian. This conjecture is the basis for interpretation of experimental data. In fact, suppose that we consider an experiment involving a certain number (preferably small) of particles. For definiteness, we can imagine that these particles are electrons and nuclei. During a typical experiment, these particles scatter from one another; then they move away in various directions. During scattering, it is difficult to describe the state of the system.

But after a long enough time, one can observe that the system breaks up into independent clusters such as molecules, atoms, ions or single electrons. The same picture is true if we go with time to $-\infty$.

In practice, it is difficult to measure the state of a system while it undergoes a scattering process. What experimentalists usually measure are the probabilities of obtaining the final configuration provided we know the initial configuration. Asymptotic completeness is a precise statement that describes all the possible initial and final configurations in very simple terms.

During scattering, system of particles can undergo all kinds of changes: clusters change their momenta, sometimes they change their composition, they break up or bind. In our example, where we treat electrons and nuclei as "elementary particles", an ion may capture an electron, an atom may become ionized, molecules may undergo a chemical reaction. It is remarkable that most of these phenomena, which comprise a large part of physics and chemistry, can be understood mathematically using the Hamiltonian (6.0.1). In particular, the existence and completeness of wave operators explains in a very satisfactory way why the standard description of scattering processes is correct.

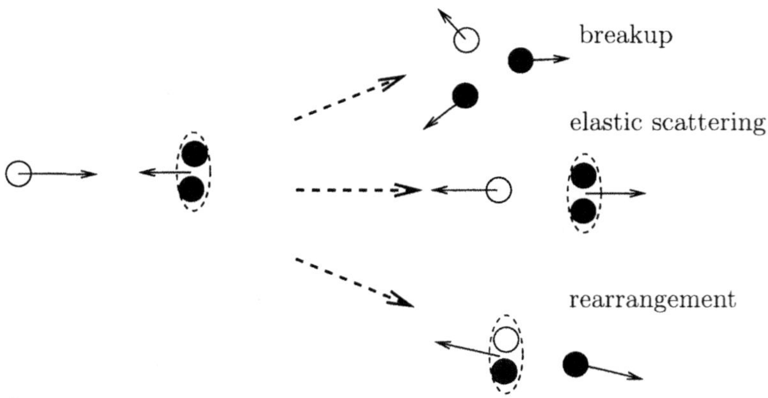

Fig. 6.1. Three possible scattering processes

We remember from Chap. 4 that, roughly speaking, typical mathematical assumptions that give a reasonable 2-body quantum scattering theory are the following:

$$\text{the short-range assumption: } |V(x)| \leq C\langle x\rangle^{-\mu}, \quad \mu > 1, \tag{6.0.2}$$

$$\text{the long-range assumption: } |\partial_x^\alpha V(x)| \leq C\langle x\rangle^{-|\alpha|-\mu}, \quad \mu > 0, \quad |\alpha| \leq 2. \tag{6.0.3}$$

In the N-body case, the situation is somewhat more complicated. The short-range assumption is sufficient to guarantee that the usual wave operators exist and are complete. The long-range assumption (6.0.2) in the N-body case with $N \geq 3$ does not guarantee the existence and completeness of modified wave operators. In fact, there are counterexamples for $\mu < 1/2$ (see [Yaf7]). The existence and completeness of wave operators for systems with an arbitrary number of particles has been shown only for $\mu > \sqrt{3} - 1 \sim 0.73$. For physical applications, this is probably sufficient, because in the nature there seem to be no potentials with a slower decay than $\mu = 1$, as in Coulomb potentials. Various other physically interesting potentials are usually short-range.

Nevertheless, N-body systems satisfying (6.0.3) have many elegant mathematical properties related to asymptotic completeness. Studying these properties, even though they may not have an immediate physical relevance, seems to clarify mathematical arguments needed to handle physically important cases.

It should be stressed, however, that N-body Hamiltonians are interesting not only because of their physical importance. We think that N-body scattering theory is also a very appealing piece of mathematics worth studying even if we disregard its physical aspects. The aim of our exposition is to present its logical structure and relationships between its various elements, and therefore in our theorems we usually try to state assumptions that are as weak as reasonably possible.

Let us now describe the contents of this chapter.

Basic definitions concerning the configuration space of N-body systems were given at the beginning of Chap. 5. In Sect. 6.1 we only recall some of them. We also define basic Hamiltonians that will be studied in this chapter.

Section 6.2 gives a short introduction to geometric methods used in the study of N-body systems. We also prove the so-called HVZ theorem, which describes the essential spectrum of H in terms of the spectra of its cluster Hamiltonians (see [RS, vol III] and references therein).

In the classical case (if the potentials are bounded), a bound on the energy yields automatically the finiteness of the velocity of trajectories. In the quantum case, there are similar results, but they are much more subtle. They are known as large velocity estimates. We studied various kinds of large velocity estimates in Chap. 4. In this chapter we will need only weak large velocity estimates, which we prove in Sect. 6.3 (see [SS1, Gr]).

One of the deepest technical results about N-body systems is the Mourre estimate. It says that, for some $C_0 > 0$,

$$\mathbb{1}_{\Delta}(H) i[H, A] \mathbb{1}_{\Delta}(H) \geq C_0 \mathbb{1}_{\Delta}(H) + \text{ compact operator}, \tag{6.0.4}$$

if Δ is an interval disjoint of thresholds of H. The Mourre estimate is used to prove various important properties of N-body Hamiltonians. Some of them, especially those related to the boundary values of the resolvent, are beyond the scope of this book (see [Mo1, PSS, JMP, Jen, CFKS, ABG]). Another consequence is the local finiteness of the pure point spectrum. We will need the Mourre estimate again in a crucial step of the proof of asymptotic completeness.

The proof of the Mourre estimate for N-body systems is given in Sect. 6.4 (see [Mo1, Mo2, PSS, FH1, CFKS]).

In Sect. 6.5 we prove the exponential decay of non-threshold eigenfunctions. The result is a consequence of the Mourre estimate. We also describe two related results. We prove that the eigenvalues do not accumulate from above at thresholds and that there are no positive eigenvalues (see [FH2, FHHO, Pe2, CFKS]).

Section 6.6 presents a number of weak propagation estimates. Using these estimates, we can show that, for long-range N-body systems satisfying (6.0.3), there exists the so-called *asymptotic velocity*

$$P^+ := \text{s}-C_\infty - \lim_{t\to\infty} \mathrm{e}^{itH}\frac{x}{t}\mathrm{e}^{-itH}.$$

We also prove that the eigenstates of P^+ with zero eigenvalue are eigenstates of H. Essentially, all the propagation estimates of Sect. 6.6 are taken from [Gr], with some modifications based on [Yaf5, De8]. The existence of the asymptotic velocity was first shown (in a different formulation) in [De6], and then reformulated to its present form in [De8]. The ideas of considering similar asymptotic observables is much older; they were used for example by Enss [E3].

With help of the results of Sect. 6.6, it is easy to show the asymptotic completeness of short-range N-body systems, which we present in Sect. 6.7. The original proof due to Sigal and Soffer [SS1] was simplified by Graf [Gr], then also by Yafaev [Yaf5]; nowadays one can argue that it is one of the most elegant and deepest pieces of mathematical physics.

Section 6.7 ends the main part of this chapter, which gives the most important and best understood elements of N-body quantum scattering theory. In fact, a full proof of asymptotic completeness for short-range N-body systems is contained in Sects. 5.1, 5.2, 6.1, 6.2, 6.3, 6.4, 6.6 and 6.7. The remaining sections of this chapter are devoted to more special topics of mathematical rather than physical interest and to the proof of asymptotic completeness for N-body long-range systems, which is technically more involved.

In Sect. 6.8 we introduce the notion of the asymptotic separation of a dynamics. We say that a dynamics is asymptotically separated with respect to the factorization $L^2(X) = L^2(X_a)\otimes L^2(X^a)$ if it can be approximated for large times by a dynamics that acts independently in $L^2(X_a)$ and $L^2(X^a)$. An example of a Hamiltonian that generates a separated dynamics is the cluster Hamiltonian H_a. Other examples are $H_{[a]}$ and $H_{a-\text{sep}}$, which are defined in this section. In the short-range case, e^{-itH} can be asymptotically approximated by a the dynamics e^{-itH_a}, $\mathrm{e}^{-itH_{[a]}}$ and $\mathrm{e}^{-itH_{a-\text{sep}}}$ on some large subspaces of the Hilbert space. The problem becomes much more difficult in the long-range case and is closely related to asymptotic completeness. As we will see later on, in the long-range case with $\mu > \sqrt{3}-1$, we will be able to use $\mathrm{e}^{-itH_{a-\text{sep}}}$ to approximate the full dynamics. This result will serve as the key element of our proof of asymptotic completeness in the long-range case.

Sometimes it is convenient to replace the full Hamiltonian H by a simpler many-body Hamiltonian with a time-decaying perturbation. This trick, which goes back to Sigal and Soffer, is described in Sect. 6.9.

N-body systems satisfying (6.0.3) have a quite well behaved scattering theory, even though asymptotic completeness may fail. This we already saw in Sect. 6.6, where we proved the existence of the asymptotic velocity and we showed some of its properties. Section 6.10 can be regarded as a supplement to Sect. 6.6. In this section we give a complete description of the joint spectrum $\sigma(P^+, H)$. We also describe some large subspaces on which asymptotic separation holds.

In Sect. 6.11 we describe asymptotic clustering. This property involves a simplified effective dynamics that commutes with the external momentum. This property was first proven by Sigal and Soffer [SS2] for $\mu = 1$; then it was extended to the $\mu > 1/2$ case [DeGe1]. It is related to Dollard wave operators (see Sects. 3.6 and 4.8). It implies the so-called asymptotic absolute continuity, which is a certain property of the spectral measure of P^+ [De6, De8].

The remaining part of this chapter is devoted to a proof of the asymptotic completeness of long-range N-body systems. The most technical elements of this proof are contained in Sects. 6.12 and 6.13. It is based on [De8]. It is technically convenient to use the framework of effective time-dependent Hamiltonians of the form $\check{H}(t) = H + W(t,x)$. We also use various special observables such as $R(x)$ and $r(x)$, which were constructed in Sect. 5.2. Characteristic for these sections is the use of functions of x/t^δ rather than of x/t, which was typical for the previous sections.

Recall from Sect. 5.2 that $r(x)$ was a deformation of the function $|x|$ that took into account the geometry of configuration space. A special role in our analysis is played by the operator

$$b_{\chi,t} = \tfrac{1}{2}\chi(H)(D\nabla r(\tfrac{x}{t^\delta}) + \nabla r(\tfrac{x}{t^\delta})D)\chi(H).$$

In Sect. 6.12 we show that in the Heisenberg picture this observable possesses a limit, which we call $\check{b}_\chi^+$, and which is equal to $|\check{P}^+|\chi^2(H)$.

In Sect. 6.13 we concentrate our attention on the states for which the asymptotic velocity is zero, and hence that belong to the kernel of $\check{b}_\chi^+$. A priori, such states may spread not faster than $o(t)$. We show that in fact they spread not faster than $O(t^\delta)$ with $\delta = 2(2+\mu)^{-1}$. Note that an analogous statement was proven in the case of classical N-body systems (see Theorem 5.7.2). Nevertheless, the result in the quantum case is more difficult to show.

In Sect. 6.14 we prove that if $\mu > \sqrt{3} - 1$, then asymptotic separation holds, that is, on a certain subspace of the Hilbert space defined in terms of the spectrum of the asymptotic velocity, there exists a relative wave operator

$$\text{s-}\lim_{t\to\infty} \mathrm{e}^{itH_{a-\mathrm{sep}}}\mathrm{e}^{-itH}, \tag{6.0.5}$$

where the Hamiltonian $H_{a-\mathrm{sep}}$ has the property

$$H_{a-\mathrm{sep}} = h_a \otimes \mathbb{1} + \mathbb{1} \otimes H^a.$$

The proof of the existence of (6.0.5) is based on the Cook method. We need to show that a certain expression of the order

$$\text{force along the trajectory} \times \text{size of the cluster}$$

is integrable in time. The force decays as $t^{-1-\mu}$ and, as follows from Sect. 6.13, the size of a cluster grows not faster than t^{δ}, where $\delta = 2(2+\mu)^{-1}$. Hence we obtain the condition

$$\mu > 2(2+\mu)^{-1},$$

which is satisfied for $\mu > \sqrt{3} - 1$.

N-body scattering theory has a long and interesting history. Below we will try to describe the main contributions to its development, focusing on the problem of the existence and completeness of wave operators.

The formulation of N-body scattering theory using wave operators and a proof of the orthogonality of channels was given by Jauch [Jau]. The existence of wave operators for a class of short-range potentials was proven by Hack [Hack], and then extended to potentials with local singularities by Hunziker [Hu1].

The reader will find a review of most results of the earlier phase of N-body scattering theory in the monograph of Reed and Simon [RS, vol III].

Whereas the existence of N-body wave operators (under some restrictive conditions, which in the long-range case involved also bound states) was relatively easy to show, the problem of asymptotic completeness for a long time remained open. The first attempts to prove it were made by Faddeev. Faddeev used clever resolvent identities (named afterwards Faddeev equations) and the stationary method to study asymptotic completeness for a certain class of short-range potentials in dimension 3 or bigger for 3-body systems [Fa]. Unfortunately, his method required to impose certain implicit assumptions on the potentials (the absence of zero-energy resonances and bound states for subsystems).

After Faddeev the stationary method was developed by Ginibre and Moulin [GM] and Thomas [Th] for 3-body systems, by Hagedorn [Ha] for 4-body systems and by Sigal [Sig1] for N-body systems. All of these papers had the same drawback, namely, implicit assumptions. The only exception was the work of Loss and Sigal [LoSig], which contained a stationary proof of the asymptotic completeness of a certain (rather small) class of 3-body systems without implicit assumptions.

One should also mention proofs of asymptotic completeness for some special N-body systems. Iorio and O'Carrol proved asymptotic completeness for small potentials in 3 or more dimensions [IoO'C]. Lavine proved asymptotic completeness for a class of repulsive N-body potentials [La2]. Note, however, that, under the conditions of these two theorems, only the free channel is open.

Scattering for the energies below the lowest 3-cluster threshold is also relatively simple and asymptotic completeness can be shown in this region quite easily, see e.g. [Comb].

With the advent of time-dependent and geometric methods, mathematical physicists have essentially abandoned the stationary approach based on resolvent identities, such as the Faddeev equation, in the study of N-body scattering.

Some of the early successes in mathematical N-body scattering theory are associated with the name of E. Mourre. He found an abstract theory, that started with a pair of operators A and H satisfying the estimate (6.0.4), and led to a number of results about the spectral properties of H and estimates on the boundary value of the resolvent of H. Mourre proved that the hypotheses of this theory were satisfied by 3-body systems [Mo1, Mo2], then Perry, Sigal and Simon showed that they were true also for an arbitrary number of particles [PSS].

The first big success in the quest for a rigorous proof of the asymptotic completeness of N-body systems was achieved by V. Enss. Using his time-dependent approach [E1, E2], he proved the asymptotic completeness of 3-body systems in the short-range and then in the long-range case with $\mu > \sqrt{3}-1$ [E5, E6]. His proofs were valid essentially for the same class of potentials that we cover here. They were based on the analysis of the phase space properties of propagation.

In 1985, Sigal and Soffer announced a proof of the asymptotic completeness of short-range systems with an arbitrary number of particles [SS1]. (Their proof was preceded by a seminal paper of Mourre and Sigal [MoSig]). The proof of Sigal and Soffer was time-dependent, but its philosophy was different from that of Enss. Enss relied heavily on the so-called RAGE theorem (see [RS, vol III]) and on the compactness of various operators. The basic idea of Sigal and Soffer, which proved very fruitful also in the further development, was the following: find a bounded observable whose Heisenberg derivative is positive. Then the expectation value of any observable dominated by this derivative is integrable in time. This idea could be traced back to Putnam and Kato (see Appendix B.4 for references). Nevertheless, it was Sigal and Soffer who first showed its great flexibility and used it to prove a variety of propagation estimates.

A new proof of the propagation theorem, which was the key ingredient of Sigal's and Soffer's proof, was given in [De2].

In 1989, a very elegant and simple proof of N-body asymptotic completeness for short-range potentials was found by Graf [Gr]. The basic strategy of Graf was that of Sigal and Soffer – to find observables with a positive Heisenberg derivative. In the case of Sigal and Soffer, those observables were mostly functions of

$$\gamma = \frac{1}{2}\left(\frac{x}{\langle x\rangle}D + D\frac{x}{\langle x\rangle}\right).$$

Graf found a new observable, which was essentially the Heisenberg derivative of a modification of x^2/t taking into account in a very clever way the geometry of the configuration space. Another feature of Graf's proof was its complete time-dependence. Sigal and Soffer used conical cutoffs and the local decay estimate

$$\int_0^\infty \|\langle x\rangle^{-\frac{1}{2}-\epsilon}\mathrm{e}^{-itH}\phi\|^2 \mathrm{d}t < \infty, \quad \epsilon > 0, \tag{6.0.6}$$

which followed from the Mourre estimate, whereas Graf used cutoffs of the form

$$J\left(\frac{x}{t}\right)$$

and did not have to use the local decay estimate.

The problem that remained open was the asymptotic completeness of long-range systems with 4 or more particles. A number of very interesting but somewhat technical results that were valid for long-range N-body Hamiltonians were obtained by Sigal and Soffer in [SS2, SS3]. One of them was the method of strong propagation estimates, that we described in Chap. 4 (where it was used in the case of 2-body Hamiltonians; it clearly generalizes to the N-body case). Another result was the so-called asymptotic clustering, proven for $\mu = 1$ in [SS2] and extended to the case $\mu > 1/2$ in [DeGe1].

It was typical for the proofs of the asymptotic completeness of short-range N-body systems that they used the short-range condition only in the last step and essentially all the propagation estimates that were needed were valid under the long-range assumption. In particular, the long-range condition was sufficient to show the existence of the limits of various observables in the Heisenberg picture, even without knowing if wave operators exist. These limits generate a natural commutative C^*-algebra. It is possible to describe the spectrum of this algebra. This approach to scattering theory, which was less satisfactory than the existence and completeness of wave operators but still preserved the basic physical picture that is associated with scattering, was proposed in [De6]. Technically, it was based on propagation estimates of [Gr] (although those of [SS1] were used in an earlier version of this work, see [De5]). Later on, in [De8], those results were reformulated using the vector of commuting self-adjoint operator P^+ called the asymptotic velocity, instead of commutative C^*-algebras.

The approach of Graf to asymptotic completeness for short-range N-body systems (or to the existence of the asymptotic velocity, as in Sect. 6.6) was based on the construction of a certain observable, which was the Heisenberg derivative of a distortion of

$$x^2/t.$$

Yafaev found a similar proof, which used the Heisenberg derivative of a distortion of $|x|$ [Yaf5]. Yafaev's proof used only time-independent observables. Because of that, unlike in [Gr], Yafaev had to use the local decay estimate (6.0.6), which can be regarded as an extra complication. On the other hand, Yafaev's approach yielded an extra bonus – it could be used in the framework of the stationary scattering theory to study eigenfunction expansions and wave operators of N-body systems [Yaf6].

One can say that the proof of the asymptotic completeness of 3-body long-range systems of [E5, E6] was based on considering a certain asymptotic effective Hamiltonian H^+, which enabled to localize "bad" states at thresholds. Sigal and Soffer tried to extend this idea to the 4-body Coulomb case in [SS3]. Nevertheless, it turned out that the right approach to prove the asymptotic completeness of long-range systems with 4 or more particles was to localize in x/t, that is, in the asymptotic velocity. This idea was applied in [De8], which contained the first proof of the asymptotic completeness of long-range N-body systems. The proof was valid for any number of particles for $\mu > \sqrt{3} - 1$, which was the same borderline as in the work of Enss. It also gave a heuristic argument why it should be difficult to extend the result to slower decaying potentials.

Let us mention other papers on the problem of the asymptotic completeness of N-body systems: [Ki6, Tam, SS5, Zie].

Apart from asymptotic completeness, there are various other problems about N-body scattering that one can study. In particular, one can ask whether the eigenfunction expansions are well defined, or (which is essentially equivalent) what are the properties of the kernel of wave operators. One can also study the regularity of the scattering matrix. These questions usually involve a study of the boundary value of the resolvent. We do not consider these questions in our monograph. Among the papers devoted to this subject, let us mention [Yaf6, Sk1, HeSk2, GeISk, I4, I5, I6, I7, Bom, Va].

3-body systems can be regarded as an intermediate case between the 2-body and N-body systems. For 3-body systems, the validity of the asymptotic completeness was pushed to $\mu > 1/2$ in [Ge4] if we impose some additional virial-type assumptions on the potentials and the spherical symmetry. The main step of the proof involves a refined analysis of a one-dimensional 2-body Hamiltonian in the presence of a time-decaying potential.

Positive potentials also give the opportunity to prove asymptotic completeness for a slower decay in 3-body systems, see [Wa4].

A counterexample to asymptotic completeness for 3-body systems with potentials satisfying (6.0.3) with $\mu < 1/2$ was given by Yafaev [Yaf7]. His counterexample involves a construction of an additional non-standard wave operator.

The traditional scattering theory regards the Laplacian as the unperturbed Hamiltonian. It is also possible to consider other types of scattering, with other exactly solvable operators serving as free Hamiltonians.

For example, N charged particles in the constant electric field $\mathbf{E}$ are described by the N-body Stark Hamiltonian

$$H := \sum_{j=1}^{N} \frac{1}{2m_j} D_j^2 + \sum_{j=1}^{N} q_j \mathbf{E} \cdot x_j + \sum_{1 \leq i < j \leq N} V_{ij}(x_i - x_j).$$

Scattering theory for such Hamiltonians was studied in [AT, HeMSk].

N charged particles in the presence of a constant magnetic field $\mathbf{B}$ are described by a Hamiltonian of the form

$$H := \sum_{j=1}^{N} \frac{1}{2m_j} (D_j + q_j A x_j)^2 + \sum_{1 \leq i < j \leq N} V_{ij}(x_i - x_j),$$

where Ax is the vector potential of $\mathbf{B}$. Scattering theory for such Hamiltonians was studied in [GeLa1, GeLa2, GeLa3].

Scattering theory for N-body Hamiltonians in combined constant magnetic and electric fields was studied in [Sk2, Sk3].

In relativistic and solid state physics, one sometimes considers the so-called dispersive N-body Hamiltonians

$$H := \sum_{j=1}^{N} \omega_j(D_j) + \sum_{1\leq i<j\leq N} V_{ij}(x_i - x_j).$$

Here $\omega_j(\xi_j)$ is the kinetic energy of the jth particle, which does not have to be a quadratic form. Scattering theory for 2-body dispersive systems is very similar to the usual non-dispersive case, described e.g. in Chap. 4. Scattering theory for N-body dispersive systems is still poorly understood, see [De3, Ge2].

6.1 Basic Definitions

Basic definitions and facts about Hilbert spaces are given in Appendix B.1. Basic definitions about the geometry of the N-body configuration space X are given in Sect. 5.1. Section 5.2 contains constructions of certain observables related to N-body systems that will be used in this chapter. This section will be devoted to some basic definitions about N-particle quantum Hamiltonians.

Most of the time, we will work in the Hilbert space $L^2(X)$, where X is a Euclidean space having the meaning of the configuration space of an N-body system. Recall from Sect. 5.1 that X is equipped with families of subspaces

$$\{X_b \mid b \in \mathcal{B}\} \subset \{X_a \mid a \in \mathcal{A}\}.$$

We will denote by D, D_a and D^a the momentum operators $i^{-1}\nabla$, $i^{-1}\nabla_a$ and $i^{-1}\nabla^a$ on $L^2(X)$ respectively. Likewise, $\Delta = -D^2$, $\Delta_a = -D_a^2$ and $\Delta^a = -(D^a)^2$ will denote the Laplacians corresponding to the variables x, x_a and x^a respectively.

We will now introduce the definition of a (generalized) quantum N-particle Hamiltonian. We assume that, for every $b \in \mathcal{B}$, we are given a real function (called a potential) $X^b \ni x^b \mapsto v^b(x^b)$. We will always assume that

$$v^b(x^b)(1-\Delta^b)^{-1} \text{ is compact on } L^2(X^b),\ b \in \mathcal{B}. \tag{6.1.1}$$

In other words, for every $b \in \mathcal{B}$, the potential $v^b(x^b)$ is a relatively compact perturbation of $-\Delta^b$. We will also assume that

$$v^{a_{\min}}(x^{a_{\min}}) = 0.$$

We set

$$V(x) := \sum_{b\in\mathcal{B}} v^b(x^b), \tag{6.1.2}$$

$$V^a(x^a) := \sum_{b\leq a} v^b(x^b).$$

Note that the hypotheses (6.1.1) implies that $V(x)$ and $V^a(x^a)$ are bounded relatively to $-\frac{1}{2}\Delta$ with the infinitesimal bound. Using Kato-Rellich's theorem (see Theorem B.1.3), we can introduce the following definition.

Definition 6.1.1
The Hamiltonian on $L^2(X)$

$$H := \frac{1}{2}D^2 + V(x)$$

with domain $H^2(X)$*, where* $V(x)$ *is given by (6.1.2), is self-adjoint and bounded from below. It is called a (generalized) many-body Hamiltonian.*

Similarly we define *clustered Hamiltonians*

$$H_a := \frac{1}{2}D^2 + V^a(x), \quad a \in \mathcal{A}.$$

Clearly, $H = H_{a_{\max}}$.

We may identify $L^2(X)$ with $L^2(X_a) \otimes L^2(X^a)$. Then we can write

$$H_a = -\frac{1}{2}\Delta_a \otimes 1 + 1 \otimes H^a,$$

where

$$H^a := -\frac{1}{2}\Delta^a + V^a(x^a)$$

is a self-adjoint operator on $L^2(X^a)$ called a *reduced clustered Hamiltonian* with domain $\mathcal{D}(\Delta^a)$.

For any $a \in \mathcal{A}$, the clustered Hamiltonian H_a and the reduced clustered Hamiltonian H^a are examples of generalized many-body Hamiltonians. Their configuration spaces are X and X^a respectively. Their families of subspaces are $\{X_b \mid b \in \mathcal{B},\ b \le a\}$ and $\{X^a \cap X_b \mid b \in \mathcal{B},\ b \le a\}$ respectively.

Note that $X^{a_{\min}} = \{0\}$, $L^2(X^{a_{\min}}) = \mathbb{C}$, $V^{a_{\min}}(x^{a_{\min}}) = 0$ and $H^{a_{\min}} = 0$.

If we set $I_a(x) := V(x) - V^a(x)$, we have:

$$H = H_a + I_a(x).$$

Obviously, instead of studying H it is sufficient to study the *reduced Hamiltonian* $H^{a_{\max}}$. Clearly, $H^{a_{\max}}$ is an N-body Hamiltonian acting on the space $L^2(X^{a_{\max}})$ associated with the Euclidean space $X^{a_{\max}}$ and the family of vector subspaces $\{X^{a_{\max}} \cap X_a \mid a \in \mathcal{A}\}$. We can always replace H with $H^{a_{\max}}$ and X with $X^{a_{\max}}$. If we do this we guarantee that

$$X_{a_{\max}} = \{0\}. \tag{6.1.3}$$

This procedure is a generalization of the separation of the center-of-mass motion for standard N-body Hamiltonians, which was described in Sect. 5.1. Most of our results are valid without the assumption (6.1.3). Sometimes it will be convenient to impose this assumption, which we will always state explicitly.

We will sometimes prove certain properties $P(a)$ of the Hamiltonians H_a using the induction with respect to the semi-lattice $\mathcal{A}$. The logical structure of such a proof is the following:

$$(\forall_{a\in\mathcal{A}}(\forall_{b<a}P(b)\Rightarrow P(a)))\Rightarrow\forall_{a\in\mathcal{A}}P(a).$$

Every $a\in\mathcal{A}$ can be renamed $a_{\max}$. Therefore it is actually sufficient to show

$$\forall_{a<a_{\max}}P(a)\Rightarrow P(a_{\max}),$$

which is notationally more convenient.

6.2 HVZ Theorem

In this section we just assume (6.1.1).

A very important role in the analysis of N-particle systems is played by the following set of energy levels, called the set of *thresholds.*

Definition 6.2.1
The set of thresholds of a subsystem $a\in\mathcal{A}$ is defined as

$$\mathcal{T}^a:=\bigcup_{b<a}\sigma_{\mathrm{pp}}(H^b).$$

The set $\mathcal{T}^{a_{\max}}$ will be simply denoted by $\mathcal{T}$. Moreover, we set $\Sigma^a:=\inf(\mathcal{T}^a)$ and $\Sigma:=\Sigma^{a_{\max}}=\inf(\mathcal{T})$.

Note that $\sigma_{\mathrm{pp}}(H^{a_{\min}})=\{0\}$. Hence $\Sigma\leq 0$.

Let us now prove the following fundamental result, known as the *Hunziker-Van Winter-Zhislin theorem.* Its proof can be considered as a good introduction to geometric methods in the study of N-particle systems.

Theorem 6.2.2
Assume (6.1.1). Let H be an N-particle Hamiltonian with $X_{a_{\max}}=\{0\}$. Then the essential spectrum of H is equal to

$$\sigma_{\mathrm{ess}}(H)=[\Sigma,\infty[.$$

From Theorem 6.2.2 we obtain the following picture of the spectrum of a typical N-particle Hamiltonian.

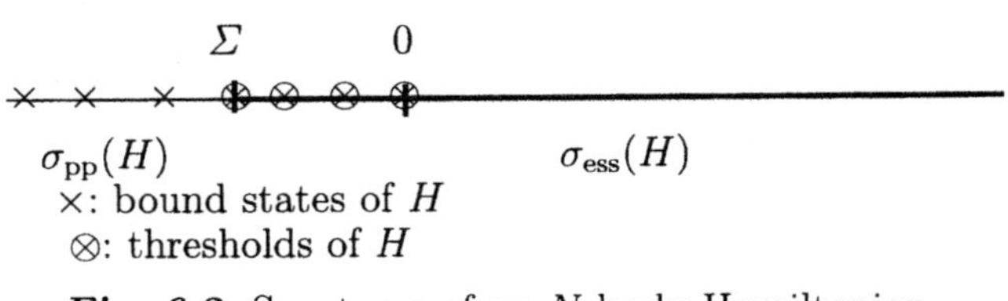

Fig. 6.2. Spectrum of an N-body Hamiltonian

Before the proof of the above theorem, let us state some general properties of N-body Hamiltonians that hold under the assumptions (6.1.1) and will be useful throughout his chapter. The first lemma describes properties that can be proven exactly as in the 2-body case.

Lemma 6.2.3
Assume (6.1.1). Let $\chi \in C_0^\infty(\mathbb{R})$ and $J \in C^\infty(X)$ such that ∇J is bounded. Then

$$\left[\chi(H), J\left(\frac{x}{t}\right)\right](H+i) \in O(t^{-1}), \tag{6.2.1}$$

$$\langle x\rangle^s \chi(H)\langle x\rangle^{-s} \text{ is bounded for any } s \in \mathbb{R}, \tag{6.2.2}$$

$$1\!\!1_{[1,\infty[}\left(\frac{|x^b|}{t}\right) v^b(x^b)(1-\Delta^b)^{-1} \in o(t^0), \quad b \in \mathcal{B}. \tag{6.2.3}$$

The next lemma says that on Y_a we can in some weak sense approximate the full Hamiltonian H by H_a.

Lemma 6.2.4
Let $\chi \in C_0^\infty(\mathbb{R})$, $\delta > 0$, $J \in C^1(X)$ such that $\mathrm{supp}J \subset Y_a^\delta$ *and $\partial_x^\alpha J$, $|\alpha| = 0, 1$ are bounded. Then*

$$(\chi(H) - \chi(H_a))\, J(\tfrac{x}{t})(H+i) \in o(t^0).$$

Proof. Let $\tilde{\chi}$ be an almost-analytic extension of χ constructed in Proposition C.2.1. Then

$$\begin{aligned}&(\chi(H) - \chi(H_a))\, J(\tfrac{x}{t})(H+i)\\ &= \tfrac{i}{2\pi}\int\limits_{\mathbb{C}} \tfrac{\partial\tilde{\chi}}{\partial\bar{z}}(z)(z-H)^{-1}I_a(x)(z-H_a)^{-1}J(\tfrac{x}{t})(H+i)\mathrm{d}z\wedge\mathrm{d}\bar{z}.\end{aligned}$$

and

$$\begin{aligned}&(z-H)^{-1}I_a(x)(z-H_a)^{-1}J(\tfrac{x}{t})\\ &= (z-H)^{-1}I_a(x)J(\tfrac{x}{t})(z-H_a)^{-1}\\ &+(z-H)^{-1}I_a(x)(z-H_a)^{-1}[H_a, J(\tfrac{x}{t})](z-H_a)^{-1}.\end{aligned} \tag{6.2.4}$$

The first term on the right of (6.2.4) is $o(t^0)$ by (6.2.3), the second is $O(t^{-1})$, which completes the proof of the lemma. □

We will often use Lemmas 6.2.3 and 6.2.4 without quoting them.

Next let us recall that in Sect. 5.2 we constructed a "smooth partition of unity" $q_a(x)$ satisfying, for some $\epsilon, \delta > 0$,

$$\begin{aligned} &\mathrm{supp} q_a \subset Z_a^{\epsilon,\delta}, \quad \sum_{a\in\mathcal{A}} q_a(x) = 1, \\ &|\partial_x^\alpha q_a(x)| \leq C_\alpha, \quad 0 \leq q_a(x) \leq 1. \end{aligned} \tag{6.2.5}$$

We will also need a closely related family of functions

$$\tilde{q}_a(x) := \frac{q_a(x)}{(\sum_{b\in\mathcal{A}} q_b^2(x))^{\frac{1}{2}}}. \tag{6.2.6}$$

They have the following properties:

$$\begin{aligned} &\mathrm{supp} \tilde{q}_a \subset Z_a^{\epsilon,\delta}, \quad \sum_{a\in\mathcal{A}} \tilde{q}_a^2(x) = 1, \\ &|\partial_x^\alpha \tilde{q}_a(x)| \leq C_\alpha, \quad 0 \leq \tilde{q}_a(x) \leq 1. \end{aligned}$$

Let B be an operator. The following formula is sometimes useful:

$$B = \sum_{a\in\mathcal{A}} \tilde{q}_a(x) B \tilde{q}_a(x) + \frac{1}{2} \sum_{a\in\mathcal{A}} [\tilde{q}_a(x), [\tilde{q}_a(x), B]]. \tag{6.2.7}$$

In particular, if $B = -\Delta$ then we get the so-called IMS localization formula [CFKS]:

$$-\Delta = -\sum_{a\in\mathcal{A}} \tilde{q}_a(x) \Delta \tilde{q}_a(x) - \frac{1}{2} \sum_{a\in\mathcal{A}} |\nabla \tilde{q}_a(x)|^2. \tag{6.2.8}$$

Proof of Theorem 6.2.2. Let us first prove that $\sigma_{\mathrm{ess}}(H) \subset [\Sigma, \infty[$.

We prove by the induction with respect to $a \in \mathcal{A}$ that, for any a,

$$\sigma_{\mathrm{ess}}(H^a) \subset [\Sigma^a, \infty[. \tag{6.2.9}$$

It is enough to consider $a_{\max}$ and to assume that (6.2.9) is known for $a < a_{\max}$.

Let $\chi \in C_0^\infty(\mathbb{R})$ with $\mathrm{supp}\chi \subset]-\infty, \Sigma[$. Let $q_a(x)$ be the partition of unity satisfying (6.2.5). We have

$$\begin{aligned} \chi(H) &= \sum_{a\in\mathcal{A}} \chi(H) q_a(\tfrac{x}{c}) \\ &= \sum_{a\in\mathcal{A}} \chi(H_a) q_a(\tfrac{x}{c}) + o(c^0), \end{aligned} \tag{6.2.10}$$

using Lemma 6.2.4.

On the other hand, by the induction assumption, for any $a < a_{\max}$, we clearly have

$$\sigma(H^a) = \sigma_{\mathrm{pp}}(H^a) \cup [\Sigma^a, \infty[\subset [\Sigma, \infty[.$$

But

$$\sigma(H_a) = \sigma(H^a) + \mathbb{R}^+,$$

where + denotes the algebraic sum of subsets of $\mathbb{R}$. Therefore,

$$\sigma(H_a) \subset [\Sigma, \infty[,$$

which implies that all the terms in the second line of (6.2.10) vanish for $a \neq a_{\max}$. The remaining term is compact. Letting c tend to ∞, we see that $\chi(H)$ is compact for any χ with supp$\chi \subset]-\infty, \Sigma[$, which shows that $\sigma_{\mathrm{ess}}(H) \subset [\Sigma, \infty[$.

Let us now prove the converse inclusion by constructing suitable Weyl's sequences for a given energy level $\lambda \geq \Sigma$. Let $a \in \mathcal{A}$, $a \neq a_{\max}$ such that $\Sigma = \inf \sigma_{\mathrm{pp}}(H^a)$.

First we note that we will find a sequence of vectors $u_n \in \mathcal{D}(H^a)$ and $\rho_n \to \infty$ such that

$$\begin{aligned} &\|u_n\| = 1, \\ &\lim_{n\to\infty} (H^a - \Sigma) u_n = 0, \qquad (6.2.11) \\ &u_n = 1\!\!1_{[0,2]}(\tfrac{|x^a|}{\rho_n}) u_n. \end{aligned}$$

In fact, since $\Sigma \in \sigma(H^a)$, there exists a sequence $\tilde{u}_n \in \mathcal{D}(H^a)$ satisfying the first two properties of (6.2.11). Let $F \in C_0^\infty(\mathbb{R})$ such that $F = 1$ on $[0,1]$ and $F = 0$ on $[2, \infty[$. Replacing $\tilde{u}_n$ by

$$c_n F(\tfrac{|x^a|}{\rho_n}) \tilde{u}_n$$

for suitable sequences $\rho_n \to \infty$ and $c_n \to 1$, we may guarantee that also the third property of (6.2.11) holds.

Next we pick $y_a \in Z_a$ such that

$$\{x \mid |x^a| \leq 2,\ |x_a - y_a| \leq 2\} \subset Y_a.$$

We claim that we will find a sequence of vectors $v_n \in \mathcal{D}(\frac{1}{2}D_a^2)$ such that

$$\begin{aligned} &\|v_n\| = 1, \\ &\lim_{n\to\infty} (-\tfrac{1}{2}D_a^2 - \lambda + \Sigma) v_n = 0, \\ &1\!\!1_{[0,2]}(|\tfrac{x_a}{\rho_n} - y_a|) v_n = v_n. \end{aligned}$$

In fact, if $\omega_a \in X_a'$ with $|\omega_a| = 1$, we can set

$$v_n(x_a) := c_n F(|\tfrac{x_a}{\rho_n} - y_a|) \mathrm{e}^{i\sqrt{(2\lambda - 2\Sigma)}\langle x_a, \omega_a \rangle},$$

where c_n is such that $\|v_n\| = 1$. We put

$$\phi_n(x) := v_n(x_a) \otimes u_n(x^a).$$

Note that $\|\phi_n\| = 1$, $\|H\phi_n\| \in O(1)$ and ϕ_n tend weakly to 0. We have

$$\begin{aligned} (H - \lambda)\phi_n &= I_a(x)\phi_n \\ &+ \Big((-\tfrac{1}{2}\Delta_a - \lambda + \Sigma) v_n\Big) \otimes u_n + v_n \otimes ((H^a - \Sigma) u_n). \end{aligned} \qquad (6.2.12)$$

Now, using (6.2.3), we get

$$I_a(x)\phi_n = I_a(x) 1\!\!1_{[0,2]}(|\tfrac{x_a}{\rho_n} - y_a|) 1\!\!1_{[0,2]}(\tfrac{|x^a|}{\rho_n}) (H+i)^{-1} (H+i)\phi_n \to 0.$$

Therefore,

$$(H-\lambda)\phi_n \to 0.$$

Consequently, ϕ_n is a Weyl's sequence for the energy level λ, which proves that $\lambda \in \sigma_{\mathrm{ess}}(H)$ (see [RS, vol. I, Thm. VII.12]). □

6.3 Weak Large Velocity Estimates

In this section we will show large velocity estimates for N-body Hamiltonians. They are similar to those proved in Sect. 4.2 for 2-body Hamiltonians.

In the 2-body case, the maximal velocity is essentially equal to $\sqrt{2H}$. This comes from the fact that, for 2-body Hamiltonians, the kinetic energy of a particle far away from the origin is very close to its total energy. For N-particle Hamiltonians, it is possible to extract more kinetic energy from the system by building bounded clusters of particles. Therefore, the maximal allowed velocity is in general larger in the N-body case.

Proposition 6.3.1
Assume (6.1.1). Let $\chi \in C_0^\infty(\mathbb{R})$ *and* $\mathrm{supp}\chi \subset]-\infty, \frac{1}{2}\theta_1^2 + \Sigma[$.
(i) Let $\theta_1 < \theta_2$. *Then*

$$\int_1^\infty \left\| \mathbb{1}_{[\theta_1,\theta_2]}\left(\frac{|x|}{t}\right)\chi(H)\phi_t \right\|^2 \frac{\mathrm{d}t}{t} \le C\|\phi\|^2.$$

(ii) We have

$$\mathrm{s-}\lim_{t\to\infty} \mathbb{1}_{[\theta_1,\infty]}\left(\frac{|x|}{t}\right)\chi(H)\mathrm{e}^{-itH} = 0.$$

The proof of the above proposition follows by exactly the same arguments as in the proof of Proposition 4.2.1 except that it uses Lemma 6.3.2 instead of Lemma 4.2.2.

Lemma 6.3.2
Assume (6.1.1). Let $\chi \in C_0^\infty(\mathbb{R})$ *with* $\mathrm{supp}\chi \subset [-\infty, \frac{1}{2}\theta^2 + \Sigma[$ *and* $0 \le \chi \le 1$. *Let* $f \in C_0^\infty(\mathbb{R})$, $0 \le f \le 1$ *and* $0 \notin \mathrm{supp}f$. *Then one has*

$$\left\| \chi(H) D\frac{x}{|x|} f\left(\frac{|x|}{t}\right) \right\| \le \theta + o(t^0). \tag{6.3.1}$$

Proof. We write

$$\begin{aligned}\chi(H)D\tfrac{x}{|x|}f(\tfrac{|x|}{t}) &= \textstyle\sum_{a\in\mathcal{A}}\chi(H)D\tfrac{x}{|x|}q_a(\tfrac{x}{\epsilon t})f(\tfrac{|x|}{t})\\ &= \textstyle\sum_{a\neq a_{\max}}\chi(H)D\tfrac{x}{|x|}q_a(\tfrac{x}{\epsilon t})f(\tfrac{|x|}{t}),\end{aligned}$$

for ϵ small enough, using the fact that $0 \notin \mathrm{supp} f$. Next we write

$$\begin{aligned}\|\chi(H)D\tfrac{x}{|x|}q_a(\tfrac{x}{\epsilon t})f(\tfrac{|x|}{t})\| &\leq \|\chi(H)D_a\tfrac{x_a}{|x|}q_a(\tfrac{x}{\epsilon t})f(\tfrac{|x|}{t})\|\\ &\quad+\|\chi(H)D^a\tfrac{x^a}{|x|}q_a(\tfrac{x}{\epsilon t})f(\tfrac{|x|}{t})\|.\end{aligned}\tag{6.3.2}$$

We have

$$\|\chi(H)D^a\tfrac{x^a}{|x|}q_a(\tfrac{x}{\epsilon t})f(\tfrac{|x|}{t})\| \leq \|\chi(H)D^a\|\|\tfrac{x^a}{|x|}q_a(\tfrac{x}{\epsilon t})\| \in O(\epsilon).$$

Using then the direct sum representation

$$H_a \simeq \int_{X'_a}^{\oplus}(\tfrac{1}{2}\xi_a^2 + H^a)\mathrm{d}\xi_a,$$

we obtain that

$$\|\chi(H_a)D_a\| \leq \theta_1$$

for some $\theta_1 < \theta$. Therefore,

$$\begin{aligned}\|\chi(H)D_a\tfrac{x_a}{|x|}q_a(\tfrac{x}{\epsilon t})f(\tfrac{|x|}{t})\| &= \|\chi(H_a)D_a\tfrac{x_a}{|x|}q_a(\tfrac{x}{\epsilon t})f(\tfrac{|x|}{t})\| + o_\epsilon(t^0)\\ &\leq \theta_1 + o_\epsilon(t^0).\end{aligned}$$

Choosing ϵ small enough in (6.3.2) this gives (6.3.1). □

6.4 The Mourre Estimate

This section is devoted to a proof of the *Mourre estimate*, which plays a very important role in the study of N-particle Hamiltonians. The Mourre estimate was first proved in [Mo1] for $N = 3$, and in [PSS] for general N. Our proof follows essentially [FH1, CFKS].

In this section we will consider N-body Hamiltonians satisfying

$$X_{a_{\max}} = \{0\},\tag{6.4.1}$$

i.e. we assume that the center-of-mass motion has been separated. We assume (6.1.1) and

$$(1-\Delta^b)^{-1}x^b\nabla_{x^b}v^b(x^b)(1-\Delta^b)^{-1} \text{ is compact on } L^2(X^b),\ b\in\mathcal{B}.\tag{6.4.2}$$

We recall from Sect. 4.3 that we denote by A the generator of dilations

$$A := \frac{1}{2}\left(\langle x, D\rangle + \langle D, x\rangle\right).$$

Note that if $a \in \mathcal{A}$, we have

$$A = A_a \otimes 1\!\!1 + 1\!\!1 \otimes A^a,$$

where

$$A^a := \frac{1}{2}\left(\langle x^a, D^a\rangle + \langle D^a, x^a\rangle\right), \quad A_a := \frac{1}{2}\left(\langle x_a, D_a\rangle + \langle D_a, x_a\rangle\right)$$

are the generators of dilations along the spaces X^a and X_a respectively. If $a = a_{\min}$, we put $A^{a_{\min}} = 0$ acting on $L^2(X^{a_{\min}}) = \mathbb{C}$.

Note also that it follows immediately from (6.1.1) and (6.4.2) that

$$(H+i)^{-1}[H, iA](H+i)^{-1} \text{ is bounded.} \tag{6.4.3}$$

The main result of this section is the following theorem.

Theorem 6.4.1
Let H be an N-particle Hamiltonian satisfying the hypotheses (6.1.1), (6.4.1), (6.4.2). Then the following results hold:
(i) For all $\lambda_1 \leq \lambda_2$ such that $[\lambda_1, \lambda_2] \cap \mathcal{T} = \emptyset$, we have

$$\dim 1\!\!1^{\mathrm{pp}}_{[\lambda_1,\lambda_2]}(H) < \infty.$$

Consequently, $\sigma_{\mathrm{pp}}(H)$ can accumulate only at $\mathcal{T}$. Moreover, $\mathcal{T}$ and $\mathcal{T} \cup \sigma_{\mathrm{pp}}(H)$ are closed countable sets.
(ii) For $\lambda \in [\Sigma, \infty[$, let

$$d(\lambda) := \inf\{\lambda - \tau \mid \tau \leq \lambda, \tau \in \mathcal{T}\}.$$

Then for any $\epsilon > 0$, $\lambda \in [\Sigma, \infty[$, there exists an open interval Δ containing λ and a compact operator K such that

$$1\!\!1_\Delta(H)[H, iA]1\!\!1_\Delta(H) \geq 2\left(d(\lambda) - \epsilon\right) 1\!\!1_\Delta(H) + K. \tag{6.4.4}$$

(iii) For any $\epsilon > 0$, $\lambda \in [\Sigma, \infty[$ there exists an open interval Δ containing λ such that

$$1\!\!1^{\mathrm{c}}_\Delta(H)[H, iA]1\!\!1^{\mathrm{c}}_\Delta(H) \geq 2\left(d(\lambda) - \epsilon\right) 1\!\!1^{\mathrm{c}}_\Delta(H). \tag{6.4.5}$$

Note the following easy consequence of Theorem 6.4.1 *(i)*:

Corollary 6.4.2

$$1\!\!1^{\mathrm{pp}}(H) = 1\!\!1_{\mathcal{T}\cup\sigma_{\mathrm{pp}}(H)}(H). \tag{6.4.6}$$

The remaining part of this section will be devoted to the proof of Theorem 6.4.1. For any $\lambda \in \mathbb{R}$ and $\kappa \geq 0$, we define

$$\Delta_\lambda^\kappa := [\lambda - \kappa, \lambda + \kappa].$$

If $a \in \mathcal{A}$, $\lambda \geq \Sigma^a$ and $\kappa \geq 0$, then we set

$$d^a(\lambda) := \inf\{\lambda - \tau \mid \tau \leq \lambda, \tau \in \mathcal{T}^a\},$$

$$d^{a,\kappa}(\lambda) := \inf\{d^a(\lambda_1) \mid \lambda_1 \in \Delta_\lambda^\kappa\},$$

$$\tilde{d}^a(\lambda) := \begin{cases} d^a(\lambda) & \lambda \in [\Sigma^a, \infty[\backslash \sigma^{\rm pp}(H^a) \\ 0 & \lambda \in [\Sigma^a, \infty] \cap \sigma^{\rm pp}(H^a) \end{cases},$$

$$\tilde{d}^{a,\kappa}(\lambda) := \inf\{\tilde{d}^a(\lambda_1) \mid \lambda_1 \in \Delta_\lambda^\kappa\}.$$

Note that $d(\lambda) = d^{a_{\max}}(\lambda)$. Likewise, we will write $d^\kappa(\lambda)$, $\tilde{d}(\lambda)$ $\tilde{d}^\kappa(\lambda)$ instead of $d^{a_{\max},\kappa}(\lambda)$, $\tilde{d}^{a_{\max}}(\lambda)$ $\tilde{d}^{a_{\max},\kappa}(\lambda)$.

Note also that $d^{a,0}(\lambda) = d^a(\lambda)$ and $\tilde{d}^{a,0}(\lambda) = \tilde{d}^a(\lambda)$.

The proof will use the induction with respect to $a \in \mathcal{A}$. Let us list the statements that we will show:

$H_1(a)$ *Let* $\epsilon > 0$, $\lambda \in [\Sigma^a, \infty[$. *Then there exists an operator* K^a *compact on* $L^2(X^a)$ *and an open interval* Δ *containing* λ *such that*

$$1\!\!1_\Delta(H^a)[H^a, iA^a]1\!\!1_\Delta(H^a) \geq 2\,(d^a(\lambda) - \epsilon)\,1\!\!1_\Delta(H^a) + K^a.$$

$H_2(a)$ *Let* $\epsilon > 0$, $\lambda \in [\Sigma^a, \infty[$. *Then there exists an open interval* Δ *containing* λ *such that*

$$1\!\!1_\Delta(H^a)[H^a, iA^a]1\!\!1_\Delta(H^a) \geq 2\left(\tilde{d}^a(\lambda) - \epsilon\right)1\!\!1_\Delta(H^a).$$

$H_3(a)$ *Fix* $\lambda_0 \in \mathbb{R}$, $\kappa > 0$ *and* $\epsilon > 0$. *Then there exists* $\delta > 0$ *such that, for any* $\lambda \in [\Sigma^a, \lambda_0[$, *we have*

$$1\!\!1_{\Delta_\lambda^\delta}(H^a)[H^a, iA^a]1\!\!1_{\Delta_\lambda^\delta}(H^a) \geq 2\left(\tilde{d}^{a,\kappa}(\lambda) - \epsilon\right)1\!\!1_{\Delta_\lambda^\delta}(H^a).$$

$S_1(a)$ $\mathcal{T}^a$ *is a closed countable set.*

$S_2(a)$ *For all* $\lambda_1 \leq \lambda_2$ *such that* $[\lambda_1, \lambda_2] \cap \mathcal{T}^a = \emptyset$, *we have*

$$\dim 1\!\!1^{\rm pp}_{[\lambda_1,\lambda_2]}(H^a) < \infty.$$

Consequently, $\sigma^{\rm pp}(H^a)$ *can accumulate only at* $\mathcal{T}^a$. *Moreover,* $\mathcal{T}^a \cup \sigma_{\rm pp}(H^a)$ *is a closed countable set.*

Note that $S_1(a_{\max})$, $S_2(a_{\max})$ and $H_1(a_{\max})$ are the statements *(i)* and *(ii)* of Theorem 6.4.1.

We will show, for any $a \in \mathcal{A}$, the following implications.

$$(S_1(a) \text{ and } (\forall_{b<a} H_3(b))) \Rightarrow H_1(a), \tag{6.4.7}$$

$$(\forall_{b<a} S_2(b)) \Rightarrow S_1(a), \tag{6.4.8}$$

$$H_1(a) \Rightarrow S_2(a), \tag{6.4.9}$$

$$H_1(a) \Rightarrow H_2(a), \tag{6.4.10}$$

$$H_2(a) \Rightarrow H_3(a). \tag{6.4.11}$$

As explained at the end of Sect. 6.1, it is enough to show these statements for $a = a_{\max}$, which we will do in the sequel.

Note that the implication (6.4.8) is obvious. The implication (6.4.9) follows exactly as in Theorem 4.3.3 using the virial theorem.

Lemma 6.4.3
Let $\lambda \in]-\infty, \Sigma[$. Then there exists an open interval Δ containing λ such that

$$\mathbb{1}_\Delta(H)[H, iA]\mathbb{1}_\Delta(H) = 0.$$

Proof. The virial theorem says that, for any $\lambda \in \mathbb{R}$,

$$\mathbb{1}_{\{\lambda\}}(H)i[H, A]\mathbb{1}_{\{\lambda\}}(H) = 0.$$

So it is enough to take Δ such that $\Delta \cap \sigma^{\mathrm{pp}}(H) = \{\lambda\}$. □

Lemma 6.4.4
The implication $H_1(a_{\max}) \Rightarrow H_2(a_{\max})$ is true.

Proof. By $H_1(a_{\max})$, we can find $\Delta_1 \ni \lambda$ and a compact operator K_1 such that

$$\mathbb{1}_{\Delta_1}(H)[H, iA]\mathbb{1}_{\Delta_1}(H) \geq 2\left(d(\lambda) - \epsilon/2\right)\mathbb{1}_{\Delta_1}(H) + K_1. \tag{6.4.12}$$

If $\lambda \notin \sigma_{\mathrm{pp}}(H)$, then s$-\lim \mathbb{1}_\Delta(H) = 0$ for $\Delta \searrow \{\lambda\}$. By the compactness of K, we will find an open interval $\Delta \ni \lambda$ such that

$$\mathbb{1}_\Delta(H)K_1\mathbb{1}(H) \geq -\epsilon\mathbb{1}_\Delta(H). \tag{6.4.13}$$

Now since $\tilde{d}(\lambda) = d(\lambda)$, (6.4.12) and (6.4.13) imply

$$\mathbb{1}_{\Delta_1}(H)[H, iA]\mathbb{1}_{\Delta_1}(H) \geq 2\left(\tilde{d}(\lambda) - \epsilon\right)\mathbb{1}_{\Delta_1}(H).$$

Now assume that $P := \mathbb{1}_{\{\lambda\}}(H) \neq 0$. Using the compactness of K_1, we pick a finite rank projection F such that $F \leq P$ and

$$\|(1-P)K_1(1-P) - (1-F)K_1(1-F)\| \leq \frac{\epsilon}{2}. \tag{6.4.14}$$

We have

$$\begin{aligned}\mathbb{1}_{\Delta_1}(H)[H,iA]\mathbb{1}_{\Delta_1}(H) &= \mathbb{1}_{\Delta_1}(H)(1-F)[H,iA](1-F)\mathbb{1}_{\Delta_1}(H)\\ &+\mathbb{1}_{\Delta_1}(H)P[H,iA]P\mathbb{1}_{\Delta_1}(H)\\ &-\mathbb{1}_{\Delta_1}(H)(1-F)P[H,iA]P(1-F)\mathbb{1}_{\Delta_1}(H)\\ &+\mathbb{1}_{\Delta_1}(H)F[H,iA](1-P)\mathbb{1}_{\Delta_1}(H)+\mathrm{hc}\\ &=: R_1+R_2+R_3+R_4+R_4^*.\end{aligned} \tag{6.4.15}$$

Composing (6.4.12) to the left and right by $(1-F)$ and using (6.4.14), we obtain

$$\begin{aligned}R_1 &\geq 2(d(\lambda)-\tfrac{\epsilon}{2})\mathbb{1}_{\Delta_1}(H)(1-F)+(1-F)K_1(1-F)\\ &\geq -\epsilon\mathbb{1}_{\Delta_1}(H)(1-F)+(1-P)K_1(1-P)-\tfrac{\epsilon}{2}.\end{aligned} \tag{6.4.16}$$

It follows then from the virial theorem (Theorem 4.3.3) that

$$R_2 = R_3 = 0. \tag{6.4.17}$$

Finally,

$$\begin{aligned}R_4+R_4^* &= F^*R_4+R_4^*F\\ &\geq -\epsilon F^*F-\epsilon^{-1}R_4^*R_4\\ &\geq -\epsilon-\epsilon^{-1}(1-P)\mathbb{1}_{\Delta_1}(H)K_2\mathbb{1}_{\Delta_1}(H)(1-P),\end{aligned} \tag{6.4.18}$$

where

$$K_2 = \mathbb{1}_{\Delta_1}(H)[H,iA]F[H,iA]\mathbb{1}_{\Delta_1}(H)$$

is a compact operator using (6.4.3) and the fact that F is a finite rank projection. Collecting (6.4.16), (6.4.17) and (6.4.18), we obtain

$$\begin{aligned}&\mathbb{1}_{\Delta_1}(H)[H,iA]\mathbb{1}_{\Delta_1}(H)\\ &\geq -\epsilon\mathbb{1}_{\Delta_1}(H)-\tfrac{\epsilon}{2}+(1-P)(K_1+K_2)(1-P).\end{aligned} \tag{6.4.19}$$

Using the compactness of K_1+K_2, the fact that $\mathbb{1}_{\Delta}(H)(1-P)$ tends strongly to 0 when Δ tends to $\{\lambda\}$, and composing (6.4.19) to the left and right by $\mathbb{1}_{\Delta}(H)$ with Δ sufficiently small, we obtain

$$\mathbb{1}_{\Delta}(H)[H,iA]\mathbb{1}_{\Delta}(H) \geq -2\epsilon\mathbb{1}_{\Delta}(H).$$

Since $\tilde{d}(\lambda)=0$, this completes the proof of the lemma. □

Next we give a uniform version of the above lemma.

Lemma 6.4.5
The implication $H_2(a_{\max}) \Rightarrow H_3(a_{\max})$ is true.

Proof. By $H_2(a_{\max})$, for every $\lambda \in \mathbb{R}$, we will find $\delta(\lambda)$ such that $\kappa > \delta(\lambda)$ and

$$\mathbb{1}_{\Delta_\lambda^{\delta(\lambda)}}(H)[H, iA]\mathbb{1}_{\Delta_\lambda^{\delta(\lambda)}}(H) \geq 2\left(\tilde{d}(\lambda) - \epsilon\right)\mathbb{1}_{\Delta_\lambda^{\delta(\lambda)}}(H). \tag{6.4.20}$$

We will find a finite sequence of λ_i, $i = 1, \ldots, N$ such that $\Delta_{\lambda_i}^{\delta_i/2}$ cover $[\Sigma, \lambda_0]$, where $\delta_i := \delta(\lambda_i)$. Thus we have

$$\mathbb{1}_{\Delta_{\lambda_i}^{\delta_i}}(H)[H, iA]\mathbb{1}_{\Delta_{\lambda_i}^{\delta_i}}(H) \geq 2\left(\tilde{d}(\lambda_i) - \epsilon\right)\mathbb{1}_{\Delta_{\lambda_i}^{\delta_i}}(H). \tag{6.4.21}$$

We set

$$\delta := \frac{1}{2}\min\{\delta(\lambda_i) \mid i = 1, \ldots, N\}.$$

Now let $\lambda \in [\Sigma, \lambda_0]$. Then for some $i = 1, \ldots, N$, we have $\Delta_\lambda^\delta \subset \Delta_{\lambda_i}^{\delta_i}$. Moreover, $\tilde{d}^\kappa(\lambda) \leq \tilde{d}(\lambda_i)$. Hence

$$\begin{aligned}\mathbb{1}_{\Delta_\lambda^\delta}(H)[H, iA]\mathbb{1}_{\Delta_\lambda^\delta}(H) &\geq 2\left(\tilde{d}(\lambda_i) - \epsilon\right)\mathbb{1}_{\Delta_\lambda^\delta}(H) \\ &\geq 2\left(\tilde{d}^\kappa(\lambda) - \epsilon\right)\mathbb{1}_{\Delta_\lambda^\delta}(H).\end{aligned}$$

□

Lemma 6.4.6
Let $\delta > 0$, $J \in C^\infty(X)$ such that $\mathrm{supp} J \subset Y_a^\delta$ *and $\partial_x^\alpha J$ are bounded. Then*

$$(1 - \Delta)^{-1}J(\tfrac{x}{R})x\nabla_x I_a(x)(1 - \Delta)^{-1} \in o(R^0).$$

Proof. Let $b \in \mathcal{B}$, $F \in C^\infty(\mathbb{R})$, $F = 0$ around 0 and $F = 1$ around ∞. Clearly,

$$[F(\tfrac{|x^b|}{R}), (1 - \Delta^b)^{-1}](1 - \Delta^b) \in O(R^{-1}).$$

Therefore,

$$\begin{aligned}&(1 - \Delta^b)^{-1}F(\tfrac{|x^b|}{R})x^b\nabla_{x^b}v^b(x^b)(1 - \Delta^b)^{-1} \\ &= F(\tfrac{|x^b|}{R})(1 - \Delta^b)^{-1}x^b\nabla_{x^b}v^b(x^b)(1 - \Delta^b)^{-1} + O(R^{-1}) \in o(R^0),\end{aligned} \tag{6.4.22}$$

using the fact that $F(\frac{|x^b|}{R})$ tends strongly to 0 when R tends to ∞.

Now,

$$\begin{aligned}&(1 - \Delta)^{-1}J(\tfrac{x}{R})x\nabla_x I_a(x)(1 - \Delta)^{-1} \\ &= \sum_{b \not\leq a}(1 - \Delta)^{-1}J(\tfrac{x}{R})x^b\nabla_{x^b}v^b(x^b)(1 - \Delta)^{-1} \\ &= \sum_{b \not\leq a}(1 - \Delta)^{-1}J(\tfrac{x}{R})F(\tfrac{|x^b|}{\epsilon R})x^b\nabla_{x^b}v^b(x^b)(1 - \Delta)^{-1}\end{aligned} \tag{6.4.23}$$

for some $\epsilon > 0$. It follows from (6.4.22) that (6.4.23) is $o(R^0)$. □

The geometric argument needed for the proof of Theorem 6.4.1 is the following expansion of the commutator $[H, iA]$. It is related to the expansion (6.2.10) used in the proof of Theorem 6.2.2.

Lemma 6.4.7
Let $\chi \in C_0^\infty(\mathbb{R})$. *Then*

$$\chi(H)[H,iA]\chi(H) = \sum_{a\in\mathcal{A}} \tilde{q}_a(\frac{x}{R})\chi(H_a)[H_a,iA]\chi(H_a)\tilde{q}_a(\frac{x}{R}) + o(R^0), \tag{6.4.24}$$

$$\chi^2(H) = \sum_{a\in\mathcal{A}} \tilde{q}_a(\frac{x}{R})\chi^2(H_a)\tilde{q}_a(\frac{x}{R}) + o(R^0). \tag{6.4.25}$$

Proof. Using (6.2.7), we have

$$\begin{aligned}\chi(H)[H,iA]\chi(H) &= \sum_{a\in\mathcal{A}} \tilde{q}_a(\tfrac{x}{R})\chi(H_a)[H_a,iA]\chi(H_a)\tilde{q}_a(\tfrac{x}{R})\\ &+ \sum_{a\in\mathcal{A}} \tilde{q}_a(\tfrac{x}{R})(\chi(H)-\chi(H_a))[H_a,iA]\chi(H_a)\tilde{q}_a(\tfrac{x}{R})\\ &+ \sum_{a\in\mathcal{A}} \tilde{q}_a(\tfrac{x}{R})\chi(H)[H_a,iA](\chi(H)-\chi(H_a))\tilde{q}_a(\tfrac{x}{R})\\ &+ \sum_{a\in\mathcal{A}} \tilde{q}_a(\tfrac{x}{R})\chi(H_a)[I_a(x),iA]\chi(H_a)\tilde{q}_a(\tfrac{x}{R})\\ &+ \sum_{a\in\mathcal{A}} [\tilde{q}_a(\tfrac{x}{R}),[\tilde{q}_a(\tfrac{x}{R}),\chi(H)[H,iA]\chi(H)]].\end{aligned} \tag{6.4.26}$$

The second and the third term on the right-hand side of (6.4.26) are $o(R^0)$ by Lemma 6.2.4. The fifth is $O(R^{-2})$. To handle the fourth term, note that

$$\begin{aligned}&\tilde{q}_a(\tfrac{x}{R})\chi(H_a)[I_a(x),iA]\chi(H_a)\\ &= \chi(H_a)\tilde{q}_a(\tfrac{x}{R})[I_a(x),iA]\chi(H_a) + O(R^{-1}),\end{aligned}$$

which is $o(R^0)$ by Lemma 6.4.6. This proves (6.4.24). To prove (6.4.25), we write

$$\begin{aligned}\chi^2(H) &= \sum_{a\in\mathcal{A}} \tilde{q}_a(\tfrac{x}{R})\chi^2(H_a)\tilde{q}_a(\tfrac{x}{R})\\ &+ \sum_{a\in\mathcal{A}} \tilde{q}_a(\tfrac{x}{R})(\chi^2(H)-\chi^2(H_a))\tilde{q}_a(\tfrac{x}{R})\\ &+ \sum_{a\in\mathcal{A}} [\tilde{q}_a(\tfrac{x}{R}),[\tilde{q}_a(\tfrac{x}{R}),\chi^2(H)]],\end{aligned} \tag{6.4.27}$$

and we use a similar argument. □

Lemma 6.4.8
The implication

$$(S_1(a_{\max}) \text{ and } (\forall_{b<a_{\max}} H_3(b))) \Rightarrow H_1(a_{\max})$$

is true.

Proof. Using the closedness of $\mathcal{T}$ (that is, the statement $S_1(a_{\max})$), we easily see that

$$d(\lambda) = \sup_{\kappa>0} d^\kappa(\lambda).$$

Hence we will find $\kappa > 0$ such that

$$d^\kappa(\lambda) \geq d(\lambda) - \frac{\epsilon}{3}.$$

Moreover, it is easy to see that

$$\min_{a<a_{\max}} \inf\{\tilde{d}^{a,\kappa}(\lambda - \lambda_1) + \lambda_1 \mid \lambda - \lambda_1 \in [\Sigma^a, \lambda]\} = d^\kappa(\lambda).$$

By $H_3(a)$, we will find $\delta > 0$ such that, for any $\lambda - \lambda_1 \in [\Sigma^a, \lambda]$, we have

$$\begin{aligned} &\mathbb{1}_{\Delta_\lambda^\delta}(H^a + \lambda_1)[H^a, iA^a]\mathbb{1}_{\Delta_\lambda^\delta}(H^a + \lambda_1) \\ &\geq 2\left(\tilde{d}^{a,\kappa}(\lambda - \lambda_1) - \tfrac{\epsilon}{3}\right)\mathbb{1}_{\Delta_\lambda^\delta}(H^a + \lambda_1). \end{aligned} \tag{6.4.28}$$

Therefore,

$$\begin{aligned} &\mathbb{1}_{\Delta_\lambda^\delta}(H_a)[H_a, iA]\mathbb{1}_{\Delta_\lambda^\delta}(H_a) \\ &= \int_{X'_a}^{\oplus} \mathbb{1}_{\Delta_\lambda^\delta}(H^a + \tfrac{1}{2}\xi_a^2)([H^a, iA^a] + \xi_a^2)\mathbb{1}_{\Delta_\lambda^\delta}(H^a + \tfrac{1}{2}\xi_a^2)\mathrm{d}\xi_a \\ &\geq \int_{X'_a}^{\oplus} \mathbb{1}_{\Delta_\lambda^\delta}(H^a + \tfrac{1}{2}\xi_a^2)2(\tilde{d}^{a,\kappa}(\lambda - \tfrac{1}{2}\xi_a^2) + \tfrac{1}{2}\xi_a^2 - \tfrac{\epsilon}{3})\mathrm{d}\xi_a \\ &\geq 2(d^\kappa(\lambda) - \tfrac{\epsilon}{3})\mathbb{1}_{\Delta_\lambda^\delta}(H_a) \geq 2(d(\lambda) - \tfrac{2\epsilon}{3})\mathbb{1}_{\Delta_\lambda^\delta}(H_a). \end{aligned}$$

Now choose $\chi \in C_0^\infty(\mathbb{R})$ such that $\chi = 1$ around λ and $\mathrm{supp}\chi \subset \Delta_\lambda^\delta$. Then, using Lemma 6.4.7, we get

$$\begin{aligned} \chi(H)[H, iA]\chi(H) &= \sum_{a\in\mathcal{A}} \tilde{q}_a(\tfrac{x}{R})\chi(H_a)[H_a, iA]\chi(H_a)\tilde{q}_a(\tfrac{x}{R}) + o(R^0) \\ &\geq (d(\lambda) - \tfrac{2\epsilon}{3}) \sum_{a\in\mathcal{A}} \tilde{q}_a(\tfrac{x}{R})\chi^2(H_a)\tilde{q}_a(\tfrac{x}{R}) + o(R^0) + K_R \\ &= (d(\lambda) - \tfrac{2\epsilon}{3})\chi^2(H) + o(R^0) + K_R. \end{aligned}$$

where

$$\begin{aligned} K_R &= q_{a_{\max}}(\tfrac{x}{R})\chi(H)[H, iA]\chi(H)q_{a_{\max}}(\tfrac{x}{R}) \\ &\quad -2(d(\lambda) - \tfrac{2\epsilon}{3})\tilde{q}_{a_{\max}}(\tfrac{x}{R})\chi^2(H)\tilde{q}_{a_{\max}}(\tfrac{x}{R}). \end{aligned}$$

Using the fact that $\tilde{q}_{a_{\max}}(x)$ is compactly supported since $X_{a_{\max}} = \{0\}$ and the boundedness of $(i + H)^{-1}[H, iA](i + H)^{-1}$, we see that K_R is a compact operator for any R. Choosing R big enough we obtain

$$\chi(H)[H, iA]\chi(H) \geq (d(\lambda) - \tfrac{2\epsilon}{3})\chi^2(H) - \tfrac{\epsilon}{3} + K_R. \tag{6.4.29}$$

Finally, we multiply both sides of (6.4.29) by $\mathbb{1}_\Delta(H)$ such that $\chi = 1$ on Δ and Δ is a neighborhood of λ, and we get (6.4.4) with

$$K = \mathbb{1}_\Delta(H)K_R\mathbb{1}_\Delta(H).$$

□

6.5 Exponential Decay of Eigenfunctions and Absence of Positive Eigenvalues

There are basically two approaches to the study of the exponential decay of eigenfunctions of N-body Hamiltonians. One of them, due to Agmon [Ag2], is applicable in the case of eigenfunctions below the essential spectrum. It gives a very good understanding of the rate of decay of eigenfunctions in various directions of the configuration space. We will not describe this approach, since it is of a limited use in the case of embedded eigenvalues.

In order to study the exponential decay of eigenfunctions with embedded eigenvalues, one needs to apply commutator techniques. Early results and references on this subject are described in [RS, vol IV]. A very precise understanding of uniform exponential upper bounds has been achieved by Froese-Herbst [FH2] (see also [CFKS]). We describe this result in Theorem 6.5.1 below. Note, however, that the understanding of direction-dependent exponential upper bounds for eigenfunctions with embedded eigenvalues is still missing (see [De1, De4] and the references therein).

Modifying the arguments of Theorem 6.5.1, one can show that eigenvalues cannot accumulate from above at thresholds (although they can accumulate from below). This result is due to Perry [Pe2]. We describe it in Theorem 6.5.3.

Finally, if we impose slightly stronger assumptions on the potentials, then, following [FH2, CFKS], we show in Theorem 6.5.4 that there are no positive eigenvalues. (See also [FHHO] for related results).

Theorem 6.5.1
Let H be an N-particle Hamiltonian satisfying the hypotheses (6.1.1), (6.4.2). Let $\psi \in H^2(X)$ such that $H\psi = E\psi$. Let

$$\tau := \sup\{\tfrac{1}{2}\theta^2 + E \mid \theta \geq 0,\ \mathrm{e}^{\theta|x|}\psi \in L^2(X)\}. \tag{6.5.1}$$

Then $\tau \in \mathcal{T} \cup \{\infty\}$. Moreover,

$$\tau = \sup\{\tfrac{1}{2}\theta^2 + E \mid \theta \geq 0,\ \mathrm{e}^{\theta|x|}\psi \in H^2(X)\}. \tag{6.5.2}$$

Definition 6.5.2
For any $\tau \in \mathcal{T} \cup \{\infty\}$, we define inductively a closed subspace $\mathcal{E}_\tau \subset \mathrm{Ran}\,\mathbb{1}^{\mathrm{pp}}(H)$ as follows. Suppose that we defined $\mathcal{E}_{\tau_1}$ for all $\tau_1 > \tau$, $\tau \in \mathcal{T} \cup \{\infty\}$. Then $\mathcal{E}_\tau$ is spanned by the vectors

$$\psi \in (\cup_{\tau_1 > \tau} \mathcal{E}_{\tau_1})^\perp \quad \textit{such that, for some } E, \quad H\psi = E\psi$$

$$\textit{and} \quad \tau = \sup\{\tfrac{1}{2}\theta^2 + E \mid \theta \geq 0,\ \mathrm{e}^{\theta|x|}\psi \in L^2(X)\}.$$

Clearly,

$$\mathcal{E}_\tau \subset \mathrm{Ran}\,\mathbb{1}^{\mathrm{pp}}_{]-\infty,\tau]}(H).$$

Moreover, by Theorem 6.5.1 we have

$$\mathrm{Ran}\,\mathbb{1}^{\mathrm{pp}}(H) = \sum_{\tau\in\mathcal{T}\cup\{\infty\}}^{\oplus} \mathcal{E}_\tau.$$

Theorem 6.5.3
Let $\lambda < \tau \in \mathcal{T}\cup\{\infty\}$. Then

$$\dim \mathbb{1}_{]-\infty,\lambda]}(H)\mathcal{E}_\tau < \infty.$$

Consequently, the eigenvalues of H can only accumulate at thresholds from below.

Note that

$$\mathcal{E}_\infty = \{\psi \in \mathrm{Ran}\,\mathbb{1}^{\mathrm{pp}}(H) \mid \mathrm{e}^{\theta|x|}\psi \in L^2(X), \quad \theta \in \mathbb{R}\}.$$

Theorem 6.5.4
Assume (6.1.1), (6.4.2). Assume, in addition, that

$$\limsup_{\lambda\to\infty} \|(\lambda - \tfrac{1}{2}\Delta)^{-\frac{1}{2}} x\nabla_x V(x)(\lambda - \tfrac{1}{2}\Delta)^{-\frac{1}{2}}\| < 1. \tag{6.5.3}$$

Then $\mathcal{E}_\infty = \{0\}$. Consequently, the Hamiltonian H has no positive eigenvalues.

The proof of Theorem 6.5.1 will be broken up into a number of steps. First let us consider $F \in C^\infty(X)$ such that

$$F(x) = f(|x|),\ f' \geq 0,\ |\partial_x^\alpha F(x)| \leq C_\alpha \langle x\rangle^{1-|\alpha|}, \quad |\alpha| \geq 0. \tag{6.5.4}$$

Let $g(x)$ be defined by

$$\nabla F(x) = xg(x),$$

so that by (6.5.4)

$$g \geq 0,\ |\partial_x^\alpha g(x)| \leq C_\alpha \langle x\rangle^{-1-|\alpha|}, \quad |\alpha| \geq 0. \tag{6.5.5}$$

Let us define the operator

$$\begin{aligned} H_F &:= \tfrac{1}{2}(D + i\nabla F)^2 + V(x) \\ &= H - \tfrac{1}{2}(\nabla F)^2 + \tfrac{i}{2}(\langle D, \nabla F\rangle + \langle \nabla F, D\rangle). \end{aligned} \tag{6.5.6}$$

Let us list some properties of H_F:

Lemma 6.5.5
(i) The operator

$$-\tfrac{1}{2}(\nabla F)^2 + \tfrac{i}{2}(\langle D, \nabla F\rangle + \langle \nabla F, D\rangle)$$

is $-\Delta$ bounded with a zero bound. Therefore H_F with domain $H^2(X)$ is a closed operator.

(ii)

$$H_F^* = H_{-F}.$$

(iii) As forms on $C_0^\infty(X)$,

$$H_F = \mathrm{e}^F H \mathrm{e}^{-F}.$$

In the following lemma we collect some properties of $\mathrm{e}^F\psi$ where ψ is an eigenfunction of H.

Lemma 6.5.6
Let $\psi \in H^2(X)$ such that $H\psi = E\psi$, and $\psi_F := \mathrm{e}^F\psi \in L^2(X)$. Then

(i) $\|\psi_F\|_{H^2(X)} \le C\|\psi_F\|$,

(ii) $H_F\psi_F = E\psi_F$,

(iii) $(\psi_F|[H, iA]\psi_F) = -\|g^{1/2}(x)A\psi_F\|^2 - (\psi_F|G(x)\psi_F)$,
where $G(x) = \frac{1}{2}\left(x\cdot\nabla(\nabla F)^2(x) - (x\cdot\nabla)^2 g(x)\right)$,

(iv) $(\psi_F|H\psi_F) = E\|\psi_F\|^2 - \frac{1}{2}(\psi_F|(\nabla F)^2\psi_F)$,

where C in (i) depends only on the constants in (6.5.4).

Proof. By Lemma 6.5.5 *(iii)*, we have, for $\phi \in C_0^\infty(X)$,

$$(H_{-F}\phi|\psi_F) = (\mathrm{e}^{-F}H\mathrm{e}^F\phi|\psi_F) = (\mathrm{e}^F\phi|H\psi) = E(\phi|\psi_F),$$

which shows that $\psi_F \in \mathcal{D}(H_{-F}^*) = \mathcal{D}(H_F) = H^2(X)$, and that *(ii)* is true.

By Lemma 6.5.5 *(i)*, we have

$$\|-\Delta\phi\| \le C(\|H_F\phi\| + \|\phi\|),$$

which together with *(ii)* implies *(i)*.

Let us show *(iii)*. Let us denote by $H_{\mathrm{c}}^2(X)$ the Fréchet space of compactly supported functions in $H^2(X)$. It follows from (6.4.2) that

$$[H, i\mathrm{e}^F A\mathrm{e}^F] = \mathrm{e}^F[H, iA]\mathrm{e}^F + \mathrm{e}^F A g A\mathrm{e}^F + \mathrm{e}^F G\mathrm{e}^F, \tag{6.5.7}$$

is true as an identity between quadratic forms on $H_{\mathrm{c}}^2(X)$.

Let now $j \in C_0^\infty(X)$ with $j = 1$ near the origin and let $\psi_m = j(\frac{x}{m})\psi$. Clearly, we have

$$\lim_{m\to\infty}\psi_m = \psi \text{ in } H^2(X), \quad \psi_m \in H_{\mathrm{c}}^2(X),$$

and using *(i)*, we also have

$$\lim_{m\to\infty} \mathrm{e}^F \psi_m = \psi_F \text{ in } H^2(X). \tag{6.5.8}$$

Using the identity (6.5.7), we have

$$(\psi_m|[H, i\mathrm{e}^F A\mathrm{e}^F]\psi_m) = (\mathrm{e}^F\psi_m|[H, iA]\mathrm{e}^F\psi_m) + \|g^{1/2}A\mathrm{e}^F\psi_m\|^2 + (\mathrm{e}^F\psi_m|G\mathrm{e}^F\psi_m).$$

We will show that

$$\begin{aligned}
&\lim_{m\to\infty} (\psi_m|[H, i\mathrm{e}^F A\mathrm{e}^F]\psi_m) = 0,\\
&\lim_{m\to\infty} (\mathrm{e}^F\psi_m|[H, iA]\mathrm{e}^F\psi_m) = (\psi_F|[H, iA]\psi_F),\\
&\lim_{m\to\infty} \|g^{1/2}A\mathrm{e}^F\psi_m\|^2 = \|g^{\frac{1}{2}}A\psi_F\|^2,\\
&\lim_{m\to\infty} (\mathrm{e}^F\psi_m|G\mathrm{e}^F\psi_m) = (\psi_F|G\psi_F),
\end{aligned} \tag{6.5.9}$$

which will imply *(iii)*.

The second, third and fourth identity of (6.5.9) follow by (6.5.8). In order to show the first identity of (6.5.9), we write

$$|(\psi_m|[H, i\mathrm{e}^F A\mathrm{e}^F]\psi_m)| \le \|\langle x\rangle \mathrm{e}^F(H-E)\psi_m\|\|\langle x\rangle^{-1}A\mathrm{e}^F\psi_m\|. \tag{6.5.10}$$

Now it follows from (6.5.8) that

$$\|\langle x\rangle^{-1}A\mathrm{e}^F\psi_m\| \le C. \tag{6.5.11}$$

We also claim that

$$\lim_{m\to\infty} \langle x\rangle \mathrm{e}^F(H-E)\psi_m = 0. \tag{6.5.12}$$

In fact, using that $(H-E)\psi = 0$, we have

$$\langle x\rangle \mathrm{e}^F(H-E)\psi_m = \langle x\rangle \mathrm{e}^F[\tfrac{1}{2}D^2, j(\tfrac{x}{m})]\psi,$$

which shows that

$$|\langle x\rangle \mathrm{e}^F(H-E)\psi_m|(x) \le C(|D\mathrm{e}^F\psi(x)| + |\mathrm{e}^F\psi(x)|) \in L^2(X),$$

Since obviously $\langle x\rangle \mathrm{e}^F(H-E)\psi_m$ goes pointwise to 0, this implies (6.5.12). Hence (6.5.10) goes to zero. This ends the proof of *(iii)*. □

The proof of Theorem 6.5.1 will use a contradiction argument. We assume that

$$\tau = \frac{1}{2}\theta_0^2 + E \notin \mathcal{T} \cup \{\infty\}.$$

Since $\tau(H)$ is closed, we can find θ_1 and $\epsilon > 0$ such that $\frac{1}{2}\theta_1^2 + E \notin \mathcal{T} \cup \{\infty\}$, $\theta_1 \le \theta_0 < \theta_1 + \epsilon$, and

$$\mathrm{e}^{\theta_1|x|}\psi \in L^2(X), \quad \mathrm{e}^{(\theta_1+\epsilon)|x|}\psi \notin L^2(X). \tag{6.5.13}$$

We fix a cutoff function $\chi \in C_0^\infty(\mathbb{R})$ equal to 1 near the origin, and put

$$F_n(x) := \left(\theta_1 + \epsilon\chi(\tfrac{|x|}{n})\right)\langle x\rangle.$$

Let us list some properties of $F_n(x)$:

$$|\partial_x^\alpha F_n(x)| \leq C_\alpha \langle x\rangle^{1-|\alpha|},\ |\alpha| \geq 0, \tag{6.5.14}$$

$$|(\nabla F_n)^2 - \theta_1^2| \in O(\epsilon), \tag{6.5.15}$$

$$|x\cdot\nabla(\nabla F_n)^2| \in O(\epsilon) + O(\langle x\rangle^{-1}). \tag{6.5.16}$$

We have $\mathrm{e}^{F_n}\psi \in L^2(X)$ but, by (6.5.13),

$$\lim_{n\to\infty} \|\mathrm{e}^{F_n}\psi\| = \infty. \tag{6.5.17}$$

We put

$$\psi_n := \frac{\mathrm{e}^{F_n}\psi}{\|\mathrm{e}^{F_n}\psi\|}.$$

Let us describe some properties of ψ_n.

Lemma 6.5.7

(i) $\lim\limits_{n\to\infty} \int_{|x|\leq R} |\partial_x^\alpha \psi_n|^2 \mathrm{d}x = 0, \quad |\alpha| \leq 1, \quad R \geq 0,$

(ii) $\mathrm{w}-\lim\limits_{n\to\infty} \psi_n = 0,$

(iii) $\|\psi_n\|_{H^2(X)} \leq C,$

(iv) $\lim\limits_{n\to\infty} (H - E - \frac{1}{2}(\nabla F_n)^2)\psi_n = 0.$

Proof. Property *(i)* follows directly from (6.5.17) and implies *(ii)*.
(iii) follows from Lemma 6.5.6 *(i)*.
Let us show *(iv)*. By Lemma 6.5.6 *(ii)*, we have

$$\left(H - E - \tfrac{1}{2}(\nabla F_n)^2\right)\psi_n = g_n A\psi_n + ix\cdot\nabla g_n\psi_n. \tag{6.5.18}$$

Using property *(i)*, the fact that $\|\psi_n\| = 1$ and

$$x\cdot\nabla g_n(x) \in O(\langle x\rangle^{-1}),$$

we obtain

$$\lim_{n\to\infty} x\cdot\nabla g_n(x)\psi_n = 0.$$

It is more delicate to handle the first term on the right-hand side of (6.5.18). Using *(iii)* and the boundedness of $(-\Delta+1)^{-1}[H, iA](-\Delta+1)^{-1}$, we obtain that

$$|(\psi_n|[H, iA]\psi_n)| \leq C,$$

which, using Lemma 6.5.6 *(iii)*, gives

$$\|g_n^{1/2}A\psi_n\| \leq C. \tag{6.5.19}$$

Using property *(i)*, (6.5.19) and the fact that

$$g_n^{1/2} \in O(\langle x\rangle^{-1/2}),$$

we obtain

$$\lim_{n\to\infty} \|g_n A\psi_n\| = 0. \tag{6.5.20}$$

This ends the proof of *(iv)*. □

Lemma 6.5.8
For every neighborhood Δ of $\frac{1}{2}\theta_1^2 + E$, one has

$$\limsup_{n\to\infty} (\mathbb{1}_\Delta(H)\psi_n|[H,iA]\mathbb{1}_\Delta(H)\psi_n) \leq O(\epsilon).$$

Proof. By (6.5.15) and Lemma 6.5.7 *(iv)*, we have

$$\limsup_{n\to\infty} \|(H - E - \tfrac{1}{2}\theta_1^2)\psi_n\| \in O(\epsilon). \tag{6.5.21}$$

This implies that if Δ is a neighborhood of $\frac{1}{2}\theta_1^2 + E$, then

$$\limsup_{n\to\infty} \|\mathbb{1}_{\mathbb{R}\setminus\Delta}(H)\psi_n\| \in O(\epsilon), \tag{6.5.22}$$

$$\limsup_{n\to\infty} \|(H+i)\mathbb{1}_{\mathbb{R}\setminus\Delta}(H)\psi_n\| \in O(\epsilon). \tag{6.5.23}$$

Using the fact that $(H+i)^{-1}[H,iA](H+i)^{-1}$ is bounded, we deduce from (6.5.23) that

$$\begin{aligned}&\limsup_{n\to\infty}(\psi_n|[H,iA]\psi_n)\\ &= \limsup_{n\to\infty}(\psi_n|\mathbb{1}_\Delta(H)[H,iA]\mathbb{1}_\Delta(H)\psi_n) + O(\epsilon).\end{aligned} \tag{6.5.24}$$

By Lemma 6.5.6 *(iii)*, we have

$$(\psi_n|[H,iA]\psi_n) \leq -(\psi_n|G_n(x)\psi_n),$$
$$\text{where } G_n(x) = \tfrac{1}{2}\left(x\cdot\nabla(\nabla F_n)^2(x) - (x\cdot\nabla)^2 g_n(x)\right).$$

Using (6.5.16), we see that

$$G_n(x) \in O(\epsilon) + O(\langle x\rangle^{-1}).$$

Therefore,

$$\limsup_{n\to\infty} |(\psi_n|G_n\psi_n)| \in O(\epsilon).$$

Hence

$$\limsup_{n\to\infty}(\psi_n|[H,iA]\psi_n) \le O(\epsilon) \tag{6.5.25}$$

Using (6.5.24), (6.5.25), we obtain the lemma. □

Lemma 6.5.9
There exists a neighborhood Δ of $\frac{1}{2}\theta_1^2 + E$ and $C_0 > 0$ such that

$$\liminf_{n\to\infty}(\mathbb{1}_\Delta(H)\psi_n|[H,iA]\mathbb{1}_\Delta(H)\psi_n) \ge C_0 + O(\epsilon). \tag{6.5.26}$$

Proof. Since $\frac{1}{2}\theta_1^2 + E \notin \tau(H)$, we can apply the Mourre estimate

$$(\mathbb{1}_\Delta(H)\psi_n|[H,iA]\mathbb{1}_\Delta(H)\psi_n) \ge C_0\|\mathbb{1}_\Delta(H)\psi_n\|^2 + (\psi_n, K\psi_n).$$

Using (6.5.22) and the fact that ψ_n tends weakly to 0, we obtain (6.5.26). □

Proof of Theorem 6.5.1. It suffices to observe that, for small enough ϵ, Lemmas 6.5.8 and 6.5.9 contradict each other. □

Proof of Theorem 6.5.3. Let ϕ_n, $n \in \mathbb{N}$ be an orthonormal sequence of eigenvectors of H with eigenvalues E_n such that $E_n \nearrow E < \tau$. Assume that $\phi_n \in \mathcal{E}_\tau$, $n \in \mathbb{N}$. By the definition of $\mathcal{E}_\tau$, we can pick $\theta < \sqrt{2(\tau - E)}$ such that

$$E + \frac{1}{2}\theta^2 \notin \mathcal{T} \quad \text{and} \quad \mathrm{e}^F\phi_n \in L^2(X) \quad \text{where} \quad F(x) := \theta\langle x\rangle.$$

We set

$$\psi_n := \frac{\mathrm{e}^F\phi_n}{\|\mathrm{e}^F\phi_n\|}.$$

First note that Lemma 6.5.6 is true for the sequence ψ_n uniformly in n.

Next we note that also Lemma 6.5.7 is true for the sequence ψ_n. This requires, however, a somewhat different proof.

The property *(iii)* of Lemma 6.5.7 follows immediately from Lemma 6.5.6 *(ii)*.

Next we will show Lemma 6.5.7 *(ii)*, or actually its stronger version

$$\mathrm{w}-\lim_{n\to\infty}(1-\Delta)\psi_n = 0. \tag{6.5.27}$$

Clearly, using the fact that ϕ_n is an orthonormal sequence and $H\phi_n = E_n\phi_n$, we see that

$$\mathrm{w}-\lim_{n\to\infty}(1-\Delta)\phi_n = 0. \tag{6.5.28}$$

Moreover, by *(iii)* we have

$$\|(1-\Delta)\psi_n\| \le C. \tag{6.5.29}$$

Now, for any $\eta \in C_0^\infty(X)$,

$$\begin{aligned}(\eta|(1-\Delta)\psi_n) &= \frac{(\mathrm{e}^F(1-\Delta)\eta|\phi_n)}{\|\mathrm{e}^F\phi_n\|} \\ &\leq (\mathrm{e}^F(1-\Delta)\eta|\phi_n).\end{aligned}$$

Hence

$$\lim_{n\to\infty}(\eta|(1-\Delta)\psi_n) = 0, \quad \eta \in C_0^\infty(X). \tag{6.5.30}$$

Now (6.5.30) and (6.5.29) imply (6.5.28).

To prove Lemma 6.5.7 *(i)*, it is enough to pick $\chi \in C_0^\infty(\mathbb{R})$ such that $\chi = 1$ on $[0,1]$ and to show that

$$\lim_{n\to\infty}(1-\Delta)^{\frac{1}{2}}\chi\left(\tfrac{|x|}{R}\right)\psi_n = 0. \tag{6.5.31}$$

But

$$(1-\Delta)^{\frac{1}{2}}\chi\left(\tfrac{|x|}{R}\right)\psi_n = K(1-\Delta)\psi_n,$$

where

$$K := (1-\Delta)^{\frac{1}{2}}\chi\left(\frac{|x|}{R}\right)(1-\Delta)^{-1}$$

is compact. Hence using (6.5.28) we see that (6.5.31) is true.

Finally, Lemma 6.5.7 *(iv)* is proven exactly as in the proof of Theorem 6.5.1.

Next, arguing as in the proof of Lemma 6.5.8 we see that, for any neighborhood Δ of $\frac{1}{2}\theta^2 + E$, we have

$$\limsup_{n\to\infty}(\mathbb{1}_\Delta(H)\psi_n|[H, iA]\mathbb{1}_\Delta(H)\psi_n) \leq 0. \tag{6.5.32}$$

Arguing as in the proof of Lemma 6.5.9 we see that there exists an open neighborhood Δ of $\frac{1}{2}\theta^2 + E$ and $C_0 > 0$ such that

$$\liminf_{n\to\infty}(\mathbb{1}_\Delta(H)\psi_n|[H, iA]\mathbb{1}_\Delta(H)\psi_n) \geq C_0. \tag{6.5.33}$$

Now, (6.5.32) and (6.5.33) contradict one another. □

Proof of Theorem 6.5.4. Let $H\psi = E\psi$. For any $\theta \in \mathbb{R}$ set $\psi_\theta = \mathrm{e}^{\theta\langle x\rangle}\psi$.

Suppose that $\psi_\theta \in L^2(X)$ for any $\theta \in \mathbb{R}$. Since V is H_0-bounded with the relative bound less than 1, we will find $C_0 > 0$ such that as quadratic forms

$$\frac{1}{2}D^2 \geq C_0 H - C.$$

Therefore, applying Proposition 6.5.6 *(iv)* to the function $F = \theta\langle x\rangle$, we obtain on the one hand

$$\begin{aligned}(\psi_\theta|\tfrac{1}{2}D^2\psi_\theta) &\geq C_0(\psi_\theta|H\psi_\theta) - C\|\psi_\theta\|^2 \\ &= C_0\theta^2(\psi_\theta|\tfrac{x^2}{\langle x\rangle^2}\psi_\theta) + (C_0E - C)\|\psi_\theta\|^2.\end{aligned} \tag{6.5.34}$$

On the other hand, it follows from (6.5.3) that

$$\frac{1}{2}D^2 \leq C_1[H, iA] + C, \text{ as forms on } H^2(X),$$

which gives

$$\begin{aligned}(\psi_\theta|\tfrac{1}{2}D^2\psi_\theta) &\leq C_1(\psi_\theta|[H, iA]\psi_\theta) + C\|\psi_\theta\|^2 \\ &\leq C_1(\psi_\theta|G(x)\psi_\theta) + C\|\psi_\theta\|^2,\end{aligned} \tag{6.5.35}$$

using Proposition 6.5.6 *(iii)*. By a direct computation, we have

$$G(x) = \theta(3\tfrac{x^4}{\langle x\rangle^5} - 2\tfrac{x^2}{\langle x\rangle^3}) - 2\theta^2\tfrac{x^2}{\langle x\rangle^4}.$$

Combining (6.5.34) and (6.5.35), we obtain

$$\int r_\theta(x)|\psi_\theta|^2(x)dx \leq 0, \tag{6.5.36}$$

for

$$\begin{aligned}r_\theta(x) &= \theta^2\left(C_0\tfrac{x^2}{\langle x\rangle^2} + 2C_1\tfrac{x^2}{\langle x\rangle^4}\right) \\ &+\theta C_1\left(2\tfrac{x^2}{\langle x\rangle^3} - 3\tfrac{x^4}{\langle x\rangle^5}\right) - C_0E - 2C.\end{aligned}$$

When θ tends to ∞, then $r_\theta(x)$ increases to ∞ except for $x = 0$. This contradicts (6.5.36) unless $\psi = 0$. Hence $\mathcal{E}_\infty = \{0\}$.

Now a simple induction argument shows that H has no positive eigenvalues. □

6.6 Asymptotic Velocity

Traditionally, wave operators and scattering operators are regarded as the most important objects in N-body scattering theory. Their existence and completeness can be proven for a large class of potentials that includes essentially all physically interesting ones. Nevertheless, it seems that from the mathematical point of view another object has a more fundamental importance. This object is the *asymptotic velocity.*

The existence of the asymptotic velocity can be shown for a very large class of potentials – much larger than the class for which asymptotic completeness has been shown. It implies easily the asymptotic completeness of short-range systems. It is also an important preparatory step in our proof of the asymptotic completeness of long-range systems.

The construction of the asymptotic velocity is described in the following theorem.

Theorem 6.6.1
Assume (6.1.1) and

$$\int_0^\infty \left\| (1-\Delta^b)^{-1} \nabla_{x^b} v^b(x^b) 1\!\!1_{[1,\infty[}\left(\frac{|x^b|}{R}\right)(1-\Delta^b)^{-1}\right\| \mathrm{d}R < \infty, \quad b \in \mathcal{B}. \quad (6.6.1)$$

Then the conclusion of Proposition 6.3.1 is true. Likewise (6.4.2) holds and hence the conclusion of Theorem 6.4.1 is true. Moreover, the following holds:
(i) There exists

$$\text{s}-C_\infty - \lim_{t\to\infty} \mathrm{e}^{itH}\frac{x}{t}\mathrm{e}^{-itH} := P^+. \quad (6.6.2)$$

The vector of commuting self-adjoint operators P^+ is densely defined and commutes with the Hamiltonian H.
(ii) If $a \in \mathcal{A}$, then

$$\text{s}-C_\infty - \lim_{t\to\infty} \mathrm{e}^{itH} D_a \mathrm{e}^{-itH} 1\!\!1_{Y_a}(P^+) = P_a^+ 1\!\!1_{Y_a}(P^+). \quad (6.6.3)$$

(iii)

$$1\!\!1_{\{X_{a_{\max}}\}}(P^+) = 1\!\!1^{\text{pp}}(H^{a_{\max}}).$$

The proof of Theorem 6.6.1 will be divided into a series of lemmas and propositions, some of them of an independent interest.

The next lemma is an obvious generalization of Lemma 4.4.2 to the case of N-body potentials.

Lemma 6.6.2
It follows from (6.1.1) and (6.6.1) that, for any $b \in \mathcal{B}$,

$$\begin{aligned} &(1-\Delta^b)^{-1}\nabla_{x^b} v^b(x^b)(1-\Delta^b)^{-1} \text{ is compact on } L^2(X^b), \\ &(1-\Delta^b)^{-1} x^b \nabla_{x^b} v^b(x^b)(1-\Delta^b)^{-1} \text{ is compact on } L^2(X^b). \end{aligned} \quad (6.6.4)$$

Moreover, let $\delta > 0$, $J \in C^\infty(X)$ such that $\partial_x^\alpha J(x)$ are bounded and $\operatorname{supp} J \subset Y_a^\delta$. *Let $\chi \in C_0^\infty(\mathbb{R})$. Then for any $a \in \mathcal{A}$,*

$$(1-\Delta)^{-1}\nabla_x I_a(x) J\left(\frac{x}{R}\right)(1-\Delta)^{-1} \in L^1(\mathrm{d}R), \quad (6.6.5)$$

$$[D_a, \chi(H)] J\left(\frac{x}{R}\right) \in o(R^0), \quad (6.6.6)$$

$$[D_a, \chi(H)]\, J\left(\frac{x}{R}\right) \in L^1(\mathrm{d}R). \quad (6.6.7)$$

Our next proposition is an important propagation estimate due to Graf [Gr].

Proposition 6.6.3
Let $a \in \mathcal{A}$, $\chi \in C_0^\infty(\mathbb{R})$, $0 < \theta$ and $\epsilon > 0$. Then

$$\int_1^\infty \left\| 1\!\!1_{[0,\theta]}\left(\frac{|x|}{t}\right) 1\!\!1_{Y_a^\epsilon}\left(\frac{x}{t}\right)\left(\frac{x_a}{t} - D_a\right)\chi(H)\phi_t\right\|^2 \frac{dt}{t} \leq C\|\phi\|^2. \qquad (6.6.8)$$

Proof. We can always increase θ so that $\mathrm{supp}\chi \subset]-\infty, \frac{1}{2}\theta^2 + \Sigma[$, in order to be able to apply Proposition 6.3.1. Let $J \in C_0^\infty(\mathbb{R})$ be a cutoff function such that $J = 1$ on $[0,\theta]$. Let $j \in C_0^\infty(\mathbb{R})$ such that $j = 1$ on $\mathrm{supp}J'$ and $\mathrm{supp}j \subset [\theta, \infty[$. Let $R(x)$ be the function defined in Sect. 5.2.

Set

$$L(t) := \tfrac{1}{2}\langle D - \tfrac{x}{t}, \nabla R(\tfrac{x}{t})\rangle + \tfrac{1}{2}\langle \nabla R(\tfrac{x}{t}), D - \tfrac{x}{t}\rangle + R(\tfrac{x}{t}),$$
$$\Phi(t) := \chi(H)J\left(\tfrac{|x|}{t}\right)L(t)J\left(\tfrac{|x|}{t}\right)\chi(H).$$

Then $\Phi(t)$ is uniformly bounded and

$$\begin{aligned}\mathbf{D}\Phi(t) &= \chi(H)\left(\mathbf{D}_0 J\left(\tfrac{|x|}{t}\right)\right)L(t)J\left(\tfrac{|x|}{t}\right)\chi(H) + \mathrm{hc}\\ &-\chi(H)\nabla_x V(x)\nabla_x R\left(\tfrac{x}{t}\right)J^2\left(\tfrac{|x|}{t}\right)\chi(H) \\ &+t^{-1}\chi(H)J\left(\tfrac{|x|}{t}\right)\langle \tfrac{x}{t} - D, \nabla^2 R\left(\tfrac{x}{t}\right)\left(\tfrac{x}{t} - D\right)\rangle J\left(\tfrac{|x|}{t}\right)\chi(H).\end{aligned} \qquad (6.6.9)$$

The first term in the right-hand side of (6.6.9) can be written as

$$t^{-1}\chi(H)j(\tfrac{|x|}{t})B(t)j(\tfrac{|x|}{t})\chi(H) + O(t^{-2}),$$

for some uniformly bounded observable $B(t)$. Using Proposition 6.3.1, we see that this term gives an integrable contribution along the evolution.

Using Lemma 5.2.7 *(ii)*, we can rewrite the second term in the right-hand side of (6.6.9) as

$$\sum_{a\in\mathcal{A}} \chi(H)\nabla_x I_a(x)\tfrac{x_a}{t}q_a(\tfrac{x}{t})J^2(\tfrac{|x|}{t})\chi(H).$$

This is integrable in norm by Lemma 6.6.2.

Finally, using Lemma 5.2.7 *(iii)*, we have

$$\begin{aligned}&t^{-1}\chi(H)J\left(\tfrac{|x|}{t}\right)\langle \tfrac{x}{t} - D, \nabla^2 R\left(\tfrac{x}{t}\right)\left(\tfrac{x}{t} - D\right)\rangle J\left(\tfrac{|x|}{t}\right)\chi(H)\\ &\geq \sum_{a\in\mathcal{A}} t^{-1}\chi(H)J\left(\tfrac{|x|}{t}\right)\left(\tfrac{x_a}{t} - D_a\right)q_a(\tfrac{x}{t})\left(\tfrac{x_a}{t} - D_a\right)J\left(\tfrac{|x|}{t}\right)\chi(H).\end{aligned}$$

Therefore, by Lemma B.4.1, for any $a \in \mathcal{A}$, we have

$$\int_1^\infty \left\|\sqrt{q_a\left(\tfrac{x}{t}\right)}\left(\tfrac{x_a}{t} - D_a\right)J\left(\tfrac{|x|}{t}\right)\chi(H)\phi_t\right\|^2 \tfrac{dt}{t} < C\|\phi\|^2. \qquad (6.6.10)$$

For any $\epsilon > 0$, we can choose $R(x)$ and $q_a(x)$ such that

$$1\!\!1_{Y_a^\epsilon}(x) \leq \sum_{a\leq b} q_b(x).$$

Therefore,

$$\begin{aligned}
&t^{-1}\left\|1\!\!1_{Y_a^\epsilon}\left(\tfrac{x}{t}\right)1\!\!1_{[0,\theta]}\left(\tfrac{|x|}{t}\right)\left(\tfrac{x_a}{t}-D_a\right)\chi(H)\phi_t\right\|^2\\
&\leq t^{-1}\left\|\sqrt{\textstyle\sum_{a\leq b}q_b\left(\tfrac{x}{t}\right)}\left(\tfrac{x_a}{t}-D_a\right)J\left(\tfrac{|x|}{t}\right)\chi(H)\phi_t\right\|^2+O(t^{-2}) \qquad (6.6.11)\\
&\leq \sum_{a\leq b}t^{-1}\left\|\sqrt{q_b\left(\tfrac{x}{t}\right)}\left(\tfrac{x_b}{t}-D_b\right)J\left(\tfrac{|x|}{t}\right)\chi(H)\phi_t\right\|^2+O(t^{-2}).
\end{aligned}$$

Now (6.6.10) and (6.6.11) yield

$$\int_1^\infty\left\|1\!\!1_{Y_a^\epsilon}\left(\frac{x}{t}\right)1\!\!1_{[0,\theta]}\left(\frac{|x|}{t}\right)\left(\frac{x_a}{t}-D_a\right)\chi(H)\phi_t\right\|^2\frac{\mathrm{d}t}{t}\leq C\|\phi\|^2,$$

as claimed. □

Lemma 6.6.4
Let $a\in\mathcal{A}$, $\epsilon>0$, $\theta>0$. Let $\chi\in C_0^\infty(\mathbb{R})$. Then

$$\mathrm{s-}\lim_{t\to\infty}1\!\!1_{[0,\theta]}\left(\frac{|x|}{t}\right)1\!\!1_{Y_a^\epsilon}\left(\frac{x}{t}\right)\left(\frac{x_a}{t}-D_a\right)\chi(H)\mathrm{e}^{-itH}=0. \qquad (6.6.12)$$

Proof. We can always enlarge θ if needed as in the proof of Proposition 6.6.3. Let $J\in C_0^\infty(X)\cap\mathcal{F}$ such that $\mathrm{supp}J\subset Y_a^{\epsilon/2}\cap\{x\mid |x|<2\theta\}$ and $J\geq 1$ on $Y_a^\epsilon\cap\{x\mid |x|<\theta\}$. Let $\chi_1\in C_0^\infty(\mathbb{R})$, $\chi_1=1$ on $\mathrm{supp}\chi$.

Using (6.6.6), we have

$$\begin{aligned}
&\mathrm{s-}\lim_{t\to\infty}J(\tfrac{x}{t})(\tfrac{x_a}{t}-D_a)\chi(H)\mathrm{e}^{-itH}\\
&=\mathrm{s-}\lim_{t\to\infty}\chi_1(H)J(\tfrac{x}{t})(\tfrac{x_a}{t}-D_a)\chi(H)\mathrm{e}^{-itH}.
\end{aligned}$$

Set

$$\Phi(t):=\chi(H)\left(\frac{x_a}{t}-D_a\right)J\left(\frac{x}{t}\right)\chi_1^2(H)J\left(\frac{x}{t}\right)\left(\frac{x_a}{t}-D_a\right)\chi(H).$$

We compute

$$\begin{aligned}
\mathbf{D}\Phi(t)\ &=-2t^{-1}\Phi(t)\\
&+t^{-1}\chi(H)\left(\tfrac{x_a}{t}-D_a\right)\langle\nabla J(\tfrac{x}{t}),D-\tfrac{x}{t}\rangle\chi_1^2(H)J(\tfrac{x}{t})\left(\tfrac{x_a}{t}-D_a\right)\chi(H)+\mathrm{hc}\\
&-\chi(H)\nabla I_a(x)J(\tfrac{x}{t})\chi_1^2(H)J(\tfrac{x}{t})\left(\tfrac{x_a}{t}-D_a\right)\chi(H)+\mathrm{hc}+O(t^{-2})\\
&=:I_1(t)+I_2(t)+I_3(t)+O(t^{-2}).
\end{aligned}$$

$I_1(t)$ is negative.

Using the fact that $J\in\mathcal{F}$, for some $\epsilon>0$, we can write

$$\begin{aligned}
&I_2(t)\\
&= \sum_{b\le a} t^{-1}\chi(H)\left(\tfrac{x_a}{t} - D_a\right)\nabla J(\tfrac{x}{t})q_b(\tfrac{x}{\epsilon t})\left(\tfrac{x_b}{t} - D_b\right)\chi_1^2(H)J(\tfrac{x}{t})\left(\tfrac{x_a}{t} - D_a\right)\chi(H)\\
&= \sum_{b\le a} t^{-1}\chi(H)\left(\tfrac{x_b}{t} - D_b\right)q_b(\tfrac{x}{\epsilon t})B_b(t)J(\tfrac{x}{t})\left(\tfrac{x_a}{t} - D_a\right)\chi(H) + O(t^{-2}),
\end{aligned}$$

where $B_b(t)$ are uniformly bounded operators. Using then Proposition 6.6.3, we see that this term is integrable along the evolution.

$I_3(t)$ is integrable in norm by Lemma 6.6.2. This shows that there exists the limit

$$\lim_{t\to\infty}(\phi_t|\Phi(t)\phi_t). \tag{6.6.13}$$

But, again by Proposition 6.6.3, we have

$$\int_1^\infty (\phi_t|\Phi(t)\phi_t)\frac{\mathrm{d}t}{t} < \infty.$$

This implies that the limit (6.6.13) is zero. □

Proposition 6.6.5
Let $J \in C_\infty(\mathrm{I\!R})$. Then there exists

$$\mathrm{s-}\lim_{t\to\infty} \mathrm{e}^{itH} J\left(\frac{x}{t}\right)\mathrm{e}^{-itH}. \tag{6.6.14}$$

Moreover, if $J(0) = 1$, then

$$\mathrm{s-}\lim_{R\to\infty}\left(\mathrm{s-}\lim_{t\to\infty}\mathrm{e}^{itH}J\left(\frac{x}{Rt}\right)\mathrm{e}^{-itH}\right) = 1. \tag{6.6.15}$$

If we define

$$\mathrm{s-}C_\infty - \lim_{t\to\infty}\mathrm{e}^{itH}\frac{x}{t}\mathrm{e}^{-itH} =: P^+, \tag{6.6.16}$$

then the vector of commuting self-adjoint operators P^+ is densely defined and commutes with the Hamiltonian H. Hence Theorem 6.6.1 (i) is true.

Proof. By density, we may assume that $J \in C_0^\infty(X) \cap \mathcal{F}$. It also suffices to prove the existence of

$$\begin{aligned}
&\mathrm{s-}\lim_{t\to\infty}\mathrm{e}^{itH}J\left(\tfrac{x}{t}\right)\chi^2(H)\mathrm{e}^{-itH}\\
&= \mathrm{s-}\lim_{t\to\infty}\mathrm{e}^{itH}\chi(H)J\left(\tfrac{x}{t}\right)\chi(H)\mathrm{e}^{-itH}
\end{aligned} \tag{6.6.17}$$

for any $\chi \in C_0^\infty(\mathrm{I\!R})$.

Set

$$\Phi(t) := \chi(H)\left(J\left(\frac{x}{t}\right) + \left\langle D - \frac{x}{t}, \nabla J\left(\frac{x}{t}\right)\right\rangle\right)\chi(H).$$

As the first step, we will show that there exists

$$\text{s-}\lim_{t\to\infty} \mathrm{e}^{itH}\Phi(t)\mathrm{e}^{-itH}. \tag{6.6.18}$$

To see this, note that

$$\begin{aligned} \mathbf{D}\Phi(t) &= -\chi(H)\nabla V(x)\nabla J\left(\tfrac{x}{t}\right)\chi(H) \\ &+t^{-1}\chi(H)\langle D-\tfrac{x}{t}, \nabla^2 J\left(\tfrac{x}{t}\right)(D-\tfrac{x}{t})\rangle\chi(H). \end{aligned} \tag{6.6.19}$$

The first term on the right of (6.6.19) is integrable in norm by Lemma 6.6.2. The second is integrable along the evolution by Proposition 6.6.3. Therefore, the limit (6.6.18) exists.

It remains to show that (6.6.18) equals (6.6.17). This follows from Lemma 6.6.4.

(6.6.15) follows from Proposition 6.3.1 *(ii)*. The fact that P^+ exists as a densely defined vector of commuting operators follows then from Proposition B.2.1. $[P^+, H] = 0$ follows from Lemma 6.2.3. □

Proposition 6.6.6
Let $g \in C_\infty(X_a)$. Then

$$\text{s-}\lim_{t\to\infty} \mathrm{e}^{itH}g(D_a)\mathrm{e}^{-itH}\mathbb{1}_{Y_a}(P^+) = g(P_a^+)\mathbb{1}_{Y_a}(P^+). \tag{6.6.20}$$

Hence Theorem 6.6.1 (ii) is true.

Proof. It is enough to assume that $g \in C_0^\infty(X_a)$ and to prove that

$$\begin{aligned} &\text{s-}\lim_{t\to\infty} \mathrm{e}^{itH}g(D_a)J\left(\tfrac{x}{t}\right)\chi(H)\mathrm{e}^{-itH} \\ &= \text{s-}\lim_{t\to\infty} \mathrm{e}^{itH}g\left(\tfrac{x_a}{t}\right)J\left(\tfrac{x}{t}\right)\chi(H)\mathrm{e}^{-itH} \end{aligned} \tag{6.6.21}$$

for any $J \in C_0^\infty(Y_a)\cap\mathcal{F}$ and $\chi \in C_0^\infty(\mathbb{R})$.

We already know that the limit on the right-hand side of (6.6.21) exists.

Next we note that, by the Baker-Campbell-Hausdorff formula (3.2.28), we have

$$\begin{aligned} &\left(g(D_a) - g(\tfrac{x_a}{t})\right) J(\tfrac{x}{t})\chi(H)\mathrm{e}^{-itH} \\ &= B(t)\left(D_a - \tfrac{x_a}{t}\right) J(\tfrac{x}{t})\chi(H)\mathrm{e}^{-itH} + O(t^{-1}), \end{aligned} \tag{6.6.22}$$

where $B(t)$ is uniformly bounded. (6.6.22) goes strongly to zero by Lemma 6.6.4. Therefore (6.6.21) is true. □

Lemma 6.6.7
Let $\lambda \in \sigma^{\mathrm{pp}}(H)\backslash\mathcal{T}$. Then the following is true:
(i) There exists C such that, for $f \in C^{0,1}(X)$,

$$\|[\mathbb{1}_{\{\lambda\}}(H), f(x)]\| \le C\|\nabla f\|_\infty.$$

(ii) If, moreover, $f(x) = 0$ on a neighborhood of 0, then

$$\|[\mathbb{1}_{\{\lambda\}}(H), D_x^\alpha f(x)]\| \leq C\|\nabla f\|_\infty, \quad |\alpha| \leq 2.$$

Proof. We recall from Theorem 6.5.1 that, since $\lambda \notin \mathcal{T}$, $\mathbb{1}_{\{\lambda\}}(H)$ is a finite rank projection on exponentially decaying eigenfunctions. Hence

$$\|\langle x\rangle \mathbb{1}_{\{\lambda\}}(H)\| < \infty.$$

Now we have

$$\begin{aligned}
&[\mathbb{1}_{\{\lambda\}}(H), f(x)] \\
&= [\mathbb{1}_{\{\lambda\}}(H)), f(0)] + \textstyle\int_0^1 [\mathbb{1}_{\{\lambda\}}(H), x\nabla_x f(sx)]\mathrm{d}s \\
&= 0 + O(\|\nabla f\|_\infty \|x\mathbb{1}_{\{\lambda\}}(H)\|).
\end{aligned}$$

This proves *(i)*.

To show *(ii)*, we use in addition the boundedness of $\|(1-\Delta)\langle x\rangle \mathbb{1}_{\{\lambda\}}(H)\|$. □

Next we will prove the so-called low velocity estimate in a version due essentially to Graf [Gr].

Proposition 6.6.8
Let $\chi \in C_0^\infty(\mathbb{R})$ such that $\mathrm{supp}\chi \cap \mathcal{T} = \emptyset$. *Then we can find $\epsilon_0 > 0$ such that*

$$\int_1^\infty \left\| \mathbb{1}_{[0,\epsilon_0]}\left(\frac{|x|}{t}\right)\chi(H)\mathbb{1}^{\mathrm{c}}(H)\phi_t\right\|^2 \frac{\mathrm{d}t}{t} \leq C\|\phi\|^2. \tag{6.6.23}$$

Proof. Let $\lambda \in \mathbb{R}\backslash\mathcal{T}$. Let

$$\epsilon_0 < \frac{d(\lambda)}{\sqrt{2(\lambda-\Sigma)}},$$

where $d(\lambda)$ was defined in Theorem 6.4.1 and Σ in Definition 6.2.1. We will show that there exists a neighborhood Δ of λ such that if $\chi \in C_0^\infty(\mathbb{R})$ and $\mathrm{supp}\chi \subset \Delta$, then (6.6.23) holds. Then we can extend the validity of (6.6.23) to χ with larger supports by the covering argument (see the proof of Proposition 4.4.7).

Let $\lambda < \lambda_1$, $d_1 < d(\lambda)$, $\epsilon_1\sqrt{2(\lambda-\Sigma)} < d_1$ and $\epsilon_0 < \epsilon_1$.

By Theorem 6.4.1 *(iii)*, we will find a function $\tilde{\chi} \in C_0^\infty(\mathbb{R})$ such that $\tilde{\chi} = 1$ on a neighborhood of λ, $\mathrm{supp}\tilde{\chi} \subset]-\infty, \lambda_1[$ and

$$\tilde{\chi}(H)\mathbb{1}^{\mathrm{c}}(H)[H, iA]\mathbb{1}^{\mathrm{c}}(H)\tilde{\chi}(H) \geq \theta_2^2\tilde{\chi}^2(H)\mathbb{1}^{\mathrm{c}}(H). \tag{6.6.24}$$

Let $\chi \in C_0^\infty(\mathbb{R})$ such that $\tilde{\chi} = 1$ on $\mathrm{supp}\chi$. We will write

$$\chi^{\mathrm{c}}(H) := \mathbb{1}^{\mathrm{c}}(H)\chi(H), \quad \tilde{\chi}^{\mathrm{c}}(H) := \mathbb{1}^{\mathrm{c}}(H)\tilde{\chi}(H).$$

We may assume that $\mathrm{supp}\chi, \mathrm{supp}\tilde{\chi} \cap \mathcal{T} = \emptyset$. By Lemma 6.6.7, for $f \in C_0^\infty(X)$, we have that

$$[f(\tfrac{x}{t}), \chi^{\mathrm{c}}(H)] \in O(t^{-1}),$$

$$\text{if, moreover, } 0 \notin \mathrm{supp} f, \text{ then} \quad [Df(\tfrac{x}{t}), \chi^{\mathrm{c}}(H)] \in O(t^{-1}) + L^1(\mathrm{d}t).$$

Of course the same is true with χ replaced with $\tilde{\chi}$.

We also choose $J, \tilde{J} \in C_0^\infty(X)$ such that $J = 1$ for $|x| \leq \epsilon_0$, $\tilde{J} = 1$ on $\mathrm{supp}J$ and $\mathrm{supp}\tilde{J}, \mathrm{supp}J \subset \{x \mid |x| < \epsilon_1\}$, $J \in \mathcal{F}$.

Set

$$\begin{aligned} M(t) &:= J\left(\tfrac{x}{t}\right) + \langle D - \tfrac{x}{t}, \nabla J\left(\tfrac{x}{t}\right)\rangle, \\ \Phi(t) &:= \chi^{\mathrm{c}}(H) M(t) \tilde{\chi}^{\mathrm{c}}(H) \tfrac{A}{t} \tilde{\chi}^{\mathrm{c}}(H) M^*(t) \chi^{\mathrm{c}}(H). \end{aligned}$$

Using the boundedness of $\langle x\rangle \tilde{\chi}^{\mathrm{c}}(H) \langle x\rangle^{-1}$, we see that $\Phi(t)$ is uniformly bounded. We compute:

$$\begin{aligned} \mathbf{D}\Phi(t) &= t^{-1}\chi^{\mathrm{c}}(H)\langle D - \tfrac{x}{t}, \nabla^2 J\left(\tfrac{x}{t}\right)(D - \tfrac{x}{t})\rangle \tilde{\chi}^{\mathrm{c}}(H)\tfrac{A}{t}\tilde{\chi}^{\mathrm{c}}(H)M^*(t)\chi^{\mathrm{c}}(H) + \mathrm{hc} \\ &\quad -\chi^{\mathrm{c}}(H)\nabla_x V(x) J(\tfrac{x}{t})\tilde{\chi}^{\mathrm{c}}(H)\tfrac{A}{t}\tilde{\chi}^{\mathrm{c}}(H)M^*(t)\chi^{\mathrm{c}}(H) + \mathrm{hc} \\ &\quad +t^{-1}\chi^{\mathrm{c}}(H)M(t)\tilde{\chi}^{\mathrm{c}}(H)\left([H, iA] - \tfrac{A}{t}\right)\tilde{\chi}^{\mathrm{c}}(H)M^*(t)\chi^{\mathrm{c}}(H) \\ &=: R_1(t) + R_2(t) + R_3(t). \end{aligned}$$

The term $R_1(t)$ can be written as as

$$\begin{aligned} R_1(t) &= \sum_{b \leq a} t^{-1}\chi^{\mathrm{c}}(H)(D_b - \tfrac{x_b}{t})q_b(\tfrac{x}{\epsilon t})B_b(t)q_b(\tfrac{x}{\epsilon t})(D_b - \tfrac{x_b}{t})\chi^{\mathrm{c}}(H) \\ &\quad +O(t^{-2}) + \langle t\rangle^{-1}L^{-1}(\mathrm{d}t) \end{aligned}$$

for certain uniformly bounded operators $B_b(t)$. Using Proposition 6.6.3, we see that $R_1(t)$ is integrable along the evolution.

The term $R_2(t)$ is integrable in norm.

Let us now estimate the term $R_3(t)$. By (6.6.24), we have

$$\begin{aligned} & t^{-1}\chi^{\mathrm{c}}(H)M(t)\tilde{\chi}^{\mathrm{c}}(H)i[H, A]\tilde{\chi}^{\mathrm{c}}(H)M^*(t)\chi^{\mathrm{c}}(H) \\ & \geq d_1 t^{-1}\chi^{\mathrm{c}}(H)M(t)(\tilde{\chi}^{\mathrm{c}})^2(H)M^*(t)\chi^{\mathrm{c}}(H) \\ & = d_1 t^{-1}\chi^{\mathrm{c}}(H)M(t)M^*(t)\chi^{\mathrm{c}}(H) + O(t^{-2}) + \langle t\rangle^{-1}L^{-1}(\mathrm{d}t). \end{aligned}$$

By Lemma 6.3.2, we have:

$$\begin{aligned} \|\tilde{J}(\tfrac{x}{t})\tilde{\chi}^{\mathrm{c}}(H)\tfrac{A}{t}\tilde{\chi}^{\mathrm{c}}(H)\tilde{J}(\tfrac{x}{t})\| &\leq \|\tfrac{|x|}{t}\tilde{J}(\tfrac{x}{t})\|\|\tilde{J}(\tfrac{x}{t})\tfrac{x}{|x|}D\tilde{\chi}(H)\| + O(t^{-1}) \\ &\leq \epsilon_1\sqrt{2(\lambda - \Sigma)} + o(t^0). \end{aligned}$$

So,

$$-\tfrac{1}{t}\chi^{\mathrm{c}}(H)M(t)\tilde{\chi}^{\mathrm{c}}(H)\tfrac{A}{t}\tilde{\chi}^{\mathrm{c}}(H)M^*(t)\chi^{\mathrm{c}}(H)$$
$$= -\tfrac{1}{t}\chi^{\mathrm{c}}(H)M(t)\tilde{J}(\tfrac{x}{t})\tilde{\chi}^{\mathrm{c}}(H)\tfrac{A}{t}\tilde{\chi}^{\mathrm{c}}(H)\tilde{J}(\tfrac{x}{t})M^*(t)\chi^{\mathrm{c}}(H) + O(t^{-2})$$
$$\geq -\epsilon_1\sqrt{2(\lambda-\Sigma)}\tfrac{1}{t}\chi^{\mathrm{c}}(H)M(t)M^*(t)\chi^{\mathrm{c}}(H) + O(t^{-2}).$$

Therefore,

$$R_3(t) \geq C_0\chi^{\mathrm{c}}(H)M(t)M^*(t)\chi^{\mathrm{c}}(H) + O(t^{-2}) + \langle t\rangle^{-1}L^{-1}(\mathrm{d}t),$$

where by $C_0 := d_1 - \epsilon_1\sqrt{2(\lambda-\Sigma)} > 0$.

We write then

$$M(t) = J\left(\tfrac{x}{t}\right) + \langle D - \tfrac{x}{t}, \nabla J\left(\tfrac{x}{t}\right)\rangle =: M_1(t) + M_2(t).$$

We use now the inequality

$$(M_1 + M_2)(M_1^* + M_2^*) \geq (1-\epsilon)M_1M_1^* + (1-\epsilon^{-1})M_2M_2^*$$

to deduce that

$$\begin{aligned} R_3(t) &\geq (1-\epsilon)C_0t^{-1}\chi^{\mathrm{c}}(H)J^2\left(\tfrac{x}{t}\right)\chi^{\mathrm{c}}(H) \\ &+(1-\epsilon^{-1})C_0\chi^{\mathrm{c}}(H)\langle D-\tfrac{x}{t}, \nabla J\left(\tfrac{x}{t}\right)\rangle\langle\nabla J\left(\tfrac{x}{t}\right), D-\tfrac{x}{t}\rangle\chi^{\mathrm{c}}(H). \end{aligned} \tag{6.6.25}$$

The second term on the right-hand side of (6.6.25) is integrable along the evolution by Proposition 6.6.3. Hence, by Proposition 4.2.1, we obtain

$$\int_1^\infty \left\| J\left(\frac{x}{t}\right)\chi^{\mathrm{c}}(H)\phi_t\right\|^2 \frac{\mathrm{d}t}{t} \leq C\|\phi\|^2.$$

□

The following proposition shows that the states with zero asymptotic velocity coincide with the bound states.

Proposition 6.6.9

$$1\!\!1_{\{X_{a_{\max}}\}}(P^+) = 1\!\!1^{\mathrm{pp}}(H^{a_{\max}}).$$

Hence Theorem 6.6.1 (iii) is true.

Proof. Using the fact that

$$H = H^{a_{\max}} + \tfrac{1}{2}D_{a_{\max}}^2,$$

we see that it suffices to consider the case when $X_{a_{\max}} = \{0\}$, $H = H^{a_{\max}}$. Let $H\phi = \tau\phi$ and $J \in C_\infty(X)$. Then

$$\begin{aligned} &\lim_{t\to\infty} \mathrm{e}^{itH}J\left(\tfrac{x}{t}\right)\mathrm{e}^{-itH}\phi \\ &= J(0)\phi + \lim_{t\to\infty}\mathrm{e}^{it(H-\tau)}\left(J\left(\tfrac{x}{t}\right) - J(0)\right)\phi = J(0)\phi. \end{aligned} \tag{6.6.26}$$

This shows that $P^+\phi = 0$ and proves

$$1\!\!1_{\{0\}}(P^+) \geq 1\!\!1^{\mathrm{pp}}(H).$$

Let us prove the opposite inequality.

Let $\chi \in C_0^\infty(\mathbb{R})$ such that $\mathrm{supp}\chi \cap (\sigma^{\mathrm{pp}}(H) \cup \mathcal{T}) = \emptyset$. Choose also $J \in C_0^\infty(X)$ such that $J(0) > 0$ and $\mathrm{supp} J \subset B(0, \epsilon)$, where $\epsilon > 0$ is given by Lemma 6.6.8. Clearly,

$$\mathrm{s-}\lim_{t\to\infty} \mathrm{e}^{itH}\chi(H)J^2\left(\frac{x}{t}\right)\chi(H)\mathrm{e}^{-itH} = \chi^2(H)J^2(P^+). \tag{6.6.27}$$

By Proposition 6.6.8,

$$\int_1^\infty \left\| J\left(\frac{x}{t}\right)\chi(H)\phi_t \right\|^2 \frac{\mathrm{d}t}{t} < \infty. \tag{6.6.28}$$

Therefore, the limit (6.6.27) vanishes, which proves that

$$1\!\!1_{\{0\}}(P^+) \leq 1\!\!1_{\mathcal{T}\cup\sigma^{\mathrm{pp}}(H)}(H). \tag{6.6.29}$$

But by (6.4.6) the right-hand side of (6.6.29) equals $1\!\!1^{\mathrm{pp}}(H)$. □

6.7 Asymptotic Completeness of Short-Range Systems

The asymptotic velocity constructed in Theorem 6.6.1 gives a classification of the states in $L^2(X)$ according to their asymptotic behavior under the evolution e^{-itH}. In fact, we clearly have

$$1\!\!1 = \sum_{a\in\mathcal{A}} 1\!\!1_{Z_a}(P^+). \tag{6.7.1}$$

Moreover, by Theorem 6.6.1,

$$1\!\!1_{\{X_{a_{\max}}\}}(P^+) = 1\!\!1_{Z_{a_{\max}}}(P^+) = 1\!\!1^{\mathrm{pp}}(H^{a_{\max}}),$$

i.e. the states with zero internal asymptotic velocity coincide with the bound states of the full Hamiltonian with a removed center-of-mass motion. However, we would like to have a better understanding of the spaces of scattering states $\mathrm{Ran}1\!\!1_{Z_a}(P^+)$.

In this section we will assume that the potential $V(x)$ satisfies the *short-range* condition. In this case, one can describe the space $\mathrm{Ran}1\!\!1_{Z_a}(P^+)$ in a very satisfactory way by constructing the *wave operators*.

Theorem 6.7.1
Assume (6.1.1) and

$$\int_0^\infty \left\| (1-\Delta^b)^{-1} v^b(x^b) \mathbb{1}_{[1,\infty[}\left(\frac{|x^b|}{R}\right)(1-\Delta^b)^{-\frac{1}{2}} \right\| \mathrm{d}R < \infty, \quad b \in \mathcal{B}. \tag{6.7.2}$$

Then assumption (6.6.1) holds and hence the conclusions of Theorem 6.6.1 hold. Moreover, for all $a \in \mathcal{A}$, there exist

$$\mathrm{s-}\lim_{t\to\infty} \mathrm{e}^{itH}\mathrm{e}^{-itH_a}\mathbb{1}^{\mathrm{pp}}(H^a), \tag{6.7.3}$$

$$\mathrm{s-}\lim_{t\to\infty} \mathrm{e}^{itH_a}\mathrm{e}^{-itH}\mathbb{1}_{Z_a}(P^+). \tag{6.7.4}$$

If we denote (6.7.3) by $\Omega^+_{\mathrm{sr},a}$, then (6.7.4) equals $\Omega^{+*}_{\mathrm{sr},a}$. The operator $\Omega^+_{\mathrm{sr},a}$ is a partial isometry such that

$$\Omega^{+*}_{\mathrm{sr},a}\Omega^+_{\mathrm{sr},a} = \mathbb{1}^{\mathrm{pp}}(H^a), \qquad \Omega^+_{\mathrm{sr},a}\Omega^{+*}_{\mathrm{sr},a} = \mathbb{1}_{Z_a}(P^+), \tag{6.7.5}$$

$$\Omega^+_{\mathrm{sr},a}H_a = H\Omega^+_{\mathrm{sr},a}, \qquad \Omega^+_{\mathrm{sr},a}D_a = P^+_a\Omega^+_{\mathrm{sr},a} = P^+\Omega^+_{\mathrm{sr},a}.$$

Remark. By (6.7.1) and (6.7.5),

$$L^2(X) = \sum_{a\in\mathcal{A}}^{\oplus} \mathrm{Ran}\,\Omega^+_{\mathrm{sr},a}. \tag{6.7.6}$$

If one notes that $\Omega^+_{\mathrm{sr},a_{\max}} = \mathbb{1}^{\mathrm{pp}}(H^{a_{\max}})$, then we can rewrite (6.7.6) as

$$\mathrm{Ran}\,\mathbb{1}^{\mathrm{c}}(H^{a_{\max}}) = \sum_{a\neq a_{\max}}^{\oplus} \mathrm{Ran}\,\Omega^+_{\mathrm{sr},a}, \tag{6.7.7}$$

which is the usual statement of asymptotic completeness. The operators $\Omega^+_{\mathrm{sr},a}$ are called the *channel wave operators.*

Remark. Let us recall that H_a, H^a and $-\frac{1}{2}\Delta_a$ can be regarded as many-body Hamiltonians satisfying the assumptions of theorem 6.6.1, acting on the Hilbert spaces X, X^a and X_a. In particular, for all these Hamiltonians, we can define the asymptotic velocities. The asymptotic velocity for H_a will be denoted $P^+_{(a)}$; the asymptotic velocity for H^a will be denoted by $P^{(a),+}$ and the asymptotic velocity for $\frac{1}{2}D_a^2$ equals D_a. Clearly,

$$\left(P^+_{(a)}\right)_a = D_a,$$
$$\left(P^+_{(a)}\right)^a = P^{(a),+}.$$

Note that we write $P^+_{(a)}$ and $P^{(a),+}$ for the asymptotic velocities of H_a and H^a, and not P^+_a and $P^{a,+}$, because the latter symbols can be understood as the x_a- and x^a-components of P^+.

Note also that

$$\mathbb{1}_{Z_a}(P^+_{(a)}) = \mathbb{1}_{Z_a}(D_a) \otimes \mathbb{1}_{\{0\}}(P^{(a),+}).$$

But

$$\mathbb{1}_{Z_a}(D_a) = 1.$$

Therefore,

$$\mathbb{1}_{Z_a}(P^+_{(a)}) = \mathbb{1}_{\{0\}}(P^{(a),+}) = \mathbb{1}^{\mathrm{pp}}(H^a). \tag{6.7.8}$$

Hence the existence of (6.7.3) and (6.7.4) are actually analogous statements obtained by exchanging the roles of H and H_a.

The fact that (6.1.1) and (6.7.2) imply (6.6.1) follows from a remark in Sect. 4.6.

The proof of Theorem 6.7.1, given what we already know, is not very difficult. It is a prototype of a certain type of reasoning that we will repeat several times later on, while studying long-range scattering. It is useful to formalize this argument by introducing the following definition.

Definition 6.7.2
Let H_i, i=1,2, be many-body Hamiltonians on $L^2(X)$ and $\Theta \subset X$ be a Borel subset. Let P_i^+ be the corresponding asymptotic velocities. We say that e^{-itH_1} is asymptotic to e^{-itH_2} on Θ if the following limits exist:

$$\mathrm{s}-\lim_{t\to\infty} \mathrm{e}^{itH_2}\mathrm{e}^{-itH_1}\mathbb{1}_\Theta(P_1^+) =: \Gamma_{21}^+, \tag{6.7.9}$$

$$\mathrm{s}-\lim_{t\to\infty} \mathrm{e}^{itH_1}\mathrm{e}^{-itH_2}\mathbb{1}_\Theta(P_2^+) =: \Gamma_{12}^+. \tag{6.7.10}$$

Proposition 6.7.3
We have $\Gamma_{12}^+ = \Gamma_{21}^{+*}$. Moreover, Γ_{21}^+ is a partial isometry satisfying

$$\Gamma_{21}^{+*}\Gamma_{21}^+ = \mathbb{1}_\Theta(P_1^+), \qquad \Gamma_{21}^+\Gamma_{21}^{+*} = \mathbb{1}_\Theta(P_2^+), \tag{6.7.11}$$

$$\Gamma_{21}^+ P_1^+ = P_2^+\Gamma_{21}^+, \qquad \Gamma_{21}^+ H_1 = H_2\Gamma_{21}^+.$$

Proof. Let $J \in C_\infty(X)$. Then

$$\begin{aligned}\Gamma_{21}^+ J(P_1^+) &= \mathrm{s}-\lim_{t\to\infty} \mathrm{e}^{itH_2} J(\tfrac{x}{t})\mathrm{e}^{-itH_1}\mathbb{1}_\Theta(P_1^+)\\ &= J(P_1^+)\Gamma_{21}^+.\end{aligned}$$

Hence $\Gamma_{21}^+ P_1^+ = P_2^+\Gamma_{21}^+$. Therefore

$$\mathrm{Ran}\Gamma_{21}^+ \subset \mathrm{Ran}\mathbb{1}_\Theta(P_2^+).$$

The same statement is true if we exchange the roles of H_1 and H_2. Therefore, (6.7.11) is true by Lemma B.5.1. □

We will now show the following theorem that, by (6.7.8), implies Theorem 6.7.1.

Theorem 6.7.4
Assume (6.1.1) and (6.7.2). Then e^{-itH} *is asymptotic to* e^{itH_a} *on* Y_a.

Proof. Let us first show the existence of the limit

$$\mathrm{s-}\lim_{t\to\infty} \mathrm{e}^{itH}\mathrm{e}^{-itH_a}1\!\!1_{Y_a}(P^+_{(a)}). \tag{6.7.12}$$

Let $\chi \in C_0^\infty(\mathbb{R})$ and $J \in \mathcal{F} \cap C_0^\infty(X)$ such that $\mathrm{supp}J \subset Y_a$. Denote

$$\begin{aligned} M(t) &:= J(\tfrac{x}{t}) + \langle D - \tfrac{x}{t}, \nabla J(\tfrac{x}{t})\rangle, \\ \Phi(t) &:= \chi(H_a)M(t)\chi(H). \end{aligned}$$

By Lemma 6.6.4, we have

$$\mathrm{s-}\lim_{t\to\infty} \mathrm{e}^{itH_a}\mathrm{e}^{-itH}J(P^+)\chi^2(H) = \mathrm{s-}\lim_{t\to\infty} \mathrm{e}^{itH_a}M(t)\chi^2(H)\mathrm{e}^{-itH}.$$

Commuting $\chi(H)$ through $M(t)$ and using (6.6.6), this is equal to

$$\mathrm{s-}\lim_{t\to\infty} \mathrm{e}^{itH_a}\Phi(t)\mathrm{e}^{-itH}. \tag{6.7.13}$$

Now,

$$\begin{aligned} \tfrac{\mathrm{d}}{\mathrm{d}t}\Phi(t)(t) + iH_a\Phi(t) - i\Phi(t)H &= \chi(H_a)\nabla_x V^a(x)\nabla J\left(\tfrac{x}{t}\right)\chi(H) \\ &\quad + i\chi(H_a)I_a(x)M(t)\chi(H) \\ &\quad + t^{-1}\chi(H_a)\langle D - \tfrac{x}{t}, \nabla^2 J\left(\tfrac{x}{t}\right)(D - \tfrac{x}{t})\rangle\chi(H). \end{aligned}$$

The first two terms on the right of the above equation are integrable in norm by (6.7.2). The third is integrable along the evolution by Proposition 6.6.3. Therefore the limit (6.7.13) exists.

Next we interchange the role of H and H_a and we prove that

$$\mathrm{s-}\lim_{t\to\infty} \mathrm{e}^{itH_a}\mathrm{e}^{-itH}1\!\!1_{Y_a}(P^+) \tag{6.7.14}$$

exists. □

6.8 Asymptotic Separation of the Dynamics I

Suppose that the Hilbert space $\mathcal{H}$ is of the form

$$\mathcal{H} = \mathcal{H}_1 \otimes \mathcal{H}_2. \tag{6.8.1}$$

Let H be a self-adjoint operator on $\mathcal{H}$. In general, the evolution e^{-itH} does not preserve the factorization (6.8.1) and "mixes" the spaces $\mathcal{H}_1$ and $\mathcal{H}_2$. Only if the Hamiltonian H has the form

$$H = H_1 \otimes \mathbb{1}_{\mathcal{H}_2} + \mathbb{1}_{\mathcal{H}_1} \otimes H_2, \tag{6.8.2}$$

does the evolution

$$\mathrm{e}^{-itH} = \mathrm{e}^{-itH_1} \otimes \mathrm{e}^{-itH_2} \tag{6.8.3}$$

act independently in $\mathcal{H}_1$ and in $\mathcal{H}_2$. We will say that a Hamiltonian H satisfying (6.8.2) and a dynamics e^{-itH} satisfying (6.8.3) are *separated* with respect to the factorization (6.8.1).

In the case of N-body systems, for any $a \in \mathcal{A}$, the Hilbert space has a natural factorization

$$L^2(X) = L^2(X_a) \otimes L^2(X^a). \tag{6.8.4}$$

In general, an N-body Hamiltonian H is not separated with respect to (6.8.4). We can ask when e^{-itH} can be approximated by a dynamics separated with respect to (6.8.4) for $t \to \infty$.

Let us now describe some separated Hamiltonians that may be used to asymptotically approximate the full dynamics. The first one we already know: it is

$$\begin{aligned} H_a &= \tfrac{1}{2}D^2 + V^a(x^a) \\ &= \tfrac{1}{2}D_a^2 \otimes \mathbb{1} + \mathbb{1} \otimes H^a. \end{aligned}$$

Theorem 6.7.4 says that on $\mathrm{Ran}\,\mathbb{1}_{Y_a}(P^+)$ the full dynamics e^{-itH} can be approximated by the dynamics e^{-itH_a}.

This has, however, two drawbacks. First of all, this result is limited to the short-range case. Moreover, the approximating dynamics is valid on a relatively small subspace. Below we will describe separated dynamics that better approximate e^{-itH}.

It will be useful to study more closely the geometry of the configuration space and to introduce some new auxiliary definitions.

Fix $a \in \mathcal{A}$. Let us note that if we set

$$\mathcal{B}^a := \{b \in \mathcal{B} \mid b \leq a\} = \{b \in \mathcal{B} \mid X^b \subset X^a\},$$

then

$$V^a(x^a) = \sum_{b \in \mathcal{B}^a} v^b(x^b).$$

Now let us define

$$\mathcal{B}_{[a]} := \{b \in \mathcal{B} \mid X^b \subset X_a\}.$$

We set

$$\begin{aligned} V_{[a]}(x_a) &:= \sum_{b \in \mathcal{B}_{[a]}} v^b(x^b) \\ h_{[a]} &:= \tfrac{1}{2}D_a^2 + V_{[a]}(x_a), \\ I_{[a]}(x) &:= \sum_{b \notin \mathcal{B}_{[a]} \cup \mathcal{B}^a} v^b(x^b). \end{aligned}$$

If we note that $\mathcal{B}^a \cap \mathcal{B}_{[a]} = a_{\min}$ and $v^{a_{\min}} = 0$, then we see that

$$V(x) = V^a(x^a) + V_{[a]}(x_a) + I_{[a]}(x).$$

The following Hamiltonian is clearly separated:

$$\begin{aligned} H_{[a]} &:= \tfrac{1}{2}D^2 + V^a(x^a) + V_{[a]}(x_a) \\ &= h_{[a]} \otimes \mathbb{1} + \mathbb{1} \otimes H^a. \end{aligned}$$

The Hamiltonians $H_{[a]}$ and $h_{[a]}$ are many-body Hamiltonians acting on $L^2(X)$ and $L^2(X_a)$ and their asymptotic velocities will be denoted by $P^+_{[a]}$ and $p^+_{[a]}$ respectively. Clearly,

$$P^+_{[a]} = p^+_{[a]} \otimes \mathbb{1} + \mathbb{1} \otimes P^{(a),+}.$$

Set

$$\begin{aligned} Y_{[a]} &:= X \backslash \bigcup_{b \notin \mathcal{B}_{[a]} \cup \mathcal{B}^a} X_b, \\ Z_{[a]} &:= X_a \backslash \bigcup_{b \notin \mathcal{B}_{[a]} \cup \mathcal{B}^a} X_b, \\ Y_{[a],\mathrm{sep}} &:= Z_{[a]} + X^a = \{x \in X \mid x_a \in Z_{[a]}\}. \end{aligned}$$

It is easy to improve Theorem 6.7.4.

Theorem 6.8.1
Assume (6.1.1) and (6.7.2). Then $\mathrm{e}^{-\mathrm{i}tH}$ *is asymptotic to* $\mathrm{e}^{\mathrm{i}tH_{[a]}}$ *on* $Y_{[a]}$.

Note that $Y_{[a]}$, in general, is considerably bigger than Y_a, and hence $\mathrm{Ran}\,\mathbb{1}_{Y_{[a]}}(P^+)$ is larger than $\mathrm{Ran}\,\mathbb{1}_{Y_a}(P^+)$. Nevertheless, we are forced to assume the short-range assumptions, because we simply dropped the interaction $I_{[a]}(x)$. Let us define a third separated effective Hamiltonian, which will work for a large class of long-range potentials (including Coulomb potentials).

For any $a \in \mathcal{A}$, we define an auxiliary Hamiltonian on $L^2(X_a)$

$$\begin{aligned} h_a &:= \tfrac{1}{2}D_a^2 + I_a(x_a) \\ &= h_{[a]} + I_{[a]}(x_a). \end{aligned}$$

(Note that the "true interaction" $I_a(x)$ depends in principle both on x_a and x^a. In the auxiliary Hamiltonian h_a, we "freeze" the variable x^a to zero.) The following Hamiltonian on $L^2(X)$

$$\begin{aligned} H_{a-\mathrm{sep}} : &= \tfrac{1}{2}D^2 + V^a(x^a) + I_a(x_a) \\ &= \tfrac{1}{2}D^2 + V^a(x^a) + V_{[a]}(x_a) + I_{[a]}(x_a) \\ &= h_a \otimes \mathbb{1} + \mathbb{1} \otimes H^a \end{aligned}$$

is clearly separated with respect to (6.8.4). Clearly, $H_{a-\mathrm{sep}}$ and h_a are many-body Hamiltonians acting on $L^2(X)$ and $L^2(X_a)$ respectively. The asymptotic velocities associated with them will be denoted by $P^+_{a-\mathrm{sep}}$ and p^+_a respectively. Clearly,

$$P^+_{a-\mathrm{sep}} = p^+_a \otimes \mathbb{1} + \mathbb{1} \otimes P^{(a),+}.$$

The following theorem will be shown in Sects. 6.12, 6.13, 6.15:

Theorem 6.8.2
Assume (6.1.1). Suppose that $\mu = \sqrt{3} - 1$ and

$$\begin{aligned} & v^b(x^b) = v^b_{\mathrm{s}}(x^b) + v^b_{\mathrm{l}}(x^b), \;\; b \in \mathcal{B}, \\ & \int_0^\infty \left\| v^b_{\mathrm{s}}(x^b) \mathbb{1}_{[1,\infty[}\left(\tfrac{|x^b|}{R}\right)(1-\Delta^b)^{-1} \right\| R^\mu \mathrm{d}R < \infty, \quad b \in \mathcal{B}, \\ & \int_0^\infty \left\| \partial^\alpha_{x^b} v^b_{\mathrm{l}}(x^a) \mathbb{1}_{[1,\infty[}\left(\tfrac{|x^b|}{R}\right) \right\| R^\mu \mathrm{d}R < \infty, \;\; |\alpha| = 1, \quad b \in \mathcal{B}. \end{aligned} \tag{6.8.5}$$

Then for all $a \in \mathcal{A}$, the dynamics e^{itH} is asymptotic to $\mathrm{e}^{-itH_{a-\mathrm{sep}}}$ on $Z_{[a]}$.

Corollary 6.8.3
Under the assumptions of Theorem 6.8.2, there exist

$$\mathrm{s}-\lim_{t\to\infty} \mathrm{e}^{itH}\mathrm{e}^{-itH_{a-\mathrm{sep}}} \mathbb{1}_{Z_a}(p^+_a) \otimes \mathbb{1}^{\mathrm{pp}}(H^a), \tag{6.8.6}$$

$$\mathrm{s}-\lim_{t\to\infty} \mathrm{e}^{itH_{a-\mathrm{sep}}}\mathrm{e}^{-itH} \mathbb{1}_{Z_a}(P^+). \tag{6.8.7}$$

Denote (6.8.6) by Ξ^+_a. Then (6.8.7) equals Ξ^{+*}_a. The operator Ξ^+_a is a partial isometry such that

$$\begin{aligned} & \Xi^{+*}_a \Xi^+_a = \mathbb{1}^{\mathrm{pp}}(H^a) \otimes \mathbb{1}_{Z_a}(p^+_a), \qquad \Xi^+_a \Xi^{+*}_a = \mathbb{1}_{Z_a}(P^+), \\ & \Xi^+_a H_{a-\mathrm{sep}} = H \Xi^+_a, \qquad \Xi^+_a p^+_a = P^+_a \Xi^+_a. \end{aligned} \tag{6.8.8}$$

Moreover,

$$L^2(X) = \sum_{a\in\mathcal{A}}^{\oplus} \mathrm{Ran}\,\Xi^+_a. \tag{6.8.9}$$

Proof of Corollary 6.8.3 given Theorem 6.8.2. To see (6.8.6), (6.8.7) and (6.8.8), it is enough to note that $Z_a \subset Z_{[a]}$ and to use Theorem 6.8.2. To see (6.8.9), it is sufficient to use (6.7.1) and

$$\mathrm{Ran}\,\mathbb{1}_{Z_a}(P^+) = \mathrm{Ran}\,\Xi^+_a.$$

□

The property (6.8.9) is closely related to asymptotic completeness for long-range systems. In fact, we will use the operators Ξ^+_a and the identity (6.8.9) in our proof of the asymptotic completeness of long-range systems. One can even say that, up to minor technical assumptions, Corollary 6.8.3 is equivalent to asymptotic completeness. The advantage of (6.8.9) and of the concept of asymptotic separation is that it uses the operators Ξ_a^+ that are intrinsically defined, whereas modified wave operators are not.

Note that, in the short-range case, one can improve Theorem 6.8.2 as follows:

Theorem 6.8.4
Assume (6.1.1) and (6.7.2). Then for all $a \in \mathcal{A}$, the dynamics e^{-itH} is asymptotic to $\mathrm{e}^{-itH_{a-\mathrm{sep}}}$ on $Y_{[a]} \cap Y_{[a]-\mathrm{sep}}$.

Nevertheless, in the short-range case, the dynamics $\mathrm{e}^{-itH_{[a]}}$ seems to be a better approximation of the full dynamics than $\mathrm{e}^{-itH_{a-\mathrm{sep}}}$, as seen from Theorem 6.8.1.

Sometimes asymptotic separation is valid only on a part of the Hilbert space. To describe this phenomenon, let us introduce the following definition:

Definition 6.8.5
Let $a \in \mathcal{A}$. It is clear that

$$\{\phi \in L^2(X) \mid \text{ there exists s}-\lim_{t\to\infty} \mathrm{e}^{itH}\mathrm{e}^{-itH_{a-\mathrm{sep}}}\phi\}$$

is a closed linear subspace. We define $Q_{a-\mathrm{sep}}^+$ to be the orthogonal projection onto this subspace.

Likewise,

$$\{\phi \in L^2(X)) \mid \text{ there exists s}-\lim_{t\to\infty} \mathrm{e}^{itH_{a-\mathrm{sep}}}\mathrm{e}^{-itH}\phi\}$$

is a closed linear subspace. The orthogonal projection onto this subspace will be denoted by Q_a^+.

Some subspaces of $L^2(X)$ contained in the range of $Q_{a-\mathrm{sep}}^+$ and Q_a^+ will be described in Theorem 6.10.1.

The following proposition is an almost immediate consequence of Definition 6.8.5 and Lemma B.5.1:

Proposition 6.8.6
Assume (6.1.1) and (6.6.1). Let Q_a^+ and $Q_{a-\mathrm{sep}}^+$ be defined as above. Then for all $a \in \mathcal{A}$, there exists

$$\text{s}-\lim_{t\to\infty} \mathrm{e}^{itH}\mathrm{e}^{-itH_{a-\mathrm{sep}}}Q_{a-\mathrm{sep}}^+ \tag{6.8.10}$$

and

$$\text{s}-\lim_{t\to\infty} \mathrm{e}^{itH_{a-\mathrm{sep}}}\mathrm{e}^{-itH}Q_a^+. \tag{6.8.11}$$

If we denote (6.8.10) with $\Xi_{Q,a}^+$, then (6.8.11) equals $\Xi_{Q,a}^{+*}$. The operator $\Xi_{Q,a}^+$ is a partial isometry such that

$$\begin{aligned} \Xi_{Q,a}^{+*}\Xi_{Q,a}^+ &= Q_{a-\mathrm{sep}}^+, & \Xi_{Q,a}^+\Xi_{Q,a}^{+*} &= Q_a^+, \\ \Xi_{Q,a}^+ H_{a-\mathrm{sep}} &= H\Xi_{Q,a}^+, & \Xi_{Q,a}^+ P_{a-\mathrm{sep}}^+ &= P^+\Xi_{Q,a}^+. \end{aligned} \tag{6.8.12}$$

Proof. The existence of the limits (6.8.10) and (6.8.11) follows immediately from the definition 6.8.5. Then we apply Lemma B.5.1. □

Let us now consider the basic example of an N-body system with pair interactions. We can identify the set $\mathcal{B}$ with the family of pairs of $\{1,\ldots,N\}$ and $\mathcal{A}$ with the family of cluster decompositions of $\{1,\ldots,N\}$. Let us fix a cluster decomposition

$$a = \{C_1,\ldots,C_m\}.$$

Without restricting the generality we can assume that $C_1 = \{1\},\ldots,C_k = \{k\}$, and C_j have at least two elements for $j = k+1,\ldots,m$. The set $\mathcal{B}^a$ consists of all pairs contained in one of the clusters C_j, $j = k+1,\ldots,m$. The set $\mathcal{B}_{[a]}$ consists of all pairs contained in $\{1,\ldots,k\}$.

In the space X_a we can use the coordinates $y_1,\ldots,y_m$, where $y_i = x_i$, $1 \leq i \leq k$ are just the positions of the corresponding particles and y_i, $k+1 \leq i \leq m$ are the centers of mass of the clusters C_i. Suppose that the total masses of the clusters C_i are M_i for $k+1 \leq i \leq m$.

Note that

$$\begin{aligned} h_{[a]} &= \sum_{j=1}^{k} \tfrac{1}{2m_j} D_j^2 + \sum_{j=k+1}^{m} \tfrac{1}{2M_j} D_j^2 + \sum_{1\leq i<j\leq k} v_{ij}(y_i - y_j), \\ h_a &= \sum_{j=1}^{k} \tfrac{1}{2m_j} D_j^2 + \sum_{j=k+1}^{m} \tfrac{1}{2M_j} D_j^2 + \sum_{1\leq i<j\leq k} v_{ij}(y_i - y_j) \\ &+ \sum_{1\leq i\leq m,\ k+1\leq j\leq m,\ i<j} v_{ij}^{\mathrm{eff}}(y_i - y_j), \end{aligned}$$

where

$$\begin{aligned} v_{ij}^{\mathrm{eff}}(y_i - y_j) &= \sum_{p\in C_j} v_{ip}(y_i - y_j),\ 0 \leq i \leq k,\ k+1 \leq j \leq m, \\ v_{ij}^{\mathrm{eff}}(y_i - y_j) &= \sum_{p\in C_i,\ q\in C_j} v_{pq}(y_i - y_j),\ k+1 \leq i < j \leq m. \end{aligned}$$

In particular, if we have a system of N charged particles with charges Z_i, $i = 1,\ldots,N$, and 2-body potentials

$$v_{ij}(y_i - y_j) = \frac{Z_i Z_j}{|y_i - y_j|},$$

and if we set

$$Z_j^{\mathrm{eff}} := \sum_{p\in C_j} Z_p,\ k+1 \leq j \leq m,$$

then the effective potentials are equal

$$\begin{aligned} v_{ij}^{\mathrm{eff}}(y_i - y_j) &= \tfrac{Z_i Z_j^{\mathrm{eff}}}{|y_i - y_j|},\ 0 \leq i \leq k,\ k+1 \leq j \leq m, \\ v_{ij}^{\mathrm{eff}}(y_i - y_j) &= \tfrac{Z_i^{\mathrm{eff}} Z_j^{\mathrm{eff}}}{|y_i - y_j|},\ k+1 \leq i,j \leq m. \end{aligned}$$

Thus Theorem 6.8.2 about asymptotic separation says that a system of simple particles $1,\ldots,k$ and"composite particles" $C_{k+1},\ldots,C_m$ can be described asymptotically by an m-body Hamiltonian h_a, provided that we avoid collisions involving "composite particles" (which means restricting to $\mathrm{Ran}\,\mathbb{1}_{Z_{[a]}}(P^+)$). In the Coulomb case, if the composite particles are neutral ($Z_j^{\mathrm{eff}} = 0$, $k+1 \leq j \leq m$), then asymptotically the composite particles do not influence the simple particles at all.

6.9 Time-Dependent N-Body Hamiltonians

When studying N-body Hamiltonians it is convenient to introduce a certain class of Hamiltonians whose potential is a sum of an N-body part and a time-dependent part. More precisely, let us assume that the Hilbert space is

$$L^2(X) \otimes \mathcal{H}_1,$$

where $\mathcal{H}_1$ describes some additional degrees of freedom. Let us introduce a class of Hamiltonians of the form

$$\check{H}(t) = H \otimes \mathbb{1}_{\mathcal{H}_1} + W(t,x),$$

where $\mathbb{R}^+ \times X \ni (t,x) \mapsto W(t,x) \in B(\mathcal{H}_1)$ satisfies

$$\int_0^\infty \sup_{x\in X} \|\nabla_x W(t,x)\|_{B(\mathcal{H}_1)} \mathrm{d}t < \infty. \tag{6.9.1}$$

We will denote by $\check{U}(t,s)$ the dynamics generated by $\check{H}(t)$. We define the following types of Heisenberg derivatives:

$$\begin{aligned} \check{\mathbf{D}}A(t) &:= \tfrac{\mathrm{d}}{\mathrm{d}t}A(t) + [\check{H}(t), iA(t)], \\ \mathbf{D}A(t) &:= \tfrac{\mathrm{d}}{\mathrm{d}t}A(t) + [H, iA(t)], \\ \mathbf{D}_0A(t) &:= \tfrac{\mathrm{d}}{\mathrm{d}t}A(t) + [\tfrac{1}{2}D^2, iA(t)]. \end{aligned} \tag{6.9.2}$$

Let us introduce the following asymptotic observables associated with the time-dependent Hamiltonian $\check{H}(t)$.

Proposition 6.9.1
Suppose that H is an N-body Hamiltonian satisfying (6.1.1) and (6.6.1). The following limits exist and define operators with dense domains:

$$\begin{aligned} \check{H}^+ &:= C_\infty\!-\lim_{t\to\infty} \check{U}(0,t) H \check{U}(t,0), \\ \check{P}^+ &= \mathrm{s}\!-\!C_\infty\!-\lim_{t\to\infty} \check{U}(0,t)\tfrac{x}{t}\check{U}(t,0). \end{aligned} \tag{6.9.3}$$

Moreover,

$$[\check{H}^+, \check{P}^+] = 0.$$

Proof. Let $\chi \in C_0^\infty(\mathbb{R})$. Using Lemma C.1.2, we get

$$\begin{aligned}\|\check{\mathbf{D}}\chi(H)\| &= \|[W(t,x), i\chi(H)]\| \\ &\le C\|[W(t,x),(H+i)^{-1}]\| \le C\|\nabla_x W(t,x)\|,\end{aligned}$$

which is integrable. This shows the existence of

$$\lim_{t\to\infty} \check{U}(0,t)\chi(H)\check{U}(t,0).$$

Therefore $\check{H}^+$ exists.

Next let us assume, in addition, that $\chi(0) = 1$. Using

$$\|\check{\mathbf{D}}\chi(T^{-1}H)\| \le CT^{-1}\|\nabla_x W(t,x)\|, \tag{6.9.4}$$

we obtain

$$\begin{aligned}&\|\lim_{t\to\infty} \check{U}(0,t)\chi(T^{-1}H)\check{U}(t,0) - \chi(T^{-1}H)\| \\ &\le CT^{-1}\textstyle\int_0^\infty \|\nabla_x W(t,\cdot)\|_\infty \mathrm{d}t.\end{aligned} \tag{6.9.5}$$

But (6.9.5) goes to zero as $T \to \infty$. Hence

$$\text{s}-\lim_{T\to\infty}\left(\lim_{t\to\infty}\check{U}(0,t)\chi(T^{-1}H)\check{U}(t,0)\right) = \mathbb{1},$$

and so $\check{H}^+$ is densely defined.

The proof of the existence of the limit defining $\check{P}^+$ is completely analogous to the one given in Sect. 6.6. Due to the presence of $W(t,x)$, additional terms arise in the Heisenberg derivative of the various propagation observables. It follows from hypothesis (6.9.1) that these terms are integrable in norm. □

Essentially all the results valid for time-independent Hamiltonians described in this chapter generalize to the case of time-dependent Hamiltonians introduced at the beginning of this section. For further reference, we will state a generalization of Proposition 6.6.8 to time-dependent perturbations.

Lemma 6.9.2
Let $\chi \in C_0^\infty(\mathbb{R})$ such that $\operatorname{supp}\chi \cap \mathcal{T} = \emptyset$. *Then there exists $\epsilon > 0$ such that*

$$\int_1^\infty \left\| \mathbb{1}_{[0,\epsilon]}\left(\frac{|x|}{t}\right)\chi(H)\mathbb{1}^{\mathrm{c}}(H)\check{U}(t,0)\phi\right\|^2 \frac{\mathrm{d}t}{t} \le C\|\phi\|^2. \tag{6.9.6}$$

Proof. The proof is very similar to the one of Proposition 6.6.8. Using the hypothesis (6.9.1) on $W(t,x)$, it is easy to see that all the additional terms in the Heisenberg derivative coming from $W(t,x)$ are integrable in norm. □

It will be convenient to extend the definition 6.7.2 to the case of time-dependent Hamiltonians.

Definition 6.9.3
Let H_i be many-body Hamiltonians on $L^2(X)$ and let $W_i(t)$ satisfy the condition 6.9.1. Define the following operators on $L^2(X)\otimes\mathcal{H}_1$:

$$\check{H}_i(t) := H_i + W_i(t,x).$$

Let $\check{U}_i(t,s)$ be the unitary dynamics generated by $\check{H}_i(t)$. Let $\check{P}_i^+$ and $\check{H}_i^+$ be the corresponding asymptotic velocities and the asymptotic Hamiltonians. Let Θ be a Borel subset of X.

We say that $\check{U}_1(t,0)$ is asymptotic to $\check{U}_2(t,0)$ on Θ if the following limits exist:

$$\text{s-}\lim_{t\to\infty} \check{U}_2(0,t)\check{U}_1(t,0)\mathbb{1}_\Theta(\check{P}_1^+), \tag{6.9.7}$$

$$\text{s-}\lim_{t\to\infty} \check{U}_1(0,t)\check{U}_2(t,0)\mathbb{1}_\Theta(\check{P}_2^+). \tag{6.9.8}$$

We have the following generalization of Proposition 6.7.3.

Proposition 6.9.4
Denote (6.9.7) by $\check{\Gamma}_{21}^+$. Then (6.9.8) equals $\check{\Gamma}_{21}^{+*}$. It is a partial isometry satisfying

$$\check{\Gamma}_{21}^{+*}\check{\Gamma}_{21}^+ = \mathbb{1}_\Theta(P_1^+), \qquad \check{\Gamma}_{21}^+\check{\Gamma}_{21}^{+*} = \mathbb{1}_\Theta(P_2^+),$$
$$\check{\Gamma}_{21}^+ P_1^+ = P_2^+\check{\Gamma}_{21}^+.$$

If moreover $H_1 = H_2$, then

$$\check{\Gamma}_{21}^+\check{H}_1^+ = \check{H}_2^+\check{\Gamma}_{21}^+.$$

It is often useful to replace the full dynamics e^{-itH} by the dynamics generated by an effective time-dependent Hamiltonian. This trick due to Sigal-Soffer [SS2] is described in the following proposition.

Proposition 6.9.5
Assume (6.1.1) and

$$v^b(x^b) = v_\mathrm{s}^b(x^b) + v_\mathrm{l}^b(x^b),\ \ b\in\mathcal{B},$$
$$\textstyle\int_0^\infty \left\|(1-\Delta^b)^{-1/2}v_\mathrm{s}^b(x^b)\mathbb{1}_{[1,\infty[}(\frac{|x^b|}{R})(1-\Delta^b)^{-1}\right\|\mathrm{d}R<\infty,\ \ b\in\mathcal{B}, \tag{6.9.9}$$
$$\textstyle\int_0^\infty \|\nabla v_\mathrm{l}^b(x^b)\mathbb{1}_{[1,\infty[}(\frac{|x^b|}{R})\|_\infty \mathrm{d}R<\infty,\ \ b\in\mathcal{B}.$$

Suppose that $a\in\mathcal{A}$, $\Theta\subset Y_a$ is compact, $\check{J}\in C_0^\infty(X)$, $\check{J}=1$ on Θ, $\mathrm{supp}\check{J}\subset Y_a$ and y is a fixed element of Y_a. We set

$$\check{I}_a(t,x) := \check{J}\left(\frac{x}{t}\right)(I_{a,\mathrm{l}}(x) - I_{a,\mathrm{l}}(ty)) + I_{a,\mathrm{l}}(ty).$$

Define

$$\check{H}_a(t) := H_a + \check{I}_a(t,x).$$

Let $\check{U}_a(t,s)$ be the dynamics generated by $\check{H}_a(t)$. Then there exist

$$\check{H}_a^+ := C_\infty - \lim_{t\to\infty} \check{U}_a(0,t) H_a \check{U}_a(t,0),$$
$$\check{H}^{a,+} := C_\infty - \lim_{t\to\infty} \check{U}_a(0,t) H^a \check{U}_a(t,0),$$
$$\check{P}_{(a)}^+ := \mathrm{s}-C_\infty - \lim_{t\to\infty} \check{U}_a(0,t) \tfrac{x}{t} \check{U}_a(t,0).$$

Moreover,

$$\check{H}_a^+ = \frac{1}{2}(\check{P}_{(a)}^+)_a^2 + \check{H}^{a,+}, \tag{6.9.10}$$

$$(\check{P}_{(a)}^+)_a = C_\infty - \lim_{t\to\infty} \check{U}_a(0,t) D_a \check{U}_a(t,0). \tag{6.9.11}$$

Furthermore, e^{-itH} *is asymptotic to $\check{U}_a(t,0)$ on Θ.*

Proof. We may identify $\check{H}_a(t)$ with $\check{H}(t)$ considered at the beginning of this section in two different ways.

First we identify

$$X^a,\ L^2(X_a),\ H^a,\ \tfrac{1}{2}D_a^2 + \check{I}_a(t,x),\ \check{H}_a(t) \quad \text{with} \quad X,\ \mathcal{H}_1,\ H,\ W(t,x),\ \check{H}(t).$$

The fact that $W(t,x)$ satisfies (6.9.1) follows from the arguments in the proof of Proposition 4.7.5. By Proposition 6.9.1, we obtain the existence of $(\check{P}_{(a)}^+)^a$, $\check{H}^{a,+}$, which are identified with $\check{P}^+$, $\check{H}^+$.

Next we identify

$$X_a,\ L^2(X^a),\ \tfrac{1}{2}D_a^2,\ H^a + \check{I}_a(t,x),\ \check{H}_a(t) \quad \text{with} \quad X,\ \mathcal{H}_1,\ H,\ W(t,x),\ \check{H}(t).$$

Note that this time $\check{H}(t)$ belongs to the class considered in Sect. 3.2 (with internal degrees of freedom). Using the results of Sect. 3.2, we obtain the existence of $(\check{P}_{(a)}^+)_a$ and the identity (6.9.11). Now (6.9.10) is easy and left to the reader.

Let us now prove the existence of the relative wave operator

$$\mathrm{s}-\lim_{t\to\infty} \mathrm{e}^{itH} \check{U}_a(t,0) 1\!\!1_\Theta(\check{P}_{(a)}^+). \tag{6.9.12}$$

Let $\chi \in C_0^\infty(\mathbb{R})$, $J \in C_0^\infty(X) \cap \mathcal{F}$ such that $\mathrm{supp} J \subset \Theta$. Set

$$M(t) := J\left(\tfrac{x}{t}\right) + \langle D - \tfrac{x}{t}, \nabla J\left(\tfrac{x}{t}\right)\rangle,$$
$$\Phi(t) := \chi(H) M(t) \chi(H_a).$$

Using Lemmas 6.2.4 and 6.6.4 applied to $\check{U}_a(t,0)$, we have

$$\chi^2(\check{H}_a^+) J(\check{P}_{(a)}^+) = \mathrm{s}-\lim_{t\to\infty} \check{U}_a(0,t) \Phi(t) \check{U}_a(t,0).$$

Hence

$$\begin{aligned}&\text{s-}\lim_{t\to\infty} \mathrm{e}^{itH}\check{U}_a(t,0)\chi^2(\check{H}_a^+)J(\check{P}_{(a)}^+)\\ &= \text{s-}\lim_{t\to\infty} \mathrm{e}^{itH}\Phi(t)\check{U}_a(t,0).\end{aligned} \tag{6.9.13}$$

Now,

$$\begin{aligned}\tfrac{\mathrm{d}}{\mathrm{d}t}\Phi(t) + iH\Phi(t) - i\Phi(t)\check{H}_a(t) &= \chi(H)\nabla_x V(x)\nabla J\left(\tfrac{x}{t}\right)\chi(H_a)\\ &+\chi(H)M(t)[\chi(H_a), i\check{I}_a(t,x)]\\ &+\chi(H)M(t)I_{a,\mathrm{s}}(x)\chi(H_a)\\ &+t^{-1}\chi(H)\langle D-\tfrac{x}{t}, \nabla^2 J\left(\tfrac{x}{t}\right)\left(D-\tfrac{x}{t}\right)\rangle\chi(H).\end{aligned} \tag{6.9.14}$$

We used here the fact that

$$I_{a,\mathrm{l}}(x) = \check{I}_a(t,x), \qquad \tfrac{x}{t} \in \mathrm{supp} J.$$

The first three terms on the right of (6.9.14) are integrable in norm. The last term is integrable along the evolution. Therefore the limit (6.9.13) and, consequently, the limit (6.9.12) exist.

The proof of the existence of the limit

$$\text{s-}\lim_{t\to\infty} \check{U}_a(0,t)\mathrm{e}^{-itH}\mathbb{1}_\Theta(P^+) \tag{6.9.15}$$

is analogous. □

6.10 Joint Spectrum of P^+ and H

The existence and completeness of wave operators gives a very satisfactory description of the asymptotic behavior of e^{-itH}. Unfortunately, it is known that the asymptotic completeness fails for $\mu < 1/2$ and is proven only for $\mu > \sqrt{3}-1$. Nevertheless, even with $\mu \geq 0$, N-body Hamiltonians are quite well behaved from the point of view of scattering theory. We saw this in Sects. 6.2, 6.3, 6.4, and especially in Sect. 6.6, where we showed the existence of the asymptotic velocity P^+. In this section we continue the study of N-body systems with $\mu \geq 0$ from the point of view of spectral properties of P^+.

If asymptotic separation (or completeness) holds, then we can fully describe the joint spectrum of (P^+, H) in terms of the collision subspaces and thresholds. Namely, if we consider the asymptotic momentum $\xi = \xi_a \in Z_a$, then the only possible open channels are those related to thresholds and bound states of H_a and hence the energy takes only the values

$$\tau + \tfrac{1}{2}\xi_a^2, \qquad \tau \in \sigma_{\mathrm{pp}}(H^a) \cup \mathcal{T}^a.$$

Our first result will say that the same description is also true under the assumption $\mu \geq 0$.

Then we will give results related to the notion of asymptotic separation. Recall from Sect. 6.8 that we introduce there the relative wave operators Ξ_a^+, which were unitary operators from $\mathrm{Ran}Q_{a-\mathrm{sep}}^+$ to $\mathrm{Ran}Q_a^+$. We will describe some large subspaces of $L^2(X)$ that are contained in $\mathrm{Ran}Q_{a-\mathrm{sep}}^+$ and $\mathrm{Ran}Q_a^+$. Moreover, we will see that asymptotic separation is true on a large subspace of the Hilbert space $L^2(X)$. This subspace can be described explicitly in terms of the joint spectral measure of H and P^+. (We will see in Sect. 6.15 that this subspace is included in the range of modified wave operators.)

Theorem 6.10.1
Assume (6.1.1) and

$$\begin{aligned} & v^b(x^b) = v_{\mathrm{s}}^b(x^b) + v_{\mathrm{l}}^b(x^b), \quad b \in \mathcal{B}, \\ & \int_0^\infty \left\|(1-\Delta^b)^{-1/2} v_{\mathrm{s}}^b(x^b) \mathbb{1}_{[1,\infty[}\left(\tfrac{|x^b|}{R}\right)(1-\Delta^b)^{-1}\right\| \mathrm{d}R < \infty, \quad b \in \mathcal{B}, \qquad (6.10.1) \\ & \int_0^\infty \|\nabla v_{\mathrm{l}}^b(x^b) \mathbb{1}_{[1,\infty[}\left(\tfrac{|x^b|}{R}\right)\|_\infty \mathrm{d}R < \infty, \quad b \in \mathcal{B}. \end{aligned}$$

Then the following is true:
(i)

$$\sigma(P^+, H) = \bigcup_{a \in \mathcal{A}} \left\{ \left(\xi_a, \tau + \tfrac{1}{2}\xi_a^2\right) \mid \xi_a \in X_a,\ \tau \in \sigma^{\mathrm{pp}}(H^a) \right\}. \qquad (6.10.2)$$

(ii) Let $\mathbb{1}_{\mathrm{reg}}^{\mathrm{pp}}(H^a)$ *denote the projection onto* $\overline{\mathrm{Ran}\mathbb{1}^{\mathrm{pp}}(H^a) \cap \mathcal{D}(\langle x^a \rangle)}$. *Then*

$$\mathbb{1}_{Z_{[a]}}(p_a^+) \otimes \mathbb{1}_{\mathbb{R}\setminus\mathcal{T}^a}^{\mathrm{pp}}(H^a) \leq \mathbb{1}_{Z_{[a]}}(p_a^+) \otimes \mathbb{1}_{\mathrm{reg}}^{\mathrm{pp}}(H^a) \leq Q_{a,\mathrm{sep}}^+.$$

(iii) Let $\Sigma_{a,\mathrm{reg}} := \{(\xi_a, \tau + \frac{1}{2}\xi_a^2) \mid \xi_a \in Z_a,\ \tau \in \sigma_{\mathrm{pp}}(H^a)\setminus\mathcal{T}^a\}$. *Then*

$$\mathbb{1}_{\Sigma_{a,\mathrm{reg}}}(P^+, H) \leq Q_a^+.$$

Remark. To some extent, the part *(i)* of this theorem is the quantum analog of Theorem 5.6.2. In the classical case, however, we could only prove that the joint image of H and P^+ contained a certain set related to the so-called regular a−trajectories (see Corollary 5.6.4), which corresponds to the $\subset$ inclusion in (6.10.2). This was due to the instability of bounded trajectories for classical systems. In the quantum case, most bound states are strongly localized as Theorem 6.5.1 shows, which is the reason for the stronger result.

The proof of the above theorem is divided into a series of lemmas.

Lemma 6.10.2

$$\mathbb{1}_{Z_{[a]}}(p_a^+) \otimes \mathbb{1}^{\mathrm{pp}}_{\mathbb{R}\setminus\mathcal{T}^a}(H^a) \leq \mathbb{1}_{Z_{[a]}}(p_a^+) \otimes \mathbb{1}^{\mathrm{pp}}_{\mathrm{reg}}(H^a) \leq Q^+_{a-\mathrm{sep}}.$$

Hence Theorem 6.10.1 (ii) is true.

Proof. The first inequality follows by the exponential decay of non-threshold eigenfunctions (see Theorem 6.5.1). Let us show the second inequality.

Let $\phi \in \mathrm{Ran}\mathbb{1}_{\{\tau\}}(H^a) \cap \mathcal{D}(\langle x^a\rangle)$. Let P_ϕ denote the projection onto ϕ. By density, it is enough to show the existence of

$$\mathrm{s-}\lim_{t\to\infty} \mathrm{e}^{itH}\mathrm{e}^{-itH_{a-\mathrm{sep}}} g(p_a^+)\chi^2(H_{a-\mathrm{sep}})P_\phi \tag{6.10.3}$$

for any $g \in C_0^\infty(X_a)$ such that $\mathrm{supp} g \subset Z_{[a]}$ and $\chi \in C_0^\infty(\mathbb{R})$.

We may assume that there exists $J \in C_0^\infty(X) \cap \mathcal{F}$ such that $J|_{X_a} = g$. Additionally, we may suppose that

$$(x_a, x^a) \in \mathrm{supp} J \Rightarrow (x_a, s x^a) \in Y_{[a]}, \quad 0 \leq s \leq 1. \tag{6.10.4}$$

Clearly,

$$g(p_a^+)P_\phi = J(P^+_{a-\mathrm{sep}})P_\phi.$$

Set

$$\begin{aligned} M(t) &:= J\left(\tfrac{x}{t}\right) + \langle D - \tfrac{x}{t}, \nabla_x J\left(\tfrac{x}{t}\right)\rangle, \\ \Phi(t) &:= \chi(H)M(t)P_\phi\chi(H_{a-\mathrm{sep}}). \end{aligned} \tag{6.10.5}$$

By Lemmas 6.2.4 and 6.6.4, (6.10.3) equals

$$\mathrm{s-}\lim_{t\to\infty} \mathrm{e}^{itH}\Phi(t)\mathrm{e}^{-itH_{a-\mathrm{sep}}}. \tag{6.10.6}$$

We have

$$\begin{aligned} \tfrac{\mathrm{d}}{\mathrm{d}t}\Phi(t) + iH\Phi(t) - i\Phi(t)H_{a-\mathrm{sep}} &= \chi(H)\nabla_x V(x)\nabla J\left(\tfrac{x}{t}\right) P_\phi\chi(H_{a-\mathrm{sep}}) \\ &+\chi(H)M(t)(I_{[a]}(x) - I_{[a]}(x_a))P_\phi\chi(H_{a-\mathrm{sep}}) \\ &+t^{-1}\chi(H)\langle D - \tfrac{x}{t}, \nabla^2 J\left(\tfrac{x}{t}\right)\left(D - \tfrac{x}{t}\right)\rangle P_\phi\chi(H_{a-\mathrm{sep}}) \\ &= I_1(t) + I_2(t) + I_3(t). \end{aligned}$$

$I_1(t)$ is integrable in norm. $I_3(t)$ is integrable along the evolution. $I_2(t)$ we can write as

$$\begin{aligned} I_2(t) &= \chi(H)M(t)(I_{[a],\mathrm{s}}(x) - I_{[a],\mathrm{s}}(x_a))\chi(H_{a-\mathrm{sep}})P_\phi \\ &+\chi(H)M(t)\textstyle\int_0^1 \langle \nabla I_{[a],\mathrm{l}}(x_a, s x^a), x^a\rangle P_\phi\chi(H_{a-\mathrm{sep}})\mathrm{d}s, \end{aligned}$$

which is integrable in norm, using $\|x^a P_\phi\| < \infty$ and

$$\int_1^\infty \|\chi(H)M(t)\nabla I_{[a],\mathrm{l}}(x_a, s x^a)\|\mathrm{d}t < C,$$

which follows from (6.10.4). Therefore the limit (6.10.3) exists. □

Lemma 6.10.3
We have

$$\sigma(P^+) = X, \tag{6.10.7}$$

$$\frac{1}{2}(P^+)^2 \mathbb{1}_{Z_{a_{\min}}}(P^+) = H \mathbb{1}_{Z_{a_{\min}}}(P^+). \tag{6.10.8}$$

Proof. One way to prove (6.10.7) is to follow the proof of Proposition 4.5.2. Using an effective time-dependent potential, it is possible to give a different proof.

Let $\Theta \subset Z_{a_{\min}}$ be compact. Let $\check{J} \in C_0^\infty(X)$ such that $\mathrm{supp}\check{J} \subset Z_{a_{\min}}$ and $\check{J} = 1$ on Θ and $y_0 \in \Theta$. We introduce $\check{I}_{a_{\min}}(t,x)$, $\check{H}_{a_{\min}}(t)$, $\check{U}_{a_{\min}}(t,s)$, $\check{P}^+_{(a_{\min})}$, $\check{H}^+_{a_{\min}}$ as in Proposition 6.9.5. Note that

$$\check{H}(t) = -\frac{1}{2}D^2 + \check{I}_{a_{\min}}(t,x),$$

hence $\check{H}(t)$ belongs to the class of Hamiltonians considered in Chap. 3. In Chap. 3 we showed that such Hamiltonians satisfy

$$\begin{aligned}
\check{P}^+_{(a_{\min})} &= C_\infty\text{-}\lim_{t\to\infty} \check{U}_{a_{\min}}(0,t) D \check{U}_{a_{\min}}(t,0), \\
\sigma(\check{P}^+_{(a_{\min})}) &= X, \\
\check{H}^+_{a_{\min}} &= \tfrac{1}{2}(\check{P}^+_{(a_{\min})})^2.
\end{aligned}$$

Moreover, e^{-itH} is asymptotic to $\check{U}_{a_{\min}}(t,0)$ on Θ. Hence

$$\begin{aligned}
\sigma\left(P^+\big|_{\mathrm{Ran}\,\mathbb{1}_\Theta(P^+)}\right) &= \Theta, \\
H\mathbb{1}_\Theta(P^+) &= \tfrac{1}{2}(P^+)^2 \mathbb{1}_\Theta(P^+).
\end{aligned}$$

Taking a sequence of compact subsets $\Theta_n \nearrow Z_{a_{\min}}$ we obtain (6.10.7) and (6.10.8). □

Lemma 6.10.4

$$\sigma(P^+, H) \supset \bigcup_{a\in\mathcal{A}} \left\{ \left(x_a, \tau + \frac{1}{2}x_a^2\right) \mid x_a \in X_a,\ \tau \in \sigma^{\mathrm{pp}}(H^a) \right\}. \tag{6.10.9}$$

Proof. By (6.10.8) applied to h_a, we obtain

$$h_a \mathbb{1}_{Z_a}(p_a^+) = \frac{1}{2}(p_a^+)^2 \mathbb{1}_{Z_a}(p_a^+).$$

Hence

$$H_{a-\mathrm{sep}}1\!\!1_{Z_a}(p_a^+)\otimes 1\!\!1_{\{\tau\}}(H^a)=(\tau+\tfrac{1}{2}(p_a^+)^2)1\!\!1_{Z_a}(p_a^+)\otimes 1\!\!1_{\{\tau\}}(H^a).$$

By Lemma 6.10.2,

$$1\!\!1_{\sigma^{\mathrm{pp}}(H^a)\backslash\mathcal{T}^a}(H^a)\otimes 1\!\!1_{Z_a}(p_a^+)\le Q_{a-\mathrm{sep}}^+.$$

Therefore,

$$\begin{aligned}&\sigma\left(P_{a-\mathrm{sep}}^+|_{\mathrm{Ran}Q_{a-\mathrm{sep}}^+},H_{a-\mathrm{sep}}|_{\mathrm{Ran}Q_{a-\mathrm{sep}}^+}\right)\\ &\supset\left\{\left(x_a,\tau+\tfrac{1}{2}x_a^2\right)\mid x_a\in Z_a,\ \tau\in\sigma^{\mathrm{pp}}(H^a)\backslash\mathcal{T}^a\right\}.\end{aligned}$$

But the operator $\Xi_{Q,a}^+$ is unitary from $\mathrm{Ran}Q_{a-\mathrm{sep}}^+$ to $\mathrm{Ran}Q_a^+$. Moreover,

$$\Xi_{Q,a}^+H_{a-\mathrm{sep}}=H\Xi_{Q,a}^+,\qquad \Xi_{Q,a}^+P_{a-\mathrm{sep}}^+=P^+\Xi_{Q,a}^+.$$

Hence, for any $a\in\mathcal{A}$,

$$\sigma\left(P^+|_{\mathrm{Ran}Q_a^+},H|_{\mathrm{Ran}Q_a^+}\right)\supset\left\{\left(x_a,\tau+\frac{1}{2}x_a^2\right)\mid x_a\in Z_a,\ \tau\in\sigma^{\mathrm{pp}}(H^a)\backslash\mathcal{T}^a\right\}.$$

Since $\sigma(P^+,H)$ is closed, this implies

$$\begin{aligned}\sigma(P^+,H)\ &\supset\overline{\cup_{a\in\mathcal{A}}\left\{\left(x_a,\tau+\tfrac{1}{2}x_a^2\right)\mid x_a\in Z_a,\ \tau\in\sigma^{\mathrm{pp}}(H^a)\backslash\mathcal{T}^a\right\}}\\ &=\cup_{a\in\mathcal{A}}\left\{\left(x_a,\tau+\tfrac{1}{2}x_a^2\right)\mid x_a\in X_a,\ \tau\in\sigma^{\mathrm{pp}}(H^a)\right\}.\end{aligned}$$

This completes the proof of the lemma. □

Lemma 6.10.5
Let $J\in C_0^\infty(X)$, $\mathrm{supp}J\subset Y_a$, $\chi\in C_0^\infty(\mathbb{R})$. Then the following is true:
(i) We have

$$\mathrm{s-}\lim_{t\to\infty}\mathrm{e}^{itH}J(\tfrac{x}{t})\chi(H^a)\mathrm{e}^{-itH}=J(P^+)\chi(H-\tfrac{1}{2}(P_a^+)^2).$$

(ii) Let $\phi\in\mathcal{D}(\langle x^a\rangle)\cap\mathrm{Ran}1\!\!1^{\mathrm{pp}}(H^a)$ and let P_ϕ be the orthogonal projection onto ϕ. Then there exists

$$\mathrm{s-}\lim_{t\to\infty}\mathrm{e}^{itH}J(\tfrac{x}{t})P_\phi\mathrm{e}^{-itH}.$$

(iii) Assume additionally, that $\mathrm{supp}\chi\cap\mathcal{T}^a=\emptyset$. Then there exists $\epsilon>0$ such that if $\mathrm{supp}J\subset\{x\mid |x^a|<\epsilon\}$, then

$$\mathrm{s-}\lim_{t\to\infty}J(\tfrac{x}{t})1\!\!1^{\mathrm{c}}(H^a)\chi(H^a)\mathrm{e}^{-itH}=0.$$

Proof. Let $J_1,\check{J}\in C_0^\infty(X)$ such that $\mathrm{supp}J_1,\mathrm{supp}\check{J}\subset Y_a$, $\Theta\subset Y_a$ such that $\mathrm{supp}J\subset\Theta$, $J_1=1$ on Θ, and $J_1\check{J}=J_1$. Using $\check{J}$, define an effective Hamiltonian

$\check{H}_a(t)$, the corresponding asymptotic velocity $\check{P}^+_{(a)}$, etc. as in Proposition 6.9.5. Let

$$\check{\Gamma}^+_{a,\Theta} := \text{s-}\lim_{t\to\infty} \mathrm{e}^{itH}\check{U}_a(t,0)\mathbb{1}_\Theta(\check{P}^+_{(a)}).$$

(i) follows from the following computations:

$$\begin{aligned}
&\text{s-}\lim_{t\to\infty} \mathrm{e}^{itH}J(\tfrac{x}{t})\chi(H^a)\mathrm{e}^{-itH}\\
&= \text{s-}\lim_{t\to\infty} \mathrm{e}^{itH}J(\tfrac{x}{t})\chi(H^a)J_1(\tfrac{x}{t})\mathrm{e}^{-itH}\\
&= \text{s-}\lim_{t\to\infty} \mathrm{e}^{itH}\check{U}_a(t,0)J(\check{P}^+_{(a)})\\
&\times \text{s-}\lim_{t\to\infty} \check{U}_a(0,t)\chi(H^a)\check{U}_a(t,0)\text{s-}\lim_{t\to\infty} \check{U}_a(0,t)\mathrm{e}^{-itH}J_1(P^+)\\
&= \check{\Gamma}^+_{a,\Theta}J(\check{P}^+_{(a)})\chi(\check{H}^{a,+})\check{\Gamma}^{+*}_{a,\Theta}J_1(P^+)\\
&= J(P^+)\check{\Gamma}^+_{a,\Theta}\chi(\check{H}^+_a - \tfrac{1}{2}(\check{P}^+_{(a)})^2_a))\Gamma^{+*}_{a,\Theta}J_1(P^+)\\
&= J(P^+)\chi(H - \tfrac{1}{2}(P^+_a)^2))J_1(P^+).
\end{aligned}$$

To prove *(ii)*, we first note that, by an obvious modification of Lemma 6.6.7 *(i)*, $\phi \in \mathcal{D}(\langle x^a\rangle)$ implies

$$[J_1(\tfrac{x}{t}), P_\phi] \in O(t^{-1}).$$

Therefore

$$\begin{aligned}
\text{s-}\lim_{t\to\infty} \mathrm{e}^{itH}J(\tfrac{x}{t})P_\phi\mathrm{e}^{-itH} &= \text{s-}\lim_{t\to\infty} \mathrm{e}^{itH}J(\tfrac{x}{t})P_\phi J_1(\tfrac{x}{t})\mathrm{e}^{-itH}\\
&= \Gamma^+_{a,\Theta}J(\check{P}^+_{(a)})\text{s-}\lim_{t\to\infty} \check{U}_a(0,t)P_\phi\check{U}_a(t,0)\Gamma^{+*}_{a,\Theta}J_1(P^+).
\end{aligned}$$

But

$$\text{s-}\lim_{t\to\infty} \check{U}_a(0,t)P_\phi\check{U}_a(t,0)$$

exists because

$$\begin{aligned}
&\tfrac{\mathrm{d}}{\mathrm{d}t}\check{U}_a(0,t)P_\phi\check{U}_a(t,0)\\
&= \check{U}_a(0,t)[\check{I}_a(t,x), P_\phi]\check{U}_a(t,0) \in O(\nabla\check{I}_a(t,x))
\end{aligned}$$

is integrable. This ends the proof of *(ii)*.

Now let us prove *(iii)*. Let $j \in C_0^\infty(\mathbb{R})$ such that $\mathrm{supp} j \subset [-\infty, \epsilon]$. Set

$$\Phi(t) := \mathbb{1}^{\mathrm{c}}(H^a)\chi(H^a)j\left(\frac{|x^a|}{t}\right)\chi(H^a)\mathbb{1}^{\mathrm{c}}(H^a).$$

We will show that

$$\text{s-}\lim_{t\to\infty} \mathrm{e}^{itH}J\left(\frac{x}{t}\right)\Phi(t)J\left(\frac{x}{t}\right)\mathrm{e}^{-itH} = 0. \tag{6.10.10}$$

Clearly, (6.10.10) equals

$$\check{\Gamma}^{+}_{a,\Theta} J(\check{P}^{+}_{(a)}) \left(\text{s}-\lim_{t\to\infty} \check{U}_a(0,t)\Phi(t)\check{U}_a(t,0)\right) \check{\Gamma}^{+*}_{a,\Theta} J(P^+). \tag{6.10.11}$$

Applying Lemma 6.9.2 to the Hamiltonian $\check{H}_a(t)$, we see that the expression in brackets in (6.10.11) is zero (see an analogous argument in the proof of Proposition 6.6.9). This ends the proof of *(iii)*. □

Now we can show the $\subset$ inclusion in (6.10.2).

Lemma 6.10.6

$$\sigma(P^+, H) \subset \bigcup_{a\in\mathcal{A}} \left\{\left(x_a, \tau + \frac{1}{2}x_a^2\right) \mid x_a \in X_a,\ \tau \in \sigma^{\text{pp}}(H^a)\right\}. \tag{6.10.12}$$

Hence Theorem 6.10.1 (i) is true.

Proof. Let $y_a \in Z_a$ and $\tau \notin \sigma_{\text{pp}}(H^a) \cup \mathcal{T}^a$. We will show that

$$(y_a, \tau + \tfrac{1}{2}y_a^2) \notin \sigma(P^+, H). \tag{6.10.13}$$

Let $\chi \in C_0^\infty(\mathbb{R})$ such that $\text{supp}\chi \cap (\sigma^{\text{pp}}(H^a) \cup \mathcal{T}^a) = \emptyset$ and $\chi(\tau) \neq 0$. Let $\epsilon > 0$ satisfy the requirements of Lemma 6.10.5 *(iii)* and let $J \in C_0^\infty(X)$ such that $\text{supp}J \subset Y_a \cap \{x \mid |x^a| < \epsilon\}$ with $J(y_a) \neq 0$. Then, by Lemma 6.10.5 *(i)*,

$$\begin{aligned} &J^2(P^+)\chi^2\left(H - \tfrac{1}{2}(P_a^+)^2\right) \\ &= \text{s}-\lim_{t\to\infty} \mathrm{e}^{itH} J(\tfrac{x}{t})\chi(H^a)J(\tfrac{x}{t})\mathrm{e}^{-itH}. \end{aligned} \tag{6.10.14}$$

But $\chi(H^a) = 1\!\!1^{\text{c}}(H^a)\chi(H^a)$, and hence, by Lemma 6.10.5 *(iii)*, if $\epsilon > 0$ is small enough, then (6.10.14) is zero. Therefore (6.10.13) is true.

Hence

$$\sigma(P^+, H) \cap Z_a \times \mathbb{R} \subset \left\{\left(x_a, \tau + \frac{1}{2}x_a^2\right) \mid x_a \in X_a,\ \tau \in \sigma^{\text{pp}}(H^a) \cup \mathcal{T}^a\right\}.$$

But $\bigcup_{a\in\mathcal{A}} Z_a = X$, and therefore

$$\begin{aligned} \sigma(P^+, H) &\subset \bigcup_{a\in\mathcal{A}} \left\{\left(x_a, \tau + \tfrac{1}{2}x_a^2\right) \mid x_a \in X_a,\ \tau \in \sigma^{\text{pp}}(H^a) \cup \mathcal{T}^a\right\} \\ &= \bigcup_{a\in\mathcal{A}} \left\{\left(x_a, \tau + \tfrac{1}{2}x_a^2\right) \mid x_a \in X_a,\ \tau \in \sigma^{\text{pp}}(H^a)\right\}. \end{aligned}$$

□

Lemma 6.10.7

$$1\!\!1_{\Sigma_{a,\text{reg}}}(P^+, H) \leq Q_a^+.$$

Proof. Let $y_a \in Z_a$, $\tau \in \sigma_{\text{pp}}(H^a)\backslash\mathcal{T}^a$, that is,

$$(y_a, \tau + \tfrac{1}{2}y_a^2) \in \Sigma_{a,\mathrm{reg}}.$$

We will show that if $J \in C_0^\infty(X)$, $\chi, \tilde{\chi} \in C_0^\infty(\mathbb{R})$, $J(y_a) \neq 0$, $\chi(\tau + \frac{1}{2}y_a^2) \neq 0$, $\tilde{\chi}(\tau) = 1$ and the supports of J, χ are sufficiently small, then

$$\mathrm{s-}\lim_{t\to\infty} \mathrm{e}^{itH_{a,\mathrm{sep}}}\mathrm{e}^{-itH} J^2(P^+)\chi^2(H)\tilde{\chi}(H - \tfrac{1}{2}(P_a^+)^2) \tag{6.10.15}$$

exists.

We may assume that $J \in \mathcal{F}$, $\mathrm{supp}J \subset Y_a$ and $(x_a, x^a) \in \mathrm{supp}J$ implies $(x_a, sx^a) \in Y_{[a]}$ for $0 \le s \le 1$. We may also assume that $\mathrm{supp}\tilde{\chi} \cap \sigma^{\mathrm{pp}}(H^a) = \{\tau\}$.

We define

$$M(t) := J^2(\tfrac{x}{t}) + \langle \nabla J^2(\tfrac{x}{t}), D - \tfrac{x}{t}\rangle.$$

Using Lemma 6.10.5 *(i)* and *(iii)*, we see that

$$\begin{aligned} J^2(P^+)\chi^2(H)\tilde{\chi}(H - \tfrac{1}{2}(P_a^+)^2) &= \mathrm{s-}\lim_{t\to\infty} \mathrm{e}^{itH}\chi(H)J(\tfrac{x}{t})\tilde{\chi}(H^a)J(\tfrac{x}{t})\chi(H)\mathrm{e}^{-itH} \\ &= \mathrm{s-}\lim_{t\to\infty} \mathrm{e}^{itH}\chi(H_{a-\mathrm{sep}})J(\tfrac{x}{t})\mathbb{1}_{\{\tau\}}(H^a)J(\tfrac{x}{t})\chi(H)\mathrm{e}^{-itH} \\ &= \mathrm{s-}\lim_{t\to\infty} \mathrm{e}^{itH}\chi(H_{a-\mathrm{sep}})\mathbb{1}_{\{\tau\}}(H^a)M(t)\chi(H)\mathrm{e}^{-itH}. \end{aligned}$$

Therefore, (6.10.15) equals

$$\mathrm{s-}\lim_{t\to\infty} \mathrm{e}^{itH_{a-\mathrm{sep}}}\chi^2(H_{a-\mathrm{sep}})\mathbb{1}_{\{\tau\}}(H^a)M(t)\chi(H)\mathrm{e}^{-itH}. \tag{6.10.16}$$

Existence of (6.10.16) follows then by the same arguments as the existence of (6.10.3) in the proof of Lemma 6.10.2. □

6.11 Asymptotic Clustering and Asymptotic Absolute Continuity

In the previous section we obtained a complete description of the joint spectrum of the energy and the asymptotic velocity for very general long-range interactions. However, this description does not say anything about the nature of the spectral measure of these observables. In this section we will prove the property of the spectral measure of P^+ called *asymptotic absolute continuity*, which was introduced in [De6, De8]. To establish asymptotic absolute continuity, we will make use of another property of N-particle systems, *asymptotic clustering*, introduced by Sigal-Soffer [SS2]. This property holds when the potentials decay as $\langle x\rangle^{-\mu}$ with $\mu > 1/2$ ([DeGe1]) and is related (and in the 2-body case identical) to the existence and completeness of Dollard wave operators.

Let us first describe asymptotic clustering in the framework of time-dependent Hamiltonians. We fix $a \in \mathcal{A}$ and a function $t \to \check{I}_a(t,x) \in L^\infty(X)$ satisfying

$$\int_1^\infty \|\nabla_x \check{I}_a(t,\cdot)\|_\infty \mathrm{d}t < \infty.$$

We also fix potentials $v^b(x^b)$ for $b \leq a$ satisfying (6.1.1) and (6.6.1). We set

$$\check{H}_a(t) := H_a + \check{I}_a(t, x).$$

We define $\check{U}_a(t, s)$ to be the evolution generated by $\check{H}_a(t)$ and

$$\check{P}^+_{(a)} := \text{s}-C_\infty-\lim_{t\to\infty} \check{U}_a(0, t), \frac{x}{t}\check{U}_a(t, 0)$$

the corresponding asymptotic velocity. Note that

$$(\check{P}^+_{(a)})_a = C_\infty-\lim_{t\to\infty} \check{U}_a(0, t) D_a \check{U}_a(t, 0).$$

We also introduce

$$\check{H}_{a,\mathrm{D}}(t) := H_a + \check{I}_a(t, tD_a, x^a).$$

We define $\check{U}_{a,\mathrm{D}}(t, s)$ to be the evolution generated by $\check{H}_{a,\mathrm{D}}(t)$ and

$$\check{P}^+_{(a),\mathrm{D}} := \text{s}-C_\infty-\lim_{t\to\infty} \check{U}_{a,\mathrm{D}}(0, t), \frac{x}{t}\check{U}_{a,\mathrm{D}}(t, 0)$$

the corresponding asymptotic velocity (which is easily seen to exist, although, strictly speaking, $\check{H}_{a,\mathrm{D}}(t)$ does not belong to the class of Hamiltonians considered in Proposition 6.9.1). Note that $\check{U}_{a,\mathrm{D}}(t, s)$ commutes with D_a. Hence

$$(\check{P}^+_{(a),\mathrm{D}})_a = D_a.$$

The following property has been called asymptotic clustering by Sigal and Soffer [SS2]:

Proposition 6.11.1
Assume that

$$\int_0^\infty \langle t\rangle^{1/2} \|\nabla_x \check{I}_a(t, \cdot)\|_\infty \mathrm{d}t < \infty. \tag{6.11.1}$$

Then $\check{U}_a(t, 0)$ is asymptotic to $\check{U}_{a,\mathrm{D}}(t, 0)$ on X.

Proof. The proposition is proven exactly as Theorem 3.6.2. □

The following theorem describes the property of the spectral measure of P^+ called asymptotic absolute continuity:

Theorem 6.11.2
Assume (6.1.1) and

$$v^b(x^b) = v^b_{\mathrm{s}}(x^b) + v^b_{\mathrm{l}}(x^b), \quad b \in \mathcal{B},$$

$$\int_0^\infty \left\|(1-\Delta^b)^{-1/2} v^b_{\mathrm{s}}(x^b) \mathbb{1}_{[1,\infty[}(\tfrac{|x^b|}{R})(1-\Delta^b)^{-1}\right\| \mathrm{d}R < \infty, \quad b \in \mathcal{B}, \tag{6.11.2}$$

$$\int_0^\infty \|\nabla v^b_{\mathrm{l}}(x^b) \mathbb{1}_{[1,\infty[}(\tfrac{|x^b|}{R})\|_\infty R^{1/2} \mathrm{d}R < \infty, \quad b \in \mathcal{B}.$$

Let $a \in \mathcal{A}$ and let $\theta \subset Z_a$ be a set of measure zero with respect to Lebesgue's measure on X_a. Then

$$\mathbb{1}_\theta(P^+) = 0.$$

Proof. We fix $a \in \mathcal{A}$. Let $\Theta \subset Y_a$ be a compact set and let $\check{I}_a(t,x)$ be the effective time-dependent potential introduced in Proposition 6.9.5. Let $\check{H}_a(t)$, $\check{U}_a(t,s)$ and $\check{P}^+_{(a)}$ be the corresponding effective time-dependent Hamiltonian, the evolution and the asymptotic velocity. Note that $\check{I}_a(t,x)$ satisfies (6.11.1).

Combining Propositions 6.11.1 and 6.9.5, we see that e^{-itH} is asymptotic to $\check{U}_{a,\mathrm{D}}(t,0)$ on Θ.

Now let $\theta \subset Z_a \cap \Theta$ be a measurable set. Clearly, $\mathbb{1}_\theta(P^+)$ is unitarily equivalent to $\mathbb{1}_\theta(\check{P}^+_{(a),\mathrm{D}})$. Moreover,

$$\mathbb{1}_\theta(\check{P}^+_{(a),\mathrm{D}}) = \mathbb{1}_\theta(D_a)\mathbb{1}_{X_a}\left((\check{P}^+_{(a),\mathrm{D}})^a\right). \tag{6.11.3}$$

Now, if θ is of Lebesgue's measure zero, then (6.11.3) and hence $\mathbb{1}_\theta(P^+)$ are zero. This proves the theorem for $\theta \subset Z_a \cap \Theta$.

Now let $\theta \subset Z_a$ be any set of Lebesgue's measure zero. We will find a sequence $\Theta_n \subset Y_a$ of compact sets such that $\Theta_n \cap \theta \nearrow \theta$. We have $\mathbb{1}_{\Theta_n \cap \theta}(P^+) \nearrow \mathbb{1}_\theta(P^+)$, and $\mathbb{1}_{\Theta_n \cap \theta}(P^+) = 0$. Hence $\mathbb{1}_\theta(P^+) = 0$. □

6.12 Improved Propagation Estimates

In this and the next section we use the notation introduced at the beginning of Sect. 6.9. More precisely, we assume that the Hilbert space is $L^2(X) \otimes \mathcal{H}_1$ and the evolution is generated by the Hamiltonian

$$\check{H}(t) := H \otimes \mathbb{1} + W(t,x),$$

where $W(t,x)$ is a function with values in $B(\mathcal{H}_1)$ and H is a N-body Schrödinger operator on $L^2(X)$. We denote by $\check{U}(t,s)$ the unitary evolution generated by $\check{H}(t)$. We assume, for some $\mu \geq 0$ and all $b \in \mathcal{B}$,

$$\int_0^\infty \|(1-\Delta^b)^{-1}\nabla_x v^b(x^b)\mathbb{1}_{[1,\infty[}\left(\frac{|x^b|}{R}\right)(1-\Delta^b)^{-1}\|\langle R\rangle^\mu \mathrm{d}R < \infty, \tag{6.12.1}$$

$$\int_0^\infty \sup_{x\in X}\|\nabla_x W(t,x)\|\mathrm{d}t < \infty. \tag{6.12.2}$$

We define the asymptotic velocity $\check{P}^+$, the asymptotic Hamiltonian $\check{H}^+$ and three kinds of the Heisenberg derivatives: $\check{\mathbf{D}}$, $\mathbf{D}$ and $\mathbf{D}_0$ as in (6.9.2). If $\phi \in L^2(X)$, then we will write ϕ_t for $\check{U}(t,0)\phi$.

The goal of this section will be to give some refined propagation estimates on the dynamics $\check{U}(t,s)$. These estimates will use functions of x/t^δ, unlike those of Sect. 6.6 that used only functions of x/t.

Let us start with a propagation estimate that is an improved version of Lemma 6.6.3. A similar estimate appeared first in [Gr], where it was one of the main tools in the proof of asymptotic completeness for short-range systems.

Proposition 6.12.1
Let $\chi \in C_0^\infty(\mathbb{R})$, $0 < \theta$, $\epsilon > 0$, and $1 \geq \delta \geq (1+\mu)^{-1}$. Then

$$\int_1^\infty \left\| 1\!\!1_{[0,\theta]}\left(\frac{|x|}{t}\right) 1\!\!1_{Y_a^\epsilon}\left(\frac{x}{t^\delta}\right)\left(\frac{x_a}{t} - D_a\right)\chi(H)\phi_t\right\|^2 \frac{dt}{t} \leq C\|\phi\|^2. \tag{6.12.3}$$

Lemma 6.12.2
Let $R(x)$ and $q_a(x)$ be the functions constructed in Sect. 5.2. Define

$$\begin{aligned} B_t &:= \tfrac{1}{2}t^{\delta-1}\left(\langle \nabla R(\tfrac{x}{t^\delta}), D\rangle + \langle D, \nabla R(\tfrac{x}{t^\delta})\rangle\right) \\ &\quad -t^{2\delta-2}R(\tfrac{x}{t^\delta}) + \delta t^{2\delta-2}\left(2R(\tfrac{x}{t^\delta}) - \tfrac{x}{t^\delta}\nabla R(\tfrac{x}{t^\delta})\right), \\ C_t &:= \langle (D - \delta\tfrac{x}{t})\nabla^2 R(\tfrac{x}{t^\delta})(D - \delta\tfrac{x}{t})\rangle \\ &\quad -(1-\delta)t^{-1+\delta}\left(\langle \nabla R(\tfrac{x}{t^\delta}), D - \delta\tfrac{x}{t}\rangle + \langle D - \delta\tfrac{x}{t}, \nabla R(\tfrac{x}{t^\delta})\rangle\right) \\ &\quad +2(\delta-1)^2 t^{-2+2\delta}R(\tfrac{x}{t^\delta}). \end{aligned}$$

Then

$$B_t = \mathbf{D}_0 t^{2\delta-1}R(\tfrac{x}{t^\delta}), \tag{6.12.4}$$

$$\begin{aligned} \mathbf{D}_0 B_t &= t^{-1}C_t + \tfrac{1}{4}t^{-1-2\delta}\Delta^2 R(\tfrac{x}{t^\delta}) \\ &\quad +t^{-3+2\delta}\delta(\delta-1)\left(2R(\tfrac{x}{t^\delta}) - \tfrac{x}{t^\delta}\nabla R(\tfrac{x}{t^\delta})\right), \end{aligned} \tag{6.12.5}$$

$$C_t \geq \sum_{a\in\mathcal{A}}(D_a - \tfrac{x_a}{t})q_a(\tfrac{x}{t^\delta})(D_a - \tfrac{x_a}{t}). \tag{6.12.6}$$

Proof. The identities (6.12.4) and (6.12.5) follow from Lemma 5.2.7.
Let us prove the inequality (6.12.6). Write C_t as

$$\begin{aligned} C_t &= t^{2\delta-2}(1-\delta)^2(k\nabla^2 R(\tfrac{x}{t^\delta})k \\ &\quad -k\nabla R(\tfrac{x}{t^\delta}) - \nabla R(\tfrac{x}{t^\delta})k + 2R(\tfrac{x}{t^\delta})), \end{aligned}$$

for

$$k = (1-\delta)^{-1}t^{1-\delta}(D - \tfrac{x}{t}).$$

Note that

$$k - \tfrac{x}{t^\delta} = (1-\delta)^{-1}t^{1-\delta}(D - \tfrac{x}{t}). \tag{6.12.7}$$

Using Lemma 5.2.7 *(iv)* and (6.12.7), we get

$$
\begin{aligned}
C_t &\geq t^{2\delta-2}(1-\delta)^{-2}\sum_{a\in\mathcal{A}}(k_a-\tfrac{x_a}{t^\delta})q_a(\tfrac{x}{t^\delta})(k_a-\tfrac{x_a}{t^\delta})\\
&= \sum_{a\in\mathcal{A}}(D_a-\tfrac{x_a}{t})q_a(\tfrac{x}{t^\delta})(D_a-\tfrac{x_a}{t}).
\end{aligned}
\tag{6.12.8}
$$

□

The assumption (6.12.2) easily implies the following estimate.

Lemma 6.12.3
Suppose that $a\in\mathcal{A}$, $\epsilon>0$, $J\in C^\infty(X)$, $\partial_x^\alpha J$ are bounded and $\mathrm{supp}J\subset Y_a^\epsilon$. *Then*

$$
(1-\Delta)^{-1}J\left(\frac{x}{t^\delta}\right)\nabla I_a(x)(1-\Delta)^{-1}\in O(t^{1-\delta-\delta\mu})L^1(\mathrm{d}t).
$$

Proof of Proposition 6.12.1. We may choose a constant θ as in Lemma 6.3.2. Suppose that $J\in C_0^\infty(\mathbb{R})$ such that $J=1$ on $[0,\theta]$ and let $j\in C_0^\infty(\mathbb{R})$ such that $j=1$ on $\mathrm{supp}J'$ and $\mathrm{supp}j\subset[\theta,\infty[$.

We consider the propagation observable

$$
\Phi(t):=\chi(H)J\left(\frac{|x|}{t}\right)B_tJ\left(\frac{|x|}{t}\right)\chi(H).
$$

Clearly, $\Phi(t)$ is uniformly bounded.

We compute, using (6.12.6) and (6.12.8),

$$
\begin{aligned}
\check{\mathbf{D}}\Phi(t) &= \chi(H)\left(\mathbf{D}J\left(\tfrac{|x|}{t}\right)\right)B_tJ\left(\tfrac{|x|}{t}\right)\chi(H)+\mathrm{hc}\\
&+t^{-1}\chi(H)J\left(\tfrac{|x|}{t}\right)(\mathbf{D}_0B_t)J\left(\tfrac{|x|}{t}\right)\chi(H)\\
&-t^{\delta-1}\chi(H)\nabla_xV(x)\nabla_xR\left(\tfrac{x}{t^\delta}\right)J^2\left(\tfrac{|x|}{t}\right)\chi(H)\\
&+[W(t,x),i\Phi(t)].
\end{aligned}
\tag{6.12.9}
$$

The first term on the right of (6.12.9) can be written as

$$
\chi(H)j\left(\frac{|x|}{t}\right)B(t)j\left(\frac{|x|}{t}\right)\chi(H)+O(t^{-2})
$$

for some uniformly bounded $B(t)$. This is integrable along the evolution by Proposition 6.3.1.

The second term is greater than or equal to

$$
\sum_{a\in\mathcal{A}}t^{-1}\chi(H)J\left(\frac{|x|}{t}\right)\left(\frac{x_a}{t}-D_a\right)q_a\left(\frac{x}{t^\delta}\right)\left(\frac{x_a}{t}-D_a\right)J\left(\frac{|x|}{t}\right)\chi(H)
$$

plus terms $O(t^{-1-2\delta})$ and $O(t^{-3+2\delta})$, which are integrable. (The term $O(t^{-3+2\delta})$ is zero if $\delta=1$).

The third term equals

$$\begin{aligned}
&\sum_{a\in\mathcal{A}} t^{-1}\chi(H)x_a\nabla_x I_a(x)q_a\left(\tfrac{x}{t^\delta}\right)J^2\left(\tfrac{|x|}{t}\right)\chi(H)\\
&=\sum_{a\in\mathcal{A}}\chi(H)\nabla_x I_a(x)q_a\left(\tfrac{x}{t^\delta}\right)\chi(H)\tfrac{x_a}{t}J^2\left(\tfrac{|x|}{t}\right)+O(t^{-2}).
\end{aligned}$$

By Lemma 6.12.3, this is $\langle t\rangle^{1-\delta-\delta\mu}L^1(\mathrm{d}t)$. This is integrable for $\delta\geq(1+\mu)^{-1}$.

The fourth term involves commutators of $W(t,x)$ with $\chi(H)$ and D, which can be estimated by $\|\nabla_x W(t,x)\|$. This is integrable by (6.12.2).

Hence, for any $a\in\mathcal{A}$, we have

$$\int_1^\infty t^{-1}\left\|\sqrt{q_a\left(\tfrac{x}{t^\delta}\right)}\left(\tfrac{x_a}{t}-D_a\right)J\left(\tfrac{|x|}{t}\right)\chi(H)\phi_t\right\|^2\mathrm{d}t<C\|\phi\|^2. \quad (6.12.10)$$

Arguing as at the end of the proof of Proposition 6.6.3 we obtain

$$\int_1^\infty t^{-1}\left\|\sqrt{\textstyle\sum_{b>a}q_b\left(\tfrac{x}{t^\delta}\right)}\left(\tfrac{x_a}{t}-D_a\right)J\left(\tfrac{|x|}{t}\right)\chi(H)\phi_t\right\|^2\mathrm{d}t\leq C\|\phi\|^2, \quad (6.12.11)$$

which implies (6.12.3). □

We introduce now certain observables that will play a very important role in this section. Let $r(x)$ be the function constructed in Sect. 5.2. We define

$$\begin{aligned}
b_t &:= \delta t^{\delta-1}r\left(\tfrac{x}{t^\delta}\right)+\tfrac12\left(\langle\nabla r\left(\tfrac{x}{t^\delta}\right),D-\delta\tfrac{x}{t}\rangle+\langle D-\delta\tfrac{x}{t},\nabla r\left(\tfrac{x}{t^\delta}\right)\rangle\right)\\
&=\tfrac12\left(\langle\nabla r\left(\tfrac{x}{t^\delta}\right),D\rangle+\langle D,\nabla r\left(\tfrac{x}{t^\delta}\right)\rangle\right)+\delta t^{\delta-1}\left(r\left(\tfrac{x}{t^\delta}\right)-\tfrac{x}{t^\delta}\nabla r\left(\tfrac{x}{t^\delta}\right)\right).\\
c_t &:= \left(D-\delta\tfrac{x}{t}\right)\nabla^2 r\left(\tfrac{x}{t^\delta}\right)\left(D-\delta\tfrac{x}{t}\right),\\
d_t &:= \delta(\delta-1)\left(r\left(\tfrac{x}{t^\delta}\right)-\tfrac{x}{t^\delta}\nabla r\left(\tfrac{x}{t^\delta}\right)\right).
\end{aligned}$$

We have

$$\begin{aligned}
&\mathbf{D}_0 t^\delta r\left(\tfrac{x}{t^\delta}\right)=b_t,\\
&\mathbf{D}_0 b_t=t^{-\delta}c_t+t^{\delta-2}d_t+t^{-3\delta}\Delta^2 r\left(\tfrac{x}{t^\delta}\right).
\end{aligned}$$

Note that, by Lemma 5.2.9, c_t is a positive operator and d_t is uniformly bounded.

For technical reasons, it will be convenient to fix an energy interval $[\lambda_1,\lambda_2]$ and a function $\chi\in C_0^\infty(\mathbb{R})$ such that $\chi=1$ on $[\lambda_1,\lambda_2]$. We set

$$\begin{aligned}
b_{\chi,t}&:=\chi(H)b_t\chi(H),\\
c_{\chi,t}&:=\chi(H)c_t\chi(H),\\
d_{\chi,t}&:=\chi(H)d_t\chi(H).
\end{aligned}$$

Note that, again by Lemma 5.2.9, the observable $b_{\chi,t}$ is uniformly bounded. The approximate positivity of the Heisenberg derivative of $b_{\chi,t}$ implies easily the existence of the asymptotic observable associated with $b_{\chi,t}$.

Proposition 6.12.4
Assume $1 \geq \delta \geq (1+\mu)^{-1}$ *and* $\delta > 1/3$. *Then there exists*

$$\mathrm{s-}\lim_{t\to\infty} \check{U}(0,t) b_{\chi,t} \check{U}(t,0) =: \check{b}_\chi^+. \tag{6.12.12}$$

Moreover

$$\check{b}_\chi^+ = |\check{P}^+| \chi^2(\check{H}^+). \tag{6.12.13}$$

Proof. Clearly, $b_{\chi,t}$ is uniformly bounded. We compute:

$$\begin{aligned}\check{\mathbf{D}} b_{\chi,t} = \; & t^{-\delta} \chi(H) c_t \chi(H) \\ & + \chi(H) \nabla_x V(x) \nabla r(\tfrac{x}{t^\delta}) \chi(H) \\ & + [W(t,x), i\chi(H) b_t \chi(H)] + O(t^{\delta-2}) + O(t^{-3\delta}).\end{aligned}$$

The first term on the right-hand side is positive.

The second term is $\langle t\rangle^{1-\delta-\delta\mu} L^1(\mathrm{d}t)$ by Lemma 6.12.3. This is integrable for $\delta \geq (1+\mu)^{-1}$.

The third term is of the order $O(\|\nabla_x W(t,\cdot)\|)$, hence integrable.

The term $O(t^{\delta-2})$ is clearly integrable for $\delta < 1$. For $\delta = 1$, it is 0, so it is also integrable in this case.

Therefore, by Lemma B.4.1, the limit (6.12.12) exists.

Let us now prove the identity (6.12.13). Consider ϕ such that $\phi = J(\check{P}^+)\phi$ for some $J \in C_0^\infty(X)$. Such ϕ are dense in $L^2(X)$. Note that, by Lemma 5.2.9, we have

$$\begin{aligned} b_t &= \langle \nabla r(\tfrac{x}{t^\delta}), D\rangle + O(t^{\delta-1}), \\ \tfrac{x}{t} \nabla r\left(\tfrac{x}{t^\delta}\right) &- \tfrac{|x|}{t} \in O(t^{\delta-1}). \end{aligned}$$

So we get

$$\check{b}_\chi^+ \phi = \lim_{t\to\infty} \tilde{U}(0,t) \chi(H) \nabla r\left(\tfrac{x}{t^\delta}\right) J\left(\tfrac{x}{t}\right) D \chi(H) \tilde{U}(t,0) \tag{6.12.14}$$

and

$$\begin{aligned} |\check{P}^+| \chi^2(\check{H}^+)\phi &= \lim_{t\to\infty} \check{U}(0,t) \chi(H) \tfrac{|x|}{t} J\left(\tfrac{x}{t}\right) \chi(H) \check{U}(t,0)\phi \\ &= \lim_{t\to\infty} \check{U}(t,0) \chi(H) \tfrac{x}{t} \nabla r\left(\tfrac{x}{t^\delta}\right) J\left(\tfrac{x}{t}\right) \chi(H) \check{U}(t,0)\phi. \end{aligned} \tag{6.12.15}$$

We subtract (6.12.15) from (6.12.14), and we obtain

$$\begin{aligned} & \check{b}_\chi^+ \phi - |\check{P}^+| \chi^2(\check{H}^+)\phi \\ &= \lim_{t\to\infty} \check{U}(0,t) \nabla r\left(\tfrac{x}{t^\delta}\right) J\left(\tfrac{x}{t}\right) \left(D - \tfrac{x}{t}\right) \chi^2(H) \check{U}(t,0)\phi. \end{aligned} \tag{6.12.16}$$

For $\epsilon > 0$ small enough, (6.12.16) equals

$$\sum_{a\in\mathcal{A}} \lim_{t\to\infty} \tilde{U}(0,t)q_a\left(\tfrac{x}{\epsilon t^\delta}\right)\nabla r\left(\tfrac{x}{t^\delta}\right)J\left(\tfrac{x}{t}\right)\left(D_a-\tfrac{x_a}{t}\right)\chi^2(H)\tilde{U}(t,0)\phi. \tag{6.12.17}$$

But by Lemma 6.12.1,

$$\int_1^\infty t^{-1}\left\|q_a\left(\frac{x}{\epsilon t^\delta}\right)\nabla r\left(\frac{x}{t^\delta}\right)J\left(\frac{x}{t}\right)\left(D_a-\frac{x_a}{t}\right)\tilde{\chi}^2(H)\tilde{U}(t,0)\phi\right\|^2 \mathrm{d}t < \infty.$$

This implies that the limit (6.12.17) vanishes, which means that

$$\breve{b}_\chi^+\phi = |\breve{P}^+|\chi^2(\breve{H}^+)\phi.$$

This completes the proof of the proposition. □

6.13 Upper Bound on the Size of Clusters

This section is devoted to a number of estimates that will lead to a proof of asymptotic completeness for a large class of long-range systems. The existence of the asymptotic velocity observable and the spectral decomposition (6.7.1) implies that any state in $\mathrm{Ran}\mathbb{1}_{Z_a}(P^+)$ separates into clusters of size $o(t)$. We will see in this section that one can get a better estimate on the size of the clusters. For instance, if $\nabla_{x^b}v^b(x^b)$ decay like $C\langle x^b\rangle^{-1-\mu}$ then the size of the clusters can be bounded by $C\langle t\rangle^{2(2+\mu)^{-1}}$.

In this section we assume that $\mu\geq 0$, $\nu\geq 0$ and the potentials $v^b(x^b)$ used to define H satisfy (6.1.1) and the following hypothesis:

$$\begin{aligned} &v^b(x^b) = v_\mathrm{s}^b(x^b)+v_\mathrm{l}^b(x^b), \quad b\in\mathcal{B},\\ &\textstyle\int_0^\infty \left\|v_\mathrm{s}^b(x^b)\mathbb{1}_{[1,\infty[}(\frac{|x^b|}{R})(1-\Delta^b)^{-1}\right\|\langle R\rangle^\mu \mathrm{d}R<\infty, \quad b\in\mathcal{B}, \\ &\textstyle\int_0^\infty \left\|\partial_{x^b}^\alpha v_\mathrm{l}^b(x^a)\mathbb{1}_{[1,\infty[}(\frac{|x^b|}{R})\right\|\langle R\rangle^\mu \mathrm{d}R<\infty, \quad b\in\mathcal{B},\ |\alpha|=1, \end{aligned} \tag{6.13.1}$$

$$\int_0^\infty \|\nabla_x W(t,x)\|\langle t\rangle^\nu \mathrm{d}t<\infty. \tag{6.13.2}$$

For a further use, let us put

$$V_l(x) = \textstyle\sum_{b\in\mathcal{B}} v_\mathrm{l}^b(x^b), \qquad H_{a,\mathrm{l}} := H_a + I_{a,\mathrm{l}}(x).$$

$$I_{a,\mathrm{s}}(x) := \textstyle\sum_{b\not\leq a} v_\mathrm{s}^b(x^b), \qquad I_{a,\mathrm{l}}(x) := \textstyle\sum_{b\not\leq a} v_\mathrm{l}^b(x^b).$$

The theorem below summarizes the main idea of this section. Note, however, that in Sect. 6.14 we will not use this theorem but more refined and technical results contained in Propositions 6.13.8 and 6.13.9.

Theorem 6.13.1
Let $1\geq\delta\geq\max(2(2+\mu)^{-1}, 1-\nu)$, $\delta>2/3$. *Assume the hypotheses (6.1.1),*

(6.13.1) and (6.13.2). Let ϕ be a vector in $\mathrm{Ran}\mathbb{1}_{\{0\}}(\check{P}^+)$. *Then there exists $\theta > 0$ such that*

$$\lim_{t\to\infty} \mathbb{1}_{[\theta,\infty[}\left(\frac{|x|}{t^\delta}\right)\phi_t = 0.$$

One of the key technical ingredients of this section is a careful estimate of the Heisenberg derivative of a function of $b_{\chi,t}$. This estimate we describe in the following proposition.

Proposition 6.13.2
Let $F \in C^\infty(\mathbb{R})$, $f \in C_0^\infty(\mathbb{R})$ with $F' = f^2$. Then one has

$$\begin{aligned}\mathbf{D}F(cb_{\chi,t}) &= cf(cb_{\chi,t})\left(t^{-\delta}c_{\chi,t} + t^{\delta-2}d_{\chi,t}\right)f(cb_{\chi,t}) + cO(t^{-3\delta})\\ &+c\langle t\rangle^{1-\delta-\delta\mu}L^1(\mathrm{dt}) + c^2\langle t\rangle^{1-2\delta-\delta\mu}L^1(\mathrm{dt})\\ &+c^3\langle t\rangle^{1-3\delta-\delta\mu}L^1(\mathrm{dt}) + c^3O(t^{-3\delta}),\end{aligned} \tag{6.13.3}$$

$$\begin{aligned}\mathbf{D}F(t^{1-\delta}b_{\chi,t}) &= \tfrac{1}{2}f(t^{1-\delta}b_{\chi,t})\left(t^{1-2\delta}c_{\chi,t} + t^{-1}d_{\chi,t} + (1-\delta)t^{-\delta}b_{\chi,t}\right)f(t^{1-\delta}b_{\chi,t})\\ &+O(t^{1-4\delta}) + \langle t\rangle^{2-2\delta-\delta\mu}L^1(\mathrm{dt}) + \langle t\rangle^{3-4\delta-\delta\mu}L^1(\mathrm{dt})\\ &+\langle t\rangle^{4-6\delta-\delta\mu}L^1(\mathrm{dt}) + O(t^{3-6\delta}).\end{aligned} \tag{6.13.4}$$

Lemma 6.13.3
Let $\chi \in C_0^\infty(\mathbb{R})$, $f_a \in C^\infty(X)$ with all $\partial_x^\alpha f_a$ bounded, $\epsilon > 0$, $\mathrm{supp} f_a \subset Y_a^\epsilon$. Then

$$(1-\Delta)^{-1}[f_a(\tfrac{x}{t^\delta})D^\beta, V](1-\Delta)^{-1} \in \langle t\rangle^{1-\delta-\delta\mu}L^1(\mathrm{dt}),\ |\beta| \le 2, \tag{6.13.5}$$

$$(1-\Delta)^{-1}[f_a(\tfrac{x}{t^\delta})D^\beta, V_{\mathrm{l}}](1-\Delta)^{-1} \in \langle t\rangle^{1-\delta-\delta\mu}L^1(\mathrm{dt}),\ |\beta| \le 3, \tag{6.13.6}$$

$$(\chi(H) - \chi(H_{a,\mathrm{l}}))f_a(\tfrac{x}{t^\delta})\langle D\rangle \in \langle t\rangle^{1-\delta-\delta\mu}L^1(\mathrm{dt}) + O(t^{-\delta}). \tag{6.13.7}$$

Proof. Let us prove (6.13.7). Let $\tilde{\chi}$ be an almost-analytic extension of χ.

$$\begin{aligned}&(\chi(H) - \chi(H_{a,\mathrm{l}})f_a(\tfrac{x}{t^\delta})\langle D\rangle\\ &= \tfrac{i}{2\pi}\int \partial_{\bar z}\tilde{\chi}(z)(z-H)^{-1}I_{a,\mathrm{s}}(x)(z-H_{a,\mathrm{l}})^{-1}f_a(\tfrac{x}{t^\delta})\langle D\rangle \mathrm{d}z\wedge\mathrm{d}\bar z\\ &= \tfrac{i}{2\pi}\int \partial_{\bar z}\tilde{\chi}(z)(z-H)^{-1}I_{a,\mathrm{s}}(x)f_a(\tfrac{x}{t^\delta})(z-H_{a,\mathrm{l}})^{-1}\langle D\rangle \mathrm{d}z\wedge\mathrm{d}\bar z + O(t^{-\delta})\\ &\in \langle t\rangle^{1-\delta-\delta\mu}L^1(\mathrm{dt}) + O(t^{-\delta}).\end{aligned}$$

□

Definition 6.13.4
The space $\Psi^{m,n}$ is the space of all differential operators of the form

$$b_t = \sum_{j \le m, k \le n} D^j t^k b_{jk}\left(\tfrac{x}{t^\delta}\right),$$

where

$$b_{jk} \in \mathcal{F}, \quad |\partial_x^\alpha b_{jk}| \le C_\alpha \quad |\alpha| \ge 0,$$

and, for some $\epsilon > 0$ and all $a \in \mathcal{A}$,

$$b_{jk}(x) D^j = b_{jk}(x_a) D_a^j, \qquad x \in X_a^\epsilon.$$

A generic operator in $\Psi^{m,n}$ will be denoted by $b_t^{m,n}$. Note that, using Lemma 5.2.9, we have

$$b_t \in \Psi^{1,0}, \quad c_t \in \Psi^{2,0}, \quad d_t \in \Psi^{0,0}. \tag{6.13.8}$$

Lemma 6.13.5
The following properties hold:

$$(i)\ [D^2, \Psi^{m,n}] \subset \Psi^{m+1,n-\delta},$$

$$(ii)\ [\Psi^{m_1,n_1}, \Psi^{m_2,n_2}] \subset \Psi^{m_1+m_2-1,n_1+n_2-\delta}.$$

Since we will work with the observables $b_{\chi,t}$ and $c_{\chi,t}$ instead of b_t and c_t, we will need some results about the commutators of operators in the spaces $\Psi^{m,0}$ and functions of the Hamiltonian. When studying such commutators, the following object is useful.

Definition 6.13.6
Let b be an operator. Let $\chi \in C_0^\infty(\mathbb{C})$ and let $\tilde{\chi}$ be its almost-analytic extension. We set

$$R_\chi(b) := \frac{i}{2\pi} \int_{\mathbb{C}} \partial_{\bar{z}} \tilde{\chi}(z)(z-H)^{-1} b (z-H)^{-1} \mathrm{d}z \wedge \mathrm{d}\bar{z}.$$

Clearly,

$$[\chi(H), b] = R_\chi([H, b]).$$

Moreover,

$$R_\chi(b_t^{n,0}) \in O(1), \qquad n \le 4,$$

$$[\chi(H), b_t^{n,0}] = R_\chi(b_t^{n+1,-\delta}) + R_\chi([V(x), b_t^{n,0}]).$$

In the following lemma we describe the behavior of various commutators. Note that it is enough to restrict our attention to the classes $\Psi^{n,0}$, since $\Psi^{n,m} = t^m \Psi^{n,0}$.

Lemma 6.13.7
Let $\chi, \chi_1 \in C_0^\infty(\mathbb{R})$. *Then*

$$(H+i)^{-1}[b_t^{n,0}, iV](H+i)^{-1} \in \langle t\rangle^{1-\delta-\delta\mu}L^1(\mathrm{d}t), \quad n \leq 2, \tag{6.13.9}$$

$$R_\chi\left([b_t^{n,0}, iV]\right) \in \langle t\rangle^{1-\delta-\delta\mu}L^1(\mathrm{d}t), \quad n \leq 2, \tag{6.13.10}$$

$$[\chi_1(H), b_t^{n,0}] \in O(t^{-\delta}) + \langle t\rangle^{1-\delta-\delta\mu}L^1(\mathrm{d}t), \quad n \leq 2, \tag{6.13.11}$$

$$[\chi_1(H), R_\chi(b_t^{n,0})] \in O(t^{-\delta}) + \langle t\rangle^{1-\delta-\delta\mu}L^1(\mathrm{d}t), \quad n \leq 3, \tag{6.13.12}$$

$$[b_t^{1,0}, R_\chi(b_t^{n,0})] \in O(t^{-\delta}) + \langle t\rangle^{1-\delta-\delta\mu}L^1(\mathrm{d}t), \quad n \leq 3. \tag{6.13.13}$$

Proof. Note that any $f_t \in \Psi^{m,n}$ can be written as

$$f_t = \sum_{j\leq m, k\leq n} D_a^j t^k f_{jk,a}(\tfrac{x}{t^\delta}), \tag{6.13.14}$$

where $f_{jk,a}(x)$ satisfies the assumptions of Lemma 6.13.3. Now (6.13.9) follows from Lemma 6.13.3. All the remaining estimates follow now easily from (6.13.9) except for (6.13.12) with $n = 3$, which requires a somewhat more careful proof.

We have

$$[\chi_1(H), R_\chi(b_t^{n,0})] = \frac{i}{2\pi}\int_{\mathbb{C}} \partial_{\bar{z}}\tilde{\chi}(z)(z-H)^{-1}[\chi_1(H), b_t^{n,0}](z-H)^{-1}\mathrm{d}z \wedge \mathrm{d}\bar{z},$$

which shows that

$$\|[\chi_1(H), R_\chi(b_t^{n,0})]\| \leq C\|(H+i)^{-1}[\chi_1(H), b_t^{n,0}](H+i)^{-1}\|. \tag{6.13.15}$$

Using (6.13.14), we see that it is enough to consider

$$\begin{aligned} &\|(H+i)^{-1}[\chi_1(H), f_a(\tfrac{x}{t^\delta})D_a^n](H+i)^{-1}\| \\ &\leq 2\|(H+i)^{-1}\left(\chi_1(H) - \chi_1(H_{a,1})\right) f_a(\tfrac{x}{t^\delta})D_a^n(H+i)^{-1}\| \\ &+\|(H+i)^{-1}[\chi_1(H_{a,1}), f_a(\tfrac{x}{t^\delta})D_a^n](H+i)^{-1}\|. \end{aligned} \tag{6.13.16}$$

By (6.13.7), the first term on the right-hand side of (6.13.16) is $\langle t\rangle^{1-\delta-\delta\mu}L^1(\mathrm{d}t) + O(t^{-\delta})$. By (6.13.6), the second term on the right-hand side of (6.13.16) is $\langle t\rangle^{1-\delta-\delta\mu}L^1(\mathrm{d}t) + O(t^{-\delta})$. □

Proof of Proposition 6.13.2 We apply Lemma C.4.1 replacing $\frac{\mathrm{d}}{\mathrm{d}t}$ with $\mathbf{D}$. For shortness, let us write χ instead of $\chi(H)$, and $R(b)$ instead of $R_{\tilde{\chi}^2}(b)$. We set

$$\begin{aligned} B(t) &:= cb_{\chi,t} = c\chi b_t\chi, \\ C_1(t) &:= c\chi[V, ib_t]\chi, \\ A_1(t) &:= c\chi(\mathbf{D}_0 b_t)\chi. \end{aligned}$$

We have

$$[B(t), A_1(t)] = c^2\chi\left([b_t, \chi^2](\mathbf{D}_0 b_t) + \chi^2[b_t, \mathbf{D}_0 b_t] - [\mathbf{D}_0 b_t, \chi^2]b_t\right)\chi.$$

We set

$$\begin{aligned} A_2(t) &:= c^2\chi(R([b_t, H_0])(\mathbf{D}_0 b_t) + \chi^2[b_t, \mathbf{D}_0 b_t] - R([\mathbf{D}_0 b_t, H_0])b_t)\chi \\ &= c^2\chi(R(b_t^{2,-\delta})b_t^{2,-\delta} + \chi^2 b_t^{2,-2\delta} - R(b_t^{3,-2\delta})b_t^{1,0})\chi =: c^2\chi\tilde{A}_2(t)\chi, \\ C_2(t) &:= c^2\chi R([b_t, iV])(\mathbf{D}_0 b_t)\chi - c^2\chi R([\mathbf{D}_0 b_t, iV])b_t\chi, \end{aligned}$$

$$\begin{aligned} &C_3(t) := [B(t), A_2(t)] \\ &= c^3\chi([b_t, \chi^2]\tilde{A}_2(t) + \chi^2[b_t, \tilde{A}_2(t)] - [\tilde{A}_2(t), \chi^2]b_t)\chi \\ &= c^3\chi([b_t^{1,0}, \chi^2]R(b_t^{2,-\delta})b_t^{2,-\delta} + [b_t^{1,0}, \chi^2]\chi^2 b_t^{2,-2\delta} - [b_t^{1,0}, \chi^2]R(b^{3,-2\delta})b_t^{1,0})\chi \\ &+c^3\chi^3([b_t^{1,0}, R(b_t^{2,-\delta})]b_t^{2,-\delta} + R(b_t^{2,-\delta})b_t^{2,-2\delta} + [b_t^{1,0}, \chi^2]b_t^{2,-2\delta} \\ &\quad +\chi^2 b_t^{2,-3\delta} - [b_t^{1,0}, R(b_t^{3,-2\delta})]b_t^{1,0} - R(b_t^{3,-2\delta})b_t^{1,-\delta})\chi \\ &-c^3\chi([R(b_t^{2,-\delta}), \chi^2]b_t^{2,-\delta}b_t^{1,0} + R(b_t^{2,-\delta})[b_t^{2,-\delta}, \chi^2]b_t^{1,0} \\ &\quad +\chi^2[\chi^2, b_t^{2,-2\delta}]b_t^{1,0} - [R(b_t^{3,-2\delta}), \chi^2]b_t^{1,0}b_t^{1,0} - R(b_t^{3,-2\delta})[b_t^{1,0}, \chi^2]b_t^{1,0})\chi. \end{aligned}$$

By Lemma 6.13.7, we have

$$\begin{aligned} C_1(t) &\in c\langle t\rangle^{1-\delta-\delta\mu}L^1(\mathrm{d}t), \\ C_2(t) &\in c^2\langle t\rangle^{1-2\delta-\delta\mu}L^1(\mathrm{d}t), \\ C_3(t) &\in c^3\langle t\rangle^{1-3\delta-\delta\mu}L^1(\mathrm{d}t) + c^3 O(t^{-3\delta}). \end{aligned}$$

This ends the proof of (6.13.3). The estimate (6.13.4) follows from (6.13.3) by

$$\check{\mathbf{D}}F(t^{1-\delta}b_{\chi,t}) = (1-\delta)t^{-\delta}b_{\chi,t}f^2(t^{1-\delta}b_{\chi,t}) + \check{\mathbf{D}}F(cb_{\chi,t})\Big|_{c=t^{1-\delta}}.$$

□

We can now start our study of the states in the kernel of $\check{b}_\chi^+$. The following proposition gives quite precise information on the behavior of the observable $b_{\chi,t}$ on such states.

Proposition 6.13.8
Assume that $1 \geq \delta \geq \max(2(2+\mu)^{-1}, 1-\nu)$, $\delta > 2/3$. *Let*

$$\theta > \limsup_{t\to\infty}\|d_{\chi,t}\|, \quad \tfrac{\theta}{1-\delta} < \theta_0 < \theta_1. \tag{6.13.17}$$

(i) Let $\chi_1 \in C^\infty(\mathbb{R})$ *such that* $\chi_1' \in C_0^\infty(\mathbb{R})$, *and* $f \in C_0^\infty(\mathbb{R})$ *such that* $\mathrm{supp} f \in [\theta_0, \theta_1[$. *Then*

$$\int_1^\infty t^{-1}\left\|f(t^{1-\delta}b_{\chi,t})\chi_1(H)\phi_t\right\|^2 \mathrm{d}t \le C\|\phi\|^2, \tag{6.13.18}$$

$$\int_1^\infty t^{1-2\delta}\left\|\sqrt{c_{\chi,t}}f(t^{1-\delta}b_{\chi,t})\chi_1(H)\phi_t\right\|^2 \mathrm{d}t \le C\|\phi\|^2. \tag{6.13.19}$$

(ii) If $\phi \in \mathrm{Ran}\mathbb{1}_{\{0\}}(\check{b}_\chi^+)$, $F_+ \in C^\infty(\mathbb{R})$, $F_+' \in C_0^\infty(\mathbb{R})$ *and* $\mathrm{supp}F_+ \in [\theta_0, \infty[$, *then*

$$\lim_{t\to\infty} F_+(t^{1-\delta}b_{\chi,t})\phi_t = 0. \tag{6.13.20}$$

Proof. Let us prove *(i)* first. Take $F \in C^\infty(\mathbb{R})$ such that $\mathrm{supp}F \subset]\theta_0, \infty[$ and $F' = f^2$. We consider the following uniformly bounded propagation observable:

$$\Phi(t) := \chi_1(H)F(t^{1-\delta}b_{\chi,t})\chi_1(H).$$

We compute, using (6.13.4):

$$\begin{aligned}\check{\mathbf{D}}\Phi(t) &= \chi_1(H)f(t^{1-\delta}b_{\chi,t})\left((1-\delta)t^{-\delta}b_{\chi,t} + t^{-1}d_{\chi,t}\right)f(t^{1-\delta}b_{\chi,t})\chi_1(H)\\ &+\chi_1(H)f(t^{1-\delta}b_{\chi,t})t^{1-2\delta}c_{\chi,t}f(t^{1-\delta}b_{\chi,t})\chi_1(H)\\ &+O(t^{1-4\delta}) + \langle t\rangle^{2-2\delta-\delta\mu}L^1(\mathrm{d}t) + \langle t\rangle^{3-4\delta-\delta\mu}L^1(\mathrm{d}t)\\ &+\langle t\rangle^{4-6\delta-\delta\mu}L^1(\mathrm{d}t) + O(t^{3-6\delta})\\ &+[W(t,x), i\chi_1(H)]F(b_{\chi,t})\chi_1(H) + \mathrm{hc}\\ &+\chi_1(H)[W(t,x), iF(t^{1-\delta}b_{\chi,t})]\chi_1(H).\end{aligned} \tag{6.13.21}$$

We have

$$2-2\delta-\delta\mu \le 0, \qquad 3-4\delta-\delta\mu \le 0,$$
$$4-6\delta-\delta\mu \le 0, \qquad 3-6\delta < -1.$$

Moreover,

$$\|[W(t,x),\chi_1(H)]\| \le C\|\nabla_x W(t,\cdot)\|_\infty \in \langle t\rangle^{-\nu}L^1(\mathrm{d}t),$$
$$\|[W(t,x),F(t^{1-\delta}b_{\chi,t})]\| \le Ct^{1-\delta}\|\nabla_x W(t,\cdot)\|_\infty \in \langle t\rangle^{1-\delta-\nu}L^1(\mathrm{d}t).$$

Therefore, all the remainder terms on in (6.13.21) are integrable in norm.

Using the fact that $\mathrm{supp}f \subset]\theta_0, \infty[$ and $d_{\chi,t} \ge -\theta$, we obtain that the first line on the right-hand side of (6.13.21) is greater than or equal to

$$C_0 t^{-1} f^2(t^{1-\delta}b_{\chi,t})$$

for $C_0 := (1-\delta)\theta_0 - \theta > 0$, which, by Lemma B.4.1, implies (6.13.18) and (6.13.19). This ends the proof of *(i)*.

Let us prove now *(ii)*. Consider an arbitrary function $F \in C^\infty(\mathbb{R})$ with $F' \in C_0^\infty(\mathbb{R})$ and $\mathrm{supp}F \subset]\theta_0, \infty[$. We claim that

$$\text{s-}\lim_{t\to\infty} \check{U}(0,t)F(t^{1-\delta}b_{\chi,t})\check{U}(t,0) \text{ exists.} \tag{6.13.22}$$

In fact, using Proposition 6.13.4 and the estimates (6.13.18), (6.13.19), we get that the Heisenberg derivative of $F(t^{1-\delta}b_{\chi,t})$ is integrable along the evolution, which proves the existence of the limit (6.13.22).

By (6.13.18), the following is also true:

$$\text{s-}\lim_{t\to\infty} \check{U}(0,t)F(t^{1-\delta}b_{\chi,t})\check{U}(t,0) = 0 \text{ if, in addition, } F \in C_0^\infty(\mathbb{R}). \tag{6.13.23}$$

Let us now assume that $F_+ \in C^\infty(]\theta_0,\infty[)$, $(F_+^2)' = f^2$. Using Proposition 6.13.2, we obtain

$$\begin{aligned}\check{\mathbf{D}}F_+^2(cb_{\chi,t}) &\geq cf(cb_{\chi,t})(t^{-\delta}c_{\chi,t} + t^{\delta-2}d_{\chi,t})f(cb_{\chi,t}) \\ &+cO(t^{-3\delta}) + c\langle t\rangle^{1-\delta-\delta\mu}L^1(\mathrm{d}t) + c^2\langle t\rangle^{1-2\delta-\delta\mu}L^1(\mathrm{d}t) \\ &+c^3t^{1-3\delta-\delta\mu}L^1(\mathrm{d}t) + c^3O(t^{-3\delta}) + [W(t,x), iF_+^2(cb_{\chi,t})] \\ &\geq c(O(t^{\delta-2}) + \langle t\rangle^{\delta-1}L^1(\mathrm{d}t)) + c^2\langle t\rangle^{2\delta-2}L^1(\mathrm{d}t) + c^3\langle t\rangle^{3\delta-3}L^1(\mathrm{d}t).\end{aligned}$$

Let now $\phi \in \mathbb{1}_{\{0\}}(\check{b}_\chi^+)$. Consider the quantity

$$k_c(t) = (\phi_t | F(cb_{\chi,t})\phi_t).$$

We have

$$\begin{aligned}&\lim_{t\to\infty} k_c(t) = 0, \\ &\tfrac{\mathrm{d}}{\mathrm{d}t}k_c(t) \geq c(O(t^{\delta-2}) + \langle t\rangle^{\delta-1}L^1(\mathrm{d}t)) + c^2\langle t\rangle^{2\delta-2}L^1(\mathrm{d}t) + c^3\langle t\rangle^{3\delta-3}L^1(\mathrm{d}t).\end{aligned}$$

Therefore

$$\begin{aligned}k_c(t) &= -\textstyle\int_t^\infty \frac{\mathrm{d}}{\mathrm{d}s}k_c(s)\mathrm{d}s \\ &\leq cO(t^{-1+\delta}) + c^2O(t^{-2+2\delta}) + c^3O(t^{-3+3\delta}).\end{aligned} \tag{6.13.24}$$

Let now $0 \leq \rho < 1$. If we set $c = \rho t^{1-\delta}$, we get from (6.13.24)

$$\lim_{t\to\infty} k_{\rho t^{1-\delta}}(t) \leq C\rho. \tag{6.13.25}$$

On the other hand, we deduce from (6.13.23) that, for $0 < \rho_1 < \rho_2 < 1$,

$$\lim_{t\to\infty} k_{\rho_1 t^{1-\delta}}(t) - k_{\rho_2 t^{1-\delta}}(t) = 0. \tag{6.13.26}$$

Therefore, from (6.13.25) and (6.13.26) we see that if $0 < \rho < 1$, then

$$\lim_{t\to\infty} k_{\rho t^{1-\delta}}(t) = 0. \tag{6.13.27}$$

This proves (6.13.20) and completes the proof of the proposition. □

Proposition 6.13.8 means that, along the evolution of a vector in $1\!\!1_{\{0\}}(\check{b}_\chi^+)$, the observable $b_{\chi,t}$ is less than $\theta_0 t^{\delta-1}$. The next proposition uses the fact that $b_{\chi,t}$ is essentially the Heisenberg derivative of $t^\delta r(x/t^\delta)$ and turns this information into an information on the size of the observable $r(x/t^\delta)$.

Proposition 6.13.9
Let μ, ν, δ satisfy the assumptions of Proposition 6.13.8. Let θ, θ_0, θ_1, f be as in Proposition 6.13.8. Let θ_2 satisfy

$$\theta_1 < \delta\theta_2. \tag{6.13.28}$$

Let $\chi_1 \in C_0^\infty([\lambda_1, \lambda_2])$. Suppose that $f_1 \in C_0^\infty(\mathbb{R})$, $F_\pm \in C^\infty(\mathbb{R})$ such that $F_\pm \geq 0$, $\pm F_\pm' \geq 0$, $\sqrt{\pm F_\pm'} \in C_0^\infty(\mathbb{R})$, $\mathrm{supp}F_- \subset]-\infty, \theta_1]$, $\mathrm{supp}F_-' \subset]\theta_0, \infty[$, and $\mathrm{supp}f_1, \mathrm{supp}F_+ \subset [\theta_2, \infty[$.
(i) The following propagation estimates are true:

$$\int_1^\infty \left\| \sqrt{\theta_1 - t^{1-\delta} b_{\chi,t}} F_-(t^{1-\delta} b_{\chi,t}) f_1\left(r\left(\frac{x}{t^\delta}\right)\right) \chi_1(H)\phi_t \right\|^2 \frac{\mathrm{d}t}{t} \leq C\|\phi\|^2, \tag{6.13.29}$$

$$\int_1^\infty t^{-1} \left\| f(t^{1-\delta} b_{\chi,t}) F_+\left(r\left(\frac{x}{t^\delta}\right)\right) \chi_1(H)\phi_t \right\|^2 \frac{\mathrm{d}t}{t} \leq C\|\phi\|^2, \tag{6.13.30}$$

$$\int_1^\infty t^{1-2\delta} \left\| \sqrt{c_{\chi,t}} f(t^{1-\delta} b_{\chi,t}) F_+\left(r\left(\frac{x}{t^\delta}\right)\right) \chi_1(H)\phi_t \right\|^2 \mathrm{d}t \leq C\|\phi\|^2. \tag{6.13.31}$$

(ii)

$$\mathrm{s-}\lim_{t\to\infty} F_-(t^{1-\delta} b_{\chi,t}) F_+\left(r\left(\frac{x}{t^\delta}\right)\right) \chi_1(H)\breve{U}(t,0) = 0. \tag{6.13.32}$$

Before proving Proposition 6.13.9, we need some auxiliary estimates.

Lemma 6.13.10
Let $F \in S(\langle s\rangle^\rho, \langle s\rangle^{-2}\mathrm{d}s^2)$ and $f \in C^\infty(\mathbb{R})$ with all $\partial_s^\alpha f$ bounded. If $\rho = 0$, then

$$\begin{aligned} &[f(r(\tfrac{x}{t^\delta})), F(t^{1-\delta} b_{\chi,t})] \in O(t^{1-2\delta}), \\ &[f(ct^\delta r(\tfrac{x}{t^\delta})), F(t^{1-\delta} b_{\chi,t})] \in cO(t^{1-\delta}). \end{aligned} \tag{6.13.33}$$

If $\rho = 1$, then

$$\begin{aligned} [f(r(\tfrac{x}{t^\delta})), F(t^{1-\delta} b_{\chi,t})] &\in O(t^{1-2\delta}) + O(t^{2-4\delta}) \\ &\quad + \langle t\rangle^{3-4\delta-\delta\mu} L^1(\mathrm{d}t), \\ [f(ct^\delta r(\tfrac{x}{t^\delta})), F(t^{1-\delta} b_{\chi,t})] &\in cO(t^{1-\delta}) + cO(t^{2-3\delta}) + c^2 O(t^{2-2\delta}) \\ &\quad + c\langle t\rangle^{3-3\delta-\delta\mu} L^1(\mathrm{d}t). \end{aligned}$$

Proof. We observe that $f(r(\frac{x}{t^\delta}))$ belongs to $\Psi^{0,0}$. Applying then Lemma 6.13.7, we obtain easily that

$$\begin{aligned} [b_{\chi,t}, f(r(\tfrac{x}{t^\delta}))] &= \chi(H) b_t^{0,-\delta} \chi(H) \\ &\quad + R_\chi(b_t^{1,-\delta}) b_t \chi(H) + \mathrm{hc} \in O(t^{-\delta}), \qquad (6.13.34) \\ [b_{\chi,t}, [b_{\chi,t}, f(r(\tfrac{x}{t^\delta}))]] &\in O(t^{-2\delta}) + \langle t\rangle^{1-2\delta-\delta\mu} L^1(\mathrm{d}t). \end{aligned}$$

Applying then Lemma C.3.2, we obtain the estimates concerning $f(r(\frac{x}{t^\delta}))$.

Likewise,

$$\begin{aligned} [b_{\chi,t}, f(ct^\delta r(\tfrac{x}{t^\delta}))] &\in cO(t^0), \\ [b_{\chi,t}, [b_{\chi,t}, f(ct^\delta r(\tfrac{x}{t^\delta}))]] &\in c^2 O(t^0) + cO(t^{-\delta}) + c\langle t\rangle^{1-\delta-\delta\mu} L^1(\mathrm{d}t), \end{aligned} \qquad (6.13.35)$$

and Lemma C.3.2 imply the estimates concerning $f(ct^\delta r(\frac{x}{t^\delta}))$. □

Lemma 6.13.11
Let χ_1, F_+ be as in Proposition 6.13.9. Then

$$\begin{aligned} \chi_1(H)\mathbf{D}_0 F_+(r(\tfrac{x}{t^\delta})) &= \chi_1(H) F_+'(r(\tfrac{x}{t^\delta}))(t^{-\delta} b_{\chi,t} - \delta t^{-1} r(\tfrac{x}{t^\delta})) \\ &\quad + O(t^{-2\delta}) + \langle t\rangle^{1-2\delta-\delta\mu} L^1(\mathrm{d}t). \end{aligned} \qquad (6.13.36)$$

$$\begin{aligned} \chi_1(H)\mathbf{D}_0 F_+(ct^\delta r(\tfrac{x}{t^\delta})) &= c\chi_1(H) F_+'(ct^\delta r(\tfrac{x}{t^\delta})) b_{\chi,t} \\ &\quad + c^2 O(t^0) + cO(t^{-\delta}) + c\langle t\rangle^{1-\delta-\delta\mu} L^1(\mathrm{d}t). \end{aligned} \qquad (6.13.37)$$

Proof. First we compute:

$$\begin{aligned} \mathbf{D}_0 F_+(r(\tfrac{x}{t^\delta})) &= F_+'(r(\tfrac{x}{t^\delta}))(t^{-\delta} b_t - \delta t^{-1} r(\tfrac{x}{t^\delta})) \\ &\quad + t^{-2\delta} F_+''(r(\tfrac{x}{t^\delta})) |\nabla r(\tfrac{x}{t^\delta})|^2 \\ &= F_+'(r(\tfrac{x}{t^\delta}))(t^{-\delta} b_t - \delta t^{-1} r(\tfrac{x}{t^\delta})) + O(t^{-2\delta}). \end{aligned} \qquad (6.13.38)$$

Using the fact that $\chi_1\chi = \chi_1$ and the estimates

$$\begin{aligned} [\chi(H), b_t] &\in O(t^{-\delta}) + \langle t\rangle^{1-\delta-\delta\mu} L^1(\mathrm{d}t), \\ [\chi(H), F_+'(r(\tfrac{x}{t^\delta}))] &\in O(t^{-\delta}), \end{aligned} \qquad (6.13.39)$$

we get

$$\begin{aligned} t^{-\delta}\chi_1(H) F_+'(r(\tfrac{x}{t^\delta})) b_t &= t^{-\delta}\chi_1(H) F_+'(r(\tfrac{x}{t^\delta})) b_{\chi,t} \\ &\quad + O(t^{-2\delta}) + \langle t\rangle^{1-2\delta-\delta\mu} L^1(\mathrm{d}t). \end{aligned} \qquad (6.13.40)$$

Putting together (6.13.38) and (6.13.40), we obtain (6.13.36).

Likewise, we have

$$\begin{aligned}\mathbf{D}_0 F_+(ct^\delta r(\tfrac{x}{t^\delta})) &= cF'_+(ct^\delta r(\tfrac{x}{t^\delta}))b_t + c^2 F''_+(ct^\delta r(\tfrac{x}{t^\delta}))|\nabla r(\tfrac{x}{t^\delta})|^2 \\ &= cF'_+(ct^\delta r(\tfrac{x}{t^\delta}))b_t + c^2 O(t^0)\end{aligned}$$

and

$$\begin{aligned}c\chi_1(H)F'_+(ct^\delta r(\tfrac{x}{t^\delta}))b_t &= c\chi_1(H)F'_+(ct^\delta r(\tfrac{x}{t^\delta}))b_{\chi,t} \\ &\quad + c^2 O(t^0) + cO(t^{-\delta}) + ct^{1-\delta-\delta\mu}L^1(\mathrm{dt}),\end{aligned}$$

which yields (6.13.37). □

Proof of Proposition 6.13.9. We put

$$\begin{aligned}f_\pm &:= \sqrt{\pm(F_\pm^2)'}, \\ \tilde{f}_+(s) &:= \sqrt{(\delta s - \theta_1)F_+(s)F'_+(s)}.\end{aligned}$$

First we are going to show *(i)*. Our basic observable will be

$$\Phi(t) := \chi_1(H)F_+\left(r\left(\frac{x}{t^\delta}\right)\right)F_-^2(t^{1-\delta}b_{\chi,t})F_+\left(r\left(\frac{x}{t^\delta}\right)\right)\chi_1(H).$$

We would like to compute its Heisenberg derivative. Using (6.13.36), we obtain

$$\begin{aligned}&-\check{\mathbf{D}}\Phi(t) \\ &= t^{-1}\chi_1(H)F'_+(r(\tfrac{x}{t^\delta}))(\theta_1 - t^{1-\delta}b_{\chi,t})F_-^2(t^{1-\delta}b_{\chi,t})F_+(r(\tfrac{x}{t^\delta}))\chi_1(H) + \mathrm{hc} \\ &+ t^{-1}\chi_1(H)F'_+(r(\tfrac{x}{t^\delta}))(\delta r(\tfrac{x}{t^\delta}) - \theta_1)F_-^2(t^{1-\delta}b_{\chi,t})F_+(r(\tfrac{x}{t^\delta}))\chi_1(H) + \mathrm{hc} \\ &+ \chi(H)F_+(r(\tfrac{x}{t^\delta}))\left(\mathbf{D}F_-^2(t^{1-\delta}b_{\chi,t})\right)F_+(r(\tfrac{x}{t^\delta}))\chi(H) \\ &+ O(t^{-2\delta}) + O(t^{1-2\delta-\delta\mu})L^1(\mathrm{dt}) - [W(t,x), i\Phi(t)]. \\ &= I_1(t) + I_2(t) + I_3(t) + I_4(t).\end{aligned}$$

$I_4(t)$ is $\langle t\rangle^{1-\delta-\nu}L^1(\mathrm{dt})$, hence integrable.

We symmetrize $I_1(t)$ and $I_2(t)$ by commuting functions of $r(\frac{x}{t^\delta})$ with functions of $t^{1-\delta}b_{\chi,t}$, using Lemma 6.13.10. We obtain that $I_1(t)$ equals

$$\begin{aligned}&t^{-1}\chi_1(H)f_+(r(\tfrac{x}{t^\delta}))(\theta_1 - t^{1-\delta}b_{\chi,t})F_-^2(t^{1-\delta}b_{\chi,t})f_+(r(\tfrac{x}{t^\delta}))\chi_1(H) \\ &+ O(t^{-2\delta}) + O(t^{1-4\delta}) + \langle t\rangle^{2-3\delta-\delta\mu}L^1(\mathrm{dt}).\end{aligned} \tag{6.13.41}$$

$I_2(t)$ equals

$$\begin{aligned}&t^{-1}\chi_1(H)\tilde{f}_+(r(\tfrac{x}{t^\delta}))F_-^2(t^{1-\delta}b_{\chi,t})\tilde{f}_+(r(\tfrac{x}{t^\delta}))\chi_1(H) \\ &+ O(t^{-2\delta}) + O(t^{1-4\delta}) + \langle t\rangle^{2-3\delta-\delta\mu}L^1(\mathrm{dt}).\end{aligned} \tag{6.13.42}$$

As in the proof of Proposition 6.13.8, using the conditions on μ, we see that, for some $C_0 > 0$, the term $I_3(t)$ is greater than or equal to

$$C_0 t^{-1}\chi_1(H)F_+(r(\tfrac{x}{t^\delta}))f_-^2(t^{1-\delta}b_{\chi,t})F_+(r(\tfrac{x}{t^\delta}))\chi_1(H)$$
$$+t^{1-2\delta}\chi(H)F_+(r(\tfrac{x}{t^\delta}))f_-(t^{1-\delta}b_{\chi,t})c_{\chi,t}f_-(t^{1-\delta}b_{\chi,t})F_+(r(\tfrac{x}{t^\delta}))\chi(H) \tag{6.13.43}$$
$$+L^1(\mathrm{d}t).$$

The expressions (6.13.41), (6.13.42) and (6.13.43) are all positive up to integrable error terms. This implies the estimates (6.13.29), (6.13.30) and (6.13.31) and ends the proof of *(i)*.

Let us now prove *(ii)*. Let $F \in C^\infty(\mathbb{R})$ such that $F' \in C_0^\infty(\mathbb{R})$ and $\mathrm{supp}F \subset]-\infty, \theta_0[$. Set

$$\Phi(t) := \chi_1(H)F(r(\tfrac{x}{t^\delta}))F_-^2(t^{1-\delta}b_{\chi,t})F(r(\tfrac{x}{t^\delta}))\chi_1(H).$$

Using Proposition 6.13.2, and the estimates (6.13.29), (6.13.30) and (6.13.31), we get that

$$\mathrm{s-}\lim_{t\to\infty} \check{U}(0,t)\Phi(t)\check{U}(t,0) \ \text{ exists.} \tag{6.13.44}$$

Moreover, it follows from (6.13.29) that

$$\text{if, in addition, } \ F \in C_0^\infty(\mathbb{R}), \ \text{ then } \mathrm{s-}\lim_{t\to\infty} \check{U}(0,t)\Phi(t)\check{U}(t,0) = 0. \tag{6.13.45}$$

For $0 \le c \le 1$, we consider the observable

$$\Phi_c(t) := \chi_1(H)F_+(ct^\delta r(\tfrac{x}{t^\delta}))F_-^2(t^{1-\delta}b_{\chi,t})F_+(ct^\delta r(\tfrac{x}{t^\delta}))\chi_1(H).$$

We have

$$\begin{aligned}
-\check{\mathbf{D}}\Phi_c(t) &= ct^{-1+\delta}\chi_1(H)F_+'(ct^\delta r(\tfrac{x}{t^\delta}))(\theta_1 - t^{1-\delta}b_{\chi,t}) \\
&\quad \times F_-^2(t^{1-\delta}b_{\chi,t})F_+(ct^\delta r(\tfrac{x}{t^\delta}))\chi_1(H) + \mathrm{hc} \\
&\quad -\theta_1 ct^{-1+\delta}\chi_1(H)F_+'(ct^\delta r(\tfrac{x}{t^\delta}))F_-^2(t^{1-\delta}b_{\chi,t})F_+(ct^\delta r(\tfrac{x}{t^\delta}))\chi_1(H) + \mathrm{hc} \\
&\quad -\chi_1(H)F_+(ct^\delta r(\tfrac{x}{t^\delta}))\left(\mathbf{D}F_-^2(t^{1-\delta}b_{\chi,t})\right)F_+(ct^\delta r(\tfrac{x}{t^\delta}))\chi_1(H) \\
&\quad -[W(t,x), i\Phi_c(t)] \\
&\quad +c^2O(t^0) + cO(t^{-\delta}) + c\langle t\rangle^{1-\delta-\delta\mu}L^1(\mathrm{d}t).
\end{aligned} \tag{6.13.46}$$

As above, we symmetrize the first term in the right hand side of (6.13.46) using Lemma 6.13.10. Hence the first term equals

$$ct^{-1+\delta}\chi_1(H)f_+(ct^\delta r(\tfrac{x}{t^\delta}))(\theta_1 - t^{1-\delta}\tilde{b}_t)F_-^2(t^{1-\delta}\tilde{b}_t)f_+(ct^\delta r(\tfrac{x}{t^\delta}))\chi_1(H)$$
$$+c^2O(t^0) + c^2O(t^{1-2\delta}) + c^2\langle t\rangle^{2-2\delta-\delta\mu}L^1(\mathrm{d}t) + c^3O(t^{1-\delta}).$$

The second term is $cO(t^{-1+\delta})$. The third term is greater than or equal to

$$C_0 t^{-1}\chi_1(H)F_+(ct^\delta r(\tfrac{x}{t^\delta}))f_-^2(t^{1-\delta}b_{\chi,t})F_+(ct^\delta r(\tfrac{x}{t^\delta}))\chi_1(H)$$
$$+t^{1-2\delta}\chi_1(H)F_+(ct^\delta r(\tfrac{x}{t^\delta}))f_-(t^{1-\delta}b_{\chi,t})\tilde{c}_t f_-(t^{1-\delta}b_{\chi,t})F_+(ct^\delta r(\tfrac{x}{t^\delta}))\chi_1(H)$$
$$+L^1(\mathrm{d}t) \ge L^1(\mathrm{d}t),$$

(see the proof of Proposition 6.13.8). The fourth term is integrable uniformly in c. Hence, for some $g \in L^1(\mathrm{d}t)$,

$$-\check{\mathbf{D}}\Phi_c(t) \geq g(t) + cO(t^{-1+\delta}) + c^2O(t^0) + c^2L^1(\mathrm{d}t) + c^3O(t^{1-\delta}).$$

Let ϕ be an arbitrary vector. Set

$$k_c(t) := (\phi_t|\Phi_c(t)\phi_t).$$

We have

$$\frac{\mathrm{d}}{\mathrm{d}t}k_c(t) \leq g(t) + cO(t^{-1+\delta}) + c^2O(t^0) + c^2L^1(\mathrm{d}t) + c^3O(t^{1-\delta}).$$

Hence, for $t_0 \leq t$, we get

$$k_c(t) \leq k_c(t_0) + \textstyle\int_{t_0}^{\infty} g(s)\mathrm{d}s + cO(t^{\delta}) + c^2O(t) + c^3O(t^{2-\delta}). \tag{6.13.47}$$

By (6.13.44), if we put $c = \rho t^{-\delta}$ with $0 < \rho \leq 1$, then the limit

$$\lim_{t\to\infty} k_{\rho t^{-\delta}}(t)$$

exists. Note that $\delta \geq 2/3$. Hence

$$\lim_{t\to\infty} k_{\rho t^{-\delta}}(t) \leq C\rho + \int_{t_0}^{\infty} g(s)\mathrm{d}s. \tag{6.13.48}$$

But for a fixed t_0,

$$\lim_{c\to 0} k_c(t_0) = 0.$$

Moreover,

$$\lim_{t_0\to\infty} \int_{t_0}^{\infty} g(s)\mathrm{d}s = 0.$$

Hence

$$\lim_{t\to\infty} k_{\rho t^{-\delta}}(t) \leq C\rho. \tag{6.13.49}$$

But we know from (6.13.45) that, for $0 \leq \rho_1 \leq \rho_2 \leq 1$,

$$\lim_{t\to\infty} \left(k_{\rho_1 t^{-\delta}}(t) - k_{\rho_2 t^{-\delta}}(t)\right) = 0. \tag{6.13.50}$$

Now (6.13.49) and (6.13.50) yield, for $0 \leq \rho \leq 1$,

$$\lim_{t\to\infty} k_{\rho t^{-\delta}}(t) = 0,$$

which means that (6.13.32) is true. □

6.14 Asymptotic Separation of the Dynamics II

In this section we are going to show Theorem 6.8.2, which says that, for a large class of long-range potentials, the dynamics is asymptotically separated. This result easily implies asymptotic completeness for a slightly smaller class of potentials.

Before considering time-independent Hamiltonians, we will prove that time-dependent Hamiltonians of the type considered in the previous two sections are in a certain sense asymptotically separated, provided that both the temporal decay of the time-dependent perturbation $W(t,x)$ and the spatial decay of the potentials $v^a(x^a)$ are fast enough. We will use the notation introduced in Sect. 6.9, and then used in Sects. 6.12 and 6.13.

Beside the Hamiltonian $\check{H}(t)$ and the corresponding dynamics $\check{U}(t,s)$, we will consider another time-dependent Hamiltonian

$$\check{H}_{\mathrm{sep}}(t) := H + W(t,0).$$

$\check{U}_{\mathrm{sep}}(t,s)$ will stand for the dynamics generated by $\check{H}_{\mathrm{sep}}(t)$. Note that

$$\check{U}_{\mathrm{sep}}(t,s) = \mathrm{e}^{-itH} \otimes T\left(\mathrm{e}^{-i\int_0^t W(s,0)\mathrm{d}s}\right).$$

Therefore

$$P^+ = \mathrm{s}{-}C_\infty{-}\lim_{t\to\infty} \check{U}_{\mathrm{sep}}(0,t)\tfrac{x}{t}\check{U}_{\mathrm{sep}}(t,0),$$
$$H = C_\infty{-}\lim_{t\to\infty} \check{U}_{\mathrm{sep}}(0,t)H\check{U}_{\mathrm{sep}}(t,0)$$

are the analogues of $\check{P}^+$ and $\check{H}^+$. Note that all the results of the previous section are valid for $\check{H}_{\mathrm{sep}}(t)$.

The following proposition describes the asymptotic separation of the dynamics $\check{U}(t,s)$:

Proposition 6.14.1
Let $1 > \mu \geq 0$ and $\nu = 2(2+\mu)^{-1}$. Assume the hypotheses (6.1.1), (6.13.1) and (6.13.2). Then $\check{U}(t,0)$ is asymptotic to $\check{U}_{\mathrm{sep}}(t,0)$ on $\{0\}$.

Proof. We will prove the existence of

$$\mathrm{s}{-}\lim_{t\to\infty} \check{U}_{\mathrm{sep}}(0,t)\check{U}(t,0)1\!\!1_{\{0\}}(\check{P}^+). \tag{6.14.1}$$

We set $\delta := \nu$. Note that δ satisfies the assumptions of the Propositions 6.13.8 and 6.13.9, that is $1 \geq \delta \geq \max(2(2+\mu)^{-1}, 1-\nu)$, $\delta > 2/3$. We define $\theta_0, \theta_1, \theta_2$ as in (6.13.17) and (6.13.28). Let us pick $F_-, F_{1-} \in C^\infty(\mathbb{R})$ such that $F_-', F_{1-}' \in C_0^\infty(\mathbb{R})$, $\mathrm{supp}F_- \subset]-\infty,\theta_1]$, $F_- = 1$ on $]-\infty,\theta_0]$ and $\mathrm{supp}F_{1-} \subset]-\infty,\theta_2]$.

It is enough to prove the existence of the limit (6.14.1) on a vector ϕ such that $\phi = \chi_1(\check{H}^+)\phi$ for some $\chi_1 \in C_0^\infty([\lambda_1,\lambda_2])$ and $\phi \in \mathrm{Ran}1\!\!1_{\{0\}}(\check{P}^+)$. Let $\chi \in C_0^\infty(\mathbb{R})$ satisfy $\chi = 1$ on $[\lambda_1,\lambda_2]$. We define $b_{\chi,t} := \chi(H)b_t\chi(H)$, etc. as in Sect. 6.12.

Note that, by Proposition 6.12.4,

$$\phi \in \mathrm{Ran}\mathbb{1}_{\{0\}}(\check{b}_\chi^+).$$

Clearly

$$\lim_{t\to\infty} \check{U}_{\mathrm{sep}}(0,t)\check{U}(t,0)\phi = \lim_{t\to\infty} \check{U}_{\mathrm{sep}}(0,t)\chi_1^2(H)\check{U}(t,0)\phi. \tag{6.14.2}$$

By Proposition 6.13.8, the limit (6.14.2) equals

$$\lim_{t\to\infty} \check{U}_{\mathrm{sep}}(0,t)\chi_1(H)F_-(t^{1-\delta}b_{\chi,t})\chi_1(H)\check{U}(t,0)\phi. \tag{6.14.3}$$

Set

$$\varPhi(t) := \chi_1(H)F_{1-}(r(\tfrac{x}{t^\delta}))F_-(t^{1-\delta}b_{\chi,t})F_{1-}(r(\tfrac{x}{t^\delta}))\chi_1(H).$$

By Proposition 6.13.9, the limit (6.14.3) equals

$$\begin{aligned}&\lim_{t\to\infty} \check{U}_{\mathrm{sep}}(0,t)\chi_1(H)F_-(t^{1-\delta}b_{\chi,t})F_{1-}(r(\tfrac{x}{t^\delta}))\chi_1(H)\check{U}(t,0)\phi\\ &= \lim_{t\to\infty} \check{U}_{\mathrm{sep}}(0,t)\varPhi(t)\check{U}(t,0)\phi.\end{aligned} \tag{6.14.4}$$

We have

$$\begin{aligned}&\tfrac{\mathrm{d}}{\mathrm{d}t}\varPhi(t) + i\check{H}_{\mathrm{sep}}(t)\varPhi(t) - i\varPhi(t)\check{H}(t) = \check{\mathbf{D}}\varPhi(t)\\ &+[\chi_1(H), iW(t,x)]F_{1-}(r(\tfrac{x}{t^\delta}))F_-(t^{1-\delta}b_{\chi,t})F_{1-}(r(\tfrac{x}{t^\delta}))\\ &+\chi_1(H)(W(t,x)-W(t,0))F_{1-}(r(\tfrac{x}{t^\delta}))F_-(t^{1-\delta}b_{\chi,t})F_{1-}(r(\tfrac{x}{t^\delta}))\chi_1(H).\end{aligned} \tag{6.14.5}$$

The first term on the right-hand side of (6.14.5) is integrable along the evolution by Propositions 6.13.8 and 6.13.9. The second term is in $\langle t\rangle^{-\nu}L^1(\mathrm{d}t)$, hence also integrable. The third term can be estimated by

$$C\|F_{1-}(r(\tfrac{x}{t^\delta}))x\|_\infty\|\nabla_x W(t,\cdot)\|_\infty \leq \langle t\rangle^{\delta-\nu}L^1(\mathrm{d}t). \tag{6.14.6}$$

But $\delta = \nu$, therefore (6.14.6) is integrable. The existence of the limit (6.14.4) follows then from Lemma B.4.2.

To prove the existence of

$$\mathrm{s-}\lim_{t\to\infty} \check{U}(0,t)\check{U}_{\mathrm{sep}}(t,0)\mathbb{1}_{\{0\}}(P^+), \tag{6.14.7}$$

it is enough to interchange $\check{H}(t)$, $\check{U}(t,s)$, $\check{P}^+$ and $\check{H}^+$ with $\check{H}_{\mathrm{sep}}(t)$, $\check{U}_{\mathrm{sep}}(t,s)$, P^+, H, and then proceed as in the proof of (6.14.1). □

Proof of Theorem 6.8.2. Let $a \in \mathcal{A}$ and let $\varTheta \subset Z_{[a]}$ be a compact subset. Let $\check{J} \in C_0^\infty(X)$ satisfy $\check{J} = 1$ on $\varTheta$ and $\mathrm{supp}\check{J} \subset Y_{[a]}$. We fix $y_0 \in \varTheta$ and set

$$\check{I}_{[a]}(t,x) := \check{J}(\tfrac{x}{t})(I_{[a]}(x) - I_{[a]}(ty_0)) + I_{[a]}(ty_0).$$

Let

$$\check{H}_{[a]}(t) := H_{[a]} + \check{I}_{[a]}(t, x).$$

Let $\check{U}_{[a]}(t, s)$ be the dynamics defined by $\check{H}_{[a]}(t)$. As in Proposition 6.9.5, we see that e^{-itH} is asymptotic to $\check{U}_{[a]}(t, 0)$ on Θ.

We also introduce

$$\check{H}_{[a],\mathrm{sep}}(t) := H_{[a]} + \check{I}_{[a]}(t, x_a).$$

Let $U_{[a],\mathrm{sep}}(t, s)$ be the dynamics generated by $\check{H}_{[a],\mathrm{sep}}(t)$. As in Proposition 6.9.5, we see that $e^{-itH_{[a],\mathrm{sep}}}$ is asymptotic to $\check{U}_{[a],\mathrm{sep}}(t, 0)$ on Θ.

Now we are ready to apply Proposition 6.14.1. We identify X^a and $L^2(X_a)$ as X and $\mathcal{H}_1$ of Sects. 6.9 and Proposition 6.14.1. We also identify H^a, $h_{[a]}+\check{I}_{[a]}(t, x)$, $\check{H}_{[a]}(t)$ with H, $W(t)$, $\check{H}(t)$. Consequently, we identify $\check{H}_{[a],\mathrm{sep}}(t)$ with $\check{H}_{\mathrm{sep}}(t)$.

Clearly, we have

$$\int_1^\infty \sup_x |\nabla_x W(t, x)| t^\mu \mathrm{d}t < \infty.$$

Therefore, the assumption (6.13.2) is satisfied with $\nu = \mu = \sqrt{3} - 1$. Thus, we can apply Proposition 6.14.1, from which we obtain that $\check{U}_{[a]}(t, 0)$ is asymptotic to $\check{U}_{[a],\mathrm{sep}}(t, 0)$ on X_a.

By the chain rule, e^{-itH} is asymptotic to $e^{-itH_{a-\mathrm{sep}}}$ on Θ. □

6.15 Modified Wave Operators and Asymptotic Completeness in the Long-Range Case

So far, we have avoided introducing modified wave operators. We preferred to speak either about usual wave operators, whose existence is restricted to the short-range case, or about the operators Ξ_a^+ introduced in Corollary 6.8.3, or $\Xi_{Q,a}^+$ introduced in Proposition 6.8.6. In fact, if we know that the operators Ξ_a^+ exist and if we assume some mild additional hypotheses on the potentials, then the existence of modified wave operators follows easily by 2-body methods, which were described in Chap. 4. This is the subject of this section.

Combining Theorem 6.8.2 with the methods of Chap. 4, we obtain the following theorem, which is the main result of this chapter. This theorem describes the existence and completeness of modified wave operators in the long-range case.

Theorem 6.15.1
Assume (6.1.1). Suppose that $\mu = \sqrt{3} - 1$ *and*

$$\begin{aligned}
&v^b(x^b) = v_{\mathrm{s}}^b(x^b) + v_{\mathrm{l}}^b(x^b), \quad b \in \mathcal{B}, \\
&\textstyle\int_0^\infty \left\| v_{\mathrm{s}}^b(x^b) \mathbb{1}_{[1,\infty[}(\frac{|x^b|}{R})(1 - \Delta^b)^{-1} \right\| \langle R \rangle^\mu \mathrm{d}R < \infty, \quad b \in \mathcal{B}, \\
&\textstyle\int_0^\infty \left\| \partial_x^\alpha v_{\mathrm{l}}^b(x^b) \mathbb{1}_{[1,\infty[}(\frac{|x^a|}{R}) \right\| \langle R \rangle^\mu \mathrm{d}R < \infty, \ |\alpha| = 1, \quad b \in \mathcal{B}.
\end{aligned} \tag{6.15.1}$$

Let $a \in \mathcal{A}$ *and*

$$S_a(t,\xi_a) := \frac{1}{2}t\xi_a^2 + \int_0^t I_{\mathrm{l},a}(s\xi_a)ds.$$

Then there exist

$$\mathrm{s-}\lim_{t\to\infty} \mathrm{e}^{itH}\mathrm{e}^{-iS_a(t,D_a)-itH^a}\mathbb{1}^{\mathrm{pp}}(H^a), \tag{6.15.2}$$

$$\mathrm{s-}\lim_{t\to\infty} \mathrm{e}^{iS_a(t,D_a)+itH^a}\mathrm{e}^{-itH}\mathbb{1}_{Z_a}(P^+). \tag{6.15.3}$$

If we denote (6.15.2) with $\Omega^+_{\mathrm{lr},a}$, then (6.15.3) equals $\Omega^{+*}_{\mathrm{lr},a}$. The operator $\Omega^+_{\mathrm{lr},a}$ is a partial isometry such that

$$\Omega^{+*}_{\mathrm{lr},a}\Omega^+_{\mathrm{lr},a} = \mathbb{1}^{\mathrm{pp}}(H^a), \qquad \Omega^+_{\mathrm{lr},a}\Omega^{+*}_{\mathrm{lr},a} = \mathbb{1}_{Z_a}(P^+),$$

$$\Omega^+_{\mathrm{lr},a}H_a = H\Omega^+_{\mathrm{lr},a}, \qquad \Omega^+_{\mathrm{lr},a}D_a = P^+_a\Omega^+_{\mathrm{lr},a}.$$

Remark. Clearly, $\mathrm{Ran}\Omega^+_{\mathrm{lr},a} = \mathrm{Ran}\mathbb{1}_{Z_a}(P^+)$. Hence

$$L^2(X) = \sum_{a\in\mathcal{A}}^{\oplus} \mathrm{Ran}\Omega^+_{\mathrm{lr},a}. \tag{6.15.4}$$

This property is called asymptotic completeness for long-range systems.

Before we prove this theorem, let us describe how it can be generalized to slower decaying potentials. Such a generalization involves two additional difficulties.

First, the Dollard modifiers that we used in (6.15.2) and (6.15.3) may be insufficient and we may be forced to use different modifiers. In the two-body case, we had an analogous problem and we know how to handle it.

The second problem is more serious. For slower decaying potentials, the asymptotic completeness and the existence of wave operators may not hold on the whole Hilbert space, but may be valid only on a certain subspace. Therefore, we will have to insert additional projections Q^+_a and $Q^+_{a,0}$ to get the existence of appropriate limits.

In the free channel a_{min}, we do not have problems with the existence and completeness of modified wave operators, even for $\mu = 0$. This fact we describe in the theorem below. It is an extension of the results of Sect. 4.7 to N-body Hamiltonians. It will be convenient to state this fact as the property of the family of Hamiltonians

$$h_a = \frac{1}{2}D_a^2 + I_a(x_a).$$

They are many-body Hamiltonians whose configuration space equals X_a and hence their free region equals Z_a. The modifiers $\tilde{S}_a(t,\xi_a)$ that we will obtain in this way will be used afterwards for modified wave operators of H corresponding to various $a \in \mathcal{A}$.

Theorem 6.15.2
Assume that $V(x)$ satisfies the hypotheses of Theorem 6.6.1 and

$$v^a(x^a) = v^a_{\mathrm{s}}(x^a) + v^a_{\mathrm{l}}(x^a), \quad a \in \mathcal{A},$$

$$\int_0^\infty \left\| (1-\Delta^a)^{-1/2} v^a_{\mathrm{s}}(x^a) \mathbb{1}_{[1,\infty[}(\tfrac{|x^a|}{R})(1-\Delta^a)^{-1} \right\| \mathrm{d}R < \infty, \quad a \in \mathcal{A}, \tag{6.15.5}$$

$$\int_0^\infty \left\| \partial_x^\alpha v^a_{\mathrm{l}}(x^a) \mathbb{1}_{[R,\infty[}(|x^a|) \right\| \langle R\rangle^{|\alpha|-1} \mathrm{d}R < \infty, \quad a \in \mathcal{A}, \ |\alpha| = 1, 2.$$

Then for all $a \in \mathcal{A}$, there exists a function

$$\mathbb{R} \times X_a \ni (t, \xi_a) \mapsto \tilde{S}_a(t, \xi_a) \in \mathbb{R}$$

such that the following limits exist:

$$\mathrm{s-}\lim_{t\to\infty} \mathrm{e}^{ith_a} \mathrm{e}^{-i\tilde{S}_a(t,D_a)}, \tag{6.15.6}$$

$$\mathrm{s-}\lim_{t\to\infty} \mathrm{e}^{i\tilde{S}_a(t,D_a)} \mathrm{e}^{-ith_a} \mathbb{1}_{Z_a}(p_a^+). \tag{6.15.7}$$

If we call (6.15.6) $\omega^+_{\mathrm{lr},a}$, then (6.15.7) equals $\omega^{+*}_{\mathrm{lr},a}$, and

$$\omega^{+*}_{\mathrm{lr},a}\omega^+_{\mathrm{lr},a} = \mathbb{1}, \qquad \omega^+_{\mathrm{lr},a}\omega^{+*}_{\mathrm{lr},a} = \mathbb{1}_{Z_a}(p_a^+),$$

$$\omega^+_{\mathrm{lr},a} D_a = p_a^+ \omega^+_{\mathrm{lr},a}, \qquad \omega^+_{\mathrm{lr},a} \tfrac{1}{2} D_a^2 = h_a \omega^+_{\mathrm{lr},a}.$$

We will see later that the functions $\tilde{S}_a(t, \xi_a)$ can be used to construct modified wave operators for the full Hamiltonian H.

Combining Theorem 6.15.2 with Proposition 6.8.6 and 6.10.1 about asymptotic separation, we obtain the following theorem. It says that even with $\mu \geq 0$ we can define modified wave operators for bound states with a sufficient decay.

Theorem 6.15.3
Assume the hypotheses of Theorem 6.15.2. Let Q_a^+ be defined as in Definition 6.8.5. Then for all $a \in \mathcal{A}$, there exists

$$\mathrm{s-}\lim_{t\to\infty} \mathrm{e}^{i\tilde{S}_a(t,D_a)+itH^a} \mathrm{e}^{-itH} Q_a^+ \mathbb{1}_{Z_a}(P^+). \tag{6.15.8}$$

Let us denote the orthogonal projection onto the range of (6.15.8) by $Q^+_{a,0}$. Then there exists

$$\mathrm{s-}\lim_{t\to\infty} \mathrm{e}^{itH} \mathrm{e}^{-i\tilde{S}_a(t,D_a)} Q^+_{a,0}. \tag{6.15.9}$$

If we call (6.15.8) $\Omega^+_{\mathrm{lr},Q,a}$, then (6.15.9) equals $\Omega^{+*}_{\mathrm{lr},Q,a}$. The operator $\Omega^+_{\mathrm{lr},Q,a}$ is a partial isometry satisfying

$$\Omega^{+,*}_{\mathrm{lr},Q,a}\Omega^+_{\mathrm{lr},Q,a} = Q_a^+ \mathbb{1}_{Z_a}(P^+) \leq \mathbb{1}_{Z_a}(P^+), \qquad \Omega^+_{\mathrm{lr},Q,a}\Omega^{+*}_{\mathrm{lr},Q,a} = Q^+_{a,0} \leq \mathbb{1}^{\mathrm{pp}}(H^a),$$

$$\Omega^+_{\mathrm{lr},Q,a} H_a = H \Omega^+_{\mathrm{lr},Q,a}, \qquad \Omega^+_{\mathrm{lr},Q,a} D_a = P_a^+ \Omega^+_{\mathrm{lr},Q,a}.$$

Moreover,

$$L^2(X_a) \otimes \mathbb{1}^{\mathrm{pp}}_{\mathbb{R}\setminus\mathcal{T}^a}(H^a) \leq L^2(X_a) \otimes \mathbb{1}^{\mathrm{pp}}_{\mathrm{reg}}(H^a) \leq Q^+_{a,0},$$

where $\mathbb{1}^{\mathrm{pp}}_{\mathrm{reg}}(H^a)$ *was defined in Theorem 6.10.1 (ii).*

The rest of this section is devoted to the proof of Theorem 6.15.2. Let us first explain how to construct the functions $\tilde{S}_a(t,\xi_a)$.

Proposition 6.15.4
Let $v^b(x^b)$ *for* $b \in \mathcal{B}$ *satisfy the assumption (6.15.5) of Theorem 6.15.2. Let* $j_a \in C_0^\infty(X_a)$ *be a cutoff function with*

$$\int j_a(y_a)\mathrm{d}y_a = 1, \quad \int y_a j(y_a)\mathrm{d}y_a = 0,$$

and let

$$\tilde{I}_a(t,x_a) = \int I_a(x_a + t^{\frac{1}{2}}y_a)j(y_a)\mathrm{d}y_a.$$

Then there exist a function $\tilde{S}_a(t,\xi_a)$ *that satisfies the following properties:*
(i) For every $\epsilon > 0$, *there exists* T_ϵ *such that*

$$\partial_t\tilde{S}_a(t,\xi_a) = \frac{1}{2}\xi_a^2 + \tilde{I}_a(t,\nabla_{\xi_a}\tilde{S}_a(t,\xi_a)), \qquad t > T_\epsilon,\ \xi_a \in Z_a^\epsilon.$$

(ii) For every $\epsilon > 0$,

$$\left|\partial_{\xi_a}^\beta\left(\tilde{S}_a(t,\xi_a) - \tfrac{1}{2}t\xi_a^2\right)\right| \leq o(t),\ |\beta| = 1,2, \qquad \xi_a \in Z_a^\epsilon,$$
$$\left|\partial_{\xi_a}^\beta\left(\tilde{S}_a(t,\xi_a) - \tfrac{1}{2}t\xi_a^2\right)\right| \leq o\left(t^{\frac{1}{2}|\beta|}\right),\ |\beta| \geq 2, \qquad \xi_a \in Z_a^\epsilon.$$

Proof. The proof consists in combining the arguments indicated in the proof of Proposition 4.7.3 and those of Theorem 5.8.2. □

Proof of Theorem 6.15.2. The proof of Theorem 6.15.2 is very similar to the proof of Theorem 4.7.1, so we will only describe its main points. Let us fix a compact set $\Theta \subset Z_a$ and repeat the construction in Sect. 6.9 with h_a as our N-particle Hamiltonian. More precisely, we fix a cutoff function $\check{j}_a \in C_0^\infty(Z_a)$ such that $\check{j}_a \equiv 1$ on a neighborhood of Θ, and also $x_0 \in \Theta$. We set

$$\check{I}_a(t,x_a) := (I_{l,a}(x_a) - I_{l,a}(t,tx_0))\,\check{j}_a\left(\frac{x_a}{t}\right) + I_{l,a}(tx_0).$$

Let us denote by $\check{u}_a(t,s)$ the dynamics generated by

$$\check{h}_a(t) := \frac{1}{2}D_a^2 + \check{I}_a(t,x_a).$$

and by $\check{p}_a^+$ the corresponding asymptotic velocity. Then e^{-ith_a} is asymptotic to $\check{u}_a(t,s)$ on Θ.

Applying the results in Sect. 4.7, we obtain the existence of the limits

$$\mathrm{s-}\lim_{t\to\infty} \breve{u}_a(0,t)\mathrm{e}^{-i\tilde{S}_a(t,D_a)}1\!\!1_\Theta(D_a) =: \breve{\omega}^+_{a,\Theta},$$

$$\mathrm{s-}\lim_{t\to\infty} \mathrm{e}^{i\tilde{S}_a(t,D_a)}\breve{u}_a(t,0)1\!\!1_\Theta(\breve{p}^+_a) = \breve{\omega}^+_{a,\Theta},$$

such that

$$\breve{\omega}^{+*}_{a,\Theta}\breve{\omega}^+_{a,\Theta} = 1\!\!1_\Theta(D_a), \qquad \breve{\omega}^+_{a,\Theta}\breve{\omega}^{+*}_{a,\Theta} = 1\!\!1_\Theta(\breve{p}^+_a),$$

$$\breve{p}^+_a\breve{\omega}^+_{a,\Theta} = \breve{\omega}^+_{a,\Theta}D_a.$$

Thus the theorem follows by the chain rule. □

Proof of Theorem 6.15.1. We proceed similarly as above, except that we use the results of Section 3.6 about the existence of the Dollard wave operators, instead of the results of Sect. 4.7. □

A. Miscellaneous Results in Real Analysis

A.1 Some Inequalities

First let us recall the well-known Gronwall inequality.

Proposition A.1.1
Suppose that $f(t)$, $g(t)$ and $z(t)$ are nonnegative functions on $[0,\infty[$. Suppose that $z(t)$ satisfies the following inequality:

$$z(t) \leq \int_s^t (f(u)z(u) + g(u))\mathrm{d}u. \tag{A.1.1}$$

Then

$$z(t) \leq \exp\left(\int_s^t f(u)\mathrm{d}u\right) \int_s^t g(u)\mathrm{d}u. \tag{A.1.2}$$

Proof. Set

$$v(t) = \int_s^t g(u) \exp\left(\int_u^t f(u_1)\mathrm{d}u_1\right) \mathrm{d}u.$$

Then

$$v(t) \leq \exp\left(\int_s^t f(u)\mathrm{d}u\right) \int_s^t g(u)\mathrm{d}u. \tag{A.1.3}$$

Moreover, $v(t)$ solves the following differential equation:

$$\begin{cases} \frac{\mathrm{d}}{\mathrm{d}t} v(t) = f(t)v(t) + g(t), \\ v(s) = 0. \end{cases}$$

Hence it also solves the integral equation

$$v(t) = \int_s^t (f(u)v(u) + g(u))\mathrm{d}u. \tag{A.1.4}$$

By subtracting (A.1.1) from (A.1.4), we obtain

$$v(t) - z(t) \geq \int_s^t (v(u) - z(u))f(u)\mathrm{d}u. \tag{A.1.5}$$

Next we put

$$n(t) := \exp\left(-\int_s^t f(u)\mathrm{d}u\right)\int_s^t (v(u)-z(u))f(u)\mathrm{d}u.$$

It satisfies

$$\begin{cases} \frac{\mathrm{d}}{\mathrm{d}t}n(t) \geq 0, \\ n(s) = 0. \end{cases}$$

Therefore, $n(t) \geq 0$ for $t \geq s$. Hence,

$$v(t) - z(t) \geq n(t) \geq 0.$$

This together with (A.1.3) implies (A.1.2). □

Next we describe a Gronwall-type lemma for solutions of second order differential inequalities.

Lemma A.1.2
Let f, g be two positive functions such that $\langle t\rangle f(t), \langle t\rangle g(t) \in L^1(\mathrm{d}t)$. Let the function $x(t) \in C^{1,1}([0,\infty[, X)$ be such that

$$\begin{cases} |\ddot{x}(t)| \leq f(t)|x(t)| + g(t), \\ \lim_{t\to\infty} \dot{x}(t) = 0. \end{cases} \tag{A.1.6}$$

Then the following is true:
(i) The limit

$$\lim_{t\to\infty} x(t)$$

exists.
(ii) For a sufficiently large time T, there exists a unique bounded function $z(t)$ that solves the following problem:

$$\begin{cases} \ddot{z}(t) + f(t)z(t) + g(t) = 0, \\ \lim_{t\to\infty} \dot{z}(t) = 0, \qquad z(T) = |x(T)|. \end{cases} \tag{A.1.7}$$

(iii) One has

$$|x(t)| \leq z(t), \quad t \geq T.$$

Proof. Define

$$\begin{aligned} r_T(t) &:= |x(T)| + \textstyle\int_T^t (s-T)g(s)\mathrm{d}s + (t-T)\int_t^\infty g(s)\mathrm{d}s, \\ \mathcal{P}_T z(t) &:= \textstyle\int_T^t (s-T)f(s)z(s)\mathrm{d}s + (t-T)\int_t^\infty f(s)z(s)\mathrm{d}s. \end{aligned}$$

Then it is easy to see that $\mathcal{P}_T$ is bounded on the Banach spaces Z_T^0 and Z_T^1 defined in Sect. 1.1. In both cases, the norm of $\mathcal{P}_T$ equals

$$\int_T^\infty (t-T)f(t)\mathrm{d}t.$$

Moreover, $\mathcal{P}_T$ maps nonnegative functions into nonnegative functions.

Next note that we can rewrite the problem (A.1.7) in the form

$$z = r_T + \mathcal{P}_T z. \tag{A.1.8}$$

Clearly, $r_T \in Z_T^0$ and, for T large enough, the norm of $\mathcal{P}_T$ is less than 1. We fix such a T. For this T, the problem (A.1.8) has a unique solution

$$z = (1-\mathcal{P})^{-1} r_T \in Z_T^0.$$

This ends the proof of *(ii)*.

Let us now compute

$$\frac{\mathrm{d}}{\mathrm{d}t}|\dot{x}(t)| \le |\ddot{x}(t)| \le f(t)|x(t)| + g(t),$$

which implies

$$|\dot{x}(t)| \le \int_t^\infty (f(s)|x(s)| + g(s))\mathrm{d}s. \tag{A.1.9}$$

Integrating (A.1.9), we obtain

$$\begin{aligned} |x(t)| - |x(T)| &\le \textstyle\int_T^t (s-T)(f(s)|x(s)| + g(s))\mathrm{d}s \\ &\quad + (t-T)\textstyle\int_t^\infty (f(s)|x(s)| + g(s))\mathrm{d}s. \end{aligned} \tag{A.1.10}$$

We can rewrite (A.1.10) as

$$|x| \le r_T + \mathcal{P}_T|x|, \tag{A.1.11}$$

Let us now set

$$x_1(t) := z(t) - |x(t)|.$$

Note that $x_1(t) \in Z_T^1$. By (A.1.11) and (A.1.8), we have

$$x_1(t) \ge \mathcal{P}_T x_1(t).$$

We also set

$$y(t) := x_1(t) - \mathcal{P}_T x_1(t).$$

This is non-negative and belongs to Z_T^1.

Because $\|\mathcal{P}_T\| < 1$ on Z_T^1, we can write

$$x_1 = (1-\mathcal{P}_T)^{-1} y = \sum_{n=0}^\infty \mathcal{P}_T^n y.$$

This is nonnegative, because $\mathcal{P}_T$ preserves the positivity. Therefore

$$|x(t)| \le z(t).$$

This completes the proof of *(iii)*.

Using the fact that $x(t)$ is bounded and that $\langle t\rangle f(t), \langle t\rangle g(t) \in L^1(\mathrm{d}t)$, we finally deduce from (A.1.6) that $|\dot{x}(t)| \in L^1(\mathrm{d}t)$, which proves that $x(t)$ has a limit at ∞. □

Next we would like to present a very simple lemma on decreasing functions.

Lemma A.1.3
Let $\mathbb{R}^+ \ni s \mapsto f(s)$ *be a positive decreasing function and* $n > -1$ *such that*

$$\int_0^\infty s^n f(s)\mathrm{d}s < \infty. \tag{A.1.12}$$

Then,

$$\lim_{t\to\infty} t^{n+1} f(t) = 0, \tag{A.1.13}$$

$$\sup_{t>0} \int_t^{t+r} s^{n+1} f(s)\mathrm{d}s \in o(r). \tag{A.1.14}$$

Proof. We have

$$\int_0^t s^n f(s)\mathrm{d}s = \frac{t^{n+1}}{n+1} f(t) - \int_0^t \frac{s^{n+1}}{n+1} f'(s)\mathrm{d}s. \tag{A.1.15}$$

where $-f'(s)\mathrm{d}s$ is a positive measure. Since f is positive, we deduce from (A.1.15) that $-\int_0^t s^{n+1} f'(s)\mathrm{d}s$ is a bounded function. Obviously, it is increasing, hence it tends to a limit when t tends to ∞. Therefore $t^{n+1} f(t)$ also has a limit at ∞, which can only be 0 by (A.1.12).

To prove (A.1.14), we note that

$$\sup_{0\le t\le r} \frac{1}{r} \int_t^{t+r} s^{n+1} f(s)\mathrm{d}s \le \int_0^{2r} \frac{s^{n+1}}{r} f(s)\mathrm{d}s$$

goes to zero as $r \to \infty$ by Lebesgue's theorem, and

$$\sup_{r\le t} \frac{1}{r} \int_t^{t+r} s^{n+1} f(s)\mathrm{d}s \le \sup_{t\ge r} t^{n+1} f(t)$$

goes to zero by (A.1.13). □

A.2 The Fixed Point Theorem

In this appendix we describe some well-known facts related to the fixed point theorem, which are useful in proofs of the existence of solutions of differential equations. We start with the fixed point theorem itself, sometimes called Banach's principle.

Theorem A.2.1
Suppose that (Z, d) is a complete metric space and $\mathcal{P} : Z \to Z$ is a contraction, that is, there exists $q < 1$ such that, for any $z, z' \in Z$,

$$\mathrm{d}(\mathcal{P}(z), \mathcal{P}(z')) \le q\mathrm{d}(z, z').$$

Then there exists a unique $z \in Z$ such that

$$z = \mathcal{P}(z). \tag{A.2.1}$$

Proof. Let z_0 be an arbitrary element of Z. Set $z_n := \mathcal{P}^n(z_0)$. We easily verify that if $j < k$, then

$$\mathrm{d}(z_j, z_k) \le (1-q)^{-1} q^j \mathrm{d}(z_0, z_1).$$

Hence the sequence z_n is Cauchy. We easily check that its limit satisfies (A.2.1) and that two distinct fixed points of (A.2.1) cannot exist. □

If the map $\mathcal{P}$ depends continuously on parameters then so does the solution of the fixed point equation (A.2.1). This fact follows from the following proposition.

Proposition A.2.2
Suppose that (Z, d) is a complete metric space, X is a topological space, $\mathcal{P} : X \times Z \to Z$ is a map that satisfies the following two properties:
(i) for any $x_0 \in X$, there exists a neighborhood U of x_0 and $q < 1$ such that, for all $x \in U$ and $z, z' \in Z$,

$$\mathrm{d}(\mathcal{P}(x, z), \mathcal{P}(x, z')) \le q\mathrm{d}(z, z');$$

(ii) for any $z \in Z$, the function

$$X \ni x \mapsto \mathcal{P}(x, z) \in Z$$

is continuous.

Let $z(x)$ be the solution of the fixed point equation

$$\mathcal{P}(x, z(x)) = z(x).$$

Then the function

$$X \ni x \mapsto z(x) \in Z$$

is continuous.

Proof. Fix $x_0 \in X$. For any $\epsilon > 0$, we will find a neighborhood U of x_0 and $q < 1$ such that, for any $x \in U$ and $z, z' \in Z$,

$$\mathrm{d}(\mathcal{P}(x, z)), \mathcal{P}(x, z')) \le q\mathrm{d}(z, z')$$

and

$$d(\mathcal{P}(x_0, z(x_0), \mathcal{P}(x, z(x_0))) \le \epsilon.$$

Then

$$\begin{aligned} d(z(x_0), z(x)) &= d(\mathcal{P}(x_0, z(x_0)), \mathcal{P}(x, z(x))) \\ &\le d(\mathcal{P}(x_0, z(x_0)), \mathcal{P}(x, z(x_0))) + d(\mathcal{P}(x, z(x_0)), \mathcal{P}(x, z(x))) \\ &\le \epsilon + q d(z(x_0), z(x)). \end{aligned}$$

Hence

$$d(z(x_0), z(x)) \le (1-q)^{-1}\epsilon.$$

This clearly implies the continuity of $z(x)$. □

Let us now state the following simple criterion for the contractivity of a map that follows immediately from the mean value theorem.

Proposition A.2.3
Let Z be a Banach space and U a convex subset of Z. Suppose that $\mathcal{P} : U \to U$ is a map such that, for any $z \in U$,

$$\|\nabla_z \mathcal{P}(z)\| \le q < 1.$$

Then $\mathcal{P}$ is a contraction.

As an application of the fixed point theorem, let us state the following theorem about the existence of the solution of the Cauchy problem for ordinary differential equations.

Proposition A.2.4
Let X be a Banach space. Suppose that a measurable function

$$[T_1, T_2] \times X \ni (t, x) \mapsto f(t, x) \in X$$

satisfies the following estimates

$$\int_{T_1}^{T_2} |f(t, 0)| dt < \infty, \tag{A.2.2}$$

$$\int_{T_1}^{T_2} \|\nabla_x f(t, \cdot)\|_\infty dt < \infty. \tag{A.2.3}$$

Then for any $(s, y) \in [T_1, T_2] \times X$, there exists a unique solution

$$[T_1, T_2] \ni t \mapsto x(t, s, y) \in X$$

of the problem

$$\begin{cases} \partial_t x(t, s, y) = f(t, x(t, s, y)), \\ x(s, s, y) = y. \end{cases} \tag{A.2.4}$$

Moreover,

$$|\nabla_y x(t,s,y)| \le C. \tag{A.2.5}$$

Proof. For any $(s,y) \in [T_1, T_2] \times X$, we set

$$z(t) := x(t,s,y) - y$$

and define

$$\mathcal{P}(z)(t) := \int_s^t f(u, y + z(u)) du.$$

Note that $z(t)$ satisfies the equation

$$z = \mathcal{P}(z). \tag{A.2.6}$$

If $T > 0$ is such that $[s-T, s+T] \subset [T_1, T_2]$, we consider the Banach space $C([s-T, s+T], X)$ with the supremum norm. Note that $\mathcal{P}$ maps this space into itself. In fact, from (A.2.2) and (A.2.3) it follows that, for any C,

$$\int_{T_1}^{T_2} \sup_{|x|\le C} |f(u,x)| du \le \int_{T_1}^{T_2} |f(u,0)| du + C \int_{T_1}^{T_2} \|\nabla_x f(u,\cdot)\|_\infty du < \infty.$$

Next let us compute the derivative of $\mathcal{P}$ with respect to z:

$$\nabla_z \mathcal{P}(z) v(t) = \int_s^t \nabla_x f(u, y + z(u)) v(u) du.$$

Therefore on $C([s-T, s+T], X)$ we have

$$\|\nabla_z \mathcal{P}(z)\| \le \int_{s-T}^{s+T} \|\nabla_x f(u,\cdot)\|_\infty du. \tag{A.2.7}$$

Clearly, (A.2.7) goes to zero as $T \to 0$. Therefore, for small enough T, the map $\mathcal{P}$ is contractive. Hence, for small enough T, there exists a unique solution of (A.2.6) in $C([s-T, s+T], X)$. This is equivalent to the existence and uniqueness of the solution of (A.2.4) on $[s-T, s+T]$. In a finite number of steps, we can extend this solution on the whole $[T_1, T_2]$.

Next we note that

$$\nabla_y \mathcal{P}(z)(t) = \int_s^t \nabla_x f(u, y + z(u)) du. \tag{A.2.8}$$

Therefore, $\nabla_y \mathcal{P}(z)$ is bounded and

$$\nabla_y z = (1 - \nabla_z \mathcal{P}(z))^{-1} \nabla_y \mathcal{P}(z)$$

implies the boundedness of $\nabla_y z(s)$ for $t \in [s-T, s+T]$. Hence (A.2.5) is true for $t \in [s-T, s+T]$. By applying the chain rule a finite number of times, we obtain (A.2.5) on the whole $[T_1, T_2]$. □

Let us remark that the choice of 0 in (A.2.2) is completely arbitrary. If we replace 0 with any $x_0 \in X$, then we obtain an equivalent set of conditions.

A.3 The Hamilton-Jacobi Equation

In this appendix we recall some rather standard facts about the Hamilton-Jacobi equation.

First of all, we will study the existence of solutions $S(t,\xi)$ of the time-dependent Hamilton-Jacobi equation with given initial conditions in a certain open subset. Such initial conditions determine a solution inside a certain "tube". This solution can be computed from a well-known formula that involves the action integral along certain trajectories.

We will also study the solution $S(t_1,t_2,x,\xi)$ of the Hamilton-Jacobi equation naturally associated with the parametrization of trajectories by the initial position and the final momentum. This solution satisfies two different Hamilton-Jacobi equations – with respect to x and with respect to ξ. It appears naturally in the semi-classical approximation of the quantum evolution. It is also useful in quantum scattering theory. (For more information on Hamilton-Jacobi equations, we refer the reader for example to [FM]).

We will consider in this appendix a time-dependent Hamiltonian $h(t,x,\xi)$ on $X \times X'$ defined for $t \in [T_1,T_2]$. We will assume that the function $h(t,x,\xi)$ is measurable and satisfies

$$\begin{aligned} &\textstyle\int_{T_1}^{T_2} |\partial_x^\alpha \partial_\xi^\beta h(t,0,0)| \mathrm{d}t < \infty, \quad |\alpha|+|\beta| = 1, \\ &\textstyle\int_{T_1}^{T_2} \|\partial_x^\alpha \partial_\xi^\beta h(t,\cdot,\cdot)\|_\infty \mathrm{d}t < \infty, \quad |\alpha|+|\beta| = 2. \end{aligned} \tag{A.3.1}$$

Note that if $V(t,x)$ is a measurable function that satisfies

$$\begin{aligned} &\textstyle\int_{T_1}^{T_2} |\partial_x^\alpha V(t,0)| \mathrm{d}t < \infty, \quad |\alpha| = 1, \\ &\textstyle\int_{T_1}^{T_2} \|\partial_x^\alpha V(t,\cdot)\|_\infty \mathrm{d}t < \infty, \quad |\alpha| = 2, \end{aligned} \tag{A.3.2}$$

then

$$h(t,x,\xi) = \frac{1}{2}\xi^2 + V(t,x)$$

satisfies (A.3.1).

It follows by Proposition A.2.4 that conditions (A.3.1) guarantee the existence of solutions of the Hamilton equations

$$\begin{cases} \partial_t x(t,s,y,\eta) = \nabla_\xi h(t, x(t,s,y,\eta), \xi(t,s,y,\eta)), \\ \partial_t \xi(t,s,y,\eta) = -\nabla_x h(t, x(t,s,y,\eta), \xi(t,s,y,\eta)), \\ x(s,s,y,\eta) = y, \qquad \xi(s,s,y,\eta) = \eta. \end{cases} \tag{A.3.3}$$

We denote by $\phi(t,s)$ the flow generated by $h(t,x,\xi)$, which is defined by

$$\phi(t,s)(y,\eta) = (x(t,s,y,\eta), \xi(t,s,y,\eta)).$$

We are now going to describe how to solve the Hamilton-Jacobi equation with given initial conditions.

Theorem A.3.1
Suppose that $s \in [T_1, T_2]$ *and* Θ_s *is an open subset in* X'. *Let* $\psi \in C^{1,1}(\Theta_s)$. *Denote, for shortness,*

$$x(t,\eta) := x(t, s, \nabla_\eta \psi(\eta), \eta),$$
$$\xi(t,\eta) := \xi(t, s, \nabla_\eta \psi(\eta), \eta).$$

For $t \in [T_1, T_2]$, *we set*

$$\Theta_t := \{\xi(t,\eta) \in X' \mid \eta \in \Theta_s\}$$

and

$$\Theta = \{(t,\xi) \in [T_1, T_2] \times X' \mid \xi \in \Theta_t\}.$$

Suppose that, for $t \in [T_1, T_2]$, *the mapping*

$$\Theta_s \ni \eta \mapsto \xi(t,\eta) \in \Theta_t \tag{A.3.4}$$

is bijective and denote by $\eta(t,\xi)$ *its inverse. Assume also that*

$$\nabla_\xi \eta(t,\xi) \in L^\infty_{\text{loc}}(\Theta). \tag{A.3.5}$$

Then there exists a unique function

$$\Theta \ni (t,\xi) \mapsto S(t,\xi) \in \mathbb{R}$$

that solves the Hamilton-Jacobi equation

$$\begin{cases} \partial_t S(t,\xi) = h(t, \nabla_\xi S(t,\xi), \xi), \\ S(s,\xi) = \psi(\xi) \end{cases} \tag{A.3.6}$$

and satisfies $\nabla^2_\xi S \in L^\infty_{\text{loc}}(\Theta)$. *The solution* $S(t,\xi)$ *is equal to*

$$S(t,\xi) := Q(t, \eta(t,\xi)), \tag{A.3.7}$$

where

$$\begin{aligned} &Q(t,\eta) \\ &:= \psi(\eta) + \textstyle\int_s^t (h(u, x(u,\eta), \xi(u,\eta)) + \langle x(u,\eta), \partial_u \xi(u,\eta)\rangle) \mathrm{d}u. \end{aligned} \tag{A.3.8}$$

Moreover,

$$\nabla_\xi S(t,\xi) = x(t, \eta(t,\xi)). \tag{A.3.9}$$

Remark. The condition (A.3.4) saying that the mapping $\eta \mapsto \xi(t,\eta)$ is invertible is, of course, the familiar condition of the absence of caustics. Namely, if Λ_s denotes the Lagrangian manifold $\{(y,\eta) \mid y = \nabla_\eta \psi(\eta),\ \eta \in \Theta_s\}$, then the condition that (A.3.4) is bijective means that, for $t \in [T_1, T_2]$, the Lagrangian manifold $\Lambda_t := \phi(t,s)\Lambda_s$ projects bijectively on the ξ variables, which implies that no caustics appear in the time evolution of Λ_s.

Proof. We first observe that, by shrinking a little the open sets Θ_s and Θ, we may assume that $\nabla_\xi^2\psi$ and $\nabla_\xi\eta$ are uniformly bounded. Let us first prove that solutions of (A.3.6) such that $\nabla_\xi^2 S \in L^\infty(\Theta)$ are unique.

Let $\tilde{S}(t,\xi)$ be a solution of (A.3.6) . We fix $\eta \in \Theta_s$ and compute:

$$\begin{aligned}
&\tfrac{\mathrm{d}}{\mathrm{d}t}(\nabla_\xi\tilde{S}(t,\xi(t,\eta)) - x(t,\eta)) \\
&= \nabla_\xi^2\tilde{S}(t,\xi(t,\eta)(\nabla_x h(t,\nabla_\xi\tilde{S}(t,\xi(t,\eta)),\xi(t,\eta)) - \nabla_x h(t,x(t,\eta),\xi(t,\eta))) \\
&+\nabla_\xi h(t,\nabla_\xi\tilde{S}(t,\xi(t,\eta)),\xi(t,\eta)) - \nabla_\xi h(t,x(t,\eta),\xi(t,\eta)).
\end{aligned}$$

Set

$$k(t) := \nabla_\xi\tilde{S}(t,\xi(t,\eta)) - x(t,\eta).$$

Then

$$\begin{aligned}
&\tfrac{\mathrm{d}}{\mathrm{d}t}|k(t)| \le \left|\tfrac{\mathrm{d}}{\mathrm{d}t}k(t)\right| \\
&\le \left(\sup_{\xi\in\Theta_t}|\nabla_\xi^2\tilde{S}(t,\xi)| + 1\right)\|\nabla_{x,\xi}^2 h(t,\cdot,\cdot)\|_\infty |k(t)| = f(t)|k(t)|,
\end{aligned}$$

where $f(t)$ is integrable. Thus

$$|k(t)| \le |k(s)|\mathrm{e}^{\int_s^t f(u)\mathrm{d}u}.$$

But $k(s) = 0$. Therefore $k(t) = 0$ for all $t \in [T_1, T_2]$, which proves that

$$\nabla_\xi\tilde{S}(t,\xi) \;=\; x(t,\eta(t,\xi)). \tag{A.3.10}$$

We consider now the quantity $\tilde{S}(t,\xi(t,\eta))$, and compute its derivative with respect to t. Using (A.3.10), we get

$$\begin{aligned}
&\tfrac{\mathrm{d}}{\mathrm{d}t}\tilde{S}(t,\xi(t,\eta)) \\
&= \partial_t\tilde{S}(t,\xi(t,\eta)) + \langle\nabla_\xi\tilde{S}(t,\xi(t,\eta)),\partial_t\xi(t,\eta)\rangle \\
&= h(t,x(t,\eta),\xi(t,\eta)) + \langle x(t,\eta),\partial_t\xi(t,\eta)\rangle.
\end{aligned} \tag{A.3.11}$$

Integrating (A.3.11) between t_0 and t, we obtain

$$\tilde{S}(t,\xi(t,\eta)) = Q(t,\eta),$$

which shows that

$$\tilde{S}(t,\xi) = Q(t,\eta(t,\xi)) = S(t,\xi).$$

This proves the uniqueness part of the theorem.

Let us now prove that $S(t,\xi)$ defined by (A.3.7) satisfies (A.3.9). To this end, we remark that, using the equations of motion, we obtain

$$\begin{aligned}\partial_t \nabla_\eta Q(t,\eta) &= \langle \nabla_x h(t,x(t,\eta),\xi(t,\eta)), \nabla_\eta x(t,\eta)\rangle \\ &+\langle \nabla_\xi h(t,x(t,\eta),\xi(t,\eta)), \nabla_\eta \xi(t,\eta)\rangle \\ &+\langle \nabla_\eta x(t,\eta), \partial_t \xi(t,\eta)\rangle + \langle x(t,\eta), \partial_t \nabla_\eta \xi(t,\eta)\rangle \\ &= \partial_t \langle x(t,\eta), \nabla_\eta \xi(t,\eta)\rangle.\end{aligned} \tag{A.3.12}$$

Clearly,

$$\nabla_\eta Q(s,\eta) = \nabla_\eta \psi(\eta) = \langle \nabla_\eta \psi(\eta), \nabla_\eta \xi(s,\eta)\rangle. \tag{A.3.13}$$

Now (A.3.12) and (A.3.13) imply

$$\nabla_\eta Q(t,\eta) = \langle x(t,\eta), \nabla_\eta \xi(t,\eta)\rangle. \tag{A.3.14}$$

Moreover, we have

$$\nabla_\xi S(t,\xi) = \langle \nabla_\eta Q(t,\eta(t,\xi)), \nabla_\xi \eta(t,\xi)\rangle. \tag{A.3.15}$$

The identities (A.3.14) and (A.3.15) imply (A.3.9).

Let us now prove that $S(t,\xi)$ solves the Hamilton-Jacobi equation. We have

$$\partial_t Q(t,\eta) = h(t,x(t,\eta),\xi(t,\eta)) + \langle x(t,\eta), \partial_t \xi(t,\eta)\rangle. \tag{A.3.16}$$

Using (A.3.7), and then (A.3.9), we obtain

$$\begin{aligned}\partial_t Q(t,\eta) &= \partial_t S(t,\xi(t,\eta)) + \langle \nabla_\xi S(t,\xi(t,\eta)), \partial_t \xi(t,\eta)\rangle \\ &= \partial_t S(t,\xi(t,\eta)) + \langle x(t,\eta), \partial_t \xi(t,\eta)\rangle.\end{aligned} \tag{A.3.17}$$

By comparing (A.3.16) and (A.3.17), we see that $S(t,\xi)$ solves the Hamilton-Jacobi equation. □

It will be convenient to state a variant of the above theorem with the interchanged role of the variables x and ξ.

Theorem A.3.2
Suppose that $s \in [T_1, T_2]$ and Θ_s is an open subset in X. Let $\psi \in C^{1,1}(\Theta_s)$. Denote, for shortness,

$$x(t,y) := x(t,s,y,\nabla_y \psi(y)),$$
$$\xi(t,y) := \xi(t,s,y,\nabla_y \psi(y)).$$

For $t \in [T_1, T_2]$, set

$$\Theta_t := \{x(t,y) \in X \mid y \in \Theta_s\},$$

and

$$\Theta := \{(t,x) \in [T_1,T_2] \times X \mid x \in \Theta_t\}.$$

Suppose that, for $t \in [T_1, T_2]$, the mapping

$$\Theta_s \ni y \mapsto x(t,y) \in \Theta_t \tag{A.3.18}$$

is bijective and denote by $y(t,x)$ its inverse. Assume also that

$$\nabla_x y \in L^\infty_{\mathrm{loc}}(\Theta).$$

Then there exists a unique function

$$\Theta \ni (t,x) \mapsto S(t,x) \in \mathbb{R}$$

that solves the Hamilton-Jacobi equation

$$\begin{cases} -\partial_t S(t,x) = h(t,x,\nabla_x S(t,x)), \\ S(s,x) = \psi(x) \end{cases} \tag{A.3.19}$$

and satisfies $\nabla_x^2 S \in L^\infty_{\mathrm{loc}}(\Theta)$. The solution $S(t,x)$ is equal to

$$S(t,x) := Q(t,y(t,x)),$$

where

$$\begin{aligned} &Q(t,y) \\ &:= \psi(y) - \textstyle\int_s^t (h(u,x(u,y),\xi(u,y)) - \langle \xi(u,y), \partial_u x(u,y)\rangle) du. \end{aligned} \tag{A.3.20}$$

Moreover,

$$\nabla_x S(t,x) = \xi(t,y(t,x)). \tag{A.3.21}$$

Proof. Consider the Hamiltonian

$$\tilde{h}(t,\tilde{x},\tilde{\xi}) := h(t,-\tilde{\xi},\tilde{x}).$$

Denote by $\tilde{\phi}(t,s)$ the flow associated with $\tilde{h}(t,\tilde{x},\tilde{\xi})$ and set $\tilde{\psi}(\tilde{\xi}) = -\psi(-\tilde{\xi})$.

Let us denote by χ the symplectic map

$$\chi : (\tilde{x},\tilde{\xi}) \mapsto (x,\xi) := (-\tilde{\xi},\tilde{x}).$$

Then

$$\tilde{h} = h \circ \chi.$$

Since χ is symplectic, we have

$$\tilde{\phi}(t,s) = \chi^{-1} \circ \phi(t,s) \circ \chi. \tag{A.3.22}$$

Using (A.3.18), (A.3.22), we see that we can apply Theorem A.3.1 to $\tilde{h}(t,\tilde{x},\tilde{\xi})$ and $\tilde{\psi}(\tilde{\xi})$, and find the function $\tilde{S}(t,\tilde{\xi})$ that solves

$$\begin{cases} \partial_t \tilde{S}(t,\tilde{\xi}) = \tilde{h}(t,\nabla_{\tilde{\xi}}\tilde{S}(t,\tilde{\xi}),\tilde{\xi}), \\ \tilde{S}(s,\tilde{\xi}) = \tilde{\psi}(\tilde{\xi}). \end{cases}$$

If we put now

$$S(t,x) := -\tilde{S}(t,-x),$$

we see that $S(t,x)$ solves (A.3.19). Finally, (A.3.20) and (A.3.21) are direct consequences of (A.3.8), (A.3.9) and of (A.3.22). □

Using Theorems A.3.1 and A.3.2, we will now solve the type of Hamilton-Jacobi equation that is encountered in the semi-classical approximation of the kernel of the unitary dynamics generated by a quantized $h(t,x,\xi)$. It is also useful in quantum scattering theory. We first introduce some notation. We set

$$Q_1(t_1,t_2,x_1,\xi_1) = \langle x_1,\xi_1\rangle$$
$$+\int_{t_1}^{t_2}(h(u,x(u,t_1,x_1,\xi_1),\xi(u,t_1,x_1,\xi_1)) + \langle x(u,t_1,x_1,\xi_1),\partial_u\xi(u,t_1,x_1,\xi_1)\rangle)du,$$

$$Q_2(t_1,t_2,x_2,\xi_2) = \langle x_2,\xi_2\rangle$$
$$+\int_{t_1}^{t_2}(h(u,x(u,t_2,x_2,\xi_2),\xi(u,t_2,x_2,\xi_2)) - \langle \partial_u x(u,t_2,x_2,\xi_2),\xi(u,t_2,x_2,\xi_2)\rangle)du$$

Note that if

$$(x_2,\xi_2) = (x(t_2,t_1,x_1,\xi_1),\xi(t_2,t_1,x_1,\xi_1))$$

or, equivalently,

$$(x_1,\xi_1) = (x(t_1,t_2,x_2,\xi_2),\xi(t_1,t_2,x_2,\xi_2)),$$

then

$$Q_1(t_1,t_2,x_1,\xi_1) = Q_2(t_1,t_2,x_2,\xi_2). \qquad \text{(A.3.23)}$$

This follows by integration by parts of the second term of the integrand in the definitions of Q_1 and Q_2.

Let us define

$$S(t_1,t_2,x_1,\xi_2) = Q_1(t_1,t_2,x_1,\xi_1(t_2,t_1,x_1,\xi_2)) \qquad \text{(A.3.24)}$$

or, equivalently,

$$S(t_1,t_2,x_1,\xi_2) = Q_2(t_1,t_2,x_2(t_2,t_1,x_1,\xi_2),\xi_2). \qquad \text{(A.3.25)}$$

Theorem A.3.3
Let $s\in[T_1,T_2]$, and let $\Omega_{s,s}\subset X\times X'$ be open. For any $t_1,t_2\in[T_1,T_2]$, denote by $\Gamma_{t_1,t_2}\subset X\times X'\times X\times X'$ the part of the graph of the flow $\phi(t_1,t_2)$ defined by

$$\Gamma_{t_1,t_2} := \{(x_1,\xi_1,x_2,\xi_2)\mid$$
$$(x_1,\xi_1)=\phi(t_1,s)(y,\eta),\ (x_2,\xi_2)=\phi(t_2,s)(y,\eta),\ (y,\eta)\in\Omega_{s,s}\}.$$

Let us also denote by π the projection

$$\pi : X\times X'\times X\times X' \to X\times X'$$
$$(x_1,\xi_1,x_2,\xi_2)\mapsto(x_1,\xi_2).$$

We will set

$$\Omega_{t_1,t_2} := \pi(\Gamma_{t_1,t_2}) = \{x(t_1,s,y,\eta), \xi(t_2,s,y,\eta)|\ (y,\eta) \in \Theta_{s,s}\}$$

and

$$\Omega := \{(t_1,t_2,x_1,\xi_2)|\ (x_1,\xi_2) \in \Omega_{t_1,t_2}\}.$$

Assume that, for any $t_1, t_2 \in [T_1, T_2]$, *the projection* π *restricted to* Γ_{t_1,t_2} *is a bijection. We denote by*

$$\Omega_{t_1,t_2} \ni (x_1,\xi_2) \mapsto (x_1, \xi_1(t_1,t_2,x_1,\xi_2), x_2(t_1,t_2,x_1,\xi_2), \xi_2) \in \Gamma_{t_1,t_2}$$

the mapping $\left(\pi\Big|_{\Gamma_{t_1,t_2}}\right)^{-1}$

Assume that

$$\nabla_{\xi_2}\xi_1(t_1,t_2,x_1,\xi_2) \in L^\infty_{\mathrm{loc}}(\Omega)$$

or, equivalently,

$$\nabla_{x_1}x_2(t_1,t_2,x_1,\xi_2) \in L^\infty_{\mathrm{loc}}(\Omega).$$

Then the function

$$\Omega \ni (t_1,t_2,x_1,\xi_2) \mapsto S(t_1,t_2,x_1,\xi_2) \in \mathbb{R}$$

is the unique solution of the Hamilton-Jacobi equation

$$\begin{cases} \partial_{t_2}S(t_1,t_2,x_1,\xi_2) = h(t_2, \nabla_{\xi_2}S(t_1,t_2,x_1,\xi_2), \xi_2), \\ S(t,t,x_1,\xi_2) = \langle x_1, \xi_2\rangle \end{cases} \tag{A.3.26}$$

such that

$$\nabla^2_{\xi_2}S(t_1,t_2,x_1,\xi_2) \in L^\infty_{\mathrm{loc}}(\Omega),$$

and it is also the unique solution of another Hamilton-Jacobi equation

$$\begin{cases} -\partial_{t_1}S(t_1,t_2,x_1,\xi_2) = h(t_1, x_1, \nabla_{x_1}S(t_1,t_2,x_1,\xi)), \\ S(t,t,x_1,\xi_2) = \langle x_1, \xi_2\rangle \end{cases} \tag{A.3.27}$$

such that

$$\nabla^2_{x_1}S(t_1,t_2,x_1,\xi_2) \in L^\infty_{\mathrm{loc}}(\Omega).$$

Moreover, the following identities are true:

$$\nabla_{\xi_2}S(t_1,t_2,x_1,\xi_2) = x_2(t_2,t_1,x_1,\xi_2), \tag{A.3.28}$$

$$\nabla_{x_1}S(t_1,t_2,x_1,\xi_2) = \xi_1(t_2,t_1,x_1,\xi_2). \tag{A.3.29}$$

Proof. First, we treat t_1, x_1 as parameters and we apply Theorem A.3.1 with $s = t_1$ and $\psi(\eta) = \langle \eta, x_1\rangle$. This implies the existence and uniqueness of the solution of (A.3.26) expressed in terms of Q_1 by the equation (A.3.24). Theorem A.3.1 also implies (A.3.28).

In (A.3.23) we have already proven the equality of (A.3.24) and (A.3.25), which means that we can express S also in terms of Q_2. Starting from (A.3.25) we apply now Theorem A.3.2 treating t_2, ξ_2 as parameters, with $s = t_2$ and $\psi(y) = \langle y, \xi_2 \rangle$. We obtain (A.3.29) and (A.3.27). □

Remark. The symplecticity of the flow $\phi(t,s)$ implies that the function

$$(x_2(t_1,t_2,x_1,\xi_2), \xi_1(t_1,t_2,x_1,\xi_2))$$

introduced in the above theorem has the following properties: $\nabla_{x_1}\xi_1(t_1,t_2,x_1,\xi_2)$ and $\nabla_{\xi_2}x_2(t_1,t_2,x_1,\xi_2)$ are symmetric and

$$\nabla_{\xi_2}\xi_1(t_1,t_2,x_1,\xi_2) = \nabla_{x_2}x_1(t_1,t_2,x_1,\xi_2).$$

In fact, the symplectic form is conserved by the flow, hence

$$\mathrm{d}x_1 \wedge \mathrm{d}\xi_1 = \mathrm{d}x_2 \wedge \mathrm{d}\xi_2.$$

Therefore,

$$\begin{aligned} &\mathrm{d}x_1 \wedge (\nabla_{x_1}\xi_1(t_1,t_2,x_1,\xi_2)\mathrm{d}x_1 + \nabla_{\xi_2}\xi_1(t_1,t_2,x_1,\xi_2)\mathrm{d}\xi_2) \\ &= (\nabla_{x_1}x_2(t_1,t_2,x_1,\xi_2)\mathrm{d}x_1 + \nabla_{\xi_2}x_2(t_1,t_2,x_1,\xi_2)\mathrm{d}\xi_2) \wedge \mathrm{d}\xi_2. \end{aligned}$$

Remark. Assuming that the hypotheses of Theorems A.3.1 and A.3.3 are both satisfied, one can express the solution $S(t,\xi)$ of (A.3.6) by

$$S(t,\xi) = \mathrm{c.v}_{x,\eta}(S(s,t,x,\xi) - \langle x, \eta \rangle + \psi(\eta)),$$

where $\mathrm{c.v}_{x,\eta}$ means the critical value with respect to the variable (x,η).

A.4 Construction of Some Cutoff Functions

It is easy to see that if $F \in C_0^\infty(\mathbb{R})$ and $F \geq 0$, then $F^{1/2}$ does not have to be smooth. In fact, if $z_0 \in \mathrm{supp}F$ and $F(z_0) = 0$, then F may fail to be smooth at z_0. In our estimates we often need non-negative cutoff functions F with the property that $F^{1/2} \in C^\infty(\mathbb{R})$ and $(F')^{1/2} \in C^\infty(\mathbb{R})$. The existence of sufficiently many of such functions is guaranteed by the following lemma.

Lemma A.4.1
Let $f \in C_0^\infty(\mathbb{R})$ and $F \in C^\infty(\mathbb{R})$ be defined as

$$f(t) := \begin{cases} \mathrm{e}^{-\frac{1}{t(1-t)}}, & t \in [0,1], \\ 0, & t \notin [0,1]. \end{cases} \quad \text{and} \quad F(t) := \int_{-\infty}^t f(s)\mathrm{d}s.$$

Then for any $\alpha > 0$, we have $F^\alpha \in C^\infty(\mathbb{R})$.

Proof. First let us show that, for t near 0,

$$F(t) = \mathrm{e}^{-\frac{1}{t(1-t)}} t^2(1 + O(t)). \tag{A.4.1}$$

Consider the change of variable $\frac{1}{x} = t(1-t)$. Clearly, it is invertible for $x \to \infty$ and

$$\tfrac{\mathrm{d}t}{\mathrm{d}x} = x^{-2} + O(x^{-3}).$$

Therefore, integrating by parts we get

$$F(\tfrac{1}{x}) = \mathrm{e}^{-x}x^{-2} + O(\mathrm{e}^{-x}x^{-3}).$$

Coming back to the original variables, we get (A.4.1).

Now let us show that $F^\alpha(t)$ is smooth at the origin. It suffices to prove that

$$\lim_{t \to 0^+} (F^\alpha)^{(n)}(t) = 0, \ n \in \mathbb{N}. \tag{A.4.2}$$

By Faa di Bruno's formula, we have

$$(F^\alpha)^{(n)}(t) = \sum_{n_1+\cdots+n_q=n} C_{n_1,\dots,n_q} F^{\alpha-q}(t) F^{(n_1)}(t) \cdots F^{(n_q)}(t). \tag{A.4.3}$$

By a direct computation, we see that

$$F^{(n)}(t) = O(t^{-2n})\mathrm{e}^{-\frac{1}{t(1-t)}}. \tag{A.4.4}$$

Using (A.4.1) and (A.4.4) in (A.4.3), we get (A.4.2), which completes the proof of the lemma. □

A.5 Propagation Estimates

The following lemma describes a certain type of reasoning that leads to estimates on observables integrated along a trajectory. These estimates go under the name of propagation estimates. They are better known in quantum scattering theory, where they turned out to be a very powerful tool in the proof of asymptotic completeness [SS1, Gr, De8].

We will introduce propagation estimates in the context of the flow generated by a rather general differential equation. Suppose that we consider solutions of the differential equation

$$\frac{\mathrm{d}}{\mathrm{d}t} x(t) = f(t, x(t)). \tag{A.5.1}$$

If $G(t, x)$ is a function on $\mathbb{R} \times X$, then we define the Liouville derivative of $G(t, x)$:

$$\mathbf{D}G(t,x) := \partial_t G(t,x) + f(t,x)\nabla_x G(t,x).$$

Note that

$$\frac{\mathrm{d}}{\mathrm{d}t}G(t,x(t)) = \mathbf{D}G(t,x(t)).$$

The following lemma should be compared with its quantum analog contained in Lemma B.4.1 *ii)*.

Lemma A.5.1
Suppose that $x(t)$ is a solution of (A.5.1). Let $\Phi(t,x)$, $B_1(t,x)$ and $B(t,x)$ be functions such that $B_1(t,x)$ and $B(t,x)$ are nonnegative, $\Phi(t,x(t))$ is bounded uniformly in t,

$$\begin{aligned}&\textstyle\int_0^\infty B_1(t,x(t))\mathrm{d}t < \infty,\\ &\mathbf{D}\Phi(t,x) \geq B(t,x) - B_1(t,x).\end{aligned}$$

Then

$$\int_0^\infty B(t,x(t))\mathrm{d}t < \infty.$$

A.6 Comparison of Two Dynamics

In classical scattering theory, it is common that we study two differential equations that are close in some sense. Given a solution of one of them we may want to find a solution of the other equation that is asymptotic to it. Below we give a proposition that describes a certain set of conditions when it is possible.

Proposition A.6.1
Let

$$\mathbb{R}^+ \times X \ni (t,x) \mapsto f_i(t,x) \in X,\ i = 1,2,$$

and let $x_1(t)$ be a solution of

$$\dot{x}_1(t) = f_1(t,x_1(t)).$$

Let $\theta > \kappa$ and

$$\begin{aligned}&\sup_{(t,x)\in\mathbb{R}_+\times X} |\nabla_x f_2(t,x)| \leq \kappa,\\ &\lim_{t\to\infty} \mathrm{e}^{t\theta} \textstyle\int_t^\infty |f_1(s,x_1(s)) - f_2(s,x_1(s))|\mathrm{d}s = 0.\end{aligned}$$

Let $x_{2,T}(t)$ be the unique solution of

$$\begin{cases} \dot{x}_{2,T}(t) = f_2(t,x_{2,T}(t)),\\ x_{2,T}(T) = x_1(T).\end{cases}$$

Then there exists the limit

$$\lim_{T\to\infty} x_{2,T}(t) =: x_2(t).$$

The function $x_2(t)$ is the unique solution of the problem

$$\begin{cases} \dot{x}_2(t) = f_2(t, x_2(t)), \\ \lim\limits_{t\to\infty} |x_1(t) - x_2(t)|e^{\theta t} = 0. \end{cases}$$

Proof. Note that $x_1(t)$ and $x_{2,T}(t)$ satisfy

$$\begin{aligned} x_1(t) &= x_1(T) - \textstyle\int_t^T f_1(s, x_1(s))ds, \\ x_{2,T}(t) &= x_1(T) - \textstyle\int_t^T f_2(s, x_{2,T}(s))ds. \end{aligned}$$

Thus

$$x_{2,T}(t) - x_1(t) = -\int_t^T (f_2(s, x_{2,T}(s)) - f_1(s, x_1(s)))ds. \tag{A.6.1}$$

Therefore, if we introduce

$$z_T(t) := x_{2,T}(t) - x_1(t),$$

then we can rewrite (A.6.1) as

$$z_T(t) = \mathcal{P}_T z_T(t), \tag{A.6.2}$$

where the operator $\mathcal{P}_T$ is defined as

$$\mathcal{P}_T z(t) := \int_t^T (f_2(s, x_1(s) + z_T(s)) - f_1(s, x_1(s))ds.$$

Let us introduce the Banach space

$$Z := \{z \in C(\mathbb{R}^+, X) \mid \lim_{t\to\infty} e^{\theta t}|z(t)| = 0\}$$

with the norm

$$\|z\| = \sup_{t\geq 0} e^{\theta t}|z(t)|.$$

Now,

$$\begin{aligned} |\mathcal{P}_T z_T(t)| &\leq \textstyle\int_t^T |f_2(s, x_1(s) + z_T(s)) - f_2(s, x_1(s))|ds \\ &\quad + \textstyle\int_t^T |f_2(s, x_1(s)) - f_1(s, x_1(s))|ds \in o(t^0)e^{-t\theta}(\|z\| + 1). \end{aligned}$$

Therefore, $\mathcal{P}_T$ is well defined for $0 \leq T \leq \infty$ as a map on Z. It depends continuously on T for $0 \leq T \leq \infty$.

Moreover,

$$
\begin{aligned}
|\mathcal{P}_T(z_1) - \mathcal{P}_T(z_2)(t)| &\leq \int_t^\infty \kappa \mathrm{e}^{-\theta s} \mathrm{d}s \|z_1 - z_2\| \\
&= \tfrac{\kappa}{\theta} \mathrm{e}^{-\theta t} \|z_1 - z_2\|,
\end{aligned}
$$

which proves that $\mathcal{P}_T$ is a contraction on Z for $\theta > \kappa$. By the fixed point theorem (see Theorem A.2.1), there exists a unique solution in Z of (A.6.2). □

Let us state a version of the above proposition for the Newton equation.

Corollary A.6.2
Let

$$
\mathbb{R}^+ \times X \ni (t, x) \mapsto F_i(t, x) \in X, \quad i = 1, 2
$$

and let $x_1(t)$ be a solution of

$$
\ddot{x}_1(t) = F_1(t, x_1(t)).
$$

Let $\kappa < \theta$ and

$$
\sup_{(t,x) \in \mathbb{R}_+ \times X} |\nabla_x F_2(t, x)| \leq \kappa^2,
$$

$$
\lim_{t \to \infty} \mathrm{e}^{t\theta} \int_t^\infty |F_1(s, x_1(s)) - F_2(s, x_1(s))| \mathrm{d}s = 0.
$$

Let $x_{2,T}$ be the unique solution of

$$
\begin{cases}
\ddot{x}_{2,T}(t) = F_2(t, x_{2,T}(t)), \\
x_{2,T}(T) = x_1(T), \qquad \dot{x}_{2,T}(T) = \dot{x}_1(T).
\end{cases}
$$

Then there exists the limit

$$
\lim_{T \to \infty} x_{2,T}(t) =: x_2(t).
$$

Moreover, the function $x_2(t)$ is the unique solution of the problem

$$
\begin{cases}
\ddot{x}_2(t) = F_2(t, x_2(t)), \\
\lim_{t \to \infty} (|x_1(t) - x_2(t)| + |\dot{x}_1(t) - \dot{x}_2(t)|) \mathrm{e}^{\theta t} = 0.
\end{cases}
$$

Proof. We introduce the variables $\tilde{x} = \kappa x$ and $\xi = \dot{x}$. Then the equations of motion can be rewritten as

$$
\begin{cases}
\dot{\tilde{x}}(t) = \kappa \xi(t), \\
\dot{\xi}(t) = F(t, \kappa^{-1} \tilde{x}(t)).
\end{cases}
\tag{A.6.3}
$$

Clearly,

$$
\|\nabla_{\tilde{x},\xi}(\kappa \xi, F(t, \kappa^{-1} \tilde{x}))\| \leq \kappa.
$$

Therefore, we can apply Proposition A.6.1. □

A.7 Schwartz's Global Inversion Theorem

It is well known that a function whose first derivative at $y_0 \in \mathbb{R}^n$ is invertible is invertible in a certain neighborhood of y_0. The following proposition, known as the Schwartz global inversion theorem, gives sufficient conditions that guarantee the global invertibility (see [Sch]).

Proposition A.7.1
Suppose that the function

$$\mathbb{R}^n \ni y \mapsto x(y) \in \mathbb{R}^n \tag{A.7.1}$$

satisfies

$$|\det \nabla_y x| \geq C_0 > 0 \tag{A.7.2}$$

and

$$\left|\partial_y^\alpha x\right| \leq C, \quad |\alpha| = 1, 2. \tag{A.7.3}$$

Then the function (A.7.1) is bijective.

Proof. First note that, by the local inversion theorem, there exist $\epsilon, \delta > 0$ such that, for any $y_0 \in \mathbb{R}^n$,

$$x(B(y_0, \epsilon)) \supset B(x(y_0), \delta). \tag{A.7.4}$$

Therefore $x(\mathbb{R}^n) = \mathbb{R}^n$.

We will now prove that the function (A.7.1) is injective. Fix $y_1, y_2 \in \mathbb{R}^n$ such that $x(y_1) = x(y_2) = x_0$.

Note that, given a curve Γ that starts at $x(y_1)$, we can find a unique curve $\tilde{\Gamma}$ that starts at y_1 and $x\left(\tilde{\Gamma}\right) = \Gamma$ (we start at y_1 and extend it by continuity).

Join y_1 and y_2 with a curve $[0,1] \ni \tau \mapsto \tilde{\Gamma}$. Set $\Gamma := x\left(\tilde{\Gamma}\right)$. The curve Γ can be continuously deformed to form a family of curves Γ_τ, $\tau \in [0,1]$, with the following properties:

$$\begin{aligned}
&\Gamma_1 = \Gamma,\\
&\Gamma_\tau(0) = \Gamma_\tau(1) = x_0, \quad 0 \leq \tau \leq 1,\\
&\Gamma_0(s) = x_0, \quad 0 \leq s \leq 1.
\end{aligned}$$

The corresponding curves $\tilde{\Gamma}_\tau$ are also continuously deformed. But $\tilde{\Gamma}_\tau(0) = y_0$ and $\tilde{\Gamma}_\tau(1) = y_1$ for all $\tau \in [0,1]$, hence $y_1 = y_2$. □

B. Operators on Hilbert Spaces

B.1 Self-adjoint Operators

In this section we collect basic concepts of the spectral theory for self-adjoint operators that will be needed in this chapter and we fix some notation. We refer the reader for example to [RS, vol I] for a more detailed exposition.

Let $\mathcal{H}$ be a separable Hilbert space. The scalar product of $\phi, \psi \in \mathcal{H}$ will be denoted by $(\phi|\psi)$.

$B(\mathcal{H})$ denotes the set of bounded operators on $\mathcal{H}$ and $CB(\mathcal{H})$ the set of compact operators on $\mathcal{H}$.

$\mathrm{Ran}B$ denotes the range of an operator B and $\mathcal{D}(B)$ denotes its domain. B^* denotes the adjoint of an operator B. We will often write $B + \mathrm{hc}$ instead of $B + B^*$.

Let H be a closed operator on $\mathcal{H}$. The spectrum of H is denoted by $\sigma(H) \subset \mathbb{C}$. If H is self-adjoint, then $\sigma(H) \subset \mathbb{R}$ and, for any Borel subset $\Delta \subset \sigma(H)$, we can define the spectral projection of H onto Δ denoted by $\mathbb{1}_\Delta(H)$. If $(A_1, \ldots, A_n)$ is an n-tuple of commuting self-adjoint operators, then $\sigma(A_1, \ldots, A_n)$ denotes their joint spectrum. If Θ is a Borel subset of $\mathbb{R}^n$, then $\mathbb{1}_\Theta(A_1, \ldots, A_n)$ denotes the spectral projection of $(A_1, \ldots, A_n)$ onto Θ.

We also define the space $\mathcal{H}_{\mathrm{pp}}(H)$ of *bound states* of H, the *pure point spectrum* $\sigma_{\mathrm{pp}}(H)$, the *continuous spectral subspace* $\mathcal{H}_{\mathrm{c}}(H)$ and the *continuous spectrum* $\sigma_{\mathrm{c}}(H)$:

$$\begin{aligned}
&\mathbb{1}^{\mathrm{pp}}(H) := \textstyle\sum_{\lambda\in\mathbb{R}} \mathbb{1}_{\{\lambda\}}(H), \quad \mathcal{H}_{\mathrm{pp}}(H) := \mathrm{Ran}\mathbb{1}^{\mathrm{pp}}(H),\\
&\sigma_{\mathrm{pp}}(H) := \{\lambda \in \mathbb{R} \mid \mathbb{1}_{\{\lambda\}}(H) \neq 0\},\\
&\mathbb{1}^{\mathrm{c}}(H) := \mathbb{1} - \mathbb{1}^{\mathrm{pp}}(H), \qquad \mathcal{H}_{\mathrm{c}}(H) := \mathrm{Ran}\,\mathbb{1}^{\mathrm{c}}(H),\\
&\sigma_{\mathrm{c}}(H) := \sigma\left(H|_{\mathcal{H}_{\mathrm{c}}(H)}\right).
\end{aligned}$$

We set

$$\mathbb{1}^{\mathrm{c}}_\Delta(H) := \mathbb{1}^{\mathrm{c}}(H)\mathbb{1}_\Delta(H), \qquad \mathbb{1}^{\mathrm{pp}}_\Delta(H) := \mathbb{1}^{\mathrm{pp}}(H)\mathbb{1}_\Delta(H).$$

It is also useful to introduce the decomposition of $\sigma(H)$ into the *discrete* and *essential* spectrum:

$$\begin{aligned}
&\sigma_{\mathrm{disc}}(H) := \{\lambda \in \mathbb{R} \mid \mathrm{rank}\mathbb{1}_{[\lambda-\epsilon,\lambda+\epsilon]}(H) \text{ is finite for some } \epsilon > 0\},\\
&\sigma_{\mathrm{ess}}(H) := \sigma(H)\backslash\sigma_{\mathrm{disc}}(H).
\end{aligned}$$

The essential spectrum is stable with respect to perturbations that have a certain compactness property.

Theorem B.1.1 Weyl Criterion
If H, H_0 are self-adjoint operators and

$$(H+i)^{-1}-(H_0+i)^{-1} \text{ is compact,} \tag{B.1.1}$$

then for any $\chi \in C_\infty(\mathbb{R})$,

$$\chi(H)-\chi(H_0) \quad \text{is compact,} \tag{B.1.2}$$

$$\sigma_{\text{ess}}(H)=\sigma_{\text{ess}}(H_0). \tag{B.1.3}$$

Proof. The function $z \mapsto (z-H)^{-1}-(z-H_0)^{-1}$ is norm continuous and compact at $\pm i$. Hence it is compact for all $z \notin \sigma(H)\cup\sigma(H_0)$.

By density, it is sufficient to assume that $\chi \in C_0^\infty(\mathbb{R})$. Let $\tilde{\chi}$ be the almost-analytic extension of χ (see Proposition C.2.1). Then

$$\chi(H)-\chi(H_0)=\frac{i}{2\pi}\int_{\mathbb{C}}\partial_{\bar{z}}\tilde{f}(z)\left((z-H)^{-1}-(z-H_0)^{-1}\right)\mathrm{d}z\wedge\mathrm{d}\bar{z}. \tag{B.1.4}$$

The integrand in (B.1.4) is compact and integrable, hence (B.1.2) is true. (B.1.3) follows from (B.1.2). □

The domain of an operator A, denoted by $\mathcal{D}(A)$, is in a natural way a normed space with the norm

$$(\|A\phi\|^2+\|\phi\|^2)^{\frac{1}{2}}=\|(A^*A+1)^{\frac{1}{2}}\phi\|.$$

An operator is called *closed*, if $\mathcal{D}(A)$ is complete with respect to this norm. Any subspace of $\mathcal{D}(A)$ dense with respect to this norm is called a *core* of A.

If $a(\cdot,\cdot)$ is a hermitian form bounded from below by $-\lambda_0$, with the domain $\mathcal{Q}(a)$, then it defines the norm

$$\sqrt{a(\phi,\phi)+(\lambda_0+1)\|\phi\|^2}. \tag{B.1.5}$$

We say that $a(\cdot,\cdot)$ is *closed* if $\mathcal{Q}(a)$ is complete with respect to this norm. In this case, there exists a unique self-adjoint operator A bounded from below by $-\lambda_0$ such that

$$a(\phi,\phi)=(\phi|A\phi),\ \ \phi\in\mathcal{Q}(a),$$

$$\mathcal{Q}(a)=\mathcal{D}((A+\lambda_0)^{\frac{1}{2}}).$$

Conversely, if the operator A is self-adjoint and bounded from below, then we can associate with it a closed form bounded from below, with the domain

$$\mathcal{Q}(a)=\mathcal{D}((A+\lambda_0)^{\frac{1}{2}}),$$

$$\text{such that} \quad a(\phi, \phi) := (\phi|A\phi).$$

We denote $\mathcal{Q}(A) := \mathcal{Q}(a)$. Any subspace of $\mathcal{Q}(A)$ dense with respect to the norm (B.1.5) is called a *form core* of A.

Definition B.1.2
*An operator B is said to be $A-$*bounded *if*

(i) $\mathcal{D}(A) \subset \mathcal{D}(B)$,

(ii) there exist C_0, C_1 such that $\|B\phi\| \leq C_0\|A\phi\| + C_1\|\phi\|, \quad \phi \in \mathcal{D}(A)$.

*The infimum of all C_0 such that (ii) holds is called the $A-$*bound *of B.*

Note that the A-bound equals

$$\lim_{\lambda\to\infty} \|B(A^*A+\lambda)^{-\frac{1}{2}}\|.$$

If A is bounded from below, then it is also equal to

$$\lim_{\lambda\to\infty} \|B(A+\lambda)^{-1}\|.$$

The following theorem is known as the Kato-Rellich theorem.

Theorem B.1.3
Let A be self-adjoint with domain $\mathcal{D}(A)$. Let B be symmetric and $A-$bounded with the $A-$bound strictly less than 1. Then
(i) $A + B$ is self-adjoint with domain $\mathcal{D}(A)$
(ii) any core for A is a core for $A+B$.

Definition B.1.4
*B is said to be A-*compact *if $B(A^*A+1)^{-\frac{1}{2}}$ is compact.*

If $\sigma(A) \neq \mathbb{C}$, then the above definition is equivalent to the compactness of $B(A-z)^{-1}$ for any $z \notin \sigma(A)$.

Proposition B.1.5
Let B be A-compact. Then B has the $A-$bound equal to 0.

Sometimes it is useful to use a form version of the Kato-Rellich theorem, sometimes called the KLMN theorem.

Definition B.1.6
*Let us assume that the operator A is bounded from below. An operator B with the form domain $\mathcal{Q}(B)$ is said to be $A-$*form bounded *if:*

(i) $\mathcal{Q}(A) \subset \mathcal{Q}(B)$,

(ii) there exist C_0, C_1 such that $|(\phi|B\phi)| \leq C_0(\phi|A\phi) + C_1(\phi|\phi), \ \phi \in \mathcal{Q}(A)$.

*The infimum of all C_0 such that (ii) holds is called the A-*form bound *of B.*

Note that the A-form bound of B equals

$$\lim_{\lambda \to \infty} \|(\lambda + A)^{-\frac{1}{2}} B (\lambda + A)^{-\frac{1}{2}}\|.$$

One has the following form version of Theorem B.1.3.

Theorem B.1.7
Suppose that A is self-adjoint and bounded from below. Let B be symmetric and A-form bounded with the A-form bound < 1. Then
(i) the sum of the quadratic forms of A and B is a closed symmetric form on $\mathcal{Q}(A)$ that is bounded below,
(ii) there exists a unique self-adjoint operator associated with this form, which is denoted by $A+B$,
(iii) any form core for A is a form core for $A+B$.

Definition B.1.8
*The operator B is called A-*form compact *if*

$$(A - z)^{-\frac{1}{2}} B (A - z)^{-\frac{1}{2}} \text{ is compact,}$$

where $z \notin \sigma(A)$.

Of course, the above definition does not depend on the choice of z.

Proposition B.1.9
If B is A-form compact, then it is A-form bounded with relative bound 0.

Note the following relationships between operator and form boundedness.

Proposition B.1.10
(i) If B is $A-$bounded with the $A-$bound a then B is $A-$form bounded with the $A-$bound $\leq a$.
(ii) If B is $A-$compact then B is $A-$form compact.

B.2 Convergence of Self-adjoint Operators

Let A_n be a sequence of operators in $B(\mathcal{H})$. Let us define the norm limit, the strong limit and the weak limit of A_n:

$$\begin{aligned} A &= \lim_{n\to\infty} A_n && \text{if} \quad \lim_{n\to\infty} \|A_n - A\| = 0, \\ A &= \text{s-}\lim_{n\to\infty} A_n && \text{if} \quad \lim_{n\to\infty} A_n\phi = A\phi, \quad \phi \in \mathcal{H}, \\ A &= \text{w-}\lim_{n\to\infty} A_n && \text{if} \quad \lim_{n\to\infty} (\psi|A_n\phi) = (\psi|A\phi), \quad \psi, \phi \in \mathcal{H}. \end{aligned}$$

From now on, within this section, we will assume B_n to be a sequence of vectors of commuting self-adjoint operators on a Hilbert space $\mathcal{H}$. More precisely,

$$B_n = (B_n^1, \dots, B_n^m), \quad [B_n^i, B_n^j] = 0, \quad 0 \le i, j \le m, \quad n = 1, 2, \dots .$$

We will not assume the boundedness of B_n. We will study various concepts of the convergence of B_n.

Proposition B.2.1
Suppose that, for every $g \in C_\infty(\mathbb{R}^m)$, there exists

$$\text{s}-\lim_{n\to\infty} g(B_n). \tag{B.2.1}$$

Then there exists a unique (possibly, non-densely defined) vector of self-adjoint operators

$$B = (B^1, \dots, B^m)$$

such that (B.2.1) equals $g(B)$. B is densely defined if, for some $g \in C_\infty(\mathbb{R}^m)$ such that $g(0) = 1$, we have

$$\text{s}-\lim_{R\to\infty} \left(\text{s}-\lim_{n\to\infty} g(R^{-1}B_n) \right) = \mathbb{1}. \tag{B.2.2}$$

Definition B.2.2
Under the assumptions of the above proposition, we will write

$$B = \text{s}-C_\infty - \lim_{n\to\infty} B_n.$$

If the limit in (B.2.1) is the norm limit, then we will write

$$B = C_\infty - \lim_{n\to\infty} B_n.$$

Remark. If B_n are bounded uniformly in n, then

$$\text{s}-\lim_{n\to\infty} B_n = \text{s}-C_\infty - \lim_{n\to\infty} B_n,$$
$$\lim_{n\to\infty} B_n = C_\infty - \lim_{n\to\infty} B_n.$$

If $m = 1$, then the strong C_∞ convergence is equivalent to the strong resolvent convergence, that is,

$$\text{s}-\lim_{n\to\infty} (B_n \pm i)^{-1} = (B \pm i)^{-1}.$$

Likewise, the norm-C_∞ convergence is equivalent to the norm resolvent convergence, that is,

$$\lim_{n\to\infty} (B_n \pm i)^{-1} = (B \pm i)^{-1}.$$

Proof of Proposition B.2.1. Denote (B.2.1) by $\gamma(g)$. Clearly, the strong limit preserves the multiplication. Moreover, we check that $\gamma(\overline{g}) = \gamma(g)^*$. Hence

$$C_\infty(X) \ni g \mapsto \gamma(g) \in B(\mathcal{H})$$

is a homomorphism of C^*-algebras.

For any open $\Theta \subset \mathbb{R}^m$, we define

$$\mathbb{1}_\Theta := \sup\{\gamma(g) |\ g \in C_\infty(X),\ 0 \leq g \leq 1,\ \ \mathrm{supp} g \subset \Theta\}.$$

In a standard way, we can extend the definition of $\mathbb{1}_\Theta$ to arbitrary Borel subsets Θ. We obtain a projection valued measure, that is, a map

$$\Theta \mapsto \mathbb{1}_\Theta$$

defined for every Borel subset $\Theta \subset \mathbb{R}^m$ that satisfies the following conditions:

(i) $\mathbb{1}_\Theta$ is an orthogonal projection,

(ii) $\mathbb{1}_\emptyset = 0$,

(iii) $\mathbb{1}_{\Theta_1}\mathbb{1}_{\Theta_2} = \mathbb{1}_{\Theta_1 \cap \Theta_2}$,

(iv) if $\Theta = \bigcup\limits_{n=1}^{\infty} \Theta_n$ and $\Theta_j \cap \Theta_n = \emptyset$ for $j \neq n$, then $\mathbb{1}_\Theta = \mathrm{s}-\lim\limits_{N\to\infty} \sum_{n=1}^N \mathbb{1}_{\Theta_n}$.

In a standard way, for any Borel function f on $\mathbb{R}^m$, we can define the integral

$$\int f(x)\mathrm{d}\mathbb{1}(x).$$

We can now set

$$B := \int x\mathrm{d}\mathbb{1}(x).$$

It is easy to check that B satisfies the requirements of our proposition. The operator B is densely defined if only if

$$\mathbb{1}_{\mathbb{R}^m} = 1,$$

which is equivalent to (B.2.2). □

Let us now describe the relationship of the spectrum of B with the spectra of B_n.

Proposition B.2.3

(i) Suppose that

$$\mathrm{s}-C_\infty-\lim_{n\to\infty} g(B_n) = g(B).$$

Let $\lambda \in \sigma(B)$. *Then there exist* $\lambda_n \in \sigma(B_n)$ *such that* $\lim_{n\to\infty} \lambda_n = \lambda$.

(ii) Suppose that

$$C_\infty-\lim_{n\to\infty} g(B_n) = g(B).$$

Then $\lambda \in \sigma(B)$ *if and only if there exist* $\lambda_n \in \sigma(B_n)$ *such that* $\lim_{n\to\infty} \lambda_n = \lambda$.

Proof. Let $\lambda \in \sigma(B)$. Let $\mathcal{U}$ be a neighborhood of λ and $g \in C_0(X)$ such that $g(\lambda) = 1$ and supp$g \subset \mathcal{U}$. Clearly,

$$\liminf_{n\to\infty} \|g(B_n)\| \geq \|\text{s-}\lim_{n\to\infty} g(B_n)\| = \|g(B)\| \geq 1.$$

Therefore, we will find λ_n such that, for $n \geq N$, we have $\lambda_n \in \sigma(B_n) \cup \mathcal{U}$. This proves *(i)* and the $\Rightarrow$ part of *(ii)*.

To show the $\Leftarrow$ part of *(ii)*, consider $\lambda_n \in \sigma(B_n)$ such that $\lim_{n\to\infty} \lambda_n = \lambda$. Let $\mathcal{U}$ be a neighborhood of λ, and $g \in C_0(X)$ such that supp$g \subset \mathcal{U}$ and, for large enough n, we have $g(\lambda_n) = 1$. Because of the norm convergence, we have

$$1 \leq \lim_{n\to\infty} \|g(B_n)\| = \|\lim_{n\to\infty} g(B_n)\| = \|g(B)\|.$$

Therefore, $\mathcal{U} \cap \sigma(B) \neq \emptyset$. □

B.3 Time-Dependent Hamiltonians

In this section we try to answer the following question: when does a time-dependent Hamiltonian generate a unitary dynamics and what does it mean. If the Hamiltonian is independent of time, the answer is simple: the Hamiltonian has to be self-adjoint. However, not every time-dependent self-adjoint Hamiltonian generates a dynamics.

The problem of finding sufficient conditions on a time-dependent Hamiltonian that guarantee the existence of a dynamics has been studied by many authors using a variety of methods. Among all these works, we would like to mention just a few. The papers [Ka1, Ka4] used the abstract theory of evolution equations in Banach spaces. A different approach by reduction to a time-independent problem has been given by Howland [How].

In the case of time-dependent Schrödinger operators, it is possible to use some special methods to construct the evolution. A method based on Fourier integral operators has been introduced by Fujiwara and developed in [Fu1, Fu2, KiK] and [Ki4]. Another approach based on the study of the integral equation satisfied by $U(t, s)$ by the perturbation of the free evolution has been introduced by Yajima [Ya2].

We need first to recall some basic facts about the measurability and integration for operator-valued functions. We will say that a function $[T_1, T_2] \ni t \to B(t) \in B(\mathcal{H})$ is *Bochner integrable* iff there exists a sequence $B_n(t)$ of measurable step functions (i.e. functions whose range is a finite set and every preimage is measurable) such that

$$\lim_{n\to\infty} \int_{T_1}^{T_2} \|B(t) - B_n(t)\| \mathrm{d}t = 0.$$

We will denote the set of Bochner integrable functions by $L^1_{\mathrm{loc}}(\mathbb{R}, B(\mathcal{H}))$. Note that, for such $B(t)$, the integral

$$\int_{t_1}^{t_2} B(s)\mathrm{d}s = \lim_{n\to\infty} \int_{t_1}^{t_2} B_n(s)\mathrm{d}s$$

is well defined and does not depend on the choice of $B_n(t)$.

We will say that $A(t) \in W^{1,1}(\mathbb{R}, B(\mathcal{H}))$ if there exists $B(t) \in L^1_{\mathrm{loc}}(\mathbb{R}, B(\mathcal{H}))$ such that, for any $t_1, t_2 \in \mathbb{R}$,

$$A(t_2) - A(t_1) = \int_{t_1}^{t_2} B(s)\mathrm{d}s.$$

We will write

$$\frac{\mathrm{d}}{\mathrm{d}t} A(t) := B(t).$$

Note that if $A_1(t), A_2(t) \in W^{1,1}(\mathbb{R}, B(\mathcal{H}))$, then $A_1(t)A_2(t) \in W^{1,1}(\mathbb{R}, B(\mathcal{H}))$ and

$$\frac{\mathrm{d}}{\mathrm{d}t} A_1(t)A_2(t) = \left(\frac{\mathrm{d}}{\mathrm{d}t} A_1(t)\right) A_2(t) + A_1(t)\frac{\mathrm{d}}{\mathrm{d}t} A_2(t). \tag{B.3.1}$$

Definition B.3.1
A unitary dynamics is a strongly continuous map $[T_1, T_2] \times [T_1, T_2] \ni (t, s) \mapsto U(t, s)$ *with values in unitary operators such that*

$$U(s,s) = 1,\ \ s \in [T_1, T_2], \qquad U(t,u)U(u,s) = U(t,s),\ \ t,u,s \in [T_1, T_2].$$

Note that we have $U(t,u) = U(u,t)^*$.

As we saw above, the definition of of a unitary dynamics is quite obvious. Unfortunately, the definition of a generator of such a dynamics is much more arbitrary.

Definition B.3.2
Suppose that B *is a positive invertible operator with a dense domain. Let* $[T_1, T_2] \ni t \mapsto H(t)$ *be a function with values in self-adjoint operators. We say that the unitary dynamics* $U(t,s)$ *is* B-regularly generated *by the time-dependent Hamiltonian* $H(s)$ *if the following conditions are satisfied:*
(i) $s \mapsto U(t,s)B^{-1}$ *belongs to* $W^{1,1}([T_1, T_2], B(\mathcal{H}))$, $\mathcal{D}(B) \subset \mathcal{D}(H(s))$ *for almost all* s, *and* $H(s)B^{-1}$ *belongs to* $L^1_{\mathrm{loc}}([T_1, T_2], B(\mathcal{H}))$;
(ii) $\partial_s U(t,s)B^{-1} = U(t,s)iH(s)B^{-1}$;
(iii) $B^{-1/2}[H(t), B]B^{-1/2}$, *originally defined as a quadratic form on* $\mathcal{D}(B^{1/2})$, *extends to an element of* $L^1_{\mathrm{loc}}([T_1, T_2], B(\mathcal{H}))$.

Note that, by conjugation, *(ii)* implies

$$\partial_t BU(t,s) = iB^{-1}H(t)U(t,s).$$

Proposition B.3.3
Suppose that $U(t,s)$ is B-regularly generated. Then it preserves $\mathcal{D}(B^{1/2})$.

Proof. Let $\epsilon > 0$ and $\psi \in \mathcal{D}(B^{1/2})$. Set

$$\begin{aligned} k_\epsilon(t) &= \|B^{\frac{1}{2}}(1+\epsilon B)^{-\frac{1}{2}}U(t,s)\psi\|^2 \\ &= \tfrac{1}{\epsilon}\|\psi\|^2 - \tfrac{1}{\epsilon}\|(1+\epsilon B)^{-\frac{1}{2}}U(t,s)\psi\|^2. \end{aligned}$$

Using *(ii)*, we obtain

$$\begin{aligned} \tfrac{\mathrm{d}}{\mathrm{d}t}k_\epsilon(t) &= \tfrac{1}{\epsilon}\left(\psi|U(s,t)[iH(t),(1+\epsilon B)^{-1}]U(t,s)\psi\right) \\ &= (\psi|U(s,t)(1+\epsilon B)^{-1}[B,iH(t)](1+\epsilon B)^{-1}U(t,s)\psi) \end{aligned}$$

Hence

$$\left|\tfrac{\mathrm{d}}{\mathrm{d}t}k_\epsilon(t)\right| \leq \left\|B^{-\frac{1}{2}}[H(t),B]B^{-\frac{1}{2}}\right\| k_\epsilon(t).$$

By Gronwall's inequality,

$$k_\epsilon(t) \leq Ck_\epsilon(0) = C\|(1+\epsilon B)^{-\frac{1}{2}}B^{\frac{1}{2}}\psi\|^2.$$

Letting ϵ go to zero we obtain that $B^{1/2}U(t,s)B^{-1/2}$ is bounded. □

Let us note another property of B-regularly generated dynamics that we will often use in this chapter.

Proposition B.3.4
Let $U_i(t,s)$, $i = 1,2$, be two dynamics B-regularly generated by two time-dependent Hamiltonians $H_i(t)$. Suppose that

$$\Phi(t) \in W^{1,1}([T_1,T_2],B(\mathcal{H})), \quad \|B^{\frac{1}{2}}\Phi(t)B^{-\frac{1}{2}}\| \leq C \textit{ for almost all } t,$$

and

$$H_2(t)\Phi(t) - \Phi(t)H_1(t)$$

originally defined as a quadratic form on $\mathcal{D}(B^{1/2})$ extends to an element of $L^1_{\mathrm{loc}}([T_1,T_2],B(\mathcal{H}))$. Set

$$_2\mathbf{D}_1\Phi(t) := \frac{\mathrm{d}}{\mathrm{d}t}\Phi(t) + iH_2(t)\Phi(t) - i\Phi(t)H_1(t).$$

Then

$$\begin{aligned} &U_2(s,t_2)\Phi(t_2)U_1(t_2,s) - U_2(s,t_1)\Phi(t_1)U_1(t_1,s) \\ &= \textstyle\int_{t_1}^{t_2} U_2(s,u)\left({}_2\mathbf{D}_1\Phi(t)\right)U_1(u)\mathrm{d}u. \end{aligned} \tag{B.3.2}$$

Proof. We can write

$$\begin{aligned}&U_2(s,t_2)\Phi(t_2)U_1(t_2,s) - U_2(s,t_1)\Phi(t_1)U_1(t_1,s)\\&= \mathrm{s-}\lim_{\epsilon\to 0} U_2(s,t_2)(1+\epsilon B)^{-1}\Phi(t_2)(1+\epsilon B)^{-1}U_1(t_2,s)\\&-\mathrm{s-}\lim_{\epsilon\to 0} U_2(s,t_1)(1+\epsilon B)^{-1}\Phi(t_1)(1+\epsilon B)^{-1}U_1(t_1,s).\end{aligned} \tag{B.3.3}$$

(B.3.3) equals the limit as $\epsilon \to 0$ of the following expression:

$$\begin{aligned}&\int_{t_1}^{t_2} U_2(s,u)(1+\epsilon B)^{-1}\left({}_2\mathbf{D}_1\Phi(t)\right)(1+\epsilon B)^{-1}U_1(u,s)\mathrm{d}u\\&+\int_{t_1}^{t_2} U_2(s,u)i[H_2(u),(1+\epsilon B)^{-1}]\Phi(u)U_1(u,s)\mathrm{d}u\\&+\int_{t_1}^{t_2} U_2(s,u)\Phi(u)[iH_1(u),(1+\epsilon B)^{-1}]U_1(u,s)\mathrm{d}u\\&=: R_1(\epsilon)+R_2(\epsilon)+R_3(\epsilon).\end{aligned}$$

Now $R_1(\epsilon)$ goes to the right-hand side of (B.3.2). Moreover, for $\psi_1, \psi_2 \in \mathcal{D}(B^{1/2})$, we have

$$\begin{aligned}&(\psi_2|R_2(\epsilon))\psi_1)\\&\leq \epsilon\int_{t_1}^{t_2} \|B^{\frac12}U_2(s,u)\psi_2\|\|B^{-\frac12}[H_2(u),B]]B^{-\frac12}\|\|B^{\frac12}\Phi(u)B^{-\frac12}\|\|B^{\frac12}U_1(u,s)\psi_1\|\mathrm{d}u,\end{aligned}$$

which clearly goes to zero. $R_3(\epsilon)$ can be handled in a similar fashion. □

So far, we have not given explicit conditions that, for a given time-dependent Hamiltonian $H(t)$, guarantee the existence of a unitary dynamics. Below we will describe one possible set of such conditions, unfortunately, quite restrictive.

Definition B.3.5
Let

$$t \to W(t)$$

belong to $L^1_{\mathrm{loc}}([T_1,T_2],B(\mathcal{H}))$. *For any* $[s,t]\subset[T_1,T_2]$, *we define the time-ordered exponential of* $W(t)$ *by the following convergent expansion:*

$$T\left(e^{\int_s^t W(u)\mathrm{d}u}\right) := \sum_{n=0}^{\infty} \int_{t\geq u_n\geq\cdots\geq u_1\geq s}\!\!\!\cdots\int W(u_n)\cdots W(u_1)\mathrm{d}u_n\ldots\mathrm{d}u_1.$$

Proposition B.3.6
Suppose that H_0 *is a fixed self-adjoint operator, and*

$$t \to V(t)$$

belongs to $L^1_{\mathrm{loc}}([T_1,T_2],B(\mathcal{H}))$. *Set*

$$H(t) := H_0 + V(t),$$

$$W(t) := \mathrm{e}^{itH_0}V(t)\mathrm{e}^{-itH_0},$$

$$U(t,s) := \mathrm{e}^{-itH_0}T\left(e^{\int_s^t W(u)\mathrm{d}u}\right)\mathrm{e}^{isH_0},$$

$$B := (H_0+1)^{\frac{1}{2}}.$$

Then $U(t,s)$ satisfies the following properties:
(i) $s \mapsto U(t,s)B^{-1}$ and $t \mapsto B^{-1}U(t,s)$ belong to $W^{1,1}([T_1,T_2],B(\mathcal{H}))$,
(ii)

$$\partial_s U(t,s)B^{-1} = U(t,s)iH(s)B^{-1},$$

$$\partial_t B^{-1}U(t,s) = -iB^{-1}H(t)U(t,s).$$

If, moreover, $V(t)$ is self-adjoint for all t, then $U(t,s)$ is a unitary dynamics in the sense of Definition B.3.1, and the conditions (i) and (ii) of Definition B.3.2 are satisfied.

B.4 Propagation Estimates

In this section we describe certain abstract arguments that are used in scattering theory to prove propagation estimates and the existence of asymptotic observables and of wave operators. These arguments do not depend on the concrete form of a time-dependent Hamiltonian.

In the first lemma we describe how two prove two types of the so-called propagation estimates. This name is usually given to various estimates on the evolution. The first type of a propagation estimate is a direct consequence of the fundamental theorem of calculus. The second one is a version of the Putnam-Kato theorem developed by Sigal and Soffer (see [RS, vol IV] and [SS1]).

Let $U(t,s)$ be the unitary evolution generated by a time-dependent Hamiltonian $H(t)$. We assume that all the conditions of Definitions B.3.1 and B.3.2 are satisfied and $U(t,s)$ is B-regularly generated. For simplicity, we will write $U(t) := U(t,0)$. Note that $U(t)$ satisfies

$$\begin{cases} \frac{\mathrm{d}}{\mathrm{d}t}U(t)\phi = -iH(t)U(t)\phi, & \phi \in \mathcal{D}(B^{\frac{1}{2}}), \\ U(0) = 1. \end{cases}$$

We will denote by $\mathbf{D}\Phi(t)$ the Heisenberg derivative associated with $U(t)$:

$$\mathbf{D}\Phi(t) := \frac{\mathrm{d}}{\mathrm{d}t}\Phi(t) + i[H(t),\Phi(t)].$$

Lemma B.4.1
Suppose that $\Phi(t)$ is a family of self-adjoint operators satisfying the conditions of Proposition B.3.4.
(i) Let $\mathbf{D}\Phi(t) \in L^1(\mathbb{R}^+, B(\mathcal{H}))$. Then

$$\|\Phi(t)U(t)\phi\| \leq \|\Phi(0)\phi\| + \int_0^t \|\mathbf{D}\Phi(s)\|\mathrm{d}s. \tag{B.4.1}$$

(ii) Suppose that $\Phi(t)$ is uniformly bounded and that there exist $C_0 > 0$ and operator valued functions $B(t)$ and $B_i(t)$, $i = 1, \ldots, n$, such that

$$\mathbf{D}\Phi(t) \geq C_0 B^*(t)B(t) - \sum_{i=1}^{n} B_i^*(t)B_i(t),$$
$$\int_1^\infty \|B_i(t)U(t)\phi\|^2 \mathrm{d}t \leq C\|\phi\|^2, \quad i = 1, \ldots, n.$$

Then there exists C_1 such that

$$\int_1^\infty \|B(t)U(t)\phi\|^2 \mathrm{d}t \leq C_1\|\phi\|^2. \tag{B.4.2}$$

Proof. *(i)* follows directly from the fact that

$$\frac{\mathrm{d}}{\mathrm{d}t}U(t)^*\Phi(t)U(t) = U(t)^*(\mathbf{D}\Phi(t))U(t).$$

To prove *(ii)*, for $0 \leq t_1 \leq t_2$, we compute

$$\begin{aligned} C_0 \textstyle\int_{t_1}^{t_2} \|B(t)U(t)\phi\|^2 \mathrm{d}t &\leq \textstyle\int_{t_1}^{t_2} (U(t)\phi|\mathbf{D}\Phi(t)U(t)\phi)\mathrm{d}t + \sum_i \int_{t_1}^{t_2} \|B_i(t)U(t)\phi\|^2 \mathrm{d}t \\ &\leq (U(t_2)\phi|\Phi(t_2)U(t_2)\phi) - (U(t_1)\phi|\Phi(t_1)U(t_1)\phi) \\ &\quad + \textstyle\sum_{i=1}^n \int_{t_1}^{t_2} \|B_i(t)U(t)\phi\|^2 \mathrm{d}t \leq C\|\phi\|^2, \end{aligned}$$

which proves the desired result. □

The observable $\Phi(t)$ used to derive (B.4.2) is called a *propagation observable.* As we saw above, the main idea of the proof of (B.4.2) is to find a propagation observable whose Heisenberg derivative is “essentially positive”.

Next we describe two methods of proving the existence of wave operators and asymptotic observables. The first one is known as Cook's method and the second one is its variation due to Kato (see [RS, vol IV] and references therein).

Let $H_1(t)$ and $H_2(t)$ be two time-dependent self-adjoint operators. Let $U_i(t)$ be the unitary evolutions generated by $H_i(t)$ in the sense of Definitions B.3.1 and B.3.2. Let ${}_2\mathbf{D}_1$ be defined as in Proposition B.3.4.

Lemma B.4.2
Suppose that $\Phi(t)$ is a uniformly bounded function with values in self-adjoint operators satisfying the conditions of Proposition B.3.2. Let $\mathcal{D}_1 \subset \mathcal{H}$ be a dense subspace.

(i) Assume that, for $\phi \in \mathcal{D}_1$,

$$\int_1^\infty \|({}_2\mathbf{D}_1\Phi(t))U_1(t)\phi\|\,\mathrm{d}t < \infty.$$

Then there exists

$$\text{s-}\lim_{t\to\infty} U_2^*(t)\Phi(t)U_1(t). \tag{B.4.3}$$

(ii) Assume that

$$\begin{aligned}
&|(\psi_2|{}_2\mathbf{D}_1\Phi(t)\psi_1)| \leq \sum_{i=1}^n \|B_{2i}(t)\psi_2\|\|B_{1i}(t)\psi_1\|,\\
&\int_1^\infty \|B_{2i}(t)U_2(t)\phi\|^2\mathrm{d}t \leq C\|\phi\|^2, \quad \phi \in \mathcal{H}, \quad i = 1,\ldots,n,\\
&\int_1^\infty \|B_{1i}(t)U_1(t)\phi\|^2\mathrm{d}t \leq C\|\phi\|^2, \quad \phi \in \mathcal{D}_1, \quad i = 1,\ldots,n.
\end{aligned}$$

Then the limit (B.4.3) exists.

The proof of *(i)* is easy and left to the reader. Let us show *(ii)*.
Let $\phi \in \mathcal{D}_1$, $\psi \in \mathcal{H}$. Then

$$\begin{aligned}
&|(\psi|U_2^*(t_2)\Phi(t_2)U_1(t_2)\phi) - (\psi|U_2^*(t_1)\Phi(t_1)U_1(t_1)\phi)|\\
&\leq \int_{t_1}^{t_2} |(\psi|U_2^*(t)({}_2\mathbf{D}_1\Phi(t))U_1(t)\phi)|\mathrm{d}t\\
&\leq \sum_{j=1}^n \left(\int_{t_1}^{t_2} \|B_{2j}(t)U_2(t)\psi\|^2\mathrm{d}t\right)^{1/2} \left(\int_{t_1}^{t_2} \|B_{1j}(t)U_1(t)\phi\|^2\mathrm{d}t\right)^{1/2}.
\end{aligned} \tag{B.4.4}$$

Therefore,

$$\begin{aligned}
&\|U_2^*(t_2)\Phi(t_2)U_1(t_2)\phi - U_2^*(t_1)\Phi(t_1)U_1(t_1)\phi\|\\
&= \sup_{\|\psi\|=1} |(\psi|U_2^*(t_2)\Phi(t_2)U_1(t_2)\phi) - (\psi|U_2^*(t_1)\Phi(t_1)U_1(t_1)\phi)|\\
&\leq \sum_{j=1}^n C \left(\int_{t_1}^{t_2} \|B_{1j}(t)U_1(t)\phi\|^2\mathrm{d}t\right)^{1/2}.
\end{aligned} \tag{B.4.5}$$

If we choose T big enough and $T \leq t_1 \leq t_2$, then we can make (B.4.5) arbitrarily small. This proves the existence of

$$\text{s-}\lim_{t\to\infty} U_2^*(t)\Phi(t)U_1(t)\phi, \quad \phi \in \mathcal{D}_1,$$

and hence it implies the existence of (B.4.3). □

B.5 Limits of Unitary Operators

The following lemma describes an argument often used in N-body scattering theory. Its easy proof is left to the reader.

Lemma B.5.1
Let $U(t)$ be a family of unitary operators. Let Q, Q_0 be a pair of orthogonal projections. Suppose that there exist

$$\Omega_0 := \mathrm{s-}\lim_{t\to\infty} U(t)Q_0, \quad \Omega := \mathrm{s-}\lim_{t\to\infty} U^*(t)Q.$$

Let us assume that

$$\mathrm{Ran}\Omega_0 \subset \mathrm{Ran}Q, \quad \mathrm{Ran}\Omega \subset \mathrm{Ran}Q_0.$$

Then Ω_0, Ω are partial isometries such that

$$\Omega_0^* = \Omega, \quad \Omega_0^*\Omega_0 = Q_0, \quad \Omega_0\Omega_0^* = Q.$$

B.6 Schur's Lemma

The next classical lemma is known as Schur's Lemma.

Lemma B.6.1
Let Y, Y' be spaces with measures $\mathrm{d}\mu, \mathrm{d}\mu'$. Let $k(\cdot,\cdot)$ be a function on $Y \times Y'$ such that

$$\mathrm{essup}_Y \int |k(y,y')|\mathrm{d}\mu' \le C, \quad \mathrm{essup}_{Y'} \int |k(y,y')|\mathrm{d}\mu \le C'.$$

Then the operator

$$K : L^2(Y, \mathrm{d}\mu) \to L^2(Y', \mathrm{d}\mu'),$$

$$Ku(y) := \int k(y,y')u(y')\mathrm{d}\mu'$$

is bounded and

$$\|K\| \le (CC')^{\frac{1}{2}}.$$

Proof. We compute

$$\begin{aligned}(Ku|v) &\le \int |k(y,y')||u(y)||v(y')|\mathrm{d}\mu\mathrm{d}\mu' \\ &\le \int |k(y,y')|^{\frac{1}{2}}|k(y,y')|^{\frac{1}{2}}|u(y)||v(y')|\mathrm{d}\mu\mathrm{d}\mu' \\ &\le \left(\int |k(y,y')||u(y)|^2\mathrm{d}\mu\mathrm{d}\mu'\right)^{\frac{1}{2}} \left(\int |k(y,y')||v(y')|^2\mathrm{d}\mu\mathrm{d}\mu'\right)^{\frac{1}{2}} \\ &\le (CC')^{\frac{1}{2}}\|u\|\|v\|,\end{aligned}$$

which proves the lemma. □

B.7 Compact Operators in $L^2(\mathbb{R}^n)$

The following criterion for compactness of operators in $L^2(\mathbb{R}^n)$ is often useful.

Proposition B.7.1
Suppose that $f, g \in L^\infty(\mathbb{R}^n)$ and

$$\lim_{|x|\to\infty} f(x) = 0, \qquad \lim_{|\xi|\to\infty} g(\xi) = 0.$$

Then the operator $f(x)g(D)$ is compact.

Proof. Set $f_n(x) := \mathbb{1}_{[0,n]}(|x|)f(x)$, $g_n := \mathbb{1}_{[0,n]}(|\xi|)g(\xi)$. Then the Hilbert-Schmidt norm of $f_n(x)g_n(D)$ equals

$$\|f_n\|_2\|g_n\|_2 < \infty.$$

Hence $f_n(x)g_n(D)$ are Hilbert-Schmidt operators. But

$$\lim_{n\to\infty} f_n(x)g_n(D) = f(x)g(D).$$

Hence $f(x)g(D)$ is compact as the limit of a sequence of compact operators. □

C. Estimates on Functions of Operators

Suppose that we know the properties of a certain operator A. In this appendix we will describe how to study the properties of its function $f(A)$. For instance, we will prove some estimates on the commutator $[B, f(A)]$. Besides, if the operator $A(s)$ depends on a parameter s, we will prove some estimates on the derivative of $f(A(s))$ with respect to s. Note that the second problem can be viewed as a generalization of the first. In fact, if we take $A(s) := \mathrm{e}^{isB} A \mathrm{e}^{-isB}$, then we have

$$\frac{\mathrm{d}}{\mathrm{d}s} f(A(s)) = i[B, f(A(s))].$$

We will use two approaches to study functions of operators. The first uses the properties of e^{itA} and the Fourier transform of the function f. The second uses the properties of the resolvent $(z - A)^{-1}$ and an almost-analytic extension of f. Both approaches yield similar results. Sometimes we will use the former approach, sometimes the latter.

It is difficult to determine who first proved similar estimates. We learned most of them from [SS1], where they were proven using the Fourier transform method. The commutator expansion lemma was proven in [SS3]. An estimate similar to that of Lemma C.4.1 played an important role in [SS1]. Its version with a more careful remainder estimate was a key ingredient of [De8].

The notion of an almost-analytic extension is due to Hörmander [Hö2, vol I]. It was applied to study functions of operators in [HS].

C.1 Basic Estimates of Commutators

Probably the simplest estimate of $[F(A), B]$ can be formulated as follows.

Lemma C.1.1
Suppose that $A = (A^1, \ldots, A^n)$ is a vector of commuting self-adjoint operators. Let B be a self-adjoint operator. Assume that the quadratic form

$$\lim_{\epsilon \to 0} \left((A\phi | B(1 + i\epsilon B)^{-1} \psi) - (B(1 - i\epsilon B)^{-1} \phi | A\psi) \right), \quad \phi, \psi \in \mathcal{D}(A), \tag{C.1.1}$$

is well defined and extends to a bounded operator that we call $[A, B]$. Suppose that $F' = f$ and $\hat{f} \in L^1(\mathbb{R}^n)$. Then

$$\|[F(A), B]\| \leq \|\hat{f}\|_1 \|[A, B]\|. \tag{C.1.2}$$

Proof. It is enough to assume that B is bounded, because otherwise we can replace B with $B(1 + i\epsilon B)^{-1}$, and then use the limit described in (C.1.1).

Both $\hat{f}$ and $\hat{F}$ are distributions in $\mathcal{S}'(\mathbb{R}^n)$. Clearly, they can be extended to test functions that are bounded together with their first derivative. Moreover, if $\phi \in \operatorname{Ran} 1\!\!1_\Theta(A)$ for some compact Θ, then $\xi \mapsto e^{i\xi A}\phi$ is bounded together with all its derivatives. Therefore, we can write the following identity:

$$F(A)\phi = (2\pi)^{-n} \int \hat{F}(\xi) e^{i\xi A} \phi \mathrm{d}\xi. \tag{C.1.3}$$

Hence, in the sense of quadratic forms on $\operatorname{Ran} 1\!\!1_\Theta(A)$, for some compact Θ, we have

$$[F(A), B] = (2\pi)^{-n} \int \hat{F}(\xi)\xi \mathrm{d}\xi \int_0^1 e^{i\tau\xi A} i[A, B] e^{i(1-\tau)\xi A} \mathrm{d}\tau, \tag{C.1.4}$$

from which the estimate (C.1.2) follows immediately. □

Sometimes one needs to regularize the commutator $[A, B]$ on the right-hand side of (C.1.2) by using the inverse of A. Below we give an example of how this can be done.

Lemma C.1.2
Suppose that $f \in C_0^\infty(\mathbb{R}^n)$. Then there exists a C that depends on f such that

$$\|[f(A), B]\| \leq C \|[(1 + A^2)^{-1}, B]\|. \tag{C.1.5}$$

Proof. We set

$$f(A) = (1 + A^2)^{-1} f_1(A) (1 + A^2)^{-1},$$

where $f_1 \in C_0^\infty(\mathbb{R}^n)$. We have

$$\begin{aligned} [f(A), B] &= (1 + A^2)^{-1} [f_1(A), B] (1 + A^2)^{-1} \\ &\quad + [(1 + A^2)^{-1}, B] f_1(A) (1 + A^2)^{-1} + \mathrm{hc} \end{aligned} \tag{C.1.6}$$

The first term on the right of (C.1.6) we treat as in the proof of Lemma C.1.1. □

C.2 Almost-Analytic Extensions

In this section we describe the concept of an almost-analytic extension of a C^∞ function. Such extensions can be used as a tool in estimates on functions of operators.

We embed $\mathbb{R}$ in $\mathbb{C}$. We denote the variable of $\mathbb{R}$ by x and the variable of $\mathbb{C}$ by $z = x + iy$.

Proposition C.2.1
Let $f \in C_0^\infty(\mathbb{R})$. Then there exist a function $\tilde{f} \in C_0^\infty(\mathbb{C})$, called an almost-analytic extension *of f, such that*

$$\tilde{f}|_{\mathbb{R}} = f, \quad |\tfrac{\partial \tilde{f}}{\partial \bar{z}}(z)| \le C_N |\mathrm{Im} z|^N, \qquad N \in \mathbb{N}. \tag{C.2.1}$$

Moreover,

$$f(x) = \frac{i}{2\pi} \int_{\mathbb{C}} \frac{\partial \tilde{f}}{\partial \bar{z}}(z)(z-x)^{-1} \mathrm{d}z \wedge \mathrm{d}\bar{z}. \tag{C.2.2}$$

Proof. Let $\chi \in C_0^\infty(\mathbb{R})$ be a cutoff function such that $\chi(x) = 1$ for $|x| \le 1$. Then it is easy to check that, for an appropriate sequence C_n, the series

$$\tilde{f}(x+iy) := \sum_{n=0}^{\infty} i^n \partial_x^n f(x) \frac{y^n}{n!} \chi\left(\frac{y}{C_n}\right) \tag{C.2.3}$$

converges uniformly with all derivatives. From

$$\frac{\partial \tilde{f}}{\partial \bar{z}}(z) = \sum_{n=0}^{\infty} i^{n+1} \partial_x^n f(x) \frac{y^n}{n! C_n} \chi'\left(\frac{y}{C_n}\right)$$

we easily see that $\tilde{f}$ satisfies (C.2.1).

Next we note that the right-hand side of (C.2.2) equals

$$\lim_{\epsilon \to 0} \tfrac{i}{2\pi} \textstyle\int_{C_\epsilon} \tfrac{\partial \tilde{f}}{\partial \bar{z}}(z)(z-x)^{-1} \mathrm{d}z \wedge \mathrm{d}\bar{z}, \tag{C.2.4}$$

where C_ϵ is the domain

$$C_\epsilon := \{z \in \mathbb{C} \mid |\mathrm{Im} z| > \epsilon, \ |z| < c\},$$

for some c large enough. Using Green's formula, we obtain that (C.2.4) equals

$$\begin{aligned} &\lim_{\epsilon \to 0} \tfrac{i}{2\pi} \textstyle\int_{\partial C_\epsilon} \tilde{f}(z)(z-x)^{-1} \mathrm{d}z \\ &= \lim_{\epsilon \to 0} \tfrac{i}{2\pi} \textstyle\int_{\mathbb{R}} (\tilde{f}(\lambda + i\epsilon)(\lambda + i\epsilon - x)^{-1} - \tilde{f}(\lambda - i\epsilon)(\lambda - i\epsilon - x)^{-1}) \mathrm{d}\lambda. \end{aligned} \tag{C.2.5}$$

Using the fact that $|(\lambda \pm i\epsilon - x)^{-1}| \le \epsilon^{-1}$, $|\tilde{f}(\lambda \pm \epsilon) - \tilde{f}(\lambda)| \le C\epsilon$, and Lebesgue's dominated convergence theorem, we see that (C.2.5) equals

$$\lim_{\epsilon \to 0} \frac{i}{2\pi} \int_{\mathbb{R}} (\tilde{f}(\lambda)(\lambda + i\epsilon - x)^{-1} - \tilde{f}(\lambda)(\lambda - i\epsilon - x)^{-1}) \mathrm{d}\lambda = f(x).$$

This completes the proof of (C.2.2). □

For $\rho \in \mathbb{R}$, we denote by $S^\rho(\mathbb{R})$ the class of functions f in $C^\infty(\mathbb{R})$ such that

$$|\partial_s^k f(s)| \leq C_k \langle s \rangle^{\rho - k}, \quad k \geq 0.$$

Below we show how to construct almost-analytic extensions of functions from $S^\rho(\mathbb{R})$.

Proposition C.2.2
Let $f \in S^\rho(\mathbb{R})$ Then there exists $\tilde{f} \in C_0^\infty(\mathbb{C})$, called an almost-analytic extension *of f, such that*

$$\tilde{f}|_{\mathbb{R}} = f, \qquad |\tfrac{\partial \tilde{f}}{\partial \bar{z}}(z)| \leq C_N \langle x \rangle^{\rho - 1 - N} |\mathrm{Im} z|^N, \qquad N \in \mathbb{N},$$
$$\mathrm{supp} \tilde{f} \subset \{x + iy \mid |y| \leq C \langle x \rangle\}$$

and (C.2.2) is true.

Proof. We easily see that we can choose a sequence C_n such that

$$\tilde{f}(x+iy) := \sum_{n=0}^{\infty} i^n \partial_x^n f(x) \frac{y^n}{n!} \chi\left(\frac{y}{\langle x \rangle C_n}\right)$$

is well defined. We also have

$$\frac{\partial \tilde{f}}{\partial \bar{z}}(z) = \sum_{n=0}^{\infty} i^{n+1} \partial_x^n f(x) \frac{y^n}{n! C_n} \chi'\left(\frac{y}{C_n \langle x \rangle}\right) \left(i - \frac{y}{\langle x \rangle} \partial_x \langle x \rangle\right),$$

which yields the desired estimate on $\frac{\partial \tilde{f}}{\partial \bar{z}}(z)$. □

C.3 Commutator Expansions I

In this section we prove a version of the commutator expansion lemma (see [SS3]).

Lemma C.3.1
Let A, B two self-adjoint operators with

$$\|\mathrm{ad}_A^j B\| < \infty, \; j \geq 1.$$

Let $\rho \geq -1$ and $f \in S^\rho(\mathbb{R})$. Let $[\rho]$ denote the integral part of ρ. Then for all $N \in \mathbb{N}$ such that $N \geq [\rho]$, we have

$$[f(A), B] = \sum_{j=1}^{N} \frac{1}{j!} f^{(j)}(A) \mathrm{ad}_A^j B + R_{N+1}(f, A, B),$$

where

$$\|(A+i)^{N-[\rho]} R_{N+1}(f, A, B)\| \leq c_N(f) \|\mathrm{ad}_A^{N+1} B\|, \tag{C.3.1}$$

and the constants $c_N(f)$ depend only on a finite number of the semi-norms of f in $S^\rho(\mathbb{R})$.

Proof. Our first aim is to show our estimate for $-1 \le \rho < 0$. In this case, this estimate is

$$\|(A+i)^{N+1}R_{N+1}(f,A,B)\| \le c_N(f)\|\mathrm{ad}_A^{N+1}B\|.$$

Let $\tilde{f}$ be an almost-analytic extension $\tilde{f}$ of constructed in Proposition C.2.2. We have

$$f(A) = \frac{i}{2\pi}\int_{\mathbb{C}} \frac{\partial \tilde{f}}{\partial \bar{z}}(z)(z-A)^{-1}\mathrm{d}z \wedge \mathrm{d}\bar{z}.$$

So,

$$\begin{aligned}
[f(A),B] &= \tfrac{i}{2\pi}\textstyle\int_{\mathbb{C}} \frac{\partial \tilde{f}}{\partial \bar{z}}(z)(z-A)^{-1}\mathrm{ad}_A(B)(z-A)^{-1}\mathrm{d}z \wedge \mathrm{d}\bar{z} \\
&= \textstyle\sum_{j=1}^N \frac{i}{2\pi}\int_{\mathbb{C}} \frac{\partial \tilde{f}}{\partial \bar{z}}(z)(z-A)^{-j-1}\mathrm{ad}_A^j(B)\mathrm{d}z \wedge \mathrm{d}\bar{z} \\
&+ \tfrac{i}{2\pi}\textstyle\int_{\mathbb{C}} \frac{\partial \tilde{f}}{\partial \bar{z}}(z)(z-A)^{-N-1}\mathrm{ad}_A^{N+1}(B)(z-A)^{-1}\mathrm{d}z \wedge \mathrm{d}\bar{z} \\
&= \textstyle\sum_{j=1}^N \frac{1}{j!}f^{(j)}(A)\mathrm{ad}_A^j(B) + R_{N+1}(f,A,B).
\end{aligned} \tag{C.3.2}$$

It is immediate to see that, for $|\mathrm{Im}z| \le c\langle \mathrm{Re}z\rangle$, one has

$$\|\tfrac{A+i}{z-A}\| \le C\langle z\rangle|\mathrm{Im}z|^{-1}. \tag{C.3.3}$$

Using then the estimates on $\frac{\partial \tilde{f}}{\partial \bar{z}}(z)$, we obtain immediately that

$$\begin{aligned}
&\|(A+i)^{N+1}R_{N+1}(f,A,B)\| \\
&\le C \int_{\mathbb{C}} \|\tfrac{A+i}{z-A}\|^{N+1}\left|\tfrac{\partial \tilde{f}}{\partial \bar{z}}(z)\right| \|(z-A)^{-1}\|\|\mathrm{ad}_A^{N+1}(B)\|\mathrm{d}z \wedge \mathrm{d}\bar{z} \\
&\le C\|\mathrm{ad}_A^{N+1}(B)\| \textstyle\int_{\mathbb{C}} \left(\frac{\langle z\rangle}{\mathrm{Im}z}\right)^{N+1} \langle z\rangle^{-1-k+\rho}|\mathrm{Im}z|^{k-1}\mathrm{d}z \wedge \mathrm{d}\bar{z}.
\end{aligned} \tag{C.3.4}$$

Setting $k = N+2$ we obtain that (C.3.4) is less than $c_N(f)\|\mathrm{ad}_A^{N+1}B\|$. This ends the proof of the lemma for $-1 \le \rho < 0$.

Now let $P_\rho(s) := (s+i)^\rho$. For $\rho = 0, 1, \ldots$, by a straightforward computation, we get

$$R_{N+1}(P_\rho, A, B) = \begin{cases} \rho!\mathrm{ad}_A^{N+1}B & \rho = N+1, \\ 0 & N+1 > \rho. \end{cases} \tag{C.3.5}$$

Finally, we consider $f_\rho \in S^\rho$ with an arbitrary $\rho \ge 0$. We set $\tilde{\rho} := \rho - [\rho] - 1 \in [-1, 0[$ and

$$f_\rho(s) =: P_{[\rho]+1}(s)f_{\tilde{\rho}}(s),$$

where $f_{\tilde{\rho}} \in S^{\tilde{\rho}}$. We have

$$\begin{aligned}
Bf_\rho(A) &= \textstyle\sum_{j=0}^{N} \frac{1}{j!} P_{[\rho]+1}^{(j)'}(A)\mathrm{ad}_A^j B f_{\tilde\rho}(A) + R_{N+1}(P_{[\rho]+1}, A, B) f_{\tilde\rho}(A) \\
&= \textstyle\sum_{j=0}^{N} \sum_{k=0}^{N-j} \frac{1}{j!k!} P_{[\rho]+1}^{(j)}(A) f_{\tilde\rho}^{(k)}(A)\mathrm{ad}_A^{j+k} B \\
&+ \textstyle\sum_{j=0}^{N} P_{[\rho]+1}^{(j)}(A) R_{N+1-j}(f_{\tilde\rho}, A, \mathrm{ad}_A^j B) + R_{N+1}(P_{[\rho]+1}, A, B) f_{\tilde\rho}(A) \\
&= \textstyle\sum_{j=0}^{N} \frac{1}{j!} f_\rho^{(j)}(A)\mathrm{ad}_A^j B + R_{N+1}(f_\rho, A, B)
\end{aligned}$$

for

$$\begin{aligned}
&R_{N+1}(f_\rho, A, B) \\
&:= \textstyle\sum_{j=0}^{\rho+1} P_{[\rho]+1}^{(j)}(A) R_{N+1-j}(f_{\tilde\rho}, A, \mathrm{ad}_A^j B) + R_{N+1}(P_{[\rho]+1}, A, B) f_{\tilde\rho}(A),
\end{aligned}$$

Now $(A+i)^{-[\rho]-1+j} P_{[\rho]+1}^{(j)}(A)$ is bounded, and $(A+i)^{N+1-j} R_{N+1-j}(f_{\tilde\rho}, A, \mathrm{ad}_A^j B)$ can be estimated by $C\|\mathrm{ad}_A^{N+1} B\|$. Using also (C.3.5), we get (C.3.1). □

Let us state the following consequence of the commutator expansion lemma.

Lemma C.3.2
Let A, B be self-adjoint operators. If $f \in S^\rho(\mathbb{R})$ with $\rho < 1$, then

$$[f(A), B] \le C\|[A, B]\|.$$

If $f \in S^1(\mathbb{R})$, then

$$[f(A), B] \le C\|[A, B]\| + C\|[A, [A, B]]\|.$$

Proof. We use Lemma C.3.1 with $N = 0$ and $N = 1$ respectively. □

C.4 Commutator Expansions II

Next we will give yet another version of the commutator expansion with a very careful remainder estimate.

Lemma C.4.1
Let $\mathbb{R} \ni t \mapsto B(t) \in \mathcal{B}(\mathcal{H})$ be a one parameter family of bounded self-adjoint operators that is C^1 in the norm sense. Let $F \in C^\infty(\mathbb{R})$ with $F' = f^2$, $f \in C_0^\infty(\mathbb{R})$. Assume that

$$\begin{aligned}
&\tfrac{\mathrm{d}}{\mathrm{d}t} B(t) = A_1(t) + C_1(t), \\
&[B(t), A_1(t)] = A_2(t) + C_2(t), \\
&[B(t), A_2(t)] = C_3(t).
\end{aligned}$$

Then there exist C depending just on f such that

$$\left\| \tfrac{\mathrm{d}}{\mathrm{d}t} F(B(t)) - f\left(B(t)\right) A_1(t) f\left(B(t)\right) \right\| \le C(\|C_1(t)\| + \|C_2(t)\| + \|C_3(t)\|).$$

Proof. We have

$$\tfrac{\mathrm{d}}{\mathrm{d}t} F(B(t)) = \tfrac{1}{2\pi} \int \int_0^1 \hat{F}(s)(-is)\mathrm{e}^{-i\tau sB}(\tfrac{\mathrm{d}}{\mathrm{d}t} B(t))\mathrm{e}^{-i(1-\tau)sB} \mathrm{d}\tau \mathrm{d}s. \tag{C.4.1}$$

Up to an error of the order $O(\|C_1\|)$, the quantity (C.4.1) equals

$$\begin{aligned} &\tfrac{1}{2\pi} \int \int_0^1 \hat{F}(s)(-is)\mathrm{e}^{-i\tau sB} A_1 \mathrm{e}^{-i(1-\tau)sB} \mathrm{d}\tau \mathrm{d}s \\ &= \tfrac{1}{4\pi} \int \hat{F}(s)(-is)(\mathrm{e}^{-isB} A_1 + A_1 \mathrm{e}^{-isB}) \mathrm{d}s \\ &+ \tfrac{1}{4\pi} \int \int_0^1 \hat{F}(s)(-is)^2(1-2\tau)\mathrm{e}^{-i\tau sB}[B, A_1]\mathrm{e}^{-i(1-\tau)sB} \mathrm{d}\tau \mathrm{d}s. \end{aligned} \tag{C.4.2}$$

The first term on the right of (C.4.2) equals

$$\begin{aligned} &\tfrac{1}{2}\left(f^2(B)A_1 + A_1 f^2(B)\right) \\ &= f(B)A_1 f(B) + \tfrac{1}{2}[f(B), [f(B), A_1]]. \end{aligned} \tag{C.4.3}$$

We have

$$[f(B), A_1] = \frac{1}{2\pi} \int \int_0^1 \hat{f}(s)(-is)\mathrm{e}^{-i\tau sB}(A_2 + C_2)\mathrm{e}^{-i(1-\tau)sB} \mathrm{d}\tau \mathrm{d}s,$$

from which we obtain that

$$\begin{aligned} \|[f(B), [f(B), A_1]]\| &\in O(\|[f(B), A_2]\|) + O(\|C_2\|) \\ &\in O(\|C_3\|) + O(\|C_2\|). \end{aligned}$$

The second term on the right-hand side of (C.4.2) up to a term of order $O(\|C_2\|)$ equals

$$\begin{aligned} &\tfrac{1}{4\pi} \int \int_0^1 (1-2\tau)\hat{F}(s)(-is)^2 \mathrm{e}^{-i\tau sB} A_2 \mathrm{e}^{-i(1-\tau)sB} \mathrm{d}\tau \mathrm{d}s \\ &= \tfrac{1}{4\pi} \int \int_0^1 (\tau - \tau^2)\hat{F}(s)(-is)^3 \mathrm{e}^{-i\tau sB}[B, A_2]\mathrm{e}^{-i(1-\tau)sB} \mathrm{d}\tau \mathrm{d}s, \end{aligned} \tag{C.4.4}$$

which is of the order $O(\|C_3\|)$. □

D. Pseudo-differential and Fourier Integral Operators

D.0 Introduction

The name "pseudo-differential operators" is usually used in two different (although related) meanings. First, it is used to denote operators on $L^2(\mathbb{R}^n)$ defined by certain integral formulas. The main ingredient of these formulas is a function that is called the symbol of a pseudo-differential operator, which encodes the phase space properties of the operator. If we use this meaning, then essentially all operators on $L^2(\mathbb{R}^n)$ are pseudo-differential operators. This point of view is taken in Sect. D.1, where we introduce the two most commonly used notions of the symbol of an operator: the Kohn-Nirenberg symbol and the Weyl symbol.

In Sect. D.2 we introduce the phase space correlation function of an operator, which is another object used to describe phase space properties of operators. In the second, probably more common meaning, the word "pseudo-differential operators" is used to denote some classes of operators that can usually be defined by describing certain properties of their symbols. There is a large variety of such classes, some of them are very useful in partial differential equations, others are less known.

Probably the most natural class of pseudo-differential operators is the algebra associated with the constant metric, which we denote by

$$\Psi(1, \mathrm{d}x^2 + \mathrm{d}\xi^2) = \Psi(1, g_0).$$

(In the literature it often appears under the name Ψ_{00}^0). Its symbols belong to the algebra of functions with all bounded derivatives, which we denote

$$S(1, \mathrm{d}x^2 + \mathrm{d}\xi^2) = S(1, g_0).$$

(In the literature it often appears under the name S_{00}^0). The algebra $\Psi(1, g_0)$ has very elegant properties, for instance, it is invariant with respect to the metaplectic group. From the point of view of applications, however, it has a big disadvantage – it does not possess a "small parameter" (a "Planck constant"), and therefore does not have an asymptotic calculus, which is so useful in practice. Nevertheless, in Chap. 3 we use this class of pseudo-differential operators. We describe the properties of symbols and operators associated with the constant metric in Sects. D.3 and D.4 respectively.

The simplest way to introduce a "Planck constant" to pseudo-differential operators is to make the operators depend on a parameter. This formalism, useful in Chap. 3, is presented in Sect. D.5. It is essentially a reformulation of the results of the preceding section.

A more refined way to introduce a "Planck constant" is to use classes of pseudo-differential operators associated with certain non-uniform metrics. There is a wide variety of such classes. The classes of operators that we need in Chap. 4 are

$$\Psi(\langle x\rangle^m, \langle x\rangle^{-2}\mathrm{d}x^2 + \mathrm{d}\xi^2) = \Psi(\langle x\rangle^m, g_1).$$

They have symbols that belong to

$$S(\langle x\rangle^m, \langle x\rangle^{-2}\mathrm{d}x^2 + \mathrm{d}\xi^2) = S(\langle x\rangle^m, g_1).$$

In this class, the quantity $\langle x\rangle^{-1}$ serves as a small parameter.

This class is probably the most popular in applications in the literature on partial differential equations, where it is denoted by $\Psi^m_{1,0}$, and the roles of x and ξ are usually switched. Basic properties of symbols in $S(\langle x\rangle^m, g_1)$ and of the operators in $\Psi(\langle x\rangle^m, g_1)$ follow easily from similar properties concerning the case of the uniform metric. They are described in Sects. D.7 and D.8.

Because of the asymptotic calculus, we can sometimes look at operators of $\Psi(\langle x\rangle^m, g_1)$ locally in phase space. Such concepts are developed in Sects. D.9 and D.10.

Properties of functions of pseudo-differential operators are studied in Sect. D.11.

In Sect. D.12 we use the so-called non-stationary phase method to describe some simple bounds on certain Fourier integral operators.

Sometimes it is convenient to approximate an operator with an integral expression that goes under the name of Fourier integral operators (FIO's). This happens especially when we consider the evolution generated by a self-adjoint operator. The last three sections are devoted to the results about FIO's that are needed in Chaps. 3 and 4. In Sect. D.13 we describe a class of FIO's whose amplitude and the second derivative of the phase belong to $S(1, g_0)$. This class of FIO's seems very natural. For instance, generically, elements of the metaplectic group belong to this class. Nevertheless, because of the absence of a small parameter, this class does not seem to be widely used in the literature. One can introduce a "Planck constant" by introducing a parameter, as we do in Sect. D.14. In Sect. D.15 we study FIO's whose amplitudes belong to $S(\langle x\rangle^m, g_1)$ and the second derivative of phases belong to $S(1, g_1)$. Note that, in the literature, a similar class is the most commonly used.

References about pseudo-differential operators include [Hö2, vol. III] and [Ta, Tre, Ro, BoCh]. Properties of functions of operators similar to those described in Proposition D.11.4 where proven in [SS1]. Fourier integral operators similar to those considered in Sect. D.13 were probably first considered in [AF, Fu1,

Fu2]. Fourier integral operators with amplitudes in $S(\langle x\rangle^m, g_1)$ considered in the literature usually have phases homogeneous with respect to one of the variables (see [Hö2, vol IV]). Fourier integral operators similar to those considered in Sect. D.15 were also considered, for example, in [Ki5, Ki4, KiK, KiYa1, KiYa2, Ya3].

D.1 Symbols of Operators

Let $\bar{\mathcal{S}}'(X)\otimes\mathcal{S}'(X)$ denote the space of sesquilinear forms on the space of Schwartz test functions $\mathcal{S}(X)$. We will view $\bar{\mathcal{S}}'(X)\otimes\mathcal{S}'(X)$ as a kind of an extension of the set of linear operators on $L^2(X)$. We will treat all the elements of this space as "pseudo-differential operators" and we will define their symbols. Note that, by Schwartz's kernel theorem (see e.g. [RS, vol I]), elements of this set can be defined by a kernel $K \in \mathcal{S}'(X \times X)$ with help of the following equation:

$$(\phi|A\psi) = \int\int K(x,y)\bar{\phi}(x)\psi(y)\mathrm{d}x\mathrm{d}y.$$

Let $A \in \bar{\mathcal{S}}'(X)\otimes\mathcal{S}'(X)$. Then we say that $a_1 \in \mathcal{S}'(X\times X')$ is the *Kohn-Nirenberg symbol* of A if, for any $\phi,\psi\in\mathcal{S}(X)$,

$$(\phi|A\psi) = (2\pi)^{-n}\int a_1(x,\xi)\bar{\phi}(x)\psi(y)\mathrm{e}^{i\langle x-y,\xi\rangle}\mathrm{d}x\mathrm{d}\xi\mathrm{d}y. \tag{D.1.1}$$

We will write

$$A = a_1(x,D). \tag{D.1.2}$$

We say that $a_2 \in \mathcal{S}'(X\times X')$ is the *Weyl symbol* of A if, for any $\phi,\psi\in\mathcal{S}(X)$,

$$(\phi|A\psi) = (2\pi)^{-n}\int a_2\left(\frac{x+y}{2},\xi\right)\bar{\phi}(x)\psi(y)\mathrm{e}^{i\langle x-y,\xi\rangle}\mathrm{d}x\mathrm{d}\xi\mathrm{d}y. \tag{D.1.3}$$

We will write

$$A = a_2^{\mathrm{w}}(x,D). \tag{D.1.4}$$

Using basic properties of the Fourier transform on $\mathcal{S}'(X\times X')$ and Schwartz's kernel theorem, we easily see that every element of $\bar{\mathcal{S}}'(X)\otimes\mathcal{S}'(X)$ possesses a unique Kohn-Nirenberg symbol and a unique Weyl symbol. Conversely, with any symbol in $\mathcal{S}'(X\times X')$, we can associate a unique Kohn-Nirenberg pseudo-differential operator and a unique Weyl pseudo-differential operator.

The following well known identity [Hö1] allows one to go from the Kohn-Nirenberg symbol to the Weyl symbol:

$$\mathrm{e}^{\frac{i}{2}\langle D_x,D_\xi\rangle}a_1 = a_2. \tag{D.1.5}$$

Note that (D.1.5) makes sense for symbols in $\mathcal{S}'(X\times X')$.

D.2 Phase–Space Correlation Functions

Properties of operators on $L^2(X)$ are closely related to the symplectic structure of the vector space $X \times X'$. The symplectic form is defined as

$$\sigma(Y, Y') := \langle y', \eta\rangle - \langle y, \eta'\rangle,$$

where $Y = (y, \eta)$ and $Y' = (y', \eta')$ are elements of $X \times X'$. The Euclidean metric on X induces a natural Euclidean metric on $X \times X'$ given by

$$|Y|^2 = |y|^2 + |\eta|^2.$$

It is also useful to introduce the notation

$$\Xi := (x, D)$$

for a vector of self-adjoint operators acting on $L^2(X)$.

Let $\phi_0 \in L^2(X)$ be the ground state of the harmonic oscillator $D^2 + x^2$:

$$\phi_0(x) := \pi^{-n/4} \mathrm{e}^{-\frac{1}{2}x^2}.$$

For $Y = (y, \eta) \in X \times X'$, we define the family of coherent states:

$$\begin{aligned}\phi_Y(x) &:= \pi^{-n/4} \mathrm{e}^{-\frac{1}{2}(x-y)^2 + i\langle x-y, \eta\rangle + \frac{i}{2}\langle y,\eta\rangle} \\ &= \mathrm{e}^{i(\langle x,\eta\rangle - \langle y, D\rangle)}\phi_0 = \mathrm{e}^{i\sigma(Y,\Xi)}\phi_0.\end{aligned}$$

Define the following families of operators:

$$\begin{aligned}P_Y &:= |\phi_Y)(\phi_Y|, \quad Y \in X \times X', \\ P_{Y',Y} &:= |\phi_{Y'})(\phi_Y|, \quad Y, Y' \in X \times X'.\end{aligned}$$

In other words, P_Y are the orthogonal projections onto ϕ_Y and $P_{Y',Y}$ are the rank one operators with the Schwartz kernel $\phi_{Y'}(x) \otimes \overline{\phi}_Y(x')$. Clearly, $P_{Y,Y} = P_Y$. It is useful to know the Weyl symbols of those operators:

$$P_Y = p_Y^w(x, D), \qquad p_Y(Z) = 2^n \mathrm{e}^{-(Z-Y)^2},$$

$$P_{Y'Y} = p_{Y',Y}^w(x, D), \qquad p_{Y',Y}(Z) = 2^n \mathrm{e}^{-i\frac{1}{2}\sigma(Y,Y')} \mathrm{e}^{i\sigma(Z,Y'-Y)} \mathrm{e}^{-\left(Z-(\frac{Y+Y'}{2})\right)^2}.$$

Note that $p_Y(Z)$ is called the *Wigner function* of ϕ_Y. Note the following property of coherent states:

$$(\phi|\psi) = \int (\phi|P_Y\psi)\mathrm{d}Y. \tag{D.2.1}$$

Let us now define the following linear mapping:

$$\begin{aligned}W &: L^2(X) \to L^2(X \times X') \\ \psi &\mapsto W\psi(Y) := (\phi_Y|\psi).\end{aligned}$$

Let us note the following immediate properties of W:

$$\begin{aligned}&(i)\quad (\psi_1|\psi_2) = (W\psi_1|W\psi_2),\\&(ii)\quad W(D+ix) = (y+i\eta)W.\end{aligned}\tag{D.2.2}$$

Let us observe that $\psi \in \mathcal{S}(X)$ if and only if $(1+Y^2)^k W\psi \in L^2(X \times X')$ for all $k \in \mathbb{N}$. This follows directly from (D.2.2) and from the fact that $\psi \in \mathcal{S}(X)$ if and only if $(x^2+D^2)^k\psi \in L^2(X)$ for all $k \in \mathbb{N}$.

For an operator $A \in \bar{\mathcal{S}}'(X) \otimes \mathcal{S}'(X)$, we define the *phase space correlation function* of A by

$$W_A(Y,Y') := (\phi_{Y'}|A\phi_Y)\,. \tag{D.2.3}$$

Note that W_A is equal to the distribution kernel of WAW^*:

$$(V|WAW^*U) = \int W_A(Y,Y')U(Y)\overline{V}(Y')\mathrm{d}Y\mathrm{d}Y',\ U,V \in L^2(X \times X').$$

An elementary computation shows that if $a \in \mathcal{S}'(X \times X')$ and $A = a^w(x,D)$, then

$$W_A(Y,Y') = \int_{X\times X'} a(Z)p_{Y,Y'}(Z)\mathrm{d}Z, \tag{D.2.4}$$

$$a(Z) = \int W_A(Y,Y')p_{Y,Y'}(Z)\mathrm{d}Y\mathrm{d}Y'. \tag{D.2.5}$$

D.3 Symbols Associated with a Uniform Metric

Definition D.3.1
Let Y be a Euclidean space with a metric $\mathrm{d}y^2$. *We define*

$$S(1,\mathrm{d}y^2) := \{a(y) \in C^\infty(Y)|\ |\partial_y^\alpha a(y)| \le C_\alpha\}.$$

$S(1,\mathrm{d}y^2)$ is a Fréchet space with a family of semi-norms

$$\|b\|_{\mathrm{d}y,N} = \sum_{|\alpha|\le N} \|\partial_y^\alpha b\|_\infty.$$

(The space $S(\mathrm{d}y^2)$ itself does not depend on the scalar product in Y, but the family of semi-norms $\|\cdot\|_{\mathrm{d}y^2,N}$ does).

Let us describe a number of properties of $S(1,\mathrm{d}y^2)$.

Proposition D.3.2
Let $a \in S(1,\mathrm{d}y^2)$.
(i) If $b \in S(1,\mathrm{d}y^2)$, then $ba \in S(1,\mathrm{d}y^2)$.
(ii) $\partial_y^\alpha a \in S(1,\mathrm{d}y^2)$.

(iii) If $Q(\cdot)$ is a quadratic form on Y', then $\mathrm{e}^{iQ(D_y)}a \in S(1, \mathrm{d}y^2)$.
(iv) Let Y_0 be a vector space and $j : Y_0 \to Y$ a linear embedding, then $a(j(\cdot)) \in S(1, \mathrm{d}y_0^2)$.
(v) Let $m_1, \ldots, m_k \in \mathbb{R}$, $Y = Y_1 \oplus \ldots \oplus Y_2$, and let $Q(\cdot)$ be a quadratic form on Y. Then

$$\langle y_1\rangle^{m_1} \ldots \langle y_k\rangle^{m_k} \mathrm{e}^{iQ(D_y)} \langle y_1\rangle^{-m_1} \langle y_k\rangle^{-m_k} a \in S(1, \mathrm{d}y^2).$$

Proof. *(i)*, *(ii)* and *(iv)* are obvious. *(iii)* follows from Lemma D.3.3 below.

Let us show *(v)*. We can assume that $m_1, \ldots, m_l \geq 0$ and $m_{l+1}, \ldots, m_k \leq 0$. It is enough to consider the case $m_1/2, \ldots, m_k/2 \in \mathbb{Z}$, and then to use the interpolation.

Set $A_i = \langle \nabla_{\xi_i} Q, D_y\rangle$. We can write

$$\begin{aligned}&\langle y_1\rangle^{m_1} \cdots \langle y_k\rangle^{m_k} \mathrm{e}^{iQ(D_y)} \langle y_1\rangle^{-m_1} \cdots \langle y_k\rangle^{-m_k} \\ &= \langle y_{l+1}\rangle^{m_{l+1}} \cdots \langle y_k\rangle^{m_k} \mathrm{e}^{iQ(D_y)} \langle y_1 + A_1\rangle^{m_1} \cdots \langle y_l + A_l\rangle^{m_l} \langle y_1\rangle^{-m_1} \cdots \langle y_k\rangle^{-m_k}\end{aligned}$$

Then we commute $\langle y_{l+1}\rangle^{-m_{l+1}} \cdots \langle y_k\rangle^{-m_k}$ to the right. Eventually we get a linear combination of terms of the form

$$\langle y_{l+1}\rangle^{m_{l+1}} y^{\alpha_{l+1}} A_{l+1}^{\beta_{l+1}} \cdots \langle y_k\rangle^{m_k} y^{\alpha_k} A_k^{\beta_k} \mathrm{e}^{iQ(D_y)} \langle y_1\rangle^{-m_1} y^{\alpha_1} A_1^{\beta_1} \cdots \langle y_l\rangle^{-m_l} y^{\alpha_l} A_l^{\beta_l}$$

where $|\alpha_j| \leq m_j$. Then we use the fact that $\mathrm{e}^{iQ(D_y)}$, $A_i^{\beta_i}$ and $\langle y_i\rangle^{-|m_i|} y_i^{\alpha_i}$ are bounded on $S(1, \mathrm{d}y^2)$. □

Lemma D.3.3
Let $R(y)$ be a non-degenerate quadratic form on Y. Let $a \in S(Y)$. Then for some N,

$$\left| \int \mathrm{e}^{iR(y)} a(y) \mathrm{d}y \right| \leq \|a\|_{\mathrm{d}y^2, N}. \tag{D.3.1}$$

Proof. It is enough to assume that $Y = \mathbb{R}$ and $R(y) = y^2$. Consider the operator $\mathcal{L}$ defined as

$$(\mathcal{L}b)(y) := (1 + 2y^2)^{-1} \left(-iy\tfrac{\mathrm{d}}{\mathrm{d}y} + 1\right) b(y).$$

Then $\mathcal{L}\mathrm{e}^{iy^2} = \mathrm{e}^{iy^2}$, and hence

$$\int \mathrm{e}^{iy^2} a(y) \mathrm{d}y = \int (\mathcal{L}^2 \mathrm{e}^{iy^2}) a(y) \mathrm{d}y = \int \mathrm{e}^{iy^2} ({}^t\mathcal{L})^2 a(y) \mathrm{d}y, \tag{D.3.2}$$

where

$${}^t\mathcal{L}b(y) := \left(i\tfrac{\mathrm{d}}{\mathrm{d}y} y + 1\right) (1 + 2y^2)^{-1} b(y).$$

But

$$\left|({}^t\mathcal{L})^2 a(y)\right| \leq C\langle y\rangle^{-2}.$$

Hence the integral (D.3.2) is finite. □

D.4 Pseudo-differential Operators Associated with a Uniform Metric

In this section a special role will be played by the class of symbols

$$S(1, \mathrm{d}x^2 + \mathrm{d}\xi^2) := \{a \in C^\infty(X \times X') \mid |\partial_x^\alpha \partial_\xi^\beta a(x,\xi)| \leq C_{\alpha,\beta}, \quad \alpha, \beta \in \mathbb{N}^n\}.$$

Since we will often use this class it is convenient to introduce the notation

$$g_0 = \mathrm{d}x^2 + \mathrm{d}\xi^2.$$

Pseudo-differential operators associated with the class of symbols $S(1, g_0)$ can be characterized in several equivalent ways, which are described in the following theorem.

Theorem D.4.1
Let $A \in \bar{S}'(X) \otimes S'(X)$. The following statements are equivalent:
(i) $W_A(Y, Y')$ satisfies

$$|W_A(Y,Y')| \leq C_k \langle Y - Y' \rangle^{-k}, \ k \in \mathbb{N};$$

(ii) A is an operator on $L^2(X)$ such that

$$\mathrm{ad}_D^\alpha \mathrm{ad}_x^\beta (A) \in B(L^2(X)), \quad \alpha, \beta \in \mathbb{N}^n;$$

(iii) the operator A is of the form

$$A = a^{\mathrm{w}}(x, D), \quad a \in S(1, g_0);$$

(iv) the operator A is of the form

$$A = a(x, D), \quad a \in S(1, g_0);$$

The set of operators satisfying any of the above conditions will be called $\Psi(1, g_0)$. It is a Fréchet space with the family of semi-norms

$$\|A\|_{g_0,N} := \sum_{|\alpha|+|\beta| \leq N} \|\mathrm{ad}_D^\alpha \mathrm{ad}_x^\beta A\|.$$

Note that, in the literature, the properties *(iii)* or *(iv)* are usually used to define this class. The implications *(iii)⇒(ii)* and *(iv)⇒(ii)* are essentially equivalent to the Calderon-Vaillancourt theorem (see [CV, Ta] and [Hö2, vol III]). The implications *(ii)⇒(iii)* and *(ii)⇒(iv)* go under the name of the Beals criterion (see [Bea]). The characterization *(i)* is due to Unterberger [Un].

Proof of Theorem D.4.1. We will only prove that *(i)*, *(ii)* and *(iii)* are equivalent. The fact that *(iii)* and *(iv)* are equivalent follows immediately from (D.1.5) and Proposition D.3.2 *(iii)*.

(iii) $\Rightarrow$ *(i)*: Let $a \in S(X \times X')$. Using (D.2.4), the identity

$$(1 + D_Z^2)e^{i\sigma(Z,Y'-Y)} = (1 + |Y - Y'|^2)e^{i\sigma(Z,Y'-Y)},$$

and integrating by parts in (D.2.4), we see that property *(i)* is satisfied.

(i) $\Rightarrow$ *(iii)*: Using the decay of $\partial_Z^\alpha p_{Y,Y'}(Z)$ and of $W_A(Y, Y')$, we deduce from (D.2.5) that $a \in S(1, g_0)$.

(i) $\Rightarrow$ *(ii)*: By a direct computation, we see that

$$(-i\mathrm{ad}_x)^\alpha(-i\mathrm{ad}_D)^\beta a^{\mathrm{w}}(x, D) = (-1)^{|\beta|}\partial_\xi^\alpha\partial_x^\beta a^w(x, D). \tag{D.4.1}$$

Hence it suffices to show that if $a \in S(1, g_0)$, then $a^{\mathrm{w}}(x, D) \in B(L^2(X))$. For $\psi_1, \psi_2 \in \mathcal{S}(X)$, using (D.2.5), we obtain

$$(\psi_1|a^{\mathrm{w}}(x, D)\psi_2) = \int W_A(Y, Y')W\psi_1(Y)\overline{W\psi_2}(Y')\mathrm{d}Y\mathrm{d}Y'.$$

It follows from condition *(i)* that the kernel $W_A(\cdot, \cdot)$ satisfies the hypotheses of Schur's lemma (see Lemma B.6.1). So we have

$$|(\psi_1|a^{\mathrm{w}}(x, D)\psi_2)| \leq C\|W\psi_1\|\|W\psi_2\| = C\|\psi_1\|\|\psi_2\|,$$

which proves that $a^{\mathrm{w}}(x, D) \in B(L^2(X))$.

(ii) $\Rightarrow$ *(i)*: Recall that Ξ denotes the (vector-valued) operator (x, D). For $A \in \bar{\mathcal{S}}'(X) \otimes \mathcal{S}'(X)$, we have

$$(Y' - Y)A = [\Xi, A] + A(\Xi - Y) - (\Xi - Y')A. \tag{D.4.2}$$

Iterating this identity, we obtain

$$\begin{aligned}&(Y' - Y)^\alpha A \\ &= \sum_{\alpha_1+\alpha_2+\alpha_3=\alpha} C(\alpha_i)(\Xi - Y')^{\alpha_1}(\mathrm{ad}_\Xi^{\alpha_2} A)(\Xi - Y)^{\alpha_3}, \quad \alpha \in \mathbb{N}^{2n}.\end{aligned} \tag{D.4.3}$$

Let now A such that $\mathrm{ad}_\Xi^\alpha A \in B(L^2(X))$ for all $\alpha \in \mathbb{N}^{2n}$. We observe that

$$\|(\Xi - Y)^\alpha\phi_Y\| \leq C_\alpha, \ \alpha \in \mathbb{N}^{2n}. \tag{D.4.4}$$

So, by (D.4.3), we get

$$|(Y - Y')^\alpha(\phi_Y, A\phi_{Y'})| \leq C_\alpha, \quad \alpha \in \mathbb{N}^{2n},$$

which proves that property *(i)* holds. □

It follows from the proof of Theorem D.4.1 that there exists M depending just on the dimension of X such that, for any N,

$$\begin{aligned} &\|a(x,D)\|_{g_0,N} \le C_N \|a\|_{g_0,N+M}, \\ &\|a^{\mathrm{w}}(x,D)\|_{g_0,N} \le C_N \|a\|_{g_0,N+M}, \\ &\|a\|_{g_0,N} \le C_N \|a(x,D)\|_{g_0,N+M}, \\ &\|a\|_{g_0,N} \le C_N \|a^{\mathrm{w}}(x,D)\|_{g_0,N+M}. \end{aligned} \tag{D.4.5}$$

Sometimes one encounters operators defined as follows.

Proposition D.4.2
Let $b(x_1, x_2, \xi) \in S(1, \mathrm{d}x_1^2 + \mathrm{d}x_2^2 + \mathrm{d}\xi^2)$. Then the operator

$$A\phi(x_1) = (2\pi)^{-n} \int\int \mathrm{e}^{i\langle x_1 - x_2, \xi\rangle} b(x_1, x_2, \xi)\phi(x_2)\mathrm{d}\xi\mathrm{d}x_2 \tag{D.4.6}$$

belongs to $\Psi(1, g_0)$. Moreover, $A = a^w(x, D)$, where

$$a(x,\xi) := \mathrm{e}^{\frac{i}{2}(\langle D_{x_1}, D_\xi\rangle + \langle D_{x_2}, D_\xi\rangle)} b(x_1, x_2, \xi)\Big|_{x:=x_1=x_2}. \tag{D.4.7}$$

The class $\Psi(1, g_0)$ is closed with respect to the multiplication.

Proposition D.4.3
(i) Let $A, B \in \Psi(1, g_0)$. Then $C := AB$ also belongs to $\Psi(1, g_0)$.
(ii) If, moreover, $A = a^w(x, D)$, $B = b^w(x, D)$ and $C = c^w(x, D)$, then

$$c(x,\xi) = \mathrm{e}^{\frac{i}{2}(\langle D_x, D_\eta\rangle - \langle D_y, D_\xi\rangle)} a(x,\xi) b(y,\eta)|_{x=y,\xi=\eta}. \tag{D.4.8}$$

(iii) If, moreover, $A = a(x, D)$, $B = b(x, D)$ and $C = c(x, D)$, then

$$c(x,\xi) = \mathrm{e}^{i\langle D_x, D_\eta\rangle} a(x,\xi) b(y,\eta)|_{x=y,\xi=\eta}. \tag{D.4.9}$$

Proof. *(i)* follows immediately from the Beals criterion. *(ii)* and *(iii)* follow by explicit calculations. □

Propositions D.4.4 and D.4.5 are easy consequences of (D.4.8) or (D.4.9) and Proposition D.3.2 *(iii)*. Note that the constants C and N in Propositions D.4.4 and D.4.5 and Theorem D.4.6 depend just on the dimension of X.

Proposition D.4.4
There exist C and N such that

$$\begin{aligned} &\|[A_1, A_2]\| \\ &\quad \le C\|[x, A_1]\|_{g_0,N}\|[D, A_2]\|_{g_0,N} + C\|[D, A_1]\|_{g_0,N}\|[x, A_2]\|_{g_0,N}. \end{aligned}$$

It is often useful to know that, in a certain sense, the symbol of the product of two pseudo-differential operators is approximately equal to the product of their symbols.

Proposition D.4.5
There exist C and N such that

$$\begin{aligned}&\|a^{\mathrm{w}}(x,D)b^{\mathrm{w}}(x,D)+b^{\mathrm{w}}(x,D)a^{\mathrm{w}}(x,D)-2(ab)^{\mathrm{w}}(x,D)\|\\&\leq C\|\nabla_{x,\xi}^2 a\|_{g_0,N}\|\nabla_{x,\xi}^2 b\|_{g_0,N}.\end{aligned}$$

The following theorem is a version of the so-called sharp Gårding inequality.

Theorem D.4.6
There exist C and N such that if

$$a(x,\xi)\geq 0,$$

then

$$a^{\mathrm{w}}(x,D)\geq -C\|\nabla_{x,\xi}^2 a\|_{g_0,N}. \tag{D.4.10}$$

Proof. We will use the coherent states ϕ_Y and the corresponding projections P_Y introduced in Sect. D.2. Set

$$\tilde{A}:=(2\pi)^{-n}\int a(Y)P_Y\mathrm{d}Y. \tag{D.4.11}$$

The Weyl symbol of $\tilde{A}$ equals the convolution $(2\pi)^{-n}a*p_0$. Now,

$$\begin{aligned}(a-(2\pi)^{-n}a*p_0)(Z)&=(2\pi)^{-n}\textstyle\int p_0(Y)(a(Z)-a(Z-Y))\mathrm{d}Y\\&=(2\pi)^{-n}\textstyle\int p_0(Y)\frac{1}{2}\int_0^1(Y)^2\nabla_Z^2a(Z-\tau Y)\mathrm{d}\tau\mathrm{d}Y.\end{aligned}$$

Hence, for some N,

$$\|\tilde{A}-a^{\mathrm{w}}(x,D)\|\leq\|\nabla_Y^2 a\|_{g_0,N}. \tag{D.4.12}$$

Now the theorem follows from (D.4.12) and the positivity of (D.4.11). □

Suppose that T_r is the operator of dilations defined by

$$(T_r\phi)(x):=r^{-\frac{n}{2}}\phi(r^{-1}x).$$

Then it is easy to see that

$$T_ra(x,D)T_{-r}=a(rx,r^{-1}D), \tag{D.4.13}$$

$$T_ra^{\mathrm{w}}(x,D)T_{-r}=a^{\mathrm{w}}(rx,r^{-1}D). \tag{D.4.14}$$

(Note that in the case of Weyl symbols a similar covariance of symbols is true for a much larger group called the *metaplectic group*).

Functions of operators in $\Psi(1, g_0)$ often belong to the same class.

Proposition D.4.7
(i) Let A be a closed operator. We consider $\sigma(A)$ as a subset of the Riemann sphere $\mathbb{C} \cup \{\infty\}$ by adding ∞ to $\sigma(A)$ if A is unbounded.

Let $z_0 \in \mathbb{C}$, $z_0 \notin \sigma(A)$ and $(z - A)^{-1} \in \Psi(1, g_0)$. Let f be a function holomorphic on the neighborhood of $\sigma(A)$ in $\mathbb{C} \cup \{\infty\}$. Then $f(A) \in \Psi(1, g_0)$.
(ii) If A is, in addition, self-adjoint, and f is a C_0^∞ function on $\sigma(A)$, then

$$f(A) \in \Psi(1, g_0).$$

Proof. First we check by the Beals criterion that if $B \in \Psi(1, g_0)$ and B is invertible, then $B^{-1} \in \Psi(1, g_0)$.

Let $z \notin \sigma(A)$. We have

$$(z - A)^{-1} = (z_0 - A)^{-1} \left(1 - (z_0 - z)(z_0 - A)^{-1}\right)^{-1}.$$

Hence $(z_0 - A)^{-1} \in \Psi(1, g_0)$ implies $(z - A)^{-1} \in \Psi(1, g_0)$.

Consider now a function f holomorphic on a neighborhood of $\sigma(A)$. Clearly,

$$f(A) = \frac{1}{2\pi i} \int_\gamma f(z)(z - A)^{-1} \mathrm{d}z,$$

where the contour γ encloses $\sigma(A)$ in $\mathbb{C} \cup \{\infty\}$. This proves *(i)*.

To show *(ii)*, we use

$$f(A) = \frac{i}{2\pi} \int_{\mathbb{C}} \partial_{\bar{z}} \tilde{f}(z)(z - A)^{-1} \mathrm{d}z \wedge \mathrm{d}\bar{z},$$

where $\tilde{f}$ is an almost-analytic extension of f and the Beals criterion. □

D.5 Symbols and Operators Depending on a Parameter

Now we would like to consider symbols and operators that depend on a parameter t. Suppose that $f(t)$, $c_x(t)$ and $c_\xi(t)$ are some non-negative functions. We define a Euclidean metric on $X \times X'$ that depends on a parameter t:

$$g(t) := c_x^2(t)\mathrm{d}x^2 + c_\xi^2(t)\mathrm{d}\xi^2.$$

The spaces $S(f(t), g(t))$, $S(o(f(t)), g(t))$ and $L^1(f(t)\mathrm{d}t, S(g(t))$ are defined to be the spaces of functions

$$t \to a(t, x, \xi) \in C^\infty(X \times X')$$

such that

$$|\partial_x^\alpha \partial_\xi^\beta a(t, x, \xi)| \le C_{\alpha\beta} f(t) c_x^{|\alpha|}(t) c_\xi^{|\beta|}(t), \quad \alpha, \beta \in \mathbb{N}^n,$$

$$\partial_x^\alpha \partial_\xi^\beta a(t, x, \xi) \in o(f(t)) c_x^{|\alpha|}(t) c_\xi^{|\beta|}(t), \quad \alpha, \beta \in \mathbb{N}^n,$$

$$\partial_x^\alpha \partial_\xi^\beta a(t, x, \xi) f(t) c_x^{-|\alpha|}(t) c_\xi^{-|\beta|}(t) \in L^1(\mathrm{d}t), \quad \alpha, \beta \in \mathbb{N}^n,$$

respectively.

Similarly, we define the algebras $\Psi(f(t), g(t))$, $\Psi(o(f(t)), g(t))$ and $L^1(f(t)\mathrm{d}t, \Psi(g(t)))$ to be the spaces of operator-valued functions

$$t \to A(t) \in B(L^2(X))$$

such that

$$\|\mathrm{ad}_D^\alpha \mathrm{ad}_x^\beta A(t)\| \le C_{\alpha\beta} f(t) c_x^{|\alpha|}(t) c_\xi^{|\beta|}(t), \quad \alpha, \beta \in \mathbb{N}^n,$$

$$\|\mathrm{ad}_D^\alpha \mathrm{ad}_x^\beta A(t)\| \in o(f(t)) c_x^{|\alpha|}(t) c_\xi^{|\beta|}(t), \quad \alpha, \beta \in \mathbb{N}^n,$$

$$\|\mathrm{ad}_D^\alpha \mathrm{ad}_x^\beta A(t)\| f(t) c_x^{-|\alpha|}(t) c_\xi^{-|\beta|}(t) \in L^1(\mathrm{d}t), \quad \alpha, \beta \in \mathbb{N}^n,$$

respectively.

In applications to scattering theory described in Chap. 3, a special role will be played by the metric

$$g_0(t) := \langle t \rangle^{-2} \mathrm{d}x^2 + \mathrm{d}\xi^2.$$

The number $c_x(t) c_\xi(t)$ has the interpretation of the "effective Planck constant". The following proposition can serve as an alternative definition of the algebras defined above if this Planck constant is bounded.

Proposition D.5.1
Suppose that

$$c_x(t) c_\xi(t) \le C. \tag{D.5.1}$$

Then $\Psi(f(t), g(t))$, $\Psi(o(f(t)), g(t))$ *and* $L^1(f(t)\mathrm{d}t, \Psi(g(t)))$ *are the sets of operator-valued functions*

$$t \to a^{\mathrm{w}}(t, x, D)$$

such that $a \in S(f(t), g(t))$, $a \in S(o(f(t)), g(t))$ *and* $a \in L^1(f(t)\mathrm{d}t, S(g(t)))$ *respectively.*

Proof. If both $c_x(t)$ and $c_\xi(t)$ are bounded, then the proposition follows immediately from Theorem D.4.1 and (D.4.1).

If not, we conjugate our operators and symbols with a t-dependent generator of dilations $T_{r(t)}(t)$, where we take

$$r(t) := c_\xi^{-1/2}(t) c_x^{1/2}(t). \tag{D.5.2}$$

By (D.4.14), the metric becomes $c_x(t)c_\xi(t)\mathrm{d}x^2 + c_x(t)c_\xi(t)\mathrm{d}\xi^2$. Thus the coefficients in the metric become bounded. Now we can apply Theorem D.4.1. □

If

$$A_i(t) \in \Psi(f_i(t), g(t)), \quad i = 1, 2,$$

then it is easy to see that

$$A_1(t)A_2(t) \in \Psi(f_1(t)f_2(t), g(t)), \tag{D.5.3}$$

and

$$[A_1(t), A_2(t)] \in \Psi(f_1(t)f_2(t)c_x(t)c_\xi(t), g(t)), \tag{D.5.4}$$

To show (D.5.3), it is enough to use the definition of Ψ (the "Beals criterion"). (D.5.4) is much deeper – one needs to use, for example, (D.4.8).

Sometimes it is possible to improve (D.5.4). Namely, the following fact follows from the proof of Proposition D.4.4.

Proposition D.5.2
Suppose that

$$[x, A_i(t)] \in \Psi(\tilde{f}_i(t)c_\xi(t), g(t)), \quad i = 1, 2,$$

$$[D, A_i(t)] \in \Psi(\tilde{f}_i(t)c_x(t), g(t)), \quad i = 1, 2.$$

Then

$$[A_1(t), A_2(t)] \in \Psi(\tilde{f}_1(t)\tilde{f}_2(t)c_x(t)c_\xi(t), g(t)). \tag{D.5.5}$$

Below we give a consequence of the proof of Proposition D.4.5.

Proposition D.5.3
Let $a_i \in S(f_i(t), g(t))$, $i = 1, 2$. *Then*

$$a_1^{\mathrm{w}}(t, x, D)a_2^{\mathrm{w}}(t, x, D) + a_2^{\mathrm{w}}(t, x, D)a_1^{\mathrm{w}}(t, x, D) - 2(a_1a_2)^{\mathrm{w}}(t, x, D)$$

belongs to $\Psi(f_1(t)f_2(t)c_x^2(t)c_\xi^2(t), g(t))$.

Finally, Theorem D.4.10 and conjugating with t-dependent dilations can be used to obtain the following version of the sharp Gårding inequality.

Proposition D.5.4
Let $a \in S(f(t), g(t))$ *and*

$$a(t, x, \xi) \geq 0.$$

Then

$$a^{\mathrm{w}}(x, D) \geq -Cf(t)c_x(t)c_\xi(t).$$

D.6 Weighted Spaces

Pseudo-differential operators are intimately related with various weighted Hilbert spaces, such as

$$\langle x\rangle^m L^2(X) \quad := \{\phi \in \mathcal{D}'(X) \mid \langle x\rangle^{-m}\phi \in L^2(X)\}.$$

The following lemma gives a criterion for the boundedness of operators on $\langle x\rangle^m L^2(X)$.

Lemma D.6.1
Suppose that A is an operator on $L^2(X)$. Then the condition (D.6.1) is equivalent to (D.6.2):

$$\begin{cases} (\mathrm{ad}_x^\alpha A)\langle x\rangle^{-|\alpha|} \in B(L^2(X)), & \alpha \in \mathbb{N}^n, \\ \langle x\rangle^{-|\alpha|}(\mathrm{ad}_x^\alpha A) \in B(L^2(X)). & \alpha \in \mathbb{N}^n. \end{cases} \tag{D.6.1}$$

$$\textit{For any } s \in \mathbb{R}, \textit{ the operator } \langle x\rangle^s A\langle x\rangle^{-s} \textit{ is bounded.} \tag{D.6.2}$$

Moreover, for any $n \in \mathbb{N}$,

$$\|\langle x\rangle^n A\langle x\rangle^{-n}\| \le C_n \sum_{|\alpha|\le n} \|(\mathrm{ad}_x^\alpha A)\langle x\rangle^{-|\alpha|}\|.$$

Proof. It is easy to see that (D.6.2) implies (D.6.1).

Assume (D.6.1). We will show that (D.6.2) is true for $s = n \in \mathbb{N}$. Clearly,

$$\|\langle x\rangle^n A\langle x\rangle^{-n}\|^2 = \|\langle x\rangle^{-n} A^*\langle x\rangle^{2n} A\langle x\rangle^{-n}\|.$$

Recall that $\langle x\rangle^{2n} = (1+x^2)^n$. We commute n copies of x through A^* to the left and n copies of x through A to the right, and we get

$$\left\|\langle x\rangle^{-n} A^*\langle x\rangle^{2n} A\langle x\rangle^{-n}\right\| \le \sum_{|\alpha|,|\beta|\le n} C_{\alpha,\beta}\|\langle x\rangle^{-|\alpha|}\mathrm{ad}_x^\alpha A^*\|\|\mathrm{ad}_x^\beta A\langle x\rangle^{-|\beta|}\|.$$

□

D.7 Symbols Associated with Some Non-uniform Metrics

Let us consider a certain class of symbols associated with non uniform metrics.

Definition D.7.1
Let $Y = Y_1 \oplus \cdots \oplus Y_k$, $\delta_1, \dots, \delta_k \in \mathbb{R}$ and let $w(y)$ be a positive function. The space

$S\left(w(y), \frac{dy_1^2}{\langle y_1\rangle^{2\delta_1}} + \cdots + \frac{dy_k^2}{\langle y_k\rangle^{2\delta_k}}\right)$ *is the Fréchet space of functions* $a(y) \in C^\infty(Y)$ *such that*

$$|\partial_{y_1}^{\alpha_1} \cdots \partial_{y_k}^{\alpha_k} a(y_1, \dots, y_k)| \le C_\alpha w(y_1, \dots, y_k)\langle y_1\rangle^{-\delta_1|\alpha_1|} \cdots \langle y_k\rangle^{-\delta_k|\alpha_k|}, \quad \alpha \in \mathbb{N}^{nk}.$$

Let us list some properties of these spaces:

Proposition D.7.2
Let $a \in S\left(\langle y_1\rangle^{m_1} \cdots \langle y_k\rangle^{m_k}, \frac{dy_1^2}{\langle y_1\rangle^{2\delta_1}} + \cdots + \frac{dy_k^2}{\langle y_k\rangle^{2\delta_k}}\right)$.
(i) We have

$$\partial_{y_1}^{\alpha_1} \cdots \partial_{y_k}^{\alpha_k} a(y_1, \dots, y_k) \in S\left(\langle y_1\rangle^{m_1-\delta_1|\alpha_1|} \cdots \langle y_k\rangle^{m_k-\delta_k|\alpha_k|}, \frac{dy_1^2}{\langle y_1\rangle^{2\delta_1}} + \cdots + \frac{dy_k^2}{\langle y_k\rangle^{2\delta_k}}\right).$$

(ii) Let $\tilde{\delta} := \min\{\delta_1, \delta_2\} \ge 0$ *and* $\tilde{a}(\tilde{y}, y_3, \dots, y_k) := a(\tilde{y}, \tilde{y}, y_3, \dots, y_k)$. *Then*

$$\tilde{a} \in S\left(\langle \tilde{y}\rangle^{m_1+m_2}\langle y_3\rangle^{m_3} \cdots \langle y_k\rangle^{m_k}, \frac{d\tilde{y}^2}{\langle \tilde{y}\rangle^{2\tilde{\delta}}} + \frac{dy_3^2}{\langle y_3\rangle^{2\delta_3}} + \cdots + \frac{dy_k^2}{\langle y_k\rangle^{2\delta_k}}\right).$$

(iii) Let $Q(\cdot)$ *be a quadratic form on* Y' *and* $\delta_1, \dots, \delta_k \ge 0$. *Then*

$$e^{iQ(D_y)}a \in S\left(\langle y_1\rangle^{m_1} \cdots \langle y_k\rangle^{m_k}, \frac{dy_1^2}{\langle y_1\rangle^{2\delta_1}} + \cdots + \frac{dy_k^2}{\langle y_k\rangle^{2\delta_k}}\right).$$

Proof. *(i)* and *(ii)* are straightforward.
Let us show *(iii)*. Set $b := e^{iQ(D_y)}a$. We write

$$\begin{aligned}
&\langle y_1\rangle^{-m_1+|\alpha_1|\delta_1} \cdots \langle y_k\rangle^{-m_k+|\alpha_k|\delta_k}\partial_{y_1}^{\alpha_1} \cdots \partial_{y_k}^{\alpha_k} b \\
&= \langle y_1\rangle^{-m_1+|\alpha_1|\delta_1} \cdots \langle y_k\rangle^{-m_k+|\alpha_k|\delta_k} e^{iQ(D_y)}\langle y_1\rangle^{m_1-|\alpha_1|\delta_1} \cdots \langle y_k\rangle^{m_k-|\alpha_k|\delta_k} \\
&\times \langle y_1\rangle^{-m_1+|\alpha_1|\delta_1} \cdots \langle y_k\rangle^{-m_k+|\alpha_k|\delta_k}\partial_{y_1}^{\alpha_1} \cdots \partial_{y_k}^{\alpha_k} a.
\end{aligned}$$

Clearly, the expression on the last line belongs to $S(1, dy^2)$. By Proposition D.3.2 *(v)*, the operator on the second line is bounded on $S(1, dy^2)$. Therefore, the above function belongs to $S(1, dy^2)$ and, in particular, is bounded. □

D.8 Pseudo-differential Operators Associated with the Metric g_1

Set

$$g_1 := \langle x\rangle^{-2}\mathrm{d}x^2 + \mathrm{d}\xi^2.$$

Define the Poisson bracket as

$$\{a_1, a_2\}(x,\xi) := \partial_x a_1(x,\xi)\partial_\xi a_2(x,\xi) - \partial_\xi a_1(x,\xi)\partial_x a_2(x,\xi).$$

In this section we describe basic properties of pseudo-differential operators with symbols in $S(\langle x\rangle^m, g_1)$. First note that the family of spaces $S(\langle x\rangle^m, g_1)$ forms a graded algebra with respect to the multiplication and a graded Lie algebra with respect to the Poisson bracket, as explained in the following proposition.

Proposition D.8.1
If $a_i \in S(\langle x\rangle^{m_i}, g_1)$, $i = 1, 2$, then $a_1 a_2 \in S(\langle x\rangle^{m_1+m_2}, g_1)$ and $\{a_1, a_2\} \in S(\langle x\rangle^{m_1+m_2-1}, g_1)$.

The following theorem describes several equivalent definitions of quantized analogs of $S(\langle x\rangle^m, g_1)$.

Theorem D.8.2
The following conditions are equivalent.
(i) A is an operator on $L^2(X)$ such that

$$\langle x\rangle^{-m+|\alpha|}\mathrm{ad}_D^\alpha \mathrm{ad}_x^\beta A \in B(L^2(X)), \quad \alpha,\beta \in \mathbb{N}^n.$$

(ii) $A = a^{\mathrm{w}}(x,D)$ with $a \in S(\langle x\rangle^m, g_1)$.
(iii) $A = a(x,D)$ with $a \in S(\langle x\rangle^m, g_1)$.

Proof. Fix $\alpha,\beta \in \mathbb{N}^n$. *(i)* implies

$$\langle x\rangle^{-m+|\alpha|}\mathrm{ad}_D^\alpha \mathrm{ad}_x^\beta A \in \Psi(1, g_0). \tag{D.8.1}$$

By Theorem D.4.1 and (D.8.1), if $A = a(x,D)$, then

$$|\langle x\rangle^{-m+|\alpha|}\partial_x^\alpha \partial_\xi^\beta a(x,\xi)| \le C_{\alpha,\beta}.$$

This shows $(i) \Rightarrow (iii)$.
(iii) implies

$$\langle x\rangle^{-m+|\alpha|}\partial_x^\alpha \partial_\xi^\beta a(x,\xi) \in S(1, g_0). \tag{D.8.2}$$

By Theorem D.4.1, it follows from (D.8.1) that

$$\langle x\rangle^{-m+|\alpha|}\mathrm{ad}_D^\alpha \mathrm{ad}_x^\beta A \in B(L^2(X)).$$

This shows $(iii) \Rightarrow (i)$.

Formula (D.1.5) and the boundedness of $e^{\pm i\langle D_x, D_\xi\rangle}$ on $S(\langle x\rangle^m, g_1)$ (Proposition D.7.2 *(iii)*) show the equivalence of *(ii)* and *(iii)*. □

Definition D.8.3
The set of operators satisfying one, and hence all of the above conditions is denoted $\Psi(\langle x\rangle^m, g_1)$. *It is called the set of pseudo-differential operators associated with the metric* g_1 *and the weight* $\langle x\rangle^m$.

Let us describe the relationship between various symbols of the same operator.

Proposition D.8.4
Let $a(x,\xi) \in S(\langle x\rangle^m, g_1)$, *and* $a^{\mathrm{w}}(x,D) = \tilde{a}(x,D)$. *Then (D.1.5) is true, and*

$$\tilde{a}(x,\xi) - \sum_{j=0}^{n} \frac{1}{j!}\left(\frac{i}{2}\langle D_x, D_\xi\rangle\right)^j a(x,\xi) \in S(\langle x\rangle^{m-n-1}, g_1). \tag{D.8.3}$$

Proof. To show (D.8.3), we note the following consequence of (D.1.5):

$$\tilde{a}(x,\xi) = \textstyle\sum_{j=0}^{n} \frac{1}{j!}\left(\frac{i}{2}\langle D_x, D_\xi\rangle\right)^j a(x,\xi) + \int_0^1 \frac{\tau^n}{n!} e^{\tau\frac{i}{2}\langle D_x, D_\xi\rangle}\left(\frac{i}{2}\langle D_x, D_\xi\rangle\right)^{n+1} a(x,\xi)\mathrm{d}\tau,$$

and we use the fact that

$$(\langle D_x, D_\xi\rangle)^{n+1} a(x,\xi) \in S(\langle x\rangle^{m-n-1}, g_1)$$

and $e^{\tau\frac{i}{2}\langle D_x, D_\xi\rangle}$ maps $S(\langle x\rangle^{m-n-1}, g_1)$ into itself. □

It will follow from the following proposition that the family $\Psi(\langle x\rangle^m, g_1)$ forms a graded algebra with respect to multiplication, and a graded Lie algebra with respect to the commutator.

Proposition D.8.5
(i) If $A_i \in \Psi(\langle x\rangle^{m_i}, g_1)$, $i = 1, 2$, *then*

$$A_1 A_2 \in \Psi(\langle x\rangle^{m_1+m_2}, g_1), \tag{D.8.4}$$

$$[A_1, A_2] \in \Psi(\langle x\rangle^{m_1+m_2-1}, g_1). \tag{D.8.5}$$

(ii) If, moreover, $A_i = a_i^w(x,D)$ *and* $A = a^w(x,D)$, *then (D.4:8) is true, and*

$$a(x,\xi) - \sum_{j=0}^{n} \left(\frac{i}{2}(\langle D_{x_2}, D_{\xi_1}\rangle - \langle D_{x_1}, D_{\xi_2}\rangle)\right)^j a_1(x_1,\xi_1)a_2(x_2,\xi_2)\Big|_{\substack{x=x_2=x_1,\\ \xi=\xi_2=\xi_1}} \tag{D.8.6}$$

belongs to $S(\langle x\rangle^{m_1+m_2-n-1}, g_1)$.
(iii) If, moreover, $A_i = a_i(x, D)$, $A = a(x, D)$, *then (D.4.9) is true, and*

$$a(x,\xi) - \sum_{j=0}^{n} \frac{1}{j!}(i\langle D_{x_2}, D_{\xi_1}\rangle)^j a_1(x_1,\xi_1)a_2(x_2,\xi_2)\Big|_{\substack{x=x_2=x_1,\\ \xi=\xi_2=\xi_1}} \tag{D.8.7}$$

belongs to $S(\langle x\rangle^{m_1+m_2-n-1}, g_1)$.

Proof. The property (D.8.4) follows easily from the Beals criterion (Theorem D.8.2 *(ii)*). In order to show the other statements of *(i)*, one has to use the symbolic calculus described either in *(ii)* or in *(iii)*. □

Occasionally one is confronted with operators given by the expressions described in the following propositions.

Proposition D.8.6
Let

$$b(x_1, x_2, \xi) \in S(\langle x_1\rangle^{m_1}\langle x_2\rangle^{m_2}\langle\xi\rangle^k, \langle x_1\rangle^{-2}\mathrm{d}x_1^2 + \langle x_2\rangle^{-2}\mathrm{d}x_2^2 + \mathrm{d}\xi^2).$$

Then the operator A defined in (D.4.6) belongs to $\Psi(\langle x\rangle^{m_1+m_2}, g_1)$. *Moreover,* $A = a^{\mathrm{w}}(x, D)$, *where a is given by (D.4.7), and*

$$a(x,\xi) - \sum_{j=0}^{n} \tfrac{1}{j!}(\tfrac{i}{2}(\langle D_{x_1}, D_\xi\rangle + \langle D_{x_2}, D_\xi\rangle))^j b(x_1,x_2,\xi)\Big|_{x=x_1=x_2} \tag{D.8.8}$$

belongs to $S(\langle x\rangle^{m_1+m_2-n-1}, g_1)$.

D.9 Essential Support of Pseudo-differential Operators

Definition D.9.1
Let $a(x,\xi) \in C^\infty(X \times X')$ *and* $\Gamma \subset X \times X'$. *Then we say that* $a \in S(\langle x\rangle^{-\infty})$ *on* Γ *if*

$$|\partial_x^\alpha \partial_\xi^\beta a(x,\xi)| \le C_{\alpha,\beta,N}\langle x\rangle^{-N}, \quad \alpha,\beta \in \mathbb{N}^n, \quad N \in \mathbb{N}, \quad (x,\xi) \in \Gamma.$$

A subset $\Gamma \subset X \times X'$ is called *conical* if $(x,\xi) \in \Gamma$ and $t > 0$ implies $(tx,\xi) \in \Gamma$.

Let $\Gamma \subset X \times X'$ be conical. Then we define the ϵ–neighborhood of Γ as follows:

$$\Gamma^\epsilon := \Big\{x,\xi \in X \times X' \ : \ \exists_{y,\eta\in\Gamma} \text{ such that } \Big|\tfrac{x}{|x|} - \tfrac{y}{|y|}\Big| < \epsilon,\ |\xi - \eta| < \epsilon\Big\}.$$

Note that Γ^ϵ is also conical.

The lemma below gives a partial justification why conical sets are useful.

Proposition D.9.2
For any conical set $\Gamma \in X \times X'$, *there exists a function* $J \in S(1, g_1)$ *such that* $J \in S(\langle x\rangle^{-\infty})$ *outside* Γ^ϵ *and* $J - 1 \in S(\langle x\rangle^{-\infty})$ *on* Γ. *More precisely, for any* $0 < R_1 < R_2$, *we can guarantee that* $\mathrm{supp} J \subset \Gamma^\epsilon \backslash B(R_1) \times X'$ *and* $J = 1$ *on* $\Gamma \backslash B(R_2) \times X'$.

Proof. Let S be the unit sphere in X. Let J_0 be the characteristic function of $\Gamma \cap S \times X'$ in $S \times X'$. We smooth it out by convolving with a C_0^∞ function of a sufficiently small support. Then we extend it by homogeneity to $X \times X'$. Finally, we smooth it out in a neighborhood of $\{0\} \times X'$. □

Lemma D.9.3
Let $a_i \in S(\langle x\rangle^{m_i}, g_1)$, $i \in \mathbb{N}$, *and* $(m_i)_{i\in\mathbb{N}}$ *a strictly decreasing sequence of real numbers. Then there exists a symbol* $a \in S(\langle x\rangle^{m_0}, \langle x\rangle^{-2}\mathrm{d}x^2 + \mathrm{d}\xi^2)$ *such that*

$$a - \sum_{i=0}^{N} a_i \in S(\langle x\rangle^{m_{N+1}}, g_1).$$

If $a_i \in S(\langle x\rangle^{-\infty})$ *on* $\Gamma \subset X \times X'$, *then* $a \in S(\langle x\rangle^{-\infty})$ *on* Γ.

Proof. Let $\chi \in C_0^\infty(X)$ such that $\chi = 1$ on a neighborhood of zero. Then we can set

$$a(x,\xi) = \sum_{i=1}^{\infty} \left(1 - \chi\left(\frac{x}{C_i}\right)\right) a_i(x,\xi)$$

for an appropriate sequence of C_i (see [Hö2, Prop. 18.1.3]). □

In some cases, one can say that pseudo-differential operators are very small in $\Gamma \subset X \times X'$. This idea is the content of the following proposition.

Proposition D.9.4
If $A \in \Psi(\langle x\rangle^m, g_1)$, $\Gamma \subset X \times X'$, *and*

$$A = a^{\mathrm{w}}(x, D) = \tilde{a}(x, D),$$

then $a \in S(\langle x\rangle^{-\infty})$ *on* Γ *if and only if* $\tilde{a} \in S(\langle x\rangle^{-\infty})$ *on* Γ.

Proof. The proposition follows from equation (D.8.3), if we note that n can be taken arbitrarily big and, for any j,

$$(\langle D_x, D_\xi\rangle)^j a(x,\xi) \in S(\langle x\rangle^{-\infty}) \quad \text{on} \quad \Gamma.$$

□

Definition D.9.5
In the situation described in Proposition D.9.4, we will say that $A \in \Psi(\langle x\rangle^{-\infty})$ *on* Γ. *If* $\Gamma = X \times X'$, *then we will simply say that* $A \in \Psi(\langle x\rangle^{-\infty})$.

Proposition D.9.6
If $A_i \in \Psi(\langle x\rangle^m, g_1)$, $\Gamma_i \subset X \times X'$ and $A_i \in \Psi(\langle x\rangle^{-\infty})$ on Γ_i, then $A_1 A_2 \in \Psi(\langle x\rangle^{-\infty})$ on $\Gamma_1 \cup \Gamma_2$.

Proof. The proposition follows from (D.8.7) with an arbitrary n, if we note that, for any j,

$$(i\langle D_{x_2}, D_{\xi_1}\rangle)^j \tilde{a}_1(x_1,\xi_1)\tilde{a}_2(x_2,\xi_2)\Big|_{\substack{x=x_2=x_1,\\ \xi=\xi_2=\xi_1}} \in S(\langle x\rangle^{-\infty}) \text{ on } \Gamma_1 \cup \Gamma_2.$$

□

Proposition D.9.7
Let $\Gamma \subset X \times X'$,
$b(x_1,x_2,\xi) \in S(\langle x_1\rangle^{m_1}\langle x_2\rangle^{m_2}, \langle x_1\rangle^{-2}\mathrm{d}x_1^2 + \langle x_2\rangle^{-2}\mathrm{d}x_2^2 + \mathrm{d}\xi^2)$, and

$$|\partial_{x_1}^{\alpha_1}\partial_{x_2}^{\alpha_2}\partial_\xi b(x,x,\xi)| \leq C_{\alpha,\beta,N}\langle x\rangle^{-N}, \quad (x,\xi) \in \Gamma, \quad \alpha_1,\alpha_2,\beta \in \mathbb{N}^n, \ N \in \mathbb{N}.$$

Let A be given by (D.4.6). Then $A \in S(\langle x\rangle^{-\infty})$ on Γ.

Proof. We use the expansion (D.8.8). □

D.10 Ellipticity

Definition D.10.1
Let $m \in \mathbb{R}$ and $\Gamma \subset X \times X'$. Let $a \in S(\langle x\rangle^m, g_1)$. We say that a is elliptic *on Γ iff there exist $R > 0$ and $C_0 > 0$ such that*

$$|a(x,\xi)| \geq C_0\langle x\rangle^m, \quad (x,\xi) \in \Gamma\backslash(B(R) \times X').$$

Proposition D.10.2
Let $m \in \mathbb{R}$ and $\Gamma \subset X \times X'$ and $A \in \Psi(\langle x\rangle^m, g_1)$. Suppose that

$$A = a^{\mathrm{w}}(x,D) = \tilde{a}(x,D).$$

Then a is elliptic on Γ iff $\tilde{a}$ is elliptic on Γ.

Proof. We use the fact that $|a(x,\xi) - \tilde{a}(x,\xi)| \leq C\langle x\rangle^{m-1}$. □

Definition D.10.3
We say that A is elliptic on Γ iff the conditions of the above proposition are satisfied.

Proposition D.10.4
If $A_i \in \Psi(\langle x\rangle^m, g_1)$, $\Gamma_i \subset X \times X'$, and A_i is elliptic on Γ_i, then A_1A_2 is elliptic on $\Gamma_1 \cap \Gamma_2$.

Proposition D.10.5
Let $\Gamma \subset X \times X'$,
$b(x_1, x_2, \xi) \in S(\langle x_1\rangle^{m_1}\langle x_2\rangle^{m_2}, \langle x_1\rangle^{-2}\mathrm{d}x_1^2 + \langle x_2\rangle^{-2}\mathrm{d}x_2^2 + \mathrm{d}\xi^2)$ and, for some $C_0 > 0$ and R,

$$|b(x, x, \xi)| \geq \langle x\rangle^{m_1+m_2}, \quad (x, \xi) \in \Gamma \backslash (B(R) \times X').$$

Let A be given by (D.4.6). Then A is elliptic on Γ.

Proposition D.10.6
Suppose that Γ is a conical subset of $X \times X'$ and $\epsilon > 0$. Let $A \in \Psi(\langle x\rangle^m, g_1)$ be elliptic on Γ^ϵ. Let $B \in \Psi(1, g_1)$ and $B \in \Psi(\langle x\rangle^{-\infty})$ outside of Γ. Then there exist $C \in \Psi(\langle x\rangle^{-m}, g_1)$, $C \in \Psi(\langle x\rangle^{-\infty})$ on Γ^ϵ, and $R_{-\infty} \in \Psi(\langle x\rangle^{-\infty})$ such that

$$B = CA + R_{-\infty}.$$

Proof. Let

$$A = a^{\mathrm{w}}(x, D), \quad B = b_0^{\mathrm{w}}(x, D).$$

Suppose that

$$|a(x, \xi)| \geq C_0\langle x\rangle^m, \quad (x, \xi) \in \Gamma^\epsilon \backslash B(R) \times X'.$$

Let $q \in S(1, g_1)$ be such that $q = 1$ on $\Gamma \backslash B(2R) \times X'$ and $\mathrm{supp} q \subset \Gamma^\epsilon \backslash B(R) \times X'$. We set

$$c_0(x, \xi) = q(x, \xi)b_0(x, \xi)a^{-1}(x, \xi).$$

Then

$$c_0^{\mathrm{w}}(x, D)a^{\mathrm{w}}(x, D) - b_0^{\mathrm{w}}(x, D) = b_1^{\mathrm{w}}(x, D)$$

with $b_1 \in S(\langle x\rangle^{-1}, g_1)$ and $b_1 \in S(\langle x\rangle^{-\infty})$ outside of Γ. Then we set

$$c_1(x, \xi) = q(x, \xi)b_1(x, \xi)a^{-1}(x, \xi).$$

We continue this way, and we obtain $c_j \in S(\langle x\rangle^{-m-j}, g_1)$ and $b_j \in S(\langle x\rangle^{-j}, g_1)$ such that

$$(c_0^{\mathrm{w}}(x, D) + \cdots + c_n^{\mathrm{w}}(x, D))a^{\mathrm{w}}(x, D) - b_0^{\mathrm{w}}(x, D) = b_{n+1}^{\mathrm{w}}(x, D)$$

Using Lemma D.9.3, we define $c(x, \xi)$ such that

$$c(x, \xi) - \sum_{j=1}^{n} c_j(x, \xi) \in S(\langle x\rangle^{-m-n-1}, g_1).$$

We set $C := c^{\mathrm{w}}(x, D)$. □

D.11 Functional Calculus for Pseudo-differential Operators Associated with the Metric g_1

Let us study functions of pseudo-differential operators. We begin with the inverse.

Lemma D.11.1
Let $A \in \Psi(1, g_1)$ such that A is invertible in $B(L^2(X))$. Then $A^{-1} \in \Psi(1, g_1)$. Moreover, if we define the semi-norms

$$m_N(B) := \sup_{|\alpha|+|\beta|\leq n} \|\langle x\rangle^{|\beta|}\mathrm{ad}_x^\alpha \mathrm{ad}_D^\beta B\|,$$

then for any N, there exist constants C_0, N_1, M such that

$$m_N(A^{-1}) \leq C_0 m_{N_1}(A)\|A^{-1}\|^M.$$

Proof. Clearly, $\mathrm{ad}_x^\alpha \mathrm{ad}_D^\beta A^{-1}$ is a linear combination of terms

$$A^{-1}(\mathrm{ad}_x^{\alpha_1}\mathrm{ad}_D^{\beta_1} A)A^{-1} \cdots A^{-1}(\mathrm{ad}_x^{\alpha_q}\mathrm{ad}_D^{\beta_q})A^{-1}$$

such that $\alpha = \alpha_1 + \cdots = \alpha_q$, $\beta = \beta_1 + \cdots + \beta_q$. Then we use Theorem D.8.2 *(i)* and Lemma D.6.1. □

Proposition D.11.2
(i) Suppose that A is a closed operator, $(z_0 - A)^{-1} \in \Psi(1, g_1)$ for some $z_0 \notin \sigma(A)$ and f is holomorphic on a neighborhood of $\sigma(A)$ in $\mathbb{C} \cup \{\infty\}$. Then $f(A) \in \Psi(1, g_1)$.
(ii) If, moreover, A is self-adjoint and $f \in C_0^\infty(\sigma(A))$, then

$$f(A) \in \Psi(1, g_1).$$

Proof. The proof of *(i)* is essentially identical to the proof of Proposition D.4.7. To prove *(ii)*, we use, in addition, Lemma D.11.1 to estimate the semi-norms of $m_N\left((z-A)^{-1}\right)$ uniformly for z in a compact set $\mathrm{supp}\tilde{f} \subset \mathbb{C}$:

$$m_N\left((z-A)^{-1}\right) \leq C|\mathrm{Im} z|^{-M}.$$

□

The following proposition follows easily by the methods of the proof of Proposition D.11.2:

Proposition D.11.3
(i) Suppose that A_i, $i = 1, 2$, are closed operators such that, for some $z_0 \notin \sigma(A_1) \cup \sigma(A_2)$,

$$(z_0 - A_1)^{-1}, (z_0 - A_2)^{-1} \in \Psi(1, g_1),$$

and for $m \geq 0$

$$A_2 - A_1 \in \Psi(\langle x \rangle^{-m}, g_1)$$

Then for any function f *holomorphic on a neighborhood of* $\sigma(A_1) \cup \sigma(A_2)$ *in* $\mathbb{C} \cup \{\infty\}$, *we have*

$$f(A_1) - f(A_2) \in \Psi(\langle x \rangle^{-m}, g_1).$$

(ii) If, moreover, A_i *are self-adjoint, then for any* $f \in C_0^\infty(\sigma(A_1) \cup \sigma(A_2))$,

$$f(A_1) - f(A_2) \in \Psi(\langle x \rangle^{-m}, g_1).$$

Proposition D.11.4
Suppose that A_j, $j = 1, 2$, *are semi-bounded self-adjoint operators such that*

$$(A_j - i)^{-1} \in \Psi(1, g_1),$$
$$V := A_1 - A_2 \in \Psi(1, g_1).$$

Let now $f_j \in C^\infty(\mathbb{R})$, $f_j' \in C_0^\infty(\mathbb{R})$, $j = 1, 2$, *such that*

$$\mathrm{dist}(\mathrm{supp} f_1, \mathrm{supp} f_2) > \|V\|.$$

Then

$$f_1(A_1) f_2(A_2) \in \Psi(\langle x \rangle^{-\infty}, g_1).$$

Proof. By splitting the functions f_j into several pieces, we can restrict ourselves to the case where f_1 is supported on the left and f_2 on the right, or the other way around. Therefore, let us assume that $\mathrm{supp} f_1 \subset]-\infty, \lambda_1[$ and $\mathrm{supp} f_2 \subset]\lambda_2, \infty[$, with $\lambda_1 + \|V\| < \lambda_2$. Let $f \in C^\infty(\mathbb{R})$ such that $0 \leq f \leq 1$, $f = 1$ on a neighborhood of $\mathrm{supp} f_1$ and $\mathrm{supp} f \subset]-\infty, \lambda_1[$. Define

$$\tilde{A}_2 := f(A_1) A_2 f(A_1).$$

Then

$$\tilde{A}_2 = f^2(A_1) A_1 + f(A_1) V f(A_1) \leq \lambda_1 + \|V\|.$$

Therefore

$$f_2(\tilde{A}_2) = 0.$$

Using the semi-boundedness of A_2 and $\tilde{A}_2$ and the fact that $f_2' \in C_0^\infty(\mathbb{R})$, we see that there exists a function $g \in C_0^\infty(\mathbb{R})$ such that

$$f_2(A_2) - f_2(\tilde{A}_2) = g(A_2) - g(\tilde{A}_2).$$

If $\tilde{g}$ is an almost-analytic extension of g, then

$$\begin{aligned} f_1(A_1)f_2(A_2) &= f_1(A_1)(f_2(A_2) - f_2(\tilde{A}_2)) \\ &= \tfrac{i}{2\pi} \textstyle\int_{\mathbb{C}} \partial_{\bar{z}}\tilde{g}(z) f_1(A_1)(z-A_2)^{-1}(A_2 - \tilde{A}_2)(z - \tilde{A}_2)^{-1} \mathrm{d}z \wedge \mathrm{d}\bar{z}. \end{aligned} \tag{D.11.1}$$

We have

$$\begin{aligned} & f_1(A_1)(z-A_2)^{-1}(A_2-\tilde{A}_2)(z-\tilde{A}_2)^{-1} \\ &= f_1(A_1)(z-A_2)^{-1}(1-f^2(A_1))A_1(z-\tilde{A}_2)^{-1} \\ &+ f_1(A_1)(z-A_2)^{-1}(1-f(A_1))V(z-\tilde{A}_2)^{-1} \\ &+ f_1(A_1)(z-A_2)^{-1}f(A_1)V(1-f(A_1))(z-\tilde{A}_2)^{-1}. \\ &= B_1(z) + B_2(z) + B_3(z). \end{aligned} \tag{D.11.2}$$

Fix $\tilde{f} \in C^\infty(\mathbb{R})$ such that $\tilde{f}' \in C_0^\infty(\mathbb{R})$ $f_1\tilde{f} = f_1$ and $\tilde{f}(1-f) = 0$. Consider, for instance, $B_2(z)$. We have

$$B_2(z) = f_1(A_1)\left(\mathrm{ad}^n_{\tilde{f}(A_1)}(z-A_2)^{-1}\right)(1-f(A_1))V(z-\tilde{A}_2)^{-1},$$

which is a sum of terms of the form

$$\begin{aligned} & f_1(A_1)(z-A_2)^{-1}\left(\mathrm{ad}^{n_1}_{\tilde{f}(A_1)}V\right)(z-A_2)^{-1} \\ & \cdots (z-A_2)^{-1}\left(\mathrm{ad}^{n_q}_{\tilde{f}(A_1)}V\right)(z-A_2)^{-1}(1-f(A_1))V(z-\tilde{A}_2)^{-1}. \end{aligned}$$

Note that $V \in \Psi(1, g_1)$ and $\tilde{f}(A_1) \in \Psi(1, g_1)$. Therefore,

$$\mathrm{ad}^n_{\tilde{f}(A_1)}V \in \Psi(\langle x\rangle^{-n}, g_1).$$

Now, using

$$\begin{aligned} \left\|\langle x\rangle^{k+n+|\alpha|}\left(\mathrm{ad}^\alpha_D \mathrm{ad}^\beta_x \mathrm{ad}^n_{\tilde{f}(A_1)}V\right)\langle x\rangle^{-k}\right\| &\le C_1, \\ \left\|\langle x\rangle^{k+|\alpha|}\left(\mathrm{ad}^\alpha_D \mathrm{ad}^\beta_x (z-A_2)^{-1}\right)\langle x\rangle^{-k}\right\| &\le C_2|\mathrm{Im} z|^{-M}, \end{aligned}$$

where C_1, C_2, M depend on α, β, k, n, we see that

$$\left\|\langle x\rangle^N \mathrm{ad}^\alpha_D \mathrm{ad}^\beta_x B_2(z)\right\| \le C_3|\mathrm{Im} z|^{-M},$$

where C_3, M depend on α, β, N.

By the same method, we obtain an analogous estimate for $B_1(z)$ and $B_3(z)$. By (D.11.1), we conclude that

$$\left\|\langle x\rangle^N \mathrm{ad}^\alpha_D \mathrm{ad}^\beta_x f_1(A_1)f_2(A_2)\right\| \le C.$$

□

D.12 Non-stationary Phase Method

One of the basic tools in the study of integral expressions with rapidly oscillating phases is the non-stationary phase method. We use this method to describe some simple but useful estimates on a certain family of Fourier integral operators.

Proposition D.12.1
Suppose that $c(x_1, x_2, \xi) \in S(\langle x_1\rangle^{m_1}\langle x_2\rangle^{m_2}, \mathrm{d}x_1^2+\mathrm{d}x_2^2+\mathrm{d}\xi^2)$, *and there exist some* $C_0 > 0$ *and* t *such that if* $(x_1, x_2, \xi) \in \mathrm{supp}c$, *then*

$$\langle x_1 - x_2\rangle \geq C_0(\langle x_1\rangle + \langle x_2\rangle) + t.$$

Let $\Psi(x_1, x_2, \xi)$ *be a real function on* $\mathrm{supp}c(x_1, x_2, \xi)$ *satisfying*

$$\partial_{x_1}^{\alpha_1}\partial_{x_2}^{\alpha_2}\partial_\xi^\beta(\Psi(x_1, x_2, \xi) - \langle x_1 - x_2, \xi\rangle) \in O(\langle x_1\rangle) + O(\langle x_2\rangle), \quad |\beta| \geq 1,$$
$$\nabla_\xi(\Psi(x_1, x_2, \xi) - \langle x_1 - x_2, \xi\rangle) \in o(\langle x_1\rangle) + o(\langle x_2\rangle).$$

Let A *be the operator with the kernel*

$$K(x_1, x_2) = (2\pi)^{-n} \int \mathrm{e}^{i\Psi(x_1,x_2,\xi)}c(x_1, x_2, \xi)\mathrm{d}\xi.$$

Then $A \in \Psi(\langle x\rangle^{-\infty})$ *and* $\|A\| \in O(t^{-\infty})$.

Proof. Define the formal operator

$$\mathcal{L} := \left(1 + (\nabla_\xi\Psi(x_1, x_2, \xi))^2\right)^{-1} (1 + \langle\nabla_\xi\Psi(x_1, x_2, \xi), D_\xi\rangle).$$

Using the identity

$$\mathcal{L}\left(\mathrm{e}^{i\Psi(x_1,x_2,\xi)}\right) = \mathrm{e}^{i\Psi(x_1,x_2,\xi)}$$

and integrating by parts in ξ, we obtain

$$\begin{aligned} K(x_1, x_2) &= (2\pi)^{-n} \textstyle\int c(x_1, x_2, \xi)\mathcal{L}^N \mathrm{e}^{i\Psi(x_1,x_2,\xi)}\mathrm{d}\xi \\ &= (2\pi)^{-n} \textstyle\int \mathrm{e}^{i\Psi(x_1,x_2,\xi)}({}^t\mathcal{L})^N c(x_1, x_2, \xi)\mathrm{d}\xi, \end{aligned}$$

where $({}^t\mathcal{L})^N c(x_1, x_2, \xi) = c_N(x_1, x_2, \xi)$ is a sum of terms of the form

$$(1 + (\nabla_\xi\Psi(x_1, x_2, \xi))^2)^{-M} \partial_\xi^{\beta_1}\nabla_\xi\Psi(x_1, x_2, \xi)$$
$$\cdots\partial_\xi^{\beta_k}\nabla_\xi\Psi(x_1, x_2, \xi)\partial_\xi^\gamma c(x_1, x_2, \xi),$$

where $N = |\beta_1| + \cdots + |\beta_k| + |\gamma| = M - k$. We have, for some $C_0 > 0$,

$$1 + (\nabla_\xi\Psi(x_1, x_2, \xi))^2 \geq C_0(\langle x_1\rangle + \langle x_2\rangle + t)^2.$$

Therefore,

$$|\partial_{x_1}^{\alpha_1}\partial_{x_2}^{\alpha_2}\partial_\xi^\beta c_N(x_1,x_2,\xi)| \le C_{\alpha,\beta,N}\langle x_1\rangle^{m_1}\langle x_2\rangle^{m_2}(\langle x_1\rangle+\langle x_2\rangle+t)^{-N}.$$

Setting

$$\tilde{c}_N(x_1,x_2,\xi) = c_N(x_1,x_2,\xi)\mathrm{e}^{i\Psi(x_1,x_2,\xi)-i\langle x_1-x_2,\xi\rangle},$$

we obtain

$$K(x_1,x_2) = (2\pi)^{-n}\int \tilde{c}_N(x_1,x_2,\xi)\mathrm{e}^{i\langle x_1-x_2,\xi\rangle}\mathrm{d}\xi,$$

where

$$|\partial_{x_1}^{\alpha_1}\partial_{x_2}^{\alpha_2}\partial_\xi^\beta \tilde{c}_N(x_1,x_2,\xi)| \le C_{\alpha,\beta,N}\langle x_1\rangle^{m_1}\langle x_2\rangle^{m_2}(\langle x_1\rangle+\langle x_2\rangle+t)^{-N+|\alpha_1|+|\alpha_2|+|\beta|}.$$

Note that, for any N_0, we can find N such that

$$|\partial_{x_1}^{\alpha_1}\partial_{x_2}^{\alpha_2}\partial_\xi^\beta \tilde{c}_N(x_1,x_2,\xi)| \le C_N\langle x_1\rangle^{-N_0}\langle x_2\rangle^{-N_0}\langle t\rangle^{-N_0}, \quad |\alpha_1|+|\alpha_2|+|\beta|\le N_0.$$

Since N_0 is arbitrary, we obtain that $K_-(x_1,x_2)$ is the kernel of an operator in $\Psi(\langle x\rangle^{-\infty})$ and its norm is $O(t^{-\infty})$. □

D.13 FIO's Associated with a Uniform Metric

In this section we recall some results about a certain class of Fourier integral operators associated with amplitudes from $S(1,g_0)$.

Definition D.13.1
Suppose that $a : X\times X' \to \mathbb{C}$ and $\Phi : \mathrm{supp}a \to \mathbb{R}$ are some functions. The Fourier integral operator with phase $\Phi(x,\xi)$ and amplitude $a(x,\xi)$ is defined as the operator given by the following formula:

$$J(\Phi,a)\phi(x) := (2\pi)^{-n}\int a(x,\xi)\mathrm{e}^{i\Phi(x,\xi)-i\langle y,\xi\rangle}\phi(y)\mathrm{d}\xi\mathrm{d}y. \tag{D.13.1}$$

In theorems that we will give below, we will give various conditions that will make the expression (D.13.1) well defined.

The main result of this section is the following theorem on the boundedness of a certain class of Fourier integral operators.

Theorem D.13.2
(i) Suppose that

$$\left|\partial_x^\alpha\partial_\xi^\beta\Phi(x,\xi)\right| \le M_{\alpha,\beta}, \quad ||\alpha|+|\beta|\ge 2, \ |\alpha|\ge 1 \tag{D.13.2}$$

and

$$|\det M| \ge C_0 > 0, \qquad M\in \mathrm{ch}\left\{\nabla_x\nabla_\xi\Phi(x,\xi) \mid (x,\xi)\in X\times X'\right\}, \tag{D.13.3}$$

where chΘ *denotes the convex hull of the set* Θ. *Let* $a_i \in S(1, g_0)$, $i = 1, 2$. *Then*

$$J(\Phi, a_1)J(\Phi, a_2)^* \in \Psi(1, g_0). \tag{D.13.4}$$

(ii) Suppose that

$$\left|\partial_x^\alpha \partial_\xi^\beta \Phi(x, \xi)\right| \le M_{\alpha,\beta}, \qquad |\alpha| + |\beta| \ge 2, \ |\beta| \ge 1. \tag{D.13.5}$$

Assume also (D.13.3). Then

$$J(\Phi, a_1)^* J(\Phi, a_2) \in \Psi(1, g_0). \tag{D.13.6}$$

(iii) Let $\Phi(x, \xi)$ *satisfy either the hypotheses of (i) or of (ii). Let* $a \in S(1, g_0)$. *Then* $J(\Phi, a)$ *is bounded on* $L^2(\mathbb{R}^n)$, *and there exists an integer* N *and a constant* C *depending on* C_0 *and* $M_{\alpha,\beta}$, $|\alpha| + |\beta| \le N$, *such that*

$$\|J(\Phi, a)\| \le C\|a\|_{g_0,N}. \tag{D.13.7}$$

Proof. Let us prove *(i)*. The kernel of $J(\Phi, a_1)J(\Phi, a_2)^*$ equals

$$\begin{aligned} K(x_1, x_2) &= (2\pi)^{-n} \textstyle\int a_1(x_1, \xi)\bar{a}_2(x_2, \xi) e^{i\Phi(x_1,\xi) - i\Phi(x_2,\xi)} d\xi \\ &= (2\pi)^{-n} \textstyle\int a_1(x_1, \xi)\bar{a}_2(x_2, \xi) e^{i \int_0^1 \langle \nabla_x \Phi(\tau x_1 + (1-\tau)x_2, \xi), x_1 - x_2\rangle d\tau} d\xi \end{aligned} \tag{D.13.8}$$

For fixed x_1, x_2, the map

$$X' \ni \xi \mapsto \eta(x_1, x_2, \xi) := \int_0^1 \nabla_x \Phi(\tau x_1 + (1-\tau)x_2, \xi) d\tau \in X'$$

satisfies the assumptions of Proposition A.7.1, hence it is invertible. Let us denote its inverse by

$$\eta \mapsto \xi(x_1, x_2, \eta).$$

We note that

$$\left|\partial_{x_1}^{\alpha_1} \partial_{x_2}^{\alpha_2} \partial_\eta^\beta \xi(x_1, x_2, \eta)\right| \le C_{\alpha,\beta} \quad |\alpha_1| + |\alpha_2| + |\beta| \ge 1. \tag{D.13.9}$$

We rewrite (D.13.8) as

$$K(x_1, x_2) = (2\pi)^{-n} \int b(x_1, x_2, \eta) e^{i\langle \eta, (x_1 - x_2)\rangle} d\eta,$$

where

$$b(x_1, x_2, \eta) := a_1(x_1, \xi(x_1, x_2, \eta))\bar{a}_2(x_2, \xi(x_1, x_2, \eta))|\det \nabla_\eta \xi(x_1, x_2, \eta)|.$$

We easily see that all the semi-norms of $b(x_1, x_2, \eta)$ in $S(1, \mathrm{d}x_1^2 + \mathrm{d}x_2^2 + \mathrm{d}\xi^2)$ can be estimated by the right-hand side of (D.13.7). Therefore (D.13.7) follows from Theorem D.4.1.

(ii) follows from *(i)* by conjugation with the Fourier transformation.

(iii) is an immediate consequence of *(i)* and *(ii)*. □

Remark. In practice, instead of (D.13.3) we will use a simpler condition. We will just assume that

$$\|\nabla_x \nabla_\xi \Phi(x, \xi) - 1\| \le C_0 < 1. \tag{D.13.10}$$

Next we describe when Fourier integral operators described in Theorem D.13.2 are invariant with respect to multiplication on the left and right with pseudo-differential operators of the class $\Psi(1, g_0)$.

Proposition D.13.3
(i) Suppose that $a(x, \xi)$ and $\Phi(x, \xi)$ satisfy the assumptions of Theorem D.13.2 (i). Let $B = b(x, D) \in \Psi(1, g_0)$. Then

$$BJ(\Phi, a) = J(\Phi, c),$$

where $c \in S(1, g_0)$.
(ii) Suppose that $a(x, \xi)$ and $\Phi(x, \xi)$ satisfy the assumptions of Theorem D.13.2 (ii). Let $B \in \Psi(1, g_0)$. Then

$$J(\Phi, a)B = J(\Phi, d),$$

where $d \in S(1, g_0)$.

Proof. To prove *(i)*, we compute the kernel of BA:

$$\begin{aligned} K(x_1, y) &= (2\pi)^{-2n} \int\int\int b(x_1, \xi_1) a(x_2, \xi_2) \mathrm{e}^{i\langle (x_1 - x_2), \xi_1\rangle + i\Phi(x_2, \xi_2) - i\langle y, \xi_2\rangle} \mathrm{d}\xi_1 \mathrm{d}x_2 \mathrm{d}\xi_2 \\ &= (2\pi)^{-n} \int c(x_1, \xi_2) e^{i\Phi(x_1, \xi_2) - i\langle \xi_2, y\rangle} \mathrm{d}\xi_2, \end{aligned}$$

where

$$\begin{aligned} c(x_1, \xi_2) &= (2\pi)^{-n} \int\int b(x_1, \xi_1) a(x_2, \xi_2) \mathrm{e}^{i\langle (x_1 - x_2), \xi_1\rangle + i\Phi(x_1, \xi_2) - i\Phi(x_2, \xi_2)} \mathrm{d}\xi_1 \mathrm{d}x_2 \\ &= (2\pi)^{-n} \int\int b(x_1, \xi_1) a(x_2, \xi_2) \mathrm{e}^{i\langle (x_1 - x_2), (\xi_1 - \int_0^1 \mathrm{d}\tau \nabla_x \Phi(\tau x_1 + (1-\tau) x_2, \xi_2))\rangle} \mathrm{d}\xi_1 \mathrm{d}x_2 \\ &= (2\pi)^{-n} \int\int b(x_1, \textstyle\int_0^1 \mathrm{d}\tau \nabla_x \Phi(x_1 + (1-\tau) z, \xi_2) + \eta) a(x_1 + z, \xi_2) \mathrm{e}^{-i\langle z, \eta\rangle} \mathrm{d}z \mathrm{d}\eta. \end{aligned}$$

Now, using Proposition D.3.2 *(iii)*, we see that $c(x, \xi) \in S(1, g_0)$.

(ii) follows by the same arguments, if we use the conjugation with the Fourier transformation. □

D.14 FIO's Depending on a Parameter

This section is parallel to Sect. D.5. In particular, we refer the reader to that section for the notation. The following proposition is a parameter-dependent version of Theorem D.13.2 *(i)*.

Proposition D.14.1
Suppose that $c_x(t)c_\xi(t) \leq C$,

$$\begin{aligned} &\left|\partial_x^\alpha \partial_\xi^\beta \Phi(t,x,\xi)\right| \leq C_{\alpha,\beta} c_x^{|\alpha|-1}(t) c_\xi^{|\beta|-1}(t), \quad |\alpha|+|\beta| \geq 2, \quad |\alpha| \geq 1, \\ &\|\nabla_x \nabla_\xi \Phi(t,x,\xi) - 1\| \leq C_0 < 1. \end{aligned} \tag{D.14.1}$$

Let $a_i(t,x,\xi) \in S(f_i(t), g(t)),\ i = 1,2.$ *Then*

$$J(\Phi(t), a_1(t)) J(\Phi(t), a_2(t))^* \in \Psi(f_1(t) f_2(t), g(t)).$$

Next let us state a parameter dependent version of Proposition D.13.3.

Proposition D.14.2
(i) Assume $c_x(t)c_\xi(t) \leq C$, *and (D.14.1). Let* $a(t,x,\xi) \in S(f_2(t), g(t))$ *and* $B(t) \in \Psi(f_1(t), g(t))$. *Then*

$$B(t) J(\Phi(t), a(t)) = J(\Phi(t), c(t))$$

for some $c(t,x,\xi) \in S(f_1(t) f_2(t), g(t))$.
(ii) If instead of (D.14.1) we assume that

$$\begin{aligned} &\left|\partial_x^\alpha \partial_\xi^\beta \Phi(t,x,\xi)\right| \leq C_{\alpha,\beta} c_x^{|\alpha|-1}(t) c_\xi^{|\beta|-1}(t), \quad |\alpha|+|\beta| \geq 2, \quad |\beta| \geq 1, \\ &\|\nabla_x \nabla_\xi \Phi(t,x,\xi) - 1\| \leq C_0 < 1, \end{aligned}$$

then

$$J(\Phi(t), a(t)) B(t) = J(\Phi(t), d(t))$$

for some

$$d(t,x,\xi) \in S(f_1(t) f_2(t), g(t)).$$

D.15 FIO's Associated with the Metric g_1

In this section we collect some results on Fourier integral operators associated with phases satisfying $\nabla_x \nabla_\xi \Phi(x,\xi) \in S(1, g_1)$. Most of them are well known (see for example [IK4]).

Theorem D.15.1
(i) Assume that

$$\begin{aligned} &\partial_x^\alpha \partial_\xi^\beta \left(\Phi(x,\xi) - \langle x,\xi\rangle\right) \in o(\langle x\rangle^0), && |\alpha| = |\beta| = 1, \\ &|\partial_x^\alpha \partial_\xi^\beta \left(\Phi(x,\xi) - \langle x,\xi\rangle\right)| \leq C_{\alpha\beta}\langle x\rangle^{1-|\alpha|}, && |\alpha| + |\beta| \geq 2, \quad |\alpha| \geq 1. \end{aligned} \tag{D.15.1}$$

Let $a_i(x,\xi) \in S(\langle x\rangle^{m_i}, g_1)$, $i = 1,2$. *Then*

$$J(\Phi, a_1)J(\Phi, a_2)^* \in \Psi(\langle x\rangle^{m_1+m_2}, g_1). \tag{D.15.2}$$

(ii) Assume, in addition,

$$\partial_x^\alpha \left(\Phi(x,\xi) - \langle x,\xi\rangle\right) \in o(\langle x\rangle^0), \ |\alpha| = 1.$$

Let $\epsilon > 0$, *and let* $\Gamma_i \subset X \times X'$ *be conical sets. If* $a_i \in S(\langle x\rangle^{-\infty})$ *on* Γ_i^ϵ, *then* $J(\Phi, a_1)J(\Phi, a_2)^* \in \Psi(\langle x\rangle^{-\infty})$ *on* $\Gamma_1 \cup \Gamma_2$.
(iii) If a_i *are elliptic on* Γ_i^ϵ, *then* $J(\Phi, a_1)J(\Phi, a_2)^*$ *is elliptic on* $\Gamma_1 \cap \Gamma_2$.

Remark. Applying Theorem D.15.1 with $a = a_1 = a_2$, we see that $J(\Phi, a)$ is a bounded operator from $\langle x\rangle^{-m}L^2(X)$ to $L^2(X)$. In particular, if $m = 0$, it is a bounded operator on $L^2(X)$.

Proof of Theorem D.15.1. Let us note that there exists R_0 such that, for $|x| > R_0$,

$$|\nabla_x \nabla_\xi \Phi(x,\xi) - 1| \leq \frac{1}{2}.$$

Let us distinguish 2 regions in the space $X \times X'$.
Let $0 < \sigma_+ < 1/2$, $R = R_0(1 - 2\sigma_1)$ and

$$\Theta_+ := \{(x_1, x_2) \in X \times X \mid |x_1 - x_2| \leq \sigma_+(|x_1| + |x_2|), \ |x_1| > R, \ |x_2| > R\}.$$

Let $\sigma_- < \sigma_+ < 1$ and $R < R_1$

$$\begin{aligned} \Theta_- := \ & \{(x_1, x_2) \in X \times X \mid |x_1 - x_2| \geq \sigma_-(|x_1| + |x_2|), \ |x_1| > R, \ |x_2| > R\} \\ & \cup \{(x_1, x_2) \in X \times X \mid |x_1| < R_1 \text{ or } |x_2| < R_1\}. \end{aligned}$$

We choose functions $p_\pm \in S(1, \langle x_1\rangle^{-2}\mathrm{d}x_1^2 + \langle x_2\rangle^{-2}\mathrm{d}x_2^2)$ such that $\mathrm{supp} p_\pm \subset \Theta_\pm$ and $p_+(x_1, x_2) + p_-(x_1, x_2) = 1$.

Explicitly, let $\chi_+ \in C_0^\infty(\mathbb{R})$ such that $\chi_+(s) = 0$ for $|s| \leq R$, $\chi_+(s) = 1$ for $s > R_1$ and let $\chi_- := 1 - \chi_+$. Let $f_\pm \in C^\infty(\mathbb{R})$ such that $f_+ + f_- = 1$, $f_+(s) = 0$ for $t \geq \sigma_+$, $f_-(s) = 0$ for $t \leq \sigma_-$. Set

$$p_+(x_1, x_2) = \chi_+(x_1)\chi_+(x_2) f_+\left(\frac{|x_1 - x_2|}{|x_1| + |x_2|}\right), \ p_-(x_1, x_2) = 1 - p_+(x_1, x_2).$$

The kernel of $J(\Phi, a_1)J(\Phi, a_2)^*$ equals

$$K(x_1, x_2) = (2\pi)^{-n} \int a_1(x_1, \xi)\bar{a}_2(x_2, \xi)\mathrm{e}^{i\Phi(x_1,\xi)-i\Phi(x_2,\xi)}\mathrm{d}\xi. \tag{D.15.3}$$

Let us split $K(x_1, x_2) = K_+(x_1, x_2) + K_-(x_1, x_2)$ as follows:

$$\begin{aligned} c_\pm(x_1, x_2, \xi) &:= p_\pm(x_1, x_2)a_1(x_1, \xi)\bar{a}_2(x_2, \xi), \\ K_\pm(x_1, x_2) &:= (2\pi)^{-n} \int c_\pm(x_1, x_2, \xi)\mathrm{e}^{i\Phi(x_1,\xi)-i\Phi(x_2,\xi)}\mathrm{d}\xi. \end{aligned}$$

Let us consider first $K_+(x_1, x_2)$. Set

$$\eta(x_1, x_2, \xi) := \int_0^1 \nabla_x\Phi(\tau x_1 + (1-\tau)x_2, \xi)\mathrm{d}\tau.$$

Since on $\mathrm{supp}p_+$

$$|\tau x_1 + (1-\tau)x_2| \geq R_0,$$

we have

$$|\nabla_\xi\eta(x_1, x_2, \xi) - 1| < 1/2.$$

By Proposition A.7.1, this implies that the map

$$X' \ni \xi \mapsto \eta(x_1, x_2, \xi)$$

is invertible for $(x_1, x_2, \xi) \in \mathrm{supp}c_+$. Let us denote its inverse by

$$\eta \mapsto \xi(x_1, x_2, \eta).$$

We note that on the support of c_+ there exist $0 < C_1 < C_2$ such that

$$C_1|x_1| < |x_2| < C_2|x_1|.$$

Using this, we obtain

$$\left|\partial_{x_1}^{\alpha_1}\partial_{x_2}^{\alpha_2}\partial_\eta^\beta\xi(x_1, x_2, \eta)\right| \leq C_{\alpha_1,\alpha_2,\beta}\langle x_1\rangle^{-|\alpha_1|}\langle x_2\rangle^{-|\alpha_2|}, \ |\alpha_1| + |\alpha_2| + |\beta| \geq 1.$$

We can rewrite K_+ as

$$K_+(x_1, x_2) = (2\pi)^{-n} \int b_+(x_1, x_2, \eta)\mathrm{e}^{i\langle\eta,(x_1-x_2)\rangle}\mathrm{d}\eta, \tag{D.15.4}$$

where

$$b_+(x_1, x_2, \eta) = c_+(x_1, x_2, \xi(x_1, x_2, \eta))|\det\nabla_\eta\xi(x_1, x_2, \eta)|.$$

We easily see that $b_+ \in S(\langle x_1\rangle^{m_1}\langle x_2\rangle^{m_2}, \langle x_1\rangle^{-2}\mathrm{d}x_1^2 + \langle x_2\rangle^{-2}\mathrm{d}x_2^2 + \mathrm{d}\xi^2)$. So, by Proposition D.8.6, $K_+(x_1, x_2)$ is the kernel of a pseudo-differential operator in $\Psi(\langle x\rangle^{m_1+m_2}, g_1)$.

Next, let us consider $K_-(x_1, x_2)$. We have

$$\begin{aligned} \nabla_\xi(\Phi(x_1, \xi) - \Phi(x_2, \xi)) &= x_1 \int_0^1 \nabla_x\nabla_\xi\Phi(\tau x_1, \xi)\mathrm{d}\tau - x_2 \int_0^1 \nabla_x\nabla_\xi\Phi(\tau x_2, \xi)\mathrm{d}\tau \\ &= x_1 - x_2 + o(\langle x_1\rangle) + o(\langle x_2\rangle). \end{aligned}$$

Moreover, on $\operatorname{supp} c_-$ we have

$$\langle x_1 - x_2\rangle \geq C_0(\langle x_1\rangle + \langle x_2\rangle)$$

for some $C_0 > 0$. Therefore

$$\langle \nabla_\xi \Phi(x_1,\xi) - \nabla_\xi \Phi(x_2,\xi)\rangle \ \geq C_1(\langle x_1\rangle + \langle x_2\rangle) \tag{D.15.5}$$

for some $C_1 > 0$. Therefore, by Proposition D.12.1, the operator with the kernel $K_-(x_1,x_2)$ belongs to $\Psi(\langle x\rangle^{-\infty})$.

To show *(ii)*, we use (D.15.4) and the fact that on the support of $b_+(x_1,x_2,\eta)$

$$\xi(x_1,x_2,\eta) - \eta \in o((\langle x_1\rangle + \langle x_1\rangle)^0).$$

□

Next we would like to consider the product of a Fourier integral operator and a pseudo-differential operator. Note that the situation is somewhat different if the pseudo-differential operator is on the left and on the right, therefore we will have two separate propositions describing these two cases.

Proposition D.15.2
(i) Assume that $\Phi(x,\xi)$ *satisfies*

$$\begin{array}{ll} \partial_\xi^\beta \left(\Phi(x,\xi) - \langle x,\xi\rangle\right) \in o(\langle x\rangle), & |\beta| = 1, \\ |\partial_x^\alpha \partial_\xi^\beta \left(\Phi(x,\xi) - \langle x,\xi\rangle\right)| \leq C_{\alpha\beta}\langle x\rangle^{1-|\alpha|}, & |\alpha| + |\beta| \geq 1, \end{array} \tag{D.15.6}$$

and that $a \in S(\langle x\rangle^{m_1}, g_1)$. *Let* $B \in \Psi(\langle x\rangle^{m_2}, g_1)$. *Then there exists* $d \in S(\langle x\rangle^{m_1+m_2}, g_1)$ *such that*

$$J(\Phi,a)B = J(\Phi,d).$$

(ii) If $a \in S(\langle x\rangle^{-\infty})$ *on* Γ_1^ϵ *and* $B \in \Psi(\langle x\rangle^{-\infty})$ *on* Γ_2^ϵ, *where* $\Gamma_1, \Gamma_2 \subset X \times X'$ *are conical sets, then* $d \in S(\langle x\rangle^{-\infty})$ *on* $\Gamma_1 \cup \Gamma_2$.

Proof. We can write $B = b(x,D)^*$ for some $b \in S(\langle x\rangle^{m_2}, g_1)$. Then the kernel of $J(\Phi,a)B$ is equal to

$$K(x_1,x_2) = (2\pi)^{-n} \int e^{i\Phi(x_1,\xi_1) - i\langle x_2,\xi_1\rangle} c(x_1,x_2,\xi_1) d\xi_1$$

for

$$c(x_1,x_2,\xi_1) \ := a(x_1,\xi_1)\bar{b}(x_2,\xi_1).$$

We choose $0 < \sigma_- < \sigma_+ < 1$ and $f_\pm \in C^\infty(\mathbb{R})$ such that $f_+ + f_- = 1$, $f_-(t) = 0$ for $t \leq \sigma_-$, $f_+(t) = 0$ for $t \geq \sigma_+$ and we set

$$\begin{aligned} c_\pm(x_1,x_2,\xi_1) &:= f_\pm\left(\tfrac{|x_1-x_2|}{\langle x_1\rangle + \langle x_2\rangle}\right) c(x_1,x_2,\xi_1), \\ K_\pm(x_1,x_2) &:= (2\pi)^{-n} \int e^{i\Phi(x_1,\xi_1) - i\langle x_2,\xi_1\rangle} c_\pm(x_1,x_2,\xi_1) d\xi_1. \end{aligned}$$

Let us study $K_+(x_1,x_2)$ first. We can write

$$K_+(x_1,x_2) = (2\pi)^{-n} \int e^{i\Phi(x_1,\xi_2)-i\langle x_2,\xi_2\rangle} d_+(x_1,\xi_2) d\xi_2$$

for

$$\begin{aligned} &d_+(x_1,\xi_2) \\ &:= (2\pi)^{-n} \textstyle\int e^{i\Phi(x_1,\xi_1)-i\Phi(x_1,\xi_2)-i\langle x_2,\xi_1-\xi_2\rangle} c_+(x_1,x_2,\xi_1) dx_2 d\xi_1. \end{aligned} \tag{D.15.7}$$

Let

$$r(x_1,\xi_1,\xi_2) = \int_0^1 \nabla_\xi \Phi(x_1,\tau\xi_1+(1-\tau)\xi_2) d\tau - x_1.$$

We have

$$\Phi(x_1,\xi_1) - \Phi(x_1,\xi_2) = \langle x_1 + r(x_1,\xi_1,\xi_2), \xi_1-\xi_2\rangle,$$

and

$$\begin{aligned} &r(x_1,\xi_1,\xi_2) \in o(\langle x_1\rangle), \\ &|\partial_{x_1}^\alpha \partial_{\xi_1}^{\beta_1} \partial_{\xi_2}^{\beta_2} r(x_1,\xi_1,\xi_2)| \leq C_{\alpha\beta}\langle x_1\rangle^{1-|\alpha|}. \end{aligned}$$

Making the change of variables

$$\tilde{x}_2 = x_2 - r(x_1,\xi_1,\xi_2)$$

in (D.15.7), we obtain

$$\begin{aligned} &d_+(x_1,\xi_2) \\ &= (2\pi)^{-n} \textstyle\int e^{i\langle x_1-\tilde{x}_2,\xi_1-\xi_2\rangle} c_+(x_1,\tilde{x}_2+r(x_1,\xi_1,\xi_2),\xi_2) d\tilde{x}_2 d\xi_1. \end{aligned} \tag{D.15.8}$$

Using the fact that on $\mathrm{supp} f_+$ we have $\langle x_1\rangle \leq C\langle x_2\rangle$, we obtain

$$\begin{aligned} &c_+(x_1,\tilde{x}_2+r(x_1,\xi_1,\xi_2),\xi_2) \\ &\in S\left(\langle x_1\rangle^{m_1}\langle \tilde{x}_2\rangle^{m_2}, \langle x_1\rangle^{-2}dx_1^2 + \langle \tilde{x}_2\rangle^{-2}dx_2^2 + d\xi_1^2 + d\xi_2^2\right). \end{aligned}$$

Using then Proposition D.7.2, we obtain that $d_+(x,\xi) \in S(\langle x\rangle^{m_1+m_2}, g_1)$.

By Proposition D.12.1, the operator with the kernel $K_-(x_1,x_2)$ belongs to $\Psi(\langle x\rangle^{-\infty})$. This ends the proof of *(i)*.

(ii) follows from (D.15.8) and the fact that $r(x_1,\xi_1,\xi_2) \in o(\langle x_1\rangle)$. □

Proposition D.15.3
(i) Assume that $\Phi(x,\xi)$ satisfies

$$|\partial_x^\alpha \partial_\xi^\beta (\Phi(x,\xi) - \langle x,\xi\rangle)| \leq C_{\alpha\beta}\langle x\rangle^{1-|\alpha|}, \quad |\alpha|+|\beta| \geq 1, \tag{D.15.9}$$

and that $a \in S(\langle x\rangle^{m_1}, g_1)$. Let $B \in \Psi(\langle x\rangle^{m_2}, g_1)$. Then there exists $d \in S(\langle x\rangle^{m_1+m_2}, g_1)$ such that

$$BJ(\Phi,a) = J(\Phi,d).$$

(ii) Assume in addition

$$\partial_x^\alpha \left(\Phi(x,\xi) - \langle x,\xi\rangle\right) \in o(\langle x\rangle^0), \qquad |\alpha| = 1. \tag{D.15.10}$$

If $\epsilon > 0$, $a \in S(\langle x\rangle^{-\infty})$ *on* Γ_1^ϵ *and* $B \in \Psi(\langle x\rangle^{-\infty})$ *on* Γ_2^ϵ, *where* $\Gamma_1, \Gamma_2 \subset X \times X'$ *are conical sets, then* $d \in S(\langle x\rangle^{-\infty})$ *on* $\Gamma_1 \cup \Gamma_2$.

Proof. Let $B = b(x, D)$. The kernel of $BJ(\Phi, a)$ is equal to

$$K(x_1,y) = (2\pi)^{-2n} \int e^{i\langle x_1 - x_2, \xi_1\rangle} e^{i\Phi(x_2,\xi_2) - i\langle y,\xi_2\rangle} c(x_1,x_2,\xi_1,\xi_2) d\xi_1 dx_2 d\xi_2,$$

where

$$c(x_1,x_2,\xi_1,\xi_2) = b(x_1,\xi_1)a(x_2,\xi_2).$$

We choose $f_\pm$ as in the proof of Proposition D.15.2 and we set

$$K_\pm(x_1,y) = (2\pi)^{-2n} \int e^{i\langle x_1 - x_2, \xi_1\rangle} e^{i\Phi(x_2,\xi_2) - i\langle y,\xi_2\rangle} c_\pm(x_1,x_2,\xi_1,\xi_2) d\xi_1 dx_2 d\xi_2,$$

where

$$c_\pm(x_1,x_2,\xi_1,\xi_2) := f_\pm\left(\tfrac{|x_1 - x_2|}{\langle x_1\rangle + \langle x_2\rangle}\right) c(x_1,x_2,\xi_1,\xi_2).$$

We can write

$$K_\pm(x_1,x_2) = (2\pi)^{-n} \int e^{i\Phi(x_1,\xi_2) - i\langle y,\xi_2\rangle} d_\pm(x_1,\xi_2) d\xi_2,$$

where

$$d_\pm(x_1,\xi_2) := (2\pi)^{-n} \int e^{i\langle x_1 - x_2, \xi_1\rangle + i\Phi(x_2,\xi_2) - i\Phi(x_1,\xi_2)} c_\pm(x_1,x_2,\xi_1,\xi_2) d\xi_1 dx_2.$$

We change the variables

$$\tilde{\xi}_1 = \xi_1 - q(x_1,x_2,\xi_2),$$

where

$$q(x_1,x_2,\xi_2) = \int_0^1 \nabla_x \Phi(\tau x_1 + (1-\tau)x_2, \xi_2) d\tau - \xi_2,$$

and get

$$d_\pm(x_1,\xi_2) = (2\pi)^{-n} \int\int e^{i\langle x_1 - x_2, \tilde{\xi}_1 - \xi_2\rangle} c_{\pm,0}(x_1,x_2,\tilde{\xi}_1,\xi_2) dx_2 d\tilde{\xi}_1, \tag{D.15.11}$$

where

$$c_{\pm,0}(x_1,x_2,\tilde{\xi}_1,\xi_2) := c_\pm(x_1,x_2,\tilde{\xi}_1 + q(x_1,x_2,\xi_2),\xi_2).$$

Let us consider first $K_+(x_1,x_2)$. Note that

$$c_{+,0}(x_1,x_2,\tilde{\xi}_1,\xi_2) \in S\left(\langle x_1\rangle^{m_1}\langle x_2\rangle^{m_2}, \langle x_1\rangle^{-2}dx_1^2 + \langle x_2\rangle^{-2}dx_2^2 + d\tilde{\xi}_1^2 + d\xi_2^2\right).$$

By Proposition D.7.2, this implies

$$d_+ \in S(\langle x\rangle^{m_1+m_2}, g_1).$$

To handle the kernel $K_-(x_1, x_2)$, we first note that

$$c_{-,0}(x_1, x_2, \tilde{\xi}_1, \xi_2) \in S\left(\langle x_1\rangle^{m_1}\langle x_2\rangle^{m_2}, \mathrm{d}x_1^2 + \mathrm{d}x_2^2 + \mathrm{d}\tilde{\xi}_1^2 + \mathrm{d}\xi_2^2\right).$$

Next we use the proposition D.12.1 to see that $d_-(x_1, \xi_2) \in S(\langle x\rangle^{-\infty})$. Therefore $K_-(x_1, y)$ is the kernel of an operator in $\Psi(\langle x\rangle^{-\infty})$.

This ends the proof of *(i)*.

Using (D.15.10), we see that on the support of $c_{+,0}(x_1, x_2, \tilde{\xi}_1, \xi_2)$

$$q(x_1, x_2, \xi_2) \in o((\langle x_1\rangle + \langle x_2\rangle)^0).$$

Using this and (D.15.11), we obtain *(ii)*. □

Let us state certain useful properties of Fourier integral operators.

Proposition D.15.4
Assume

$$\begin{aligned}
&\partial_x^\alpha \partial_\xi^\beta \left(\Phi(x,\xi) - \langle x,\xi\rangle\right) \in o(\langle x\rangle^{1-|\alpha|}), \qquad |\alpha| = 0, 1, \quad |\beta| = 1, \\
&|\partial_x^\alpha \partial_\xi^\beta \left(\Phi(x,\xi) - \langle x,\xi\rangle\right)| \leq C_{\alpha\beta}\langle x\rangle^{1-|\alpha|}, \qquad |\alpha| + |\beta| \geq 1.
\end{aligned} \tag{D.15.12}$$

and $a \in S(1, g_1)$. Then the following holds:
(i) The following "Beals criterion" is satisfied:

$$\mathrm{ad}_D^\alpha \mathrm{ad}_x^\beta J(\Phi, a)\langle x\rangle^{-|\alpha|} \in B(L^2(X)).$$

$$\langle x\rangle^{-|\alpha|}\mathrm{ad}_D^\alpha \mathrm{ad}_x^\beta J(\Phi, a) \in B(L^2(X)).$$

(ii) For any $m, k \in \mathbb{R}$, the operator $\langle x\rangle^m\langle D\rangle^k J(\Phi, a)\langle x\rangle^{-m}\langle D\rangle^{-k}$ is bounded,
(iii) If $0 < \delta_0 < \delta_1$, then

$$\mathbb{1}_{[0,\delta_0]}(\tfrac{|x|}{t}) J(\Phi, a) \mathbb{1}_{[\delta_1,\infty]}(\tfrac{|x|}{t}) \in O(t^{-\infty}).$$

Proof. *(i)* follows by the direct calculation and it implies *(ii)* by Lemma D.6.1. *(iii)* follows from Proposition D.12.1. □

References

[AbMa] Abraham, R. Marsden, J.: *Foundations of Mechanics*, Benajmin-Cummings, New-York 1978

[AT] Adachi, T., Tamura, H.: Asymptotic completeness for long-range many-particle systems with Stark effect, II, Comm. Math. Phys. (1996)

[Ag1] Agmon, S.: Some new results in spectral and scattering theory of differential operators on $L^2(\mathbb{R}^n)$, Séminaire Goulaouic-Schwartz 1978-79, Centre de Mathématiques, Ecole Polytechnique

[Ag2] Agmon, S.: *Lectures on exponential decay of solutions of second order elliptic equations*, Lect. Notes in Math. **29** Princeton University Press, Princeton 1982

[AgHö] Agmon, S., Hörmander, L.: Asymptotic properties of solutions to differential equations with simple characteristics, J. An. Math. **30** (1976) 1-38

[AlKa] Alsholm, P.K., Kato, T.: Scattering with long range potentials, in *Partial Differential Equations*, Proc. Symp. Pure. Math. **23** (1973) 393-399

[ABG] Amrein, W.O., Boutet de Monvel, A., Georgescu, I.V.: *C_0-groups, commutator methods and spectral theory of N-body hamiltonians*, Birkhäuser 1996

[AGM] Amrein, W.O., Georgescu, I.V., Martin, P.A.: Approche algébrique de la théorie non-relativiste de la diffusion aux canaux multiples, in *Physical Reality and Mathematical Description*, Enz/Mehra eds. (1974) 255-276

[AMM] Amrein, W.O., Martin, P.A., Misra, B.: On the asymptotic condition of scattering theory, Helv. Phys. Acta **43** (1970) 313-344

[Ar] Arai, M.: Eigenfunction expansion associated with the Schrödinger operators with long-range potentials, Publ. RIMS Kyoto Univ. **16** (1980) 33-59

[AF] Asada, K., Fujiwara, D.: On some oscillatory integral transformations in $L^2(\mathbb{R}^n)$, Japan J. Math. **4** (1978) 299-361

[Bea] Beals, R.: Weighted distribution spaces and pseudodifferential operators, J. An. Math. **39** (1981) 130-187

[BC] Berthier, A., Collet, P.: Wave operators for momentum dependent long-range potentials, Ann. I. H. P. **26** (1977) 279-293

[Bom] Bommier, A.: Propriétés de la matrice de diffusion 2-amas 2-amas, pour les problèmes à N corps à longue portée, Ann. I. H. P. **59** (1993) 237-267

[BoCh] Bony, J.M., Chemin, J.Y.: Espaces fonctionnels associés àu calcul de Weyl-Hörmander **122** Bull. Soc. Math. Fr. **122** (1995) 77-118

[BG] Boutet de Monvel, A., Georgescu, I.V.: Spectral and scattering theory by the conjuguate operator method, Algebra and Analysis **4** (1992) 73-116

[BuMa] Buslaev, V.S., Matveev, V.B.: Wave operators for the Schrödinger equation with a slowly decreasing potential, Theor. Math. Phys. **2** (1970) 266-274

[CV] Calderon, A.P., Vaillancourt, R.: A class of bounded pseudodifferential operators, Proc. Nat. Acad. Sci **69** (1972) 1185-1187

[Cha] Chazy, J.: Sur l'allure du mouvement dans le problème des trois corps quand le temps croît indéfiniment, Ann. Sci. de l'ENS, **39** (1923) 29-130

[Comb] Combes, J.M.: Time-dependent approach to nonrelativistic multichannel scattering, Nuovo Cimento A **64** (1969) 111-144

[Com1] Combescure, M.: Spectral and scattering theory for a class of strongly oscillating potentials, Comm. Math. Phys. **73** (1980) 43-62

[Com2] Combescure, M.: Long-range scattering for two-body Schrödinger equation with Hörmander-like potentials, Comm. in P.D.E. **11** (1986) 123-165

[Co1] Cook, J.M.: Convergence of the Moeller wave matrix, J. Math. Phys. **36** (1957) 82-87

[Co2] Cook, J.M.: Banach Algebras and Asymptotic Mechanics, in Cargèse Lecture Notes in Theoretical Physics 1965, ed. F.Lurcat, Gordon Breach New York 1967

[CFKS] Cycon, H.L., Froese, R., Kirsch, W., Simon, B.: *Schrödinger Operators with applications to Quantum Mechanics and Global Geometry*, Springer 1987

[Dav] Davies, E.B.: On Enss's approach to scattering theory, Duke Math. J. **47** (1980) 171-185

[DKY] Deich, V.G., Korotyaev, E.L., Yafaev, D.R.: Theory of potential scattering taking into account spatial anisotropy, J. of Sov. Math. **34** (1986) 2040-2050

[DSim] Deift, P., Simon, B.: A time-dependent approach to the completeness of multiparticle quantum systems, Comm. Math. Phys. **30** (1983) 573-411

[De1] Dereziński, J.: Exponential bounds in cones for eigenfunctions of N-body Schrödinger operators, Rep. Math. Phys. **24** (1986) 219-228

[De2] Dereziński, J.: A new proof of the propagation theorem for N-body quantum systems, Comm. Math. Phys. **122** (1989) 203-231

[De3] Dereziński, J.: The Mourre estimate for dispersive N-body Schrödinger operators, Trans. of AMS **317** (1990) 773-798

[De4] Dereziński, J.: Exponential bounds in cones for eigenfunctions of N-body Schrödinger operators II, Commentationes Mathematicae **30** (1991) 269-275

[De5] Dereziński, J.: Asymptotic observables in the long-range N-body scattering theory, A. Boutet de Monvel et al. eds, *Recent developments in Quantum Mechanics* 243-255, 1991 Kluwer Academic Publishers

[De6] Dereziński, J.: Algebraic approach to the $N-$body long range scattering, Rev. in Math. Phys. **3** (1991) 1-62

[De7] Dereziński, J.: Large time behavior of classical $N-$body systems, Comm. Math. Phys. **148** (1992) 503-520

[De8] Dereziński, J.: Asymptotic Completeness for $N-$particle long-range quantum systems, Ann. of Math. **138** (1993) 427-476

[De9] Dereziński, J: Asymptotic completeness in the non-relativistic field theory, in preparation

[DeGe1] Dereziński, J., Gérard, C.: A remark on the asymptotic clustering of N-body systems, in Schrödinger Operators, *The Quantum Mechanical Many-Body Problem*, Balslev (ed), Proc. of Workshop, Aarhus,1991, Lecture Notes in Phys. **403**, Springer 1992

[DeGe2] Dereziński, J., Gérard, C.: Long-range scattering in the position representation, preprint Ecole Polytechnique 1996

[Do1] Dollard, J.: Asymptotic convergence and Coulomb interaction, J. Math. Phys. **5** (1964) 729-738

[Do2] Dollard, J.: Scattering into cones, I . Potential scattering, Comm. Math, Phys. **12** (1969) 193-203

[E1] Enss, V.: Asymptotic completeness for quantum-mechanical potential scattering, I Short range potentials, Comm. Math. Phys. **61** (1978) 285-291

[E2] Enss, V.: Asymptotic completeness for quantum-mechanical potential scattering, II Singular and long-range potentials, Ann. Phys. **119** (1979) 117-132

[E3] Enss, V.: Asymptotic observables on scattering states, Comm. Math. Phys. **89** (1983) 245-268

[E4] Enss, V.: Propagation properties of quantum scattering states, J. Func. Anal. **52** (1983) 219-251

[E5] Enss, V.: Quantum scattering theory for two- and three-body systems with potentials of short and long range, in: *Schrödinger Operators*, S.Graffi ed., Springer Lect. Notes in Math. **1218**, Berlin, 1986

[E6] Enss, V.: Long-range scattering of two- and three-body quantum systems, Journées "Equations aux dérivées partielles", Saint Jean de Monts, Juin 1989, Publications Ecole Polytechnique, Palaiseau 1989

[Fa] Faddeev L.: *Mathematical Aspects of the Three Body Problem in Quantum Scattering Theory*, Steklov Institute, 1963 (Russian); English translation: Israel Program for Scientific Translation, 1965

[FM] Fedoriuk, M.V., Maslov, V.P.: *Semiclassical Approximation in Quantum Mechanics*, D.Reidel Publishing Company Dordrecht Holland 1981

[FH1] Froese, R., Herbst, I.: A new proof of the Mourre estimate, Duke Math. J. **49** (1982) 1075-1085

[FH2] Froese, R., Herbst, I.: Exponential bounds and absence of positive eigenvalues for $N-$body Schrödinger operators, Comm. Math. Phys. **87** (1982) 429-447

[FHHO] Froese, R., Herbst, I., Hoffman-Ostenhoff, T., Hoffman-Ostenhoff, M.: L^2-exponential lower bounds to solutions of the Schrödinger equation, Comm. Math. Phys. **87** (1982) 265-286

[Fu1] Fujiwara, D.: A construction of the fundamental solution for the Schrödinger equation, J. An. Math. **35** (1979) 41-96

[Fu2] Fujiwara, D.: Remarks on convergence of the Feynman path integrals, Duke Math. J. **47** (1980) 559-600

[Geo] Georgescu, V.I.: Sur l'existence des opérateurs d'onde dans la théorie algébrique de la diffusion, Ann. I.H.P. **227** (1977) 9-29

[Ge1] Gérard, C.: Semiclassical resolvent estimates for two and three-body Schrödinger operators, Comm. in P. D. E. **15** (1990) 1161-1178

[Ge2] Gérard, C.: the Mourre estimate for regular dispersive systems, Ann. I. H. P. **54** (1991) 59-88

[Ge3] Gérard, C.: Asymptotic behavior of scattering trajectories for classical 3-body problems, Preprint Ecole Polytechnique 1993

[Ge4] Gérard, C.: Asymptotic completeness for 3-particle long-range systems, Invent. Math. **114** (1993) 333-397

[GeISk] Gérard, C., Isozaki, H., Skibsted, E.: Commutator algebra and resolvent estimates, Adv. Studies in Pure Math. **23**, (1994) 69-82

[GeLa1] Gérard, C., Laba, I.: Scattering theory for N-particle systems in constant magnetic fields, Duke Math. J., **76** (1994) 433-465

[GeLa2] Gérard, C., Laba, I.: Scattering theory for N-particle systems in constant magnetic fields. II. Long-range interactions, Comm. in P.D.E. **20** (1996) 1791-1830

[GeLa3] Gérard, C., Laba, I.: Scattering theory for 3-particle systems in constant magnetic fields: dispersive case, to appear in Ann. Inst. Fourier 1996

[GeMa1] Gérard, C., Martinez, A.: Principe d'absorption limite pour des opérateurs de Schrödinger à longue portée, C.R. Acad. Sci. Paris, **306** série I (1988) 121-123

[GeMa2] Gérard, C., Martinez, A.: Prolongement méromorphe de la matrice de scattering pour des problèmes à deux corps à longue portée, Ann. I.H.P. **51** (1989) 81-110

[GeSj] Gérard, C., Sjöstrand, J.: Semiclassical resonances generated by a closed trajectory of hyperbolic type, Comm. Math. Phys. **108** (1987) 391-421

[Gv] Gerver, J.L.: The existence of pseudocollisions in the plane, J. Diff. Equ. **89** (1991) 1-68

[GM] Ginibre, J., Moulin, M.: Hilbert space approach to the quantum mechanical three-body problem, Ann. I. H. P. A, **21** (1974) 97-145

[Gr] Graf, G.M.: Asymptotic completeness for $N-$body short range quantum systems : A new proof, Comm. Math. Phys. **132** (1990) 73-101

[GGNT] Grosse, H., Grüm, H.R., Narnhofer, H., Thirring, W.: Algebraic theory of Coulomb scattering, A. Phys. Austr. **40** (1974) 97-103

[Hack] Hack: Wave operators in multichannel scattering, Nuovo Cimento Ser. X **13** (1959) 231-236

[Ha] Hagedorn, G. Asymptotic completeness for classes of two, three and four particle Schrödinger operators, Trans. AMS **258** (1980) 1-75

[HS] Helffer, B., Sjöstrand, J.: Equation de Schrödinger avec champ magnétique et equation de Harper, Springer Lecture Notes in Physics **345** (1989) 118-197

[He] Herbst, I.: Classical scattering with long range forces, Comm. Math. Phys. **35** (1974) 193-214

[HeSk1] Herbst, I., Skibsted, E.: Time-dependent approach to radiation condition, Duke Math. J. **64** (1991) 119-147

[HeSk2] Herbst, I., Skibsted, E.: Free channel Fourier transform in the long-range $N-$body problem, J. An. Math. **65** (1995) 297-332

[HeMSk] Herbst, Møller, J.S., I., Skibsted, E.: Asymptotic completeness of N-body Stark Hamiltonians, Comm. Math. Phys. (1996)

[Hö1] Hörmander, L.: The existence of wave operators in scattering theory, Math. Z. **146** (1976) 69-91

[Hö2] Hörmander, L.: *The analysis of linear partial differential operators*, vols I-IV, Springer Verlag Berlin Heidelberg New York (1985)

[Hö3] Hörmander, L.: *Existence of solutions to partial differential equations*, l'Enseignement mathématique 1970

[How] Howland, J.: Stationary scattering theory for time-dependent Hamiltonians, Math. Ann. **207** (1974) 315-335

[Hu1] Hunziker, W.: Time-dependent scattering theory for singular potentials, Helv. Phys. Acta **40** (1967) 1052-1062

[Hu2] Hunziker, W.: The $S-$Matrix in Classical Mechanics, Comm. Math. Phys. **8** (1968) 282-299

[Hu3] Hunziker, W.: Scattering in Classical Mechanics, in *Scattering Theory in Mathematical Physics*, J.A.LaVita, J.P.Marchand eds, D.Reidel Publishing Company Dordrecht Holland 1974

[IoO'C] Iorio R., O'Carroll, M.: Asymptotic completeness for multiparticle Schrödinger Hamiltonians with weak potentials, Comm. Math. Phys. **27** (1972) 137-145

[Ike1] Ikebe, T.: Spectral representations for Schrödinger operators with long-range potentials, J. Func. Anal **20** (1975) 158-177

[Ike2] Ikebe, T.: Spectral representations for Schrödinger operators with long-range potentials, II Perturbations by short-range potentials, Publ. RIMS Kyoto Univ. **11** (1976) 551-558

[IkeI1] Ikebe, T., Isozaki, H.: Completeness of modified wave operators for long-range potentials, Publ. RIMS Kyoto Univ. **15** (1979) 679-718

[IkeI2] Ikebe, T., Isozaki, H.: A stationary approach to the existence and completeness of long-range wave operators, Int. Eq. and Op. Th. **5** (1982) 18-49

[I1] Isozaki, H.: On the long-range stationary wave operator, Publ. RIMS Kyoto Univ. **13** (1977) 589-626

[I2] Isozaki, H.: Eikonal equation and spectral representations for long-range Schrödinger Hamiltonians, J. Math. Kyoto Univ. **20** (1980) 243-261

[I3] Isozaki, H.: On the differentiability properties of generalized Fourier transforms, J. Math. Kyoto Univ. **25** (1985) 789-820

[I4] Isozaki, H.: Structures of S-matrices for three body Schrödinger operators, Comm. Math. Phys. **146** (1992) 241-258

[I5] Isozaki, H.: Asymptotic properties of generalized eigenfunctions for three body Schrödinger operators, Comm. Math. Phys. **153** (1993) 1-21

[I6] Isozaki, H.: A generalization of the radiation condition of Sommerfeld for N-body Schrödinger operators, Duke Math. J. **74** (1994) 557-584

[I7] Isozaki, H.: On N-body Schrödinger operators, Proc. Indian Acad. Sci (Math. Sci), **104** (1994) 667-703

[IK1] Isozaki, H., Kitada, H.: Modified wave operators with time-independent modifiers, J. Fac. Sci. Univ. Tokyo, Sec1A **32** (1985) 77-104

[IK2] Isozaki, H., Kitada, H.: Microlocal resolvent estimates for two-body Schrödinger operators, J. Func. Anal. **57** (1984) 270-300

[IK3] Isozaki, H., Kitada, H.: Scattering matrices for two-body Schrödinger operators, Sci. Pap. of the Coll. of Arts and Sci., Tokyo Univ. **35** (1985) 81-107

[IK4] Isozaki, H., Kitada, H.: A remark on the microlocal resolvent estimates for two-body Schrödinger operators, Publ. RIMS Kyoto Univ. **21** (1986) 889-910

[Jau] Jauch, J.: Theory of scattering operator, II, Helv. Phys. Acta **31** (1958) 661-684

[Jen] Jensen, A.: Propagation estimates for Schrödinger type operators, Trans. AMS, **291** (1985) 129-144

[JMP] Jensen, A., Mourre, E., Perry, P.: Multiple commutator estimates and resolvent smoothness in quantum scattering theory, Ann. I.H.P. **41** (1984) 207-225

[JN] Jensen, A., Nakamura, S.: On the mapping properties of the wave operators in scattering theory, in *Mathematical Physics X* K. Schmüdgen ed., Leipzig 1991, Springer-Verlag

[Kako] Kako, T.: Existence and equivalence of two types of long-range modified wave operators, J. Fac. Sci. Univ. Tokyo Sec. IA **25** (1978) 133-147

[Kal] Kalf, H.: The quantum mechanical virial theorem and the absence of positive energy bound states of Schrödinger operators, Isr. J. Math. **20** (1975) 57-69

[Ka1] Kato, T.: Integration of the equation of evolution in a Banach space, J. Math. Soc. Jap. **5** (1953) 208-234

[Ka2] Kato, T.: Wave operators and similarity for some non-self-adjoint operators, Math. Ann. **162** (1966) 258-279

[Ka3] Kato, T.: Smooth operators and commutators, Studia Math. **31** (1968) 535-546

[Ka4] Kato, T.: Linear evolution equation of "hyperbolic type", J. Fac. Sci. Univ. Tokyo Sect IA **17** (1970) 241-258
Linear evolution equation of "hyperbolic type" II, J. Math. Soc. Jap. **25** (1973) 648-666

[Ki1] Kitada, H.: On the completeness of modified wave operators, Proc. Japan Acad. Sci **52** (1976) 409-412

[Ki2] Kitada, H.: Scattering theory for Schrödinger operators with long-range potentials, I. Abstract theory, J. Math. Soc. Jap. **29** (1977) 665-691

[Ki3] Kitada, H.: Scattering theory for Schrödinger operators with long-range potentials, II. J. Math. Soc. Jap. **30** (1978) 603-682

[Ki4] Kitada, H.: On a construction of the fundamental solution for Schrödinger equations, J. Fac. Sci. Univ. Tokyo Sect IA **27** (1980) 193-226

[Ki5] Kitada, H.: Scattering theory for Schrödinger equation with time-dependent potentials of long-range type, J. Fac. Sci. Univ. Tokyo Sec IA **29** (1982) 353-369

[Ki6] Kitada, H.: Asymptotic completeness of N-body wave operators. I. Short-range quantum systems, Rev. Math. Phys. **3** (1991) 101-124

[KiK] Kitada, H., Kumano-go, H.: A family of Fourier integral operators and the fundamental solution for a Schrödinger equation, Osaka J. Math, **18** (1981) 291-360

[KiYa1] Kitada, H. ,Yajima, K.: A scattering theory for time-dependent long-range potentials, Duke Math. J. **49** (1982) 341-376

[KiYa2] Kitada, H., Yajima, K.: Remarks on our paper "A scattering theory for time-dependent long-range potentials", Duke Math. J. **50** (1983) 1005-1016

[Ku1] Kuroda, S.: Scattering theory for differential operators, I. Operator theory, J. Math. Soc. Japan **25** (1973) 75-104

[Ku2] Kuroda, S.: Scattering theory for differential operators, II. Self-adjoint elliptic operators, J. Math. Soc. Japan **25** (1973) 222-234

[KuM] Kuroda, S., Morita, H.: An estimate for solutions of Schrödinger equations with time-dependent potentials and associated scattering theory, J. Fac. Sci. Univ. Tokyo Sect IA **24** (1977) 459-475

[La1] Lavine, R.: Scattering theory for long-range potentials, J. Func. Anal. **5** (1970) 368-382

[La2] Lavine, R.: Commutators and scattering theory, I. Repulsive interactions, Comm. Math. Phys. **20** (1971) 301-323

[La3] Lavine, R.: Commutators and scattering theory, II. A class of one body problems, Indiana Univ. Math. J. **21** (1972) 643-656

[LoSig] Loss, M., Sigal, I.M.: The three-body problem with threshold singularities, ETH preprint

[LooSt] Loomis, L:, Sternberg, S.: *Advanced Calculus*, Reading Mass. Addison-Wesley 1968

[MarSa] Marchal, C., Saari, D.G.: On the final evolution of the $N-$body problem, J. Diff. Equ. **20** (1976) 150-186

[Mø] Møller, C.: General properties of the characteristic matrix in the theory of elementary particles, Danske. Vid. Selsk. Mat.-Fys. Medd. **23** (1945) 1-48

[Mo1] Mourre, E.: Absence of singular continuous spectrum for certain selfadjoint operators, Comm. Math. Phys. **78** (1981) 519-567

[Mo2] Mourre, E.: Opérateurs conjugués et propriétés de propagation, Comm. Math. Phys. **91** (1983) 279-300

[MoSig] Mourre, E., Sigal, I.M., Phase space analysis and scattering theory for N-particle systems (Micro-local analysis of propagation of particles), preprint 1984

[Pe1] Perry, P.: *Scattering Theory by the Enss Method*, Harwood Academic London 1983

[Pe2] Perry, P.: Exponential bounds and semi-finiteness of point spectrum for N-body Schrödinger operators, Comm. Math. Phys. **92** (1984) 481-483

[PSS] Perry, P., Sigal, I.M., Simon, B.: Spectral analysis of $N-$body Schrödinger operators, Ann. of Math. **114** (1981) 519-567

[Pro] Prosser, R.T.: On the asymptotic behavior of Certain Dynamical Systems, J. Math. Phys. **13** (1972) 186-201

[Put] Putnam, C.R.: *Commutation Properties of Hilbert Space Operators and Related topics*, Springer, Berlin, Heidelberg 1967

[RaS] Radin, C., Simon, B.: Invariant domains for the time-dependent Schrödinger equation, J. Diff. Eq. **29** (1978) 289-296

[Ro] Robert, D.: *Autour de l'approximation semiclassique*, Birkhäuser, 1983

[RS] Reed, M., Simon, B.: *Methods of Modern Mathematical Physics*, vol I and II, 1976, vol. III, 1979, vol. IV, 1978, Academic Press, London

[Sa1] Saari, D.G.: $N-$body collisions and singularities, Lect. Notes in Pure and Appl. Math. **70** (1981)

[Sa2] Saari, D.G.: Expanding gravitational systems, Trans. A. M. S. **156** (1971) 219-240

[SaXi] Saari, D.G., Xia, Z.: The existence of oscillatory and superhyperbolic motion in Newtonian systems, J. Diff. Equ. **82** (1989) 342-355

[Sai1] Saito, Y.: Eigenfunction expansions for the Schrödinger operators with long-range potentials $Q(y) = O(|y|^{-\epsilon})$, $\epsilon > 0$, Osaka J. Math. **14** (1977) 37-53

[Sai2] Saito, Y.: *Spectral representation for Schrödinger operators with long-range potentials*, Lect. Notes in Math. **727**, Springer, Berlin, Heidelberg, New York 1979

[Sch] Schwartz, J.T.: *Non linear functional analysis*, Gordon Breach 1969

[Sie] Siegel, C.L.: *Vorlesungen über Himmels-Mechanik*, Springer Verlag Berlin 1956

[Sig1] Sigal, I.M.: Asymptotic completeness of many body short-range systems, Lett. Math. Phys. **8** (1984) 181-188

[Sig2] Sigal, I.M.: On the long range scattering, Duke. Math. J. **60** (1990) 473-492

[SS1] Sigal, I.M., Soffer, A.: The N−particle scattering problem: asymptotic completeness for short-range quantum systems, Ann. of Math. **125** (1987) 35-108

[SS2] Sigal, I.M., Soffer, A.: Long-range many-body scattering: Asymptotic clustering for Coulomb-type potentials, Invent. Math. **99** (1990) 115-143

[SS3] Sigal, I.M., Soffer, A.: Local decay and propagation estimates for time-dependent and time-independent Hamiltonians, Preprint Princeton University 1988

[SS4] Sigal, I.M., Soffer, A.: Asymptotic completeness for four-body Coulomb systems, Duke Math. J. **71** (1993) 243-298

[SS5] Sigal, I.M., Soffer, A.: Asymptotic completeness of N-particle long-range scattering, J. AMS. **7** (1994) 307-334

[Sim1] Simon, B.: Wave operators for classical particle scattering, Comm. Math. Phys. **23** (1971) 37-48

[Sim2] Simon, B.: Phase space analysis of simple scattering systems: extensions of some work of Enss, Duke Math. J. **46** (1979) 119-168

[Sk1] Skibsted, E.: Smoothness of N-body scattering amplitudes, Rev. Math. Phys. **4** (1992) 279-300

[Sk2] Skibsted, E.: Asymptotic completeness for particles in combined constant electric and magnetic fields, in Partial Differential Equations and Mathematical Physics. The Danish-Swedish Analysis Seminar, 1995, Birkhauser Boston 1996

[Sk3] Skibsted, E.: Asymptotic completeness for particles in combined constant electric and magnetic fields, II, preprint, Aarhus University

[Tam] Tamura, H.: Asymptotic completeness for N-body Schrödinger operators with short-range interactions, Comm. P.D.E. **16** (1991) 1129-1154

[Ta] Taylor, M.: *Pseudodifferential Operators*, Princeton University Press, 1981

[Th] Thomas L. E.: Asymptotic completeness in two- and three-particle quantum mechanical scattering, Ann. Phys. **90** (1975) 127-165

[Tre] Trèves, F.: *Introduction to Pseudodifferential Operators and Fourier Integral Operators*, Plenum, New York, 1980

[Un] Unterberger, A.: *Les opérateurs métadifférentiels*, Springer Lecture Notes in Physics **126** (1980) 205-241

[Yaf1] Yafaev, D.R.: On the violation of unitarity in time-dependent potential scattering, Soviet Math. Doklady **19** (1978) 1517-1521

[Yaf2] Yafaev, D.R.: On the breakdown of asymptotic completeness of wave operators in potential scattering, Comm. Math. Phys. **65** (1979) 167-179

[Yaf3] Yafaev, D.R. Wave operators for the Schrödinger equation, Theor. Math. Phys. **45** (1980) 224-234

[Yaf4] Yafaev, D.R. *Mathematical Scattering Theory. General Theory* AMS, Providence 1992

[Yaf5] Yafaev, D.R.: Radiation conditions and scattering theory for N-particle hamiltonians, Comm. Math. Phys. **154** (1993) 523-554

[Yaf6] Yafaev, D.R.: Resolvent estimates and scattering matrix for N-particle hamiltonians, preprint, 1993

[Yaf7] Yafaev, D.R.: New channels of scattering in three-body quantum systems with long-range potentials, to appear in Duke Math. J.

[Yaf8] Yafaev, D.: New channels in two-body scattering, to appear in Mathematical Journal of Saint-Petersbourg (1996)

[Ya1] Yajima, K.: The surfboard Schrödinger equation, Comm. Math. Phys. **96** (1984) 349-360

[Ya2] Yajima, K.: Existence of solutions for Schrödinger evolution equations, Comm. Math. Phys. **110** (1987) 415-426

[Ya3] Yajima, K.: Schrödinger evolution equation and associated smoothing effect, in *Rigorous Results in Quantum Dynamics*, Proceedings Liblice 1990

[Va] Vasy, A.: Structure of the resolvent for three-body potentials, preprint 1996

[Wa1] Wang, X.P.: Time-delay operator in the semiclassical limit II: short-range potentials, Trans. AMS **322** (1990) 395-415

[Wa2] Wang, X.P.: Semiclassical resolvent estimates for N-body Schrödinger operators, J. Func. Anal. **96** (1991) 466-483

[Wa3] Wang, X.P.: Microlocal estimates for N-body Schrödinger operators, J. Fac. Sci. Univ. Tokyo Sec. IA, Math. 40 (1993) 337-385

[Wa4] Wang, X.P.: On the three-body long-range scattering problems, Rep. Math. Phys. **25** (1992) 267-276

[Wei] Weidmann, J.: The virial theorem and its application to the spectral theory of Schrödinger operators, Bull. AMS **73** (1967) 452-456

[Xia] Xia, Z.: The existence of non-collision singularities in Newtonian systems, Ann. of Math. **135** (1992) 411-468

[Zie] Zieliński, L.: A proof of asymptotic completeness for N-body Schrödinger operators, Comm. in P.D.E. **19** (1994) 455-522

Subject Index

Printing: Mercedesdruck, Berlin
Binding: Buchbinderei Lüderitz & Bauer, Berlin

9 783540 620662

GPSR Compliance

The European Union's (EU) General Product Safety Regulation (GPSR) is a set of rules that requires consumer products to be safe and our obligations to ensure this.

If you have any concerns about our products, you can contact us on ProductSafety@springernature.com

In case Publisher is established outside the EU, the EU authorized representative is:

Springer Nature Customer Service Center GmbH
Europaplatz 3
69115 Heidelberg, Germany

Batch number: 10584518

Printed by Printforce, the Netherlands